Higher Dimensional Complex Varieties

Higher Dimensional Complex Varieties

Proceedings of the International Conference
held in Trento, Italy, June 15–24, 1994

Editors

Marco Andreatta
Thomas Peternell

Walter de Gruyter · Berlin · New York 1996

Editors

Marco Andreatta
Dipartimento di Matematica
Università di Trento
I-38050 Povo (Trento)
Italy

Thomas Peternell
Mathematisches Institut
Universität Bayreuth
D-95440 Bayreuth
Germany

1991 Mathematics Subject Classification: 14–06

Keywords: Calabi-Yau manifolds, fundamental group, Kähler manifolds, linear systems, minimal model theory, singularities, vector bundles

♾ Printed on acid-free paper which falls within the guidelines of the ANSI to ensure permanence and durability.

Library of Congress Cataloging-in-Publication-Data

Higher dimensional complex varieties : proceedings of the international conference held in Trento, Italy, June 15–24, 1994 / editors, Marco Andreatta, Thomas Peternell.
p. cm.
Proceedings of the International School – Conference on Higher Dimensional Complex Geometry.
ISBN 3-11-014503-0
1. Complex manifolds – Congresses. 2. Algebraic varieties – Congresses. I. Andreatta, Marco, 1958– . II. Peternell, Th. (Thomas), 1954– . III. International School – Conference on Higher Dimensional Complex Geometry (1994 ; Trento, Italy)
QA613.H54 1996
516.3'53–dc20 96-7246
CIP

Die Deutsche Bibliothek – Cataloging-in-Publication-Data

Higher dimensional complex varieties : proceedings of the international conference, held in Trento, Italy, June 15–24, 1994 / ed. Marco Andreatta ; Thomas Peternell. – Berlin : New York : de Gruyter, 1996
ISBN 3-11-014503-0
NE: Andreatta, Marco [Hrsg.]

Printed in Germany. Typeset using the author's TEX files: J. Zimmermann, Freiburg.
Printing: Gerike GmbH, Berlin. Binding: Lüderitz & Bauer-GmbH, Berlin.
Cover design: Thomas Bonnie, Hamburg.

Preface

Since the late seventies, the theory of complex algebraic varieties of higher dimensions, i.e. of dimension at least three, has seen a tremendeous development; we think, for instance, of the wonderful results obtained in minimal model theory, in projective classification theory and in the theory of linear series, the new developments around Calabi–Yau manifolds, mirror symmetry and quantum cohomology, the theory of moduli and the topology of algebraic varieties. In this rapidly growing field of complex geometry it becomes more and more important to bring together the experts from different angles of the field and to lead young mathematicians to the frontier of research.

One of the main concerns of the International School–Conference on Higher Dimensional Complex Geometry held in Trento from June 15 to June 24, 1994, was to combine these two aspects. The members of the scientific and organising committee were Marco Andreatta (University of Trento), Giorgio Bolondi (University of Trento), Thomas Peternell (University of Bayreuth) and Edoardo Sernesi (University of Roma). The meeting took place under the auspices of C.I.R.M. (Centro Internazionale per la Ricerca Matematica-Trento), the University of Trento, the network Europroj and the Italian C.N.R.; in the nine days of the event around 100 mathematicians came to Trento.

The first days (June 15–20) were devoted to a Summer School with two main lecturers, Frédéric Campana, lecturing on "Kodaira dimensions and growth of Kähler groups" and David Morrison, lecturing on "Quantum Cohomology and Mirror Symmetry". In addition, some senior participants gave introductory talks to some of the topics of the Conference. In the last four days (June 21–24) the regular conference took place, the proceedings of which constitute this volume, together with the lecture notes of Campana.

All articles are research articles of invited speakers and participants; all contributions have been refereed. Some of the topics covered in this volume are:

- topology of algebraic varieties
- projective classification theory
- minimal model theory
- moduli of algebraic varieties
- toric varieties
- Calabi–Yau threefolds
- mirror symmetry.

We warmly thank the contributors to these proceedings, all the lecturers of the school and of the conference making the meeting a success. Last but not least we thank the referees of the articles who have done a careful and anonymous work during the last months.

Marco Andreatta
Thomas Peternell

Table of Contents

Moduli spaces $M_{g,n}(W)$ for surfaces

Valery Alexeev

Contents

0. Introduction

0.1. We recall the following coarse moduli spaces in the case of curves:

1. M_g, parameterizing nonsingular curves of genus $g \geq 2$ and its compactification $\overline{M_g}$, parameterizing Mumford–Deligne moduli-stable curves, see Mumford [20],
2. spaces $M_{g,n}$, $2g-2+n>0$, for stable n-pointed curves, see Knudsen [9],
3. a moduli space $M_{g,n}(W)$ of stable maps from reduced curves to a variety W, see Kontsevich [15].

It is well known that $\overline{M_g}$ and $M_{g,n}$ are projective, M_g is quasi-projective.

0.2. For surfaces, Gieseker [6] established the existence of a quasi-projective scheme parameterizing surfaces with at worst Du Val singularities, ample canonical class K and fixed K^2, this is a straightforward analog of M_g and we will denote it by M_{K^2}. A geometrically meaningful compactification of this space, $\overline{M^{sm}_{K^2}}$, was constructed by Kollár and Shepherd-Barron in [14] as a separated algebraic space. It is a moduli space of smoothable stable (not in the G.I.T. sense) surfaces of general type. In [11] Kollár has shown that if the class of smoothable stable surfaces with a fixed K^2 is bounded then $\overline{M^{sm}_{K^2}}$ is in fact a projective scheme (Corollary 5.6). Finally, the boundedness was proved by the author in [2].

0.3. The main purpose of this paper is to construct analogs of $M_{g,n}$ and $M_{g,n}(W)$ in the case of surfaces, and to prove their projectiveness. After this is done, we touch on a connection between our moduli spaces and the standard moduli spaces of K3 and Abelian surfaces.

0.4. An idea of "$M_{g,n}$ for surfaces" occured to me when I mentioned that my boundedness theorem 9.2 [2] is strictly stronger than what was used for $\overline{M^{sm}_{K^2}}$. Looking at the definition of $M_{g,n}(W)$ in [15] I also realized that it is simply a relative version of the same scheme, and can be done for surfaces too.

0.5. The basic construction of a moduli space as an algebraic space used here is the same as in [21], [10], [25] and elsewhere. For the hardest question involved, proof of local closedness, we refer to a result of Kollár [13].

For proving that our moduli spaces are projective schemes, rather than mere algebraic spaces, we use Kollár's Ampleness Lemma 3.9 [11], which can be applied in a straightforward way to a variety of complete moduli problems.

0.6. Kontsevich and Manin [16] use the moduli spaces $M_{g,n}(W)$ to define Gromov–Witten classes of varieties in the "quantum cohomology" theory. Hence one distant application of "$M_{g,n}(W)$ for surfaces" might be "higher" GW-classes of schemes.

Notation. All schemes are assumed to be of finite type and defined over a fixed algebraically closed field k of characteristic zero. Obstacles to extending the results to positive characteristic are discussed in the last section. In most cases, we prefer the additive notation to the multiplicative one, for divisors and line bundles. All moduli spaces in this paper are coarse.

Acknowledgment. It is a pleasure to acknowledge useful discussions that I had with F. Campana, Y. Kawamata, J. Kollár, Yu. Manin, E. Sernesi and V. V. Shokurov while working on this paper.

1. Overview

1.1. One possible approach to solving an algebro-geometric moduli problem goes through the following steps:

1. defining the objects that we are trying to parameterize,
2. giving the right definition for a moduli functor,

3. establishing properties of this functor,
4. constructing a moduli space in some fashion,
5. proving finer facts about this space.

In our treatment, we will follow two guiding principles well understood in the field:

Principle 1. Most moduli spaces exist in the category of algebraic spaces.

Principle 2. Most complete and separated moduli spaces are projective.

1.2. Moduli spaces of nonsingular curves M_g and of Deligne–Mumford stable curves $\overline{M_g}$ of genus g provide a textbook illustration of how this works in practice. Since nonsingular curves can degenerate into singular ones in an uncorrectable way, M_g is not complete. There are many different ways to compactify it but the one we are interested in here is adding more curves and trying to solve an enlarged moduli problem. It turns out that the curves one has to add are Deligne–Mumford moduli-stable curves which are defined as connected and complete reduced curves with ordinary nodes only such that every smooth rational irreducible component intersects others in at least 3 points and every irreducible component of arithmetical genus one intersects the rest in at least 1 point. The latter condition has two equivalent meanings:

1. the automorphism group $\mathrm{Aut}(X)$ is finite (and this property is a must for the Geometric Invariant Theory),
2. the dualizing sheaf ω_X is ample.

To arrive at this answer, one can look at the way the good limits are obtained. One considers a family $\mathcal{X}$ of curves over a marked curve, or the spectrum of a DVR, $(\mathcal{S}, 0)$ with a nonsingular general fibre and a degenerate special fibre. Then by the Semistable Reduction Theorem, after making a finite base change $\mathcal{S}' \to \mathcal{S}$ and resolving the singularities of $\mathcal{X}'$, the central fibre will be a reduced curve with simple nodes. Following (1) above one should contract all the rational curves E in the central fiber that intersect the rest only at 1 or 2 points. These have self-intersection numbers $E^2 = -1$ and $E^2 = -2$ respectfully. Contracting (-1)-curves leaves the ambient space, which is a surface, nonsingular. Contracting (-2)-curves introduces very simple surface singularities, called Du Val or rational double. The central fiber has nodes only.

1.3. One can recognize that the above is a field of the Minimal Model Program. In fact, we have just constructed the canonical model, in dimension 2, of $\mathcal{X}'$ over $\mathcal{S}'$. So, to generalize $\overline{M_g}$ to the surfaces of general type we have to apply the Minimal Model Program in dimension 3. This was done by Kollár and Shepherd-Barron in [14]. By that time, the end of 1980-s, all the necessary for this construction tools from MMP in dimension 3 were already available. The new objects that one has to add are defined as connected reduced surfaces with semi-log canonical singularities and ample tensor power of the dualizing sheaf $(\omega_X^N)^{**}$, where ** means taking the self-dual. Using the additive notation, we say that a $\mathbb{Q}$-Cartier divisor K_X is ample.

1.4. At the present time, the *log* Minimal Model Program in dimension 3 is in a pretty good shape. Let us see what kind of statements we can get using its principles. Keeping in line with what we did before, we now consider pairs (X, B) of surfaces X and divisors $B = \sum_{j=1}^{n} B_j$ with ample $K_X + B$. A construction very similar to the one above, with Semistable Reduction Theorem and, this time *log* canonical model, shows that we again get a complete moduli functor. What about singularities of the pair (X, B)? Why, they ought to be semi-log canonical, of course!

What is the analog of this in dimension one? That is easy to answer and we get something very familiar. The divisor $B = \sum_{j=1}^{n} B_j$ becomes a set of marked points. Semi-log canonical means that the curve has nodes only, and that marked points are distinct and lie in the nonsingular part. These are exactly the n-marked semistable curves of Knudsen [9].

1.5. Another possible generalization would be looking not at absolute curves (or surfaces) X (or pairs (X, B)) but doing it in the relative setting. In other words, let us consider maps $X \to W$ to a fixed projective scheme W with K_X (resp. $K_X + B$) *relatively* ample. The only modification in the above construction will be that we have to apply the relative version of the (log) Minimal Model Program over $\mathcal{S}' \times W$ instead of over $\mathcal{S}'$. What we get for curves is the moduli space $M_{g,n}(W)$ introduced by Kontsevich in [15].

1.6. Now that we have outlined the objects we will be dealing with, let us return back to the basic example of $\overline{M}_g$. We recall two different approaches to constructing it.

Approach 1 (G.I.T.). One first proves that moduli-stable curves are asymptotically Hilbert-stable, [20]. Then the standard G.I.T. machinery produces a quasi-projective moduli space. Since it is complete, it is actually projective.

Approach 2. Using a fairly general argument ([21], p. 172) one proves the existence of a moduli space in the category of algebraic spaces. To a family of curves $\pi : \mathcal{X} \to \mathcal{S}$ one can in a natural way associate line bundles on $\mathcal{S}$ which are defined as $\det(\pi_* \omega^k)$. They descend to ($\mathbb{Q}$-)line bundles λ_k on $\overline{M}_g$ and one can further show that λ_k are ample for $k \geq 1$.

1.7. As mentioned in [23] and [11], for surfaces of general type the first approach fails. By [20] 3.19 in order to be asymptotically Chow- or Hilbert-stable a surface has to have singularities of multiplicities at most 7. On the other hand, the semi-stable limits described above have semi-log canonical singularities and it looks like they must be included in any reasonable complete moduli problem. These semi-log canonical singularities include all quotient singularities, for example, and can have arbitrarily high multiplicities.

1.8. The second approach is what we will be using here. After establishing the existence as an algebraic space, we will use Kollár's Ampleness Lemma [11] to prove that it is projective. The Ampleness Lemma is a general scheme that shows projectiveness once some good properties of the moduli functor are established: local closedness, completeness, separateness, semipositiveness, finite reduced automorphism groups, and, crucially, boundedness. The projectiveness will be the only "finer" property of the obtained moduli spaces that we will consider.

2. The objects

2.1. The main objects into the consideration will be *stable maps of pairs* $g : (X, B) \to W$, where

1. $W \subset \mathbb{P}$ is a fixed projective scheme,
2. X is a connected reduced projective surface,
3. $B = \sum_{j=1}^{n} B_j$ is a divisor on X, B_j are reduced but not necessarily irreducible,
4. the pair (X, B) has semi-log canonical singularities,
5. the divisor $K_X + B$ is relatively g-ample.

The precise definitions follow.

2.2. For a normal variety X, K_X or simply K will always denote the class of linear equivalence of the canonical Weil divisor. The corresponding reflexive sheaf $\mathcal{O}_X(K_X)$ is defined as $i_*(\Omega_U^{\dim X})$, where $i : U \to X$ is the embedding of the nonsingular part of X.

Definition 2.3. An $\mathbb{R}$-*divisor* $D = \sum d_j D_j$ is a linear combination of prime Weil divisors with real coefficients, i.e. an element of $N^1 \otimes \mathbb{R}$. An $\mathbb{R}$-divisor is said to be $\mathbb{R}$-Cartier if it is a combination of Cartier divisors with real coefficients, i.e. if it belongs to the image of $\mathrm{Div}(X) \otimes \mathbb{R} \to N^1(X) \otimes \mathbb{R}$ (this map is of course injective for normal varieties). The $\mathbb{Q}$-divisors and $\mathbb{Q}$-Cartier divisors are defined in a similar fashion.

Definition 2.4. Consider an $\mathbb{R}$-divisor $K + B = K_X + \sum b_j B_j$ and assume that

1. $K + B$ is $\mathbb{R}$-Cartier,
2. $0 \leq b_j \leq 1$.

For any resolution $f : Y \to X$ look at the natural formula

$$K_Y + B^Y = f^*(K_X + \sum b_j B_j) = K_Y + \sum b_j f^{-1} B_j + \sum b_i F_i \tag{2.1}$$

or, equivalently,

$$K_Y + \sum b_j f^{-1} B_j + \sum F_i = f^*(K_X + \sum b_j B_j) + \sum a_i F_i \tag{2.2}$$

Here $f^{-1}B_j$ are the proper preimages of B_j and F_i are the exceptional divisors of $f: Y \to X$.

The coefficients b_i, b_j are called codiscrepancies, the coefficients $a_i = 1-b_i$, $a_j = 1-b_j$ log discrepancies.

Remark 2.5. In fact, $K + B$ is not a usual $\mathbb{R}$-divisor but rather a special gadget consisting of a linear class of a Weil divisor K (or a corresponding reflexive sheaf) and an honest $\mathbb{R}$-divisor B. This, however, does not cause any confusion.

Definition 2.6. A pair (X, B) (or a divisor $K + B$) is said to be

1. log canonical, if the log discrepancies $a_k \geq 0$,
2. Kawamata log terminal, if $a_k > 0$,
3. canonical, if $a_k \geq 1$,
4. terminal, if $a_k > 1$,

for every resolution $f: Y \to X$, $\{k\} = \{i\} \cup \{j\}$.

2.7. The notion of *semi-log canonical* is a generalization of *log canonical* to the case of varieties that are singular in codimension 1. The basic observation here is that for a curve with a simple node the definition of the log discrepancies still makes sense and gives $a_1 = a_2 = 0$, so it can also be considered to be (semi-)log canonical. No new Kawamata semi-log terminal singularities appear, however.

Recall that according to Serre's criterion normal is equivalent to Serre's condition S_2 and regularity in codimension 1. So, if we do allow singularities in codimension 1, S_2 will be exactly what we will need to keep.

Definition 2.8. Let X be a reduced (but not necessarily irreducible) equidimensional scheme which satisfies Serre's condition S_2 and is Gorenstein in codimension 1. Let $B = \sum b_j B_j$, $0 \leq b_j \leq 1$ be a linear combination with real coefficients of codimension 1 subvarieties none of irreducible components of which is contained in the singular locus of X. Denote by $\mathcal{O}(K_X)$ the reflexive sheaf $i_*(\omega_U)$, where $i: U \to X$ is the open subset of Gorenstein points of X and ω_U is the dualizing sheaf of U. We can again consider a formal combination of K_X and an $\mathbb{R}$-divisor B, and there is a good definition for $K_X + \sum b_j B_j$ to be $\mathbb{R}$-Cartier. It means that in a neighborhood of any point $P \in X$ we can choose a section s of $\mathcal{O}(K_X)$ such that the divisor $(s) + \sum b_j B_j$ is a formal combination with real coefficients of Cartier divisors with no components entirely in the singular set.

A pair (X, B) (or a divisor $K_X + B$) is said to be semi-log canonical if, similar to the above,

1. $K_X + B$ is $\mathbb{R}$-Cartier,
2. for any morphism $f: Y \to X$ which is birational on every irreducible component, and with a nonsingular Y, in the natural formula

$$f^*(K_X + B) = K_Y + f^{-1}B + \sum b_i F_i$$

with F_i being irreducible components of the exceptional set, all $b_i \leq 1$ (resp. $a_i = 1 - b_i \geq 0$).

As before, the coefficients b_i, b_j are called codiscrepancies, the coefficients a_i, a_j log discrepancies.

Remark 2.9. In the case when (X, B) has a good semi-resolution (for example, for surfaces) this definition is equivalent to that of [14], [5] chapter 12. In our opinion, it is more natural to give a definition which is independent of the existence of a semi-resolution.

Remark 2.10. For surfaces the condition S_2 is of course equivalent to Cohen–Macaulay.

Definition 2.11. By the Kleiman's criterion, the ampleness for proper schemes is a numerical condition, hence it extends to $\mathbb{R}$-Cartier divisors. If coefficients of B are rational, $K_X + B$ is g-ample iff for some positive integer n the divisor $n(K_X + B)$ is Cartier and g-ample in the usual sense.

Remark 2.12. Below we will only consider the case when B is reduced, i.e. all the coefficients $b_j = 1$. See the last section for the discussion on non-integral coefficients.

Example 2.13. If X is a curve then (X, B) is semi-log canonical iff the only singularities of X are simple nodes and B consists of distinct points lying in the nonsingular part of X. $K_X + B$ is relatively ample iff every smooth rational component of X mapping to a point on W has at least 3 points of intersection with the rest of X, or points in B_j, and every component of arithmetical genus 1 has at least 1 such point. In the absolute case, i.e. when W is a point, this is the usual definition of a stable curve with marked points. Every B_j can also be considered as a group of unordered points.

Example 2.14. By the previous example, the only codimension 1 semi-log canonical singularities are normal crossings.

Example 2.15. If X is a nonsingular surface then (X, B) is semi-log canonical iff B has only normal intersections.

Example 2.16. For the case when X is a surface and B is empty the semi-log canonical singularities over $\mathbb{C}$ were classified in [14]. They are (modulo analytic isomorphism): nonsingular points, Du Val singularities, cones over nonsingular elliptic curves, cusps or degenerate cusps (which are similar to cones over singular curves of arithmetical genus 1), double normal crossing points $xy = 0$, pinch points $x^2 = y^2 z$, and all cyclic quotients of the above. If B is nonempty then the singularities of X are from the same list and, in addition, there are different ways B can pass through them. For normal X the list could be found in [1] for example.

2.17. The following describes an easy reduction of semi-log canonical singularities to log canonical, cf. [5] 12.2.4.

Lemma 2.18. *Let* (X, B) *be as in the definition* 2.8 *and denote by* $\nu : X^\nu \to X$ *its normalization. Assume that* $K_X + B$ *is* $\mathbb{R}$*-Cartier. Then* (X, B) *is semi-log canonical iff* $\left(X^\nu, \nu^{-1}B + \operatorname{cond}(\nu)\right)$ *is log canonical, and they have the same log discrepancies.*

Proof. Clear from the definition. □

2.19. The next theorem explains how semi-log canonical surfaces appear in families (cf. [14] 5.1). But first we will need an auxiliary definition.

Definition 2.20. Let $f : (\mathcal{X}, \mathcal{B}) \to \mathcal{S}$ be a 3-dimensional one-parameter family. Let $\mathcal{B} = \sum b_j \mathcal{B}_j$ with $0 \le b_j \le 1$ be an $\mathbb{R}$-divisor and assume that $\mathcal{X}$ and all $\mathcal{B}_j$ are flat over $\mathcal{S}$, $\mathcal{X}$ satisfies S_2 and that $K_\mathcal{X} + \mathcal{B}$ is $\mathbb{R}$-Cartier. We will say that the pair $(\mathcal{X}, \mathcal{B})$ (or the divisor $K_\mathcal{X} + \mathcal{B}$) is *$f$-canonical* if in the definition of log discrepancies for all exceptional divisors with $f(F_i)$ a closed point on $\mathcal{S}$ one has for the corresponding log discrepancy $a(F_i) \ge 1$ (resp. $b(F_i) \le 0$). This condition does not say anything about log discrepancies of divisors that map surjectively onto $\mathcal{S}$.

Theorem 2.21. *Let* $f : (\mathcal{X}, \mathcal{B}) \to \mathcal{S}$ *be a* 3*-dimensional one-parameter family over a pointed curve or a spectrum of a DVR (a discrete valuation ring). Let* $\mathcal{B} = \sum b_j \mathcal{B}_j$ *with* $0 \le b_j \le 1$ *be an* $\mathbb{R}$*-divisor and assume that* $\mathcal{X}$ *is irreducible,* $\mathcal{X}$ *and all* $\mathcal{B}_j$ *are flat over* $\mathcal{S}$ *and that the fibers satisfy Serre's condition* S_2 *and are Gorenstein in codimension* 1 *(note that this implies that* $\mathcal{X}$ *itself is Cohen–Macaulay and is Gorenstein in codimension* 1*). Further assume that* $K_\mathcal{X} + \mathcal{B}$ *is* $\mathbb{R}$*-Cartier. Then the following is true:*

1. *If* $K_{\mathcal{X}_0} + \mathcal{B}_0$ *is semi-log canonical then* $K_\mathcal{X} + \mathcal{B}$ *is log canonical and* f*-canonical.*
2. *Under assumptions of* (1)*, the general fiber is also semi-log canonical.*
3. *Suppose that there exists a birational morphism* $\mu : \mathcal{Y} \to \mathcal{X}$ *with a nonsingular* $\mathcal{Y}$ *such that all exceptional divisors of* μ *and strict preimages of* $\mathcal{B}_i$ *have normal crossings and such that the central fiber is reduced. Then the opposite to* (1) *is true.*

Proof. The proof is an application of the adjunction formula.

(1) The log adjunction theorem [5] 17.12 and 2.18 imply that $K_{\mathcal{X}_0} + \mathcal{B}_0$ is semi-log canonical iff $K_\mathcal{X} + \mathcal{B} + \mathcal{X}_0$ is. Now, the connection between the log discrepancies of the divisors $K_\mathcal{X} + \mathcal{B}$ and $K_\mathcal{X} + \mathcal{B} + \mathcal{X}_0$ is clear. For a divisor E mapping onto $\mathcal{S}$ the log discrepancies are the same. For E mapping to a central point of $\mathcal{S}$ the difference is the coefficient of E in the central fiber of $\mathcal{Y} \to \mathcal{S}$, and so is at least 1.

(3) Here the differences between the log discrepancies over the central fiber are all equal to 1.

(2) is by adjunction. □

2.22. Finally, we show how to pass from a relatively ample $K + B$ to an ample divisor.

Lemma 2.23. *Let $g : (X, B) \to W \subset \mathbb{P}$ be a map, where X is a projective surface and $B = \sum b_j B_j$ is an $\mathbb{R}$-divisor on X. Assume that $K_X + B$ is semi-log canonical and is relatively g-ample. Then $K_X + B + 4H$ is ample, where $H = g^*\mathcal{O}(1)$.*

Proof. It is enough to prove that the restriction on the normalization of X is ample, therefore by 2.18 we can assume that X is normal and that (X, B) is log canonical.

We show that $K_X + B + 3H$ is nef (numerically effective) and this implies the statement. Indeed, $K_X + B + MH$ is ample for $M \gg 0$ and $K_X + B + 4H$ is a weighted average of the above two divisors.

Assume that $K_X + B + 3H$ is not nef. Then the Cone Theorem, which holds for arbitrary normal surfaces, tells us that there exists an irreducible curve C generating an extremal ray and such that

$$(K_X + B + 3H)C < 0$$

This is possible only if C does not map to a point. But then $C \cdot 3H \geq 3$ and $(K_X + B)C \geq -3$ by a theorem on the length of extremal curves, see [18].

In dimension 2 the latter statement is very elementary. Let $f : Y \to X$ be a minimal resolution of singularities of X. Then, if $X \neq \mathbb{P}^2$, one necessarily has $(f^{-1}C)^2 \leq 0$ and

$$(K_X + B)C \geq (K_Y + f^{-1}B)f^{-1}C \geq$$
$$(K_Y + f^{-1}C)f^{-1}C = 2p_a(f^{-1}C) - 2 \geq -2.$$

And the case of $X = \mathbb{P}^2$ is clear. □

3. Definition and properties of a moduli functor

3.1. Below we give a few general definitions for moduli functors. They are fairly standard (see e.g. [25], [11]) but we need to make slight modifications to adapt them to our situation.

3.2. The moduli functor for a moduli problem of polarized schemes is normally constructed in the following way. One fixes a class $\mathcal{C}$ of schemes X/k with a polarization, i.e. an ample line bundle, L and some extra structure and subject to some conditions. Then for an arbitrary scheme $\mathcal{S}/k$ one defines $\mathcal{MC}(\mathcal{S})$ as the set of all (relatively) polarized flat families over $\mathcal{S}$ with all geometric fibers from $\mathcal{C}$ and, possibly, subject to more conditions. The families are considered modulo an equivalence relation. Usually it is an isomorphism between $\mathcal{X}_1/\mathcal{S}$ and $\mathcal{X}_2/\mathcal{S}$ with whatever extra structure they have and a fiber-wise linear equivalence between $\mathcal{L}_1$ and $\mathcal{L}_2$. In other cases it is an algebraic or a numerical, or a numerical up to a scalar equivalence, instead of linear.

Sometimes, it is also useful considering a $\mathbb{Q}$-polarization L on X, i.e a reflexive sheaf such that $(L^{\otimes N})^{**}$ is an ample line bundle.

3.3. The above definition is intentionally vague since extra structures and conditions vary greatly from one moduli problem to another. Instead of trying to cover all future generalizations, we will formulate general principles and, when nontrivial, say exactly how they specialize to our situation.

Definition 3.4. The class $\mathcal{S}$ is said to be *bounded* if there exists a scheme $(\mathcal{X}, \mathcal{L})$ with an extra structure, and a morphism $F : \mathcal{X} \to \mathcal{S}$ to a scheme $\mathcal{S}$ of finite type such that all elements of $\mathcal{C}$ appear as geometric fibers of F, not necessarily in a one-to-one way.

There are two important variations of this definition. There is the *polarized* boundedness, when one requires F to be projective and $\mathcal{L}$ to restrict to the given polarization L on a fiber, versus *non-polarized*. One can also consider boundedness *in the narrow sense*, requiring that all fibers of F belong to $\mathcal{C}$, or *in the wide sense*, asking only for some of the fibers to be from $\mathcal{C}$.

Here we make the choice of the polarized boundedness in the wide sense.

Definition 3.5. A moduli functor $\mathcal{MC}$ is said to be *separated* if every one-parameter family in $\mathcal{MC}(\mathcal{S}_{\text{gen}})$, where $\mathcal{S}_{\text{gen}}$ is a generic point of a DVR, has at most one extension to $\mathcal{S}$.

Definition 3.6. A moduli functor $\mathcal{MC}$ is said to be *complete* if every one-parameter family in $\mathcal{MC}(\mathcal{S}_{\text{gen}})$, where $\mathcal{S}_{\text{gen}}$ is a generic point of a DVR, has at least one extension after a finite cover $\mathcal{S}' \to \mathcal{S}$.

Definition 3.7. A class $\mathcal{C}$ is said to be *locally closed* if for every flat family $F : (\mathcal{X}, \mathcal{L}) \to \mathcal{S}$ with an extra structure there exist locally closed subschemes $\mathcal{S}_l \subset \mathcal{S}$ with the following universal property:

- A morphism of schemes $\mathcal{T} \to \mathcal{S}$ factors through $\coprod \mathcal{S}_l$ iff $(\mathcal{X}, \mathcal{L}) \underset{\mathcal{S}}{\times} \mathcal{T} \to \mathcal{T}$ belongs to $\mathcal{MC}(\mathcal{T})$.

Definition 3.8. A class $\mathcal{C}$ is said to *have finite reduced automorphisms* if every object in $\mathcal{C}$ has a finite and reduced (the latter is automatic in characteristic 0) group of automorphisms.

Definition 3.9. A moduli functor $\mathcal{MC}$ is said to be *functorially polarizable* if for every family $(\mathcal{X}, \mathcal{L})$ in $\mathcal{MC}(\mathcal{S})$ there exists an equivalent family $(\mathcal{X}, \mathcal{L}^c)$ such that

1. if $(\mathcal{X}_1, \mathcal{L}_1)$ and $(\mathcal{X}_2, \mathcal{L}_2)$ are equivalent, then $(\mathcal{X}_1, \mathcal{L}_1^c)$ and $(\mathcal{X}_2, \mathcal{L}_2^c)$ are isomorphic,
2. for any base chance $h : \mathcal{S}' \to \mathcal{S}$, $(\mathcal{X}', \mathcal{L}'^c)$ and $\left(\mathcal{X}', h^*(\mathcal{L}^c)\right)$ are isomorphic.

The main example of a functorial polarization is delivered by the polarization $\omega_{\mathcal{X}/\mathcal{S}}$ for canonically polarized manifolds.

Definition 3.10. A functorial polarization $\mathcal{L}^c$ is said to be *semipositive* if there exists a fixed k_0 such that whenever $\mathcal{S}$ is a complete smooth curve and $f : (\mathcal{X}, \mathcal{L}) \to \mathcal{S}$ an element in $\mathcal{MC}(\mathcal{S})$, then for all $k \geq k_0$ the vector bundles $f_*(k\mathcal{L}^c)$ are semipositive, i.e. all their quotients have nonnegative degrees.

This definition will be slightly modified for our purposes, we will also require semipositiveness of restrictions of $\mathcal{L}^c$ to certain divisors $\mathcal{B}_j$ on $\mathcal{X}$.

3.11. The following is the class that we will be considering from now on.

Definition 3.12. The elements of the class $\mathcal{C}^N = \mathcal{C}^N_{(K+B)^2,(K+B)H,H^2}$ are *stable maps of pairs* $g : (X, B, L_N) \to W$, where

1. $W \subset \mathbb{P}$ is a fixed projective scheme,
2. X is a connected projective surface,
3. $B = \sum_{j=1}^n B_j$ is a divisor on X, B_j are reduced but not necessarily irreducible,
4. the pair (X, B) has semi-log canonical singularities,
5. the divisor $K_X + B$ is relatively g-ample,
6. $(K_X + B)^2 = C_1$, $(K_X + B)H = C_2$, $H^2 = C_3$ are fixed,
7. $L_N = \mathcal{O}(N(K_X + B + 5H))$, where $H = g^*\mathcal{O}_W(1)$. Here N is a positive integer such that for every map as above L_N is a line bundle. The existence of at least one (and hence infinitely many) such N is established by 3.14 and 3.16. The divisor $K_X + B + 5H$ is ample by 2.23.

Remark 3.13. The classes $\mathcal{C}^N$ and $\mathcal{C}^M$ for different N, M are in a one-to-one correspondence between each other, and the only difference is the polarizations.

As a consequence, the polarization in our functor plays a secondary role. We will switch from a polarization L_N to its multiple L_M when it will be convenient.

When we do not emphasize the polarization, the class will be denoted simply by $\mathcal{C}$.

Theorem 3.14. *For some $M > 0$ the class $\mathcal{C}^M$ is bounded.*

Proof. We start with the boundedness theorem and its corollary which give what we want in the absolute case.

Theorem 3.15 ([2], 9.2). *Fix a constant C and a set $\mathcal{A}$ satisfying the descending chain condition. Consider all surfaces X with an $\mathbb{R}$-divisor $B = \sum b_j B_j$ such that the pair (X, B) is semi-log canonical, $K_X + B$ is ample, $b_j \in \mathcal{A}$ and $(K_X + B)^2 = C$. Then the class $\{(X, \sum b_j B_j)\}$ is bounded.*

Corollary 3.16. *Let (X, B) be as in the previous theorem. Assume in addition that all b_j are rational numbers with a fixed denominator. Then there exists a positive integer N so that for every such surface X the divisor $N(K_X + B)$ has integral coefficients and is Cartier.*

Proof. Let $X^\nu \to X$ be the normalization as in lemma 2.18 and let

$$\left(X^\nu, \nu^{-1}B + \operatorname{cond}(\nu)\right) = \bigsqcup (X_l, B_l).$$

The index N is the LCM of indices N_l for each $K_{X_l} + B_l$. Each family (X_l, B_l) is also bounded (see [2] for details), so we can assume that X is normal and has log canonical singularities only.

The singularities of X where the pair $(X, 0)$ is not log terminal all have indices 1, 2, 3, 4 or 6, by the classification (cf. [1] f.e.). All the other singularities are log terminal, and in dimension 2 and characteristic 0 these are precisely the quotient singularities.

Since the deformation of a quotient singularity is again quotient, we can assume that we have a quasi-projective family over a scheme of finite type, and every fiber has only quotient singularities. But then it is clear that only finitely many indices occur. □

Apply the last corollary with the set $\mathcal{A} = \{1\}$ to $K_X + B + D$, where D is a general member of the linear system $|4H|$. Since this linear system is base point free, the pair $(X, B + D)$ also has semi-log canonical singularities. Therefore, all pairs (X, B) satisfying the conditions of the theorem can be embedded by a linear system $|M(K_X + B + 4L)|$ for a fixed large divisible M in a fixed projective space $\mathbb{P}^{d_1}$. Every map $g : X \to W$ is defined by its graph Γ_g. Consider a Veronese embedding of W by $|\mathcal{O}_W(M)|$ in some $\mathbb{P}^{d_2}$ and then look at the graphs Γ_g in a Segre embedding $\mathbb{P}^{d_1} \times \mathbb{P}^{d_2} \subset \mathbb{P}^{d_3}$. Note that $\mathcal{O}_{\mathbb{P}^{d_3}}(1)$ restricted on $X \simeq \Gamma_g$ is $L_M = M(K_X + B + 5H)$.

L^2 is fixed, hence by the boundedness theorem 3.15 above there are only finitely many possibilities for Hilbert polynomials $\chi\left(\mathcal{O}_{\Gamma_g}(t)\right)$. By the same theorem, there are also only finitely many possibilities for Hilbert polynomials $\chi\left(\mathcal{O}_{B_j}(t)\right)$. Therefore, all elements of our class $g : (X, B) \to W$ are parameterized by finitely many products of Hilbert schemes. In each product, we have to extract a subscheme parameterizing subschemes of $\mathbb{P}^{d_1} \times W$ and with fixed $\mathcal{O}_{\mathbb{P}^{d_1}}(1)^2$, $\mathcal{O}_{\mathbb{P}^{d_1}}(1) \cdot \mathcal{O}_{\mathbb{P}^{d_2}}(1)$ and $\mathcal{O}_{\mathbb{P}^{d_2}}(1)^2$, and these are obviously closed algebraic conditions. We also need to extract the graphs, i.e subschemes mapping isomorphically to $\mathbb{P}^{d_1}$, and this is an open condition.

The resulting scheme will parameterize the maps, including all maps from the class $\mathcal{C}^M$. This proves the theorem. □

3.17. We won't need the boundedness of the class $\mathcal{C}^N$ itself, although it will follow from the proof of the local closedness 3.27.

Definition 3.18. There are several ways to define the moduli functor for our class. The first one is most straightforward (cf. [14], [25] in the absolute case with $B = \emptyset$). For any scheme $\mathcal{S}/k$, $\mathcal{MC}^N = \mathcal{MC}^N_{(K+B)^2,(K+B)H,H^2}$ is given by

$$\mathcal{MC}^N(\mathcal{S}) = \left\{ \begin{array}{l} \text{all families } f : (\mathcal{X}, \mathcal{L}) \to \mathcal{S} \text{ with a divisor } \mathcal{B} = \sum_{j=1}^N \mathcal{B}_j \text{ on } X, \\ \text{a map } g : \mathcal{X} \to W \text{ and a line bundle } \mathcal{L} \text{ such that every} \\ \text{geometric fiber belongs to } \mathcal{C}^N,\ X \text{ and all } \mathcal{B}_j \text{ are flat over } \mathcal{S}. \end{array} \right\}$$

Two families over $\mathcal{S}$ are equivalent if they are isomorphic fiber-wise.

In this functor we consider a sub-functor $\mathcal{MC}'^N$, requiring in addition that for each s there exists a 1-dimensional family from $\mathcal{MC}^N$ with the central fiber $\mathcal{X}_s$ and a general fiber $\mathcal{X}_g$ such that the pair $(\mathcal{X}_g, 0)$ is (Kawamata) log terminal.

This is similar to the smoothability condition for $\overline{M^{sm}_{K^2}} \subset \overline{M_{K^2}}$ (see [11]) and is necessary due to the technical reasons. Consider a one parameter family of maps. Then we would like the 3-fold to be normal since MMP is not developed for non-normal varieties yet. We would also want the 3-fold to have log terminal singularities because they are Cohen–Macaulay in characteristic 0.

3.19. A little disadvantage of the above definition is that even though $\mathcal{MC}^{N,\text{irr}}$ and, say, $\mathcal{MC}^{2N,\text{irr}}$ are the same on the closed points, the corresponding moduli spaces can potentially have different scheme structures, the second one could be larger. So, in fact, we have not one but infinitely many moduli spaces. It would be nicer if we had a formula for the minimal N in terms of $(K+B)^2$, $(K+B)H$, H^2. We know, however, only that such an N exists.

3.20. A different solution was suggested (again, in the absolute case with $B = \emptyset$) by Kollár in [11], [13]. In a sense, it produces a moduli space with the "minimal" scheme structure.

We introduce some necessary notation first.

Definition 3.21. Let $F : \mathcal{X} \to \mathcal{S}$ be a projective family of graphs of maps $(X, B) \to W$. Assume that every fiber$\mathcal{X}_s$ is Gorenstein in codimension 1 and satisfies Serre's condition S_2. Denote by $i : \mathcal{U} \hookrightarrow \mathcal{X}$ the open subset where f is Gorenstein and the divisors $\mathcal{B}_j$ are Cartier. Note that on every fiber one has $\operatorname{codim}_{\mathcal{X}_s}(\mathcal{X}_s - \mathcal{U}_s) \geq 2$. Define the sheaves $\mathcal{L}_{\mathcal{U},k}$ and $\mathcal{L}_k$ by

$$\mathcal{L}_{\mathcal{U},k} = \mathcal{O}_{\mathcal{U}}\left(k\left(K_{\mathcal{U}/\mathcal{S}} + \mathcal{B} + g^*\mathcal{O}_W(5)\right)\right)$$

$$\mathcal{L}_k = i_*\mathcal{L}_{\mathcal{U},k}$$

It follows that the sheaves $\mathcal{L}_k$ on $\mathcal{X}$ are coherent.

Notation 3.22. Let $f : \mathcal{X} \to \mathcal{S}$ be a morphism of schemes, $i : \mathcal{U} \hookrightarrow \mathcal{X}$ be the immersion of an open set and $\mathcal{F}$ be a coherent sheaf on $\mathcal{U}$ which is flat over $\mathcal{S}$. For a base change $h : \mathcal{S}' \to \mathcal{S}$ we obtain $\mathcal{X}^h := \mathcal{X} \underset{\mathcal{S}}{\times} \mathcal{S}'$, $\mathcal{U}^h := \mathcal{U} \underset{\mathcal{S}}{\times} \mathcal{S}'$ etc. Denote the induced morphism $\mathcal{U}^h \to \mathcal{U}$ by $h_{\mathcal{U}}$ and set $\mathcal{F}^h := h_{\mathcal{U}}^*\mathcal{F}$. The induced morphism $\mathcal{X}^h \to \mathcal{X}$ is denoted by h_X.

One says that *the push forward of $\mathcal{F}$ commutes with a base change $h : \mathcal{S}' \to \mathcal{S}$* if the natural map $h_X^*(i_*\mathcal{F}) \to i_*^h\mathcal{F}^h$ is an isomorphism.

Definition 3.23. Define $\mathcal{MC}^{\text{all}} = \mathcal{MC}^{\text{all}}_{(K+B)^2,(K+B)H,H^2}$ by

$$\mathcal{MC}^{\text{all}}(\mathcal{S}) = \left\{ \begin{array}{l} \text{all families } f : \mathcal{X} \to \mathcal{S} \text{ with a divisor } \mathcal{B} = \sum_{j=1}^N \mathcal{B}_j \text{ on } \mathcal{X} \text{ and} \\ \text{a map } g : \mathcal{X} \to W \text{ such that every geometric fiber belongs to } \mathcal{C}, \\ \mathcal{X} \text{ and all } \mathcal{B}_j \text{ are flat over } \mathcal{S}, \\ \text{and for each } k,\ i_*\mathcal{L}_{\mathcal{U},k} \text{ commutes with arbitrary base changes.} \end{array} \right\}$$

As above, one considers a sub-functor $\mathcal{MC}'^{\text{all}} \subset \mathcal{MC}^{\text{all}}$.

3.24. One can see that if we require that $i_*\mathcal{L}_{\mathcal{U},k}$ commutes with arbitrary base changes only for $k = N$ instead of all positive k, then we get the previous definition of the moduli functor $\mathcal{MC}^N$. Indeed, if a line bundle $\mathcal{L}$ exists, then $\mathcal{L}_N = \mathcal{L} + f^*\mathcal{E}$ for some invertible sheaf $\mathcal{E}$ on $\mathcal{S}$. Then for every $h : \mathcal{S}' \to \mathcal{S}$ the two sheaves $i_*^h\mathcal{L}_{\mathcal{U},N}^h$ and $h_X^*(i_*\mathcal{L}_{\mathcal{U},N}) = h_X^*(\mathcal{L}_N)$ on $\mathcal{X}'$ are both reflexive and coincide on $h_X^{-1}(\mathcal{U})$, hence everywhere.

Vice versa, if $i_*\mathcal{L}_{\mathcal{U},N}$ commutes with base changes, then $\mathcal{L}_N$ is flat and for every closed point $s \in \mathcal{S}$

$$\mathcal{L}_N|_{\mathcal{X}_s} = \mathcal{O}_{\mathcal{X}_s}\left(N\left(K + B + g^*\mathcal{O}_W(5)\right)\right).$$

Since the latter restriction is locally free for every s and the sheaves $\mathcal{O}_{\mathcal{X}}$, $\mathcal{L}_N$ are coherent and flat over $\mathcal{S}$, it follows by [17] 22.5, 22.3 that $\mathcal{L}_N$ is locally free.

3.25. Now let us show that our moduli functors have all the good properties listed above. All the proofs for the functors $\mathcal{MC}^N$ and $\mathcal{MC}^{\text{all}}$ are basically the same.

We start with the local closedness. The main technical result we will be using is the following theorem.

Theorem 3.26 (Kollár [13]). *With the above notations, assume that $f : \mathcal{X} \to \mathcal{S}$ is projective, $i_*\mathcal{F}$ is coherent and that for every point $s \in \mathcal{S}$ the sheaf $\mathcal{F}_s$ on the fiber $\mathcal{U}_s$ satisfies Serre's condition S_2. Then there exist locally closed subschemes $\mathcal{S}_l \subset \mathcal{S}$ such that for any morphism $h : \mathcal{T} \to \mathcal{S}$ the following are equivalent:*

1. *h factors through $\mathcal{T} \to \coprod \mathcal{S}_l \to \mathcal{S}$,*
2. *$i_*^h\mathcal{F}^h$ commutes with all future base changes.*

Theorem 3.27. *The functors $\mathcal{MC}^N(\mathcal{MC}'^N)$ and $\mathcal{MC}^{\text{all}}(\mathcal{MC}'^{\text{all}})$ are locally closed.*

Proof. We first prove the theorem for the functor $\mathcal{MC}^N(\mathcal{MC}'^N)$.

Let $F : \mathcal{X} \to \mathcal{S}$ be an arbitrary projective family of graphs of maps $(X, B) \to W$. First of all, after the flattening decomposition (see [19] lecture 8) of $\mathcal{S}$ into locally closed subschemes, we can assume that $\mathcal{X}$ and $\mathcal{B}_j$ are flat over $\mathcal{S}$ if they are not so already.

The property of a local ring to be Cohen–Macaulay is open ([17] 24.5). This implies that for a closed (i.e. projective) morphism over a smooth base the property of a fiber to be Cohen–Macaulay is open. The openness for an arbitrary base of finite type follows quite easily by a stratification argument and Noetherian induction. Therefore, if $\mathcal{X}_0$ is Cohen–Macaulay then there exists an open neighborhood of 0 in $\mathcal{S}$ that contains exactly the points over which the fibers are Cohen–Macaulay.

The property of a local ring to be Gorenstein is also open ([17] 24.6) and by the same argument there exists a closed subset Z of non-Gorenstein points in $\mathcal{X}$. Give it the structure of a reduced scheme. Then we have to throw away all fibers on which the Hilbert polynomial of $\mathcal{O}_Z \underset{\mathcal{O}_S}{\otimes} k(s)$ has degree ≥ 1. There are only finitely many possible Hilbert polynomials and the condition on the degree is obviously closed.

At this point we use the previous theorem 3.26 to the sheaf $\mathcal{L}_{\mathcal{U},N}$ to conclude that there exist locally closed subschemes $\mathcal{S}_l \subset \mathcal{S}$ such that every map $h : \mathcal{T} \to \mathcal{S}$ with $\mathcal{X} \underset{\mathcal{S}}{\times} \mathcal{T} \in \mathcal{MC}(\mathcal{T})$ factors through $\coprod \mathcal{S}_l$. $\mathcal{S}_l$ are disjoint, so we can concentrate on one of them. If P is a point of $\mathcal{S}$ and some h as in the definition does not factor through $\mathcal{S} - P$, then the fiber of F over P has to be a pair (X, B) from our class. The sheaf $\mathcal{L}_N$ on $X \underset{\mathcal{S}}{\times} S_l$ is flat over S_l and its restriction to the fiber over P is locally free. Hence, it has to be locally free in a neighborhood of the fiber. Therefore, for each S_l if we denote by $U_l \subset S_l$ the open set over which $\mathcal{L}_N$ is locally free, then $h : \mathcal{T} \to \mathcal{S}$ has to factor through $\coprod \mathcal{U}_l$. Now we can apply 2.21(2) to conclude that there exist open subsets $\mathcal{V}_l \subset \mathcal{U}_l$ containing all the points over which the fibers have semi-log canonical singularities. Also, $\mathcal{MC}'^N \subset \mathcal{MC}^N$ is evidently closed and we end up with a disjoint union of locally closed subschemes.

There is one more thing one has to take care of: the polarization $\mathcal{O}_{\mathbb{P}^{d_3}}(1)$ on the fibers has to coincide with $\mathcal{L}_N$ or its fixed multiple $\mathcal{L}_M$. Note that the two sheaves are locally free and have the same Hilbert polynomials. Standard semi-continuity theorems for h^0 in flat families show that there exists a closed subset where the two sheaves are the same. One can also define the scheme structure on it, see lemma 1.26 of [25].

This proves the theorem for the functor $\mathcal{MC}^N(\mathcal{MC}'^N)$. To prove it for the functor $\mathcal{MC}^{\text{all}}(\mathcal{MC}'^{\text{all}})$, one has to apply 3.26 several more times, with $k = 1, \dots, N-1$. □

Lemma 3.28. *For the functors $\mathcal{MC}^N(\mathcal{MC}'^N)$, resp. $\mathcal{MC}^{\text{all}}(\mathcal{MC}'^{\text{all}})$, the polarization $\mathcal{L}_N$, resp. all $\mathbb{Q}$-polarizations $\mathcal{L}_k$, are functorial.*

Proof. $K_{\mathcal{U}/\mathcal{S}}$ of a flat family commutes with base changes, and so do $\mathcal{O}(\mathcal{B}_j)$ and $g^*\mathcal{O}_W(1)$. Therefore, $\mathcal{L}_{\mathcal{U},k}$ are functorial. By the definitions of the functors the same is true for $\mathcal{L}_N$ (resp. for every $\mathcal{L}_k$). □

Theorem 3.29. *The functors $\mathcal{MC}'^N$ and $\mathcal{MC}'^{\text{all}}$ are*

1. *separated,*
2. *complete,*
3. *have finite and reduced automorphisms.*

Proof. The first two properties have code names in the Minimal Model Program: "uniqueness and existence of the log canonical model". Since we consider only subfunctors $\mathcal{MC}' \subset \mathcal{MC}$, it is enough to check them in the case when the general fiber is irreducible and has log terminal singularities.

(1) Let $\mathcal{S}$ be a spectrum of a DVR or a pointed curve. Two families in $\mathcal{MC}(\mathcal{S})$ that coincide outside of 0 are birationally isomorphic. 2.21(1) implies that they are both

log canonical and both are relative log canonical models over $\mathcal{S} \times W$ for the same divisor, hence isomorphic. If $\mathcal{Y} \to \mathcal{S}$ is a common resolution then the divisor is

$$K_{\mathcal{Y}} + f^{-1}\mathcal{B} + \sum \mathcal{E}_i$$

where $\mathcal{E}_i$ are exceptional divisors that do not map to the central point $0 \in \mathcal{S}$.

(2) If there is a family over $\mathcal{S} - 0$, we can complete it over 0 somehow. Then by a variant of the Semistable Reduction Theorem, after a finite base change, there is a resolution $\mathcal{Y}$ of singularities such that the central fiber is reduced and all exceptional divisors and $\mathcal{B}_j$ have normal crossings. Consider the log canonical model for the same divisor as above, relative over $\mathcal{S} \times W$. It exists by standard results from the log Minimal Model in dimension 3. This log canonical model, denote it $\mathcal{Z}$, has the same fibers as $(\mathcal{X}, \mathcal{B})$ outside 0. The restriction of $\mathcal{B}$ to each fiber is $\mathbb{Q}$-Cartier, hence $\mathcal{B}$ itself is $\mathbb{Q}$-Cartier. Since the pair $(\mathcal{Z}, \mathcal{B})$ is F-canonical and the restriction on the general fiber is log terminal, the pair $(\mathcal{Z}, 0)$ is log terminal. The log terminal singularities are Cohen–Macaulay in dimension 3 and characteristic 0. Therefore, the central fiber is also Cohen–Macaulay and it is from our class $\mathcal{C}$ by 2.21(3).

According to the definition, we also have to show that the push forward $i_*\mathcal{L}_k$ commutes with arbitrary base changes. In our situation it reduces to showing that it commutes with the morphism $j : 0 \to \mathcal{S}$, i.e. that

$$(i_*\mathcal{L}_k)|_{\mathcal{X}_0} = i_{0*}(\mathcal{L}_k|_{\mathcal{X}_0})$$

Since the two sheaves coincide outside of a codimension 2 subset, the equality holds if the sheaf $(i_*\mathcal{L}_k)|_{\mathcal{X}_0}$ is reflexive, i.e. Cohen–Macaulay. This, in turn, is equivalent to saying that the sheaf $\mathcal{L}_k$ on the 3-fold $\mathcal{Z}$ is Cohen–Macaulay.

Let us show that the sheaf $\mathcal{O}_{\mathcal{Z}}(k(K_{\mathcal{Z}} + \mathcal{B}))$ is Cohen–Macaulay (hence the same is true for $\mathcal{L}_k$). This is a local question. Let $\mathcal{Z}' \to \mathcal{Z}$ be a local cover corresponding to the torsion divisors $\mathcal{B}$ and $K_{\mathcal{Z}}$. Then the preimage $\mathcal{B}'$ of $\mathcal{B}$ is Cartier and $\mathcal{Z}'$ is Gorenstein and canonical. The sheaf $\mathcal{O}_{\mathcal{Z}'}\left((K_{\mathcal{Z}'} + \mathcal{B}')\right)$ is invertible and therefore Cohen–Macaulay. On the other hand, by the covering trick the sheaf $\mathcal{O}_{\mathcal{Z}}(k(K_{\mathcal{Z}} + \mathcal{B}))$ is its direct summand, hence it is Cohen–Macaulay too.

(3) In the absolute case, the fact that $K + B$ is ample and log canonical implies that the automorphism group is finite by [7]. In the relative case we apply the same theorem to $K_X + B + D$, $D \in |4H|$ general, which is ample by lemma 2.23. We are working in characteristic 0 and so the group scheme Aut X is reduced. □

Theorem 3.30. *The functors $\mathcal{MC}^N$ ($\mathcal{MC}'^N$) and $\mathcal{MC}^{\text{all}}$($\mathcal{MC}'^{\text{all}}$) are semipositive.*

Proof. One has the following

Theorem 3.31 (Kollár [11] 4.12). *Let Z be a complete variety over a field of characteristic zero. Assume that Z satisfies Serre's condition S_2 and that it is Gorenstein in codimension one. Let $Z \to C$ be a map onto a smooth curve. Assume that the general fiber of f has only semi-log canonical singularities, and further that K of the general fiber is ample. Then $f_*\mathcal{O}(kK_{Z/C})$ is semipositive for $k \geq 1$.*

For the sheaves $\mathcal{L}_N = \mathcal{O}_{\mathcal{X}}\left(N\left(K_{\mathcal{X}/\mathcal{S}} + \mathcal{B}\right)\right)$ with empty $\mathcal{B}$ in the absolute case this is exactly what we need. Analyzing the proof of 3.31 shows that it applies in the case of a non-empty reduced $\mathcal{B}$ too, with trivial changes. In the relative case instead of $K_{\mathcal{X}/\mathcal{S}} + \mathcal{B}$ we consider $K_{\mathcal{X}/\mathcal{S}} + \mathcal{B} + 5H$, $H = g^*\mathcal{O}_W(1)$. We can think of $5H$ simply as of an additional component of the boundary $\mathcal{B}$. If a member of the linear system $|5H|$ is chosen generically, the pair $(\mathcal{X}, \mathcal{B} + 5H)$ will still be semi-log canonical.

For the positiveness of the sheaves $\mathcal{L}_N\Big|_{\mathcal{B}_j}$ we use the log adjunction formula, see [24] or [5] chapter 16. We get the following semi-log canonical divisors on $\mathcal{B}_j$:

$$K_{\mathcal{X}} + \mathcal{B}\Big|_{\mathcal{B}_j} = K_{\mathcal{B}_j} + \sum (1 - 1/m_k)\mathcal{M}_k$$

for some Weil divisors $\mathcal{M}_k$ on $\mathcal{B}_j$ and $m_k \in \mathbb{N} \cup \{\infty\}$. So, here we need a more general semipositiveness theorem, with nonempty $\mathcal{B}$ that has fractional coefficients. The situation is saved by the fact that the relative dimension of $\mathcal{B}_j$ over $\mathcal{S}$ equals 1, and the semipositiveness for this case is proved in [11] 4.7. □

4. Existence and projectivity of a moduli space

Theorem 4.1. *The functors $\mathcal{MC} = \mathcal{MC}'^N$ and $\mathcal{MC}'^{\text{all}}$ are coarsely represented by proper separated algebraic spaces of finite type $\mathbf{MC} = \mathbf{MC}'^N$ and $\mathbf{MC}'^{\text{all}}$ respectively.*

Proof. The proof is essentially the same as in [21], p. 172. We remind that we are working in characteristic zero, and over $\mathbb{C}$ the argument is easier.

The class $\mathcal{C}^M$ is bounded, and we can embed all graphs Γ_g of the maps g by a linear system $|M(K_X + B + 5H)|$ in $\mathbb{P}^{d_1} \times \mathbb{P}^{d_2} \subset \mathbb{P}^{d_3}$ as in 3.14 for a large divisible M. By taking M even larger we can assume that all $X = \Gamma_g$ and all $B_j \subset X$ are projectively normal, $h^0(M(K_X + B + 5H))$ is locally constant and there are no higher cohomologies. (Γ_g, B) are parameterized, not in a one-to-one way, by some scheme that we will denote by $\mathcal{H}$. For any family in $\mathcal{MC}^N(\mathcal{T})$, the embedding by a relatively very ample linear system $|M(K_X + B + 4H)|$ defines a non-unique map $\mathcal{T} \to \mathcal{H}$. By 3.27 there exists a disjoint union of locally closed subschemes $\mathcal{S} = \coprod \mathcal{S}_l \hookrightarrow \mathcal{H}$ with a universal property, and $\mathcal{T} \to \mathcal{H}$ factors through $\mathcal{S}$. We conclude that the coarse moduli space $\mathbf{MC}$ is a categorial quotient of $\mathcal{S}$ by an equivalence relation R, described as follows.

R is a set of pairs (h, G), where $h \in \mathcal{S}$ and G corresponds to a different embedding of X in $\mathbb{P}^{d_1}$, i.e. G varies in a group $\mathrm{PGL}(d_1 + 1)$. There is a natural map $F : R \to \mathcal{S} \times \mathcal{S}$. Every fiber of $\pi_1 \circ F$ is isomorphic to $\mathrm{PGL}(d_1 + 1)$ and this map is obviously smooth. The map F is quasi-finite and unramified because its fibers are automorphism groups of objects in $\mathcal{C}$, and these are finite reduced. The fact that $\mathcal{MC}$ is also proper implies that F is finite.

The rest of the proof is the same as in [21], p. 172 verbatim. By taking the transversal sections locally the question is reduced to the case of a finite equivalence relation dominated by a map $F' : R \to \mathcal{H}' \times \mathcal{H}'$ with $\pi_1 \circ F'$ étale, and then the quotient is easily constructed as an algebraic space.

Finally, since $\mathcal{MC}$ is proper, so is **MC**. □

Theorem 4.2. *The moduli spaces* $\mathbf{MC} = \mathbf{MC}'^{N}$ *and* $\mathbf{MC}'^{\mathrm{all}}$ *are projective.*

Proof. The proof follows the general scheme of [11]. By the very construction of **MC**, there exists a subscheme $\mathcal{S} \subset \mathcal{H}$ of a product of Hilbert schemes, with the corresponding universal family $V_{\mathcal{S}} \to \mathcal{S}$, that maps to **MC**. One starts by constructing a *finite* morphism from a scheme $Y \to \mathbf{MC}$ with a universal family $f : V_Y \to Y$. This is done locally by cutting $\mathcal{S} \to \mathbf{MC}$ transversally, then adding more copies of these sections, so that the automorphisms do not obstruct gluing the local pieces together, see [11] 2.7. The only properties of the class $\mathcal{C}$ used in this construction are boundedness and finiteness of automorphisms, which we have.

Next step is to consider the line bundles

$$\lambda_M = \det(f_*\mathcal{L}_M \oplus f_*\mathcal{L}_M|_{\mathcal{B}_j})$$

on Y for M large divisible, where

$$\mathcal{L}_M = \mathcal{O}_V\left(M\left(K_{V/Y} + \mathcal{B} + g^*\mathcal{O}_W(5)\right)\right).$$

These line bundles do not descend to **MC** because of automorphisms, but since the objects of $\mathcal{C}$ have finite groups of automorphisms and $\mathcal{C}$ is bounded, for every M there is a finite power of λ_M that does come from a line bundle on **MC**. To prove that **MC** is projective it is enough to show that one of λ_M is ample, which is achieved by the following theorem. For simplicity we formulate it only in characteristic 0.

Notation 4.3. Let Y be a scheme and let W be a vector bundle of rank w with structure group $\rho : G \to \mathrm{GL}_w$. Let $q : W \to Q$ be a quotient vector bundle of rank k. Let $Gr(w,k)/G$ denote the set of G-orbits on the k-dimensional quotients of a w-dimensional vector space. The natural map of sets

$$u_{Gr} : \{\text{ closed points of } X\} \to Gr(w,k)/G$$

is called the *classifying map*.

One says that the classifying map is *finite* if

1. every fiber of u_{Gr} is finite, and
2. for every $y \in Y$ only finitely many elements of G leave $\ker q_y$ invariant.

Theorem 4.4 (Kollár's Ampleness Lemma, [11] 3.9). *Let Y be a proper algebraic space and let W be a semipositive vector bundle with structure group G. Let Q be a quotient vector bundle of W. Assume that*

1. *G is reductive,*

2. *the classifying map is finite.*

Then det Q *is ample. In particular, Y is projective.*

This is what it translates to in our situation. The sheaves are

$$W = \mathrm{Sym}^j(f_*\mathcal{L}_M) \oplus \mathrm{Sym}^j(f_*\mathcal{L}_M|_{B_j})$$

and

$$Q = f_*L_{jM} \oplus f_*L_{jM}|_{B_j},$$

q is the multiplication map.

By 3.30 we already know that Q is semipositive, and so is W since symmetric powers of a semipositive sheaf are semipositive.

Recall that the universal family U_Y over Y is embedded into a product of Y and

$$\mathbb{P}^{d_1} \times W \subset \mathbb{P}^{d_1} \times \mathbb{P}^{d_2} \subset \mathbb{P}^{d_3}$$

and that the sheaf L_M is the restriction of $\mathcal{O}_{P^{d_3}}(1)$ in this embedding.

The group G acting on W is $\mathrm{GL}_{d_1+1} \times \mathrm{GL}_1$. If every fiber $\mathcal{X} = \Gamma_g$ together with all B_j can be uniquely reconstructed from the map $W_s \to Q_s$, then the fibers of u_{Gr} will be exactly the same as fibers of $Y \to$ **MC**, hence finite.

For this to be true we need the following:

1. every fiber in $\mathbb{P}^{d_3}$ is set-theoretically defined by degree $\leq j$ equations,
1. the multiplication maps $\mathrm{Sym}^j(f_*\mathcal{L}_M) \to f_*L_{jM}$ and $\mathrm{Sym}^j(f_*\mathcal{L}_M|_{B_J}) \to f_*L_{jM}|_{B_J}$ are surjective.

(1) holds if j is large enough. (2) is satisfied because we have chosen M so large that all X and B_j are projectively normal in $\mathbb{P}^{d_3}$.

Finally, the second condition in the definition of finiteness of the classifying map is satisfied because all graphs $(\Gamma_g, B) = (X, B)$ in $\mathbb{P}^{d_3}$ have finite groups of automorphisms. □

5. Related questions

5.1. Let us see how our moduli spaces are related to some others. For example, consider the moduli space $\mathcal{M}_{L^2}$ of K3 surfaces X with a polarization L with a fixed square. Compare it with $\mathcal{M}_{(K+B)^2}$, where $W = pt$, $B = B_1$ is one reduced divisor and $(K + B)^2 = L^2$ is the same number. $\mathcal{M}_{(K+B)^2}$ contains an open subset U parameterizing K3 surfaces with reduced divisors having normal intersections only, and we have a map $F : U \to \mathcal{M}_{L^2}$. A well-known result (Saint-Donat [22]) says that every ample linear system $|L|$ on a K3 surface contains at least one reduced divisor with normal intersections, therefore F is surjective. In fact, $\mathcal{M}_{L^2}$ is a quotient of

U modulo an obvious equivalence relation R: $(X_1, B_1) \sim R(X_2, B_2)$ iff X_1, X_2 are isomorphic and B_1, B_2 are linearly equivalent.

There is a natural map $G : R \to U \times U$. $\pi_1 \circ G$ is smooth and its fibers are open subsets in $\mathbb{P}^{h^0(H)-1}$. The situation is very similar to what we had in theorem 4.1, except this time the quotient U/R is not proper. The obvious way to try to obtain a compactification of $\mathcal{M}_{H^2}$ is to consider the closure $\overline{U}$ of U in $\mathcal{M}_{(K+B)^2}$, then somehow define the closure $\overline{R}$ of R, and ask if it has good enough properties enabling one to construct $\overline{U}/\overline{R}$ and to prove that it is projective. Alternatively, one can ask if the closure of $\overline{G}(R)$ in $\overline{U} \times \overline{U}$ has good properties.

The situation resembles what happens for elliptic curves. The natural compactification of the moduli space $\mathcal{M}_1 = \mathbb{A}^1_k$ is $\mathbb{P}^1_k$, and the infinite point corresponds not to one but to many degenerations: wheels of rational curves of lengths $1 \dots n$ if we consider $\mathcal{M}_1$ as a factor of $\mathcal{M}_{1,n}$. Similarly, the boundary points of $\mathcal{M}_{H^2}$ should correspond to many different degenerations of smooth K3 surfaces with geometric divisors, properly identified.

The first thing to ask on this way is:

Question 5.2. Is it possible to define an equivalence relation $\overline{G} : \overline{R} \to \overline{U} \times \overline{U}$, so that the morphism $\pi_1 \circ \overline{G}$ is smooth or at least flat?

Even if this is done, there are problems with taking the quotient. There does not seem to exist in the literature a ready-to-use method that would cover our situation. There is, on one hand, a theorem of M. Artin (see [3] 7.1, [4] 6.3) that shows that if $\overline{G} : \overline{R} \to \overline{U} \times \overline{U}$ were a monomorphism (which it is not) with flat projections, then the quotient would be defined as an algebraic space. In this case it would also easily follow that the quotient is actually projective.

On the other hand, there is the method of [21], p. 172 that we used in the previous section, in which the equivalence relation is smooth, and the map $\overline{G}$ is finite. Natural degenerations of K3 surfaces can have infinite groups of automorphisms, however.

I think that the question deserves a more detailed consideration.

5.3. Similar to K3 surfaces, for any principally polarized Abelian variety A with a theta divisor Θ the pair (A, Θ) has log canonical singularities, see [12]. So, the previous discussion applies to principally polarized Abelian surfaces too. However, the situation here is much simpler, since now $h^0(A, \Theta) = 1$ and the equivalence relation R becomes trivial. Thus, we obtain a canonical compactification for the moduli space of principally polarized Abelian surfaces. There is a hope that a similar construction would work in arbitrary dimension. This will be considered in detail in a forthcoming paper.

5.4. It goes without saying that the projectivity theorem 4.2 applies in the case of curves, with significant simplifications. Therefore, the moduli spaces $M_{g,n}(W)$ of [15] are also projective.

5.5. Most spaces $\mathbf{MC}_{(K+B)^2,(K+B)H,H^2}$ definitely are not irreducible and not even connected. They are subdivided according to various invariants, such as the numerical or homological type of $g(X)$ and $g(B_j)$, intersection numbers $(K+B)B_j$ etc.

One can also get by fixing only one number, $(K+B+4H)^2$. Then there are only finitely many possibilities for other invariants.

5.6. The boundedness theorem 3.15 is in fact even stronger than what we used here: it applies to the case when the coefficients b_j belong to an arbitrary set $\mathcal{A}$ that satisfies the descending chain condition. One, perhaps, would want to define even more general moduli spaces. There are two obstacles, however. First, the semipositiveness theorem 3.31 for the case of fractional coefficients seems to be quite hard to prove, but probably still possible. The second obstacle is a fundamental one: for proving the semipositiveness theorems for $\mathcal{L}_k|_{B_j}$ we used the log adjunction formula. It basically just says $K+B|_B = K_B$, and here the coefficient 1 of B is important.

5.7. The places where assumption about the characteristic 0 was used:

1. MMP in dimension 3. This is not serious since we worked in the situation of the relative dimension 2. For surfaces log MMP is characteristic free, and perhaps it is true for families of surfaces in generality needed. For the case $B=\emptyset$ see [8].
2. The semipositiveness theorem 3.31 requires characteristic 0. Since we work with the case of relative dimension 2 only, this also probably can be dealt with.
3. A group scheme in characteristic 0 is reduced, hence smooth. This was used in the proof of 4.1. Perhaps, the argument could be strengthened.
4. The argument of [21] p. 172 is a whole lot more complicated in characteristic $p>0$.

5.8. It should be possible to prove the semipositiveness theorems and the Ampleness Lemma, as well as the [21] p. 172 argument, entirely in the relative situation over W, without appealing to absolutely ample divisors. W then wouldn't have to be projective, and it wouldn't even have to be a scheme. The moduli spaces obtained should be relatively projective over W.

References

[1] V. Alexeev, Log canonical surface singularities: Arithmetical approach, in: Flips and Abundance for Algebraic Threefolds [5], pp. 47–58.

[2] —, Boundedness and K^2 for log surfaces, Internat. J. Math. 5 (1994), 779–810.

[3] M. Artin, The implicit function theorem in algebraic geometry, Algebraic Geometry, Papers presented at the Bombay Colloquium (1969), 13–34, Bombay–Oxford.

[4] —, Versal deformations and algebraic stacks, Invent. Math. 27 (1974), 165–189.

[5] J. Kollár et al., Flips and Abundance for Algebraic Threefolds, Astérisque 211, 1–258.

[6] D. Gieseker, Global moduli for surfaces of general type, Invent. Math. 43 (1977), 233–282.

[7] S. Iitaka, Algebraic geometry. An introduction to birational geometry of algebraic varieties, Grad. Texts in Math. 76, Springer-Verlag, New York 1982.

[8] Y. Kawamata, Semistable minimal models of threefolds in positive or mixed characteristic, J. Algebraic Geometry 3 (1994), 463–491.

[9] F. F. Knudsen, The projectivity of the moduli space of stable curves, II: the stacks $M_{g,n}$, Math. Scand. 52 (1983), 161–199.

[10] J. Kollár, Toward moduli of singular varieties, Compositio Math. 56 (1985), 369–398.

[11] —, Projectivity of complete moduli, J. Differential Geom. 32 (1990), 235–268.

[12] —, Shafarevich maps and automorphic forms, Invent. Math 113 (1993), 177–215.

[13] —, Push forward and base change for open immersions, Lectures at Utah Summer Workshop on Moduli of Surfaces (1994).

[14] J. Kollár and N. Shepherd-Barron, Threefolds and deformations of surface singularities, Invent. Math. 91 (1988), 299–338.

[15] M. Kontsevich, Enumeration of rational curves via torus action, Preprint (1994).

[16] M. Kontsevich and Yu. Manin, Gromov–Witten classes, quantum cohomology, and enumerative geometry, Preprint (1994).

[17] H. Matsumura, Commutative ring theory, Cambridge Stud. Adv. Math. 8, Cambridge University Press, 1986.

[18] Y. Miyaoka and S. Mori, A numerical criterion for uniruledness, Ann. of Math. 124 (1986), 65–69.

[19] D. Mumford, Lectures on curves on algebraic surface, Ann. of Math. Stud. 59, Princeton Univ. Press, Princeton, 1966.

[20] —, Stability of projective varieties, Conférences de l'Union Mathématique Internationale, no.5, vol. 24, L'Enseignement Mathématique, 1977.

[21] —, Geometric invariant theory, Ergeb. Math. Grenzgeb.(2) 34, Springer-Verlag, second enlarged edition, 1982.

[22] B. Saint-Donat, Projective models of K3 surfaces, Amer. J. Math. 96 (1974), 602–639.

[23] N. I. Shepherd-Barron, The Birational Theory of Degenerations, pp. 33–84, Birkhäuser, Boston, 1983.

[24] V. V. Shokurov, 3-fold log flips, Math. USSR–Izv. 56 (1992), 105–203.

[25] E. Viehweg, Quasi-projective Moduli for Polarized Manifolds, Ergeb. Math. Grenzgeb. (3) 30, Springer-Verlag, 1995.

On the geometry and Brill–Noether theory of spanned vector bundles on curves

E. Ballico and B. Russo

This paper is a collection of results on the geometry of sections of vector bundles on curves. Most of such results are contained in the very recent and unpublished preprints [BR1], [BR2], [B5], which will not be published elsewhere also because they were in a very preliminary form. In most cases we will consider vector bundles spanned by their global sections; hence we will be essentially studying the geometry of curves in a Grassmannian.

A key part of these topics is of course the Brill–Noether–Petri theory for vector bundles on curves. We are interested in families of such bundles (described by "measures of stability" or by cohomological conditions) and in the way in which different families fit together (see §4). In this analysis we principally referred to [Hi1], [Hi2], [Hi3], [L1], [L2], [LN], [Pe2] and the updated problem list on Brill–Noether theory [N].

Now we will briefly describe the contents of this paper.

The first section (which is essentially [BR1], §5) is on the rank-2 case; its main (and most useful) result is Proposition 1.3 (i.e. [BR1], Theorem 5.2).

The general aim is to study relations between the Brill–Noether theory for line bundles on a smooth curve C and the corresponding theory for higher rank vector bundles on C. In particular we will focus our attention on bundles which induce imbeddings or birational maps on Grassmannians, (which is essentially [BR2], §2). For example we will try to find the smallest number d such that there is a spanned, rank-2 vector bundle E of degree d on C, inducing a birational map of C to a Grassmannian.

The second section (i.e. essentially [B5], §1) contains a construction of possible "degeneration" of an embedding of a smooth curve C contained in a Grassmannian (see Theorems 2.3 and 2.4). Then we apply Theorem 2.3 to the study of the derived bundles (introduced in [Pe1] and [Pe2]) of a spanned vector bundle on C (see Theorems 2.5 and 2.6).

In the third section (i.e. essentially [B5], §2) we will study these derived bundles (and in particular their ranks) for indecomposable vector bundles on an elliptic curve (see Theorem 3.4).

In the last section (which is contained in [BR1], §6, [BR2], §3 and [B5], §3) we will deal with the following situation. Let us consider the universal family of extensions introduced by H. Lange ([L3], §4). As in the case of rank-2 extensions (see [LN]), let us suppose that the parameter space of this universal family is the product of C with an integral variety parameterizing pairs of vector bundles on C. Consider the strata which are identified by the invariants introduced by H. Lange in [L1]. We will investigate when a stratum is in the closure of an other (Proposition 4.2). As a corollary we will obtain numerical conditions for particular subsets of the space of rank-3 stable bundles, to be one in the closure of the other (Corollary 4.5). As a similar kind of situation arises in the study of irreducibility for several other moduli problems (e.g. bundles on higher dimensional varieties and some Hilbert spaces of smooth curves in $\mathbb{P}^n$), we think it worthwhile to try and obtain further and deeper results on this topic.

This research was partially supported by MURST and GNSAGA of CNR (Italy). This paper is in final form and no part of it will be published elsewhere.

§0.

Unless otherwise stated in this paper we will use the following notations.

Let us fix an algebraically closed base field $\boldsymbol{k}$ and a smooth complete curve C; set $g := p_a(C)$, $\mathcal{O} := \mathcal{O}_C$ and $K := K_C$.

For every sheaf U on C, let us write $H^i(U)$ and $h^i(U)$ instead of $H^i(C, U)$ and $h^i(C, U)$. If $V \subseteq H^0(T)$ is a vector space spanning a rank t vector bundle T on C with $w := \dim(V)$, let $h_V : C \to G(t, w)$ be the morphism to the Grassmannian $G(t, w)$ induced by V; if $V = H^0(T)$, set $h_T := h_V$.

Let us fix an integer $r \geq 1$. Let E be a rank r vector bundle on C. Let $\{s_i\}$, $1 \leq i \leq r$, be the set of invariant associated to ruled subvarieties of codimension $1 \leq i \leq r$ in the Chow ring of $P(E)$:

$$s_i(P(E)) := \inf_{Q \subset P(E)} ib^r + r \deg Q$$

with i the codimension of Q in $P(E)$, $b = [\mathcal{O}_{P(E)}(1)]$ in the Chow ring of $P(E)$ and $\deg Q$ the degree of Q defined by the intersection formula in the Chow ring of $P(E)$, (see [L1], §1). By Lemma 1.2 of [L1] they coincide with the degrees of stability of E which are the integers $\{ri(\mu(E) - \mu(H_i))\}_{1 \leq i \leq r}$, with H_i any rank i subbundle of maximal degree of C and $\mu(F) := \deg(F)/\operatorname{rank}(F)$ for any vector bundle F.

§1.

In this section we investigate under which hypothesis a rank 2 vector bundle is spanned. Unlike the usual Clifford theory for rank 2 bundles (see e.g. [Te1], [Te2], [Ta]) we are interested also in a geometric condition "spanned" which in the rank 1 case coincides with the usual one, as it will be clear in what will follows. Furthermore we are interested in results on every smooth curve, not just on a sufficiently general one (e.g. on curves for which it holds the dimensional equality of Brill Noether Griffiths Harris theorem for special divisors, as in [Ta]). Moreover we do not want to restrict ourselves to stable or semistable bundles.

(1.1.) Assume now that E is a vector bundle of rank 2 given by an extension of the form

$$0 \to A \to E \to B \to 0. \tag{1}$$

Let us consider the long exact sequence of the cohomology and let us write J for the image of $H^0(E)$ in $H^0(B)$.

Remark 1.1. If E is spanned then obviously:

i) B is spanned by the image J of $H^0(E)$,

ii) if $h^0(B) = 2$ then $\dim J = 2$,

iii) if $\dim J = 2$ then $h^0(A) > 0$ unless $E = \mathcal{O}^{\oplus 2}$.

With $\mathrm{Exs}(B, A)$ we will denote the following subspace of $\mathrm{Ext}(B, A)$:

$$\mathrm{Exs}(B, A) := \{h^0(B) = 2, \quad B \text{ spanned}\}.$$

If, moreover, $h^1(B) > 0$, the existence of such a B with a precise degree b concerns the classical rank 1 Clifford–Brill–Noether theory and then reflects a very important part of the geometry of C.

Remark 1.2. From the long exact sequence of cohomology associated with (1), it is trivial that if $H^1(A) = 0$ then $J = H^0(B)$ (e.g. $H^1(A) = 0$ when $a > 2g - 2$ or $a \geq g$ and A is a general line bundle). However in the general case a useful result is the following proposition.

Proposition 1.3. *For a general* $e \in \mathrm{Ext}(B, A)$ *we have*

$$\dim(J) = \max\{0, h^0(B) - h^1(A)\}.$$

In particular we have:

(i) *If* $h^1(A) > 0$, *then* $J \neq H^0(B)$ *and if* $e \in \mathrm{Exs}(B, A)$ *the bundle E associated to e is not spanned.*

(ii) *If* $e \in \mathrm{Exs}(B, A)$ *and* $h^1(A) \geq 2$, *then* $H^0(A) \cong H^0(E)$.

Proof. Let us discard the trivial cases. We may assume $J \neq 0$. Let $\mathcal{E}(B, A)$ be the universal extension of $p_2{}^*B$ and $p_2{}^*A$ on $\mathrm{Ext}(B, A) \times C$ (see [L3]); therefore there is the exact sequence

$$0 \to p_2{}^*A \to \mathcal{E}(B, A) \to p_2{}^*B \to 0$$

on $\mathrm{Ext}(B, A) \times C$ such that for every k-rational point e of $\mathrm{Ext}(B, A)$ the restriction of $\mathcal{E}(B, A)$ to $C \times \{e\}$ is just the vector bundle represented by e.

Let us call α the morphism induced on the zero-cohomology of $\mathcal{E}(\mathcal{B}, \mathcal{A})$ to the zero-cohomology of $p_2{}^*B$ and $\tilde{\alpha}_E$ the map between $H^0(E)$ and $H^0(Q)$ for a fixed extension E, then the map

$$E \to \alpha_e(H^0(C \times \{e\}, \mathcal{E}(B, A)_E)) = \tilde{\alpha}_E(H^0(C, E))$$

is an upper semicontinuos function on $\operatorname{Ext}(B, A)$. Let us observe that the lowest value of this map is $h^0(B) - h^1(A)$. So by semicontinuity the set in which this value is reached is open. Then we have only to show that it is nonempty. Let E be an extension of B with A and let us consider the following elementary transformation of E. Choose a general point $P \in C$ such that there is a vector v in the fiber E_P of E at P, v not in the image of the fiber A_P of A, and such that there is $z \in H^0(E)$ with $z(v) \neq 0$; set

$$0 \to E' \to E \to \boldsymbol{k}_p \to 0$$

with $\boldsymbol{k}_p$ generated by v, an elementary transformation of E with respect to v. Then we obtain a sequence

$$0 \to A \to E' \to B' \to 0 \qquad (2)$$

with $\deg(B') = \deg(B) - 1$, E' subsheaf of E and B' subsheaf of B.

Let us observe that z does not induce a section of P', since z does not induce a section on E'; then $h^0(E) = h^0(E') + 1$ and since $\deg(B') = \deg(B) - 1$ $h^0(B) = h^0(B') + 1$. Hence if we set $J' := \operatorname{Im}(H^0(E')) \subseteq H^0(B')$, we have $J' \subset J$ and $\dim(J') = \dim(J) - 1$.

Let us repeat this construction so to obtain a vector bundle E'' of rank two and the exact sequence:

$$0 \to A \to E'' \to B'' \to 0$$

with $\deg B'' = \deg B - 2$, $c_1(E'') = c_1(E) - 2$, $\dim J'' = \dim J - 2$.

With a twist of E'' by a general point $Q \in C$ we can determine a vector bundle with $c_1(G) = c_1(E)$, $h^0(G) = h^0(E) + 1$. Then $\dim J_G = \dim J - 1$. The quotient of G , with respect to A, is a line bundle with the same degree of B, but is not in general isomorphic to B. Let us iterate this procedure until $\dim J_n = h^0(B) - h^1(A)$. Though we reached the right dimension for the J_n we could obtain a line quotient B_n a priori not isomorphic to B. To resolve this final problem we proceed in this way. B_n and B differ by an element in Pic^0, so thank to the surjectivity of the map $S^t(C) \to \operatorname{Pic}^t(C)$ for $t \geq g$, (where $S^t(C)$ is the symmetric product of C) we can choose at least g points on C such that we can recover B by B_n.

Then we repeat the initial construction with respect to each of the point chosen such that $h^0(E_n)$ does not vary. □

(1.2.) A vector bundle E of rank r is called ample (resp. very ample) if $\mathcal{O}_{\mathbb{P}(E)}(1)$ (the tautological line bundle) is ample (resp. very ample). Moreover E is spanned iff $\mathcal{O}_{\mathbb{P}(E)}(1)$ is spanned.

Then we adopt the following

Definition 1.4. A bundle E is called to be *geometrically very ample* (abbreviated gva) if the pair $(E, H^0(E))$ induces an embedding in the appropriate Grassmannian.

More generally, if $V \subseteq H^0(E)$ is a vector space, we define the pair (E, V) gva if it induces an embedding in the appropriate Grassmannian.

Let us observe that both gva and very ampleness imply that E is spanned; though gva does not imply that E is very ample but only that $\det(E)$ is very ample; indeed the pair $\big(\det(E), \bigwedge^r(V)\big)$ induces an embedding of C in a projective space. Here we use the fact that the associated tautological quotient line bundle for the Plücker embedding of the Grassmannian G is $\det(Q)$ with Q being the tautological quotient vector bundle of G.

Now assume E spanned; then E is geometrically very ample if and only if for every $P \in C$, $H^0(E(-P))$ has no full fiber in its base locus, i.e. if and only if for every $Q \in C$ (the case $P = Q$ being allowed), we have $h^0\big(E(-P-Q)\big) < h^0\big(E(-P)\big)$.

Example 1.5. The notions of gva and very ampleness are different, even for ample and spanned rank 2 vector bundles which are direct sum of two line bundles. This is the case of the following example.

If E is very ample, then every quotient bundle F of E is very ample and indeed we have the very ampleness of the pair (F, M) with $M \subseteq H^0(F)$ image of $H^0(E)$ under the quotient map. Take $A', A'' \in \mathrm{Pic}(C)$ ample and spanned with A' very ample; since A'' is a quotient of $E := A' \oplus A''$, the bundle E is very ample if and only if A'' is very ample, while it is always geometrically very ample.

Let E be a rank 2 vector bundle which is the middle term of an exact sequence like (1). Let us assume that E is spanned; with the notation of the Proposition 1.3 we recall the following easy results.

If J induces an embedding of C, then E is geometrically very ample.

If A is very ample, then E is geometrically very ample.

If $B \cong \mathcal{O}$, then E is geometrically very ample if and only if A is very ample.

Remark 1.6. Since $\mathbb{P}(E)$ is smooth of dimension r and $\mathrm{Pic}\big(\mathbb{P}(E)\big) \neq \mathbb{Z}$, if E is very ample we have $h^0(E) \geq 2r$ by a theorem of Barth ([Ba] or see [Ha] in positive characteristic). If $r = 2$ and $h^0(E) = 4$, by the classification of smooth ruled surfaces in $\mathbb{P}^3$ we have $C \cong \mathbb{P}^1$ and $E \cong \mathcal{O}(1) \oplus \mathcal{O}(1)$.

Remark 1.7. Assume $r = 2$, $h^0(E) = 3$ and E geometrically very ample. Then, just by the definition of geometrically very ample, the curve C is isomorphic to a smooth plane curve X and $E \cong T\mathbb{P}^2(-1)|X$. It is known (and very easy) that $T\mathbb{P}^2(-1)|A$ is ample if and only if X is not a line.

Example 1.8. Let us suppose $r = 2$, $h^0(E) = 4$ and E geometrically very ample. In this case we define:

- E is "non-degenerate" if $h_E(C)$ is not contained in a plane of the family of planes of $G(2, 4)$ which is defined by the universal subbundle of $G(2, 4)$, or equivalently E has $h^0 > 3$.
- E is "linearly normal as bundle" if $h_E(C)$ does not come as a projection from $G(2, 5)$ to $G(2, 4)$, or equivalently $h^0(E) < 5$.

For $g \leq 1$ it is easy to give a complete classification of bundles with $h^0(E) = 4$; moreover such description shows the difference between the notion of non degeneracy for E and for $\det(E)$, while for higher genera it is easier to give examples of linearly normal vector bundle (for example one can find smooth complete intersections on the Klein quadric which do not come as a projection from $G(2, 5)$).

Proposition 1.9. (Classification of the gva bundles on curves of genus $g \leq 1$.) *Let E be a gva vector bundle of rank 2 with $h^0(E) = 4$.*

If $g = 0$ then either $E \cong \mathcal{O} \oplus \mathcal{O}(2)$, or $E \cong \mathcal{O}(1) \oplus \mathcal{O}(1)$.

If $g = 1$ and E is decomposable, then either $E \cong \mathcal{O} \oplus L$ for every $L \cong \det(E) \in \mathrm{Pic}^3(C)$ or $E \cong A \oplus B$ with $A, B \in \mathrm{Pic}^2(C)$ and $A \not\cong B$. If $g = 1$ and E is indecomposable, then $\deg(E) = 4$ and E is uniquely determined by the choice of $\det(E) \in \mathrm{Pic}^4(C)$

Proof. The case $g = 0$ is essentially the splitting criterion for vector bundle on $\mathbb{P}^1$.

If $g = 1$ and E is decomposable, the only non-obvious case is a direct consequence of the Riemann–Roch theorem on line bundles: $A \cong B$ iff the morphism h_E is not an embedding at the 4 points (or two or one in characteristic 2) P such that $A \cong \mathcal{O}(2P)$.

By the Atiyah classification of indecomposable vector bundles on an elliptic curve (see [At], Lemma 15) and Riemann–Roch theorem (see [At] Lemma 15), we have $\deg(E) = 4$, E semistable, $h^1(E(-P)) = 0$ for every P (hence E spanned) and $h^0\big(E(-P-Q)\big) < h^0\big(E(-P)\big)$ for every P, Q, (hence E is gva); then any indecomposable E with $\deg(E) = 4$ gives a gva bundle, moreover it is uniquely determined by the (arbitrary) choice of $\det(E) \in \mathrm{Pic}^4(C)$ (see [At], Theorem 6, 7). □

Remark 1.10. The existence of a gva bundle E on C with $r = 2$, $h^0 = 4$ and with a desired degree of instability (i.e. with a fixed $s_1 < 0$), is linked to the Brill–Noether theory for vector bundle of rank 1; indeed let (1) be an exact sequence such that A is the highest degree line subbundle of E (hence A unique since E is unstable). Since E has more than 3 sections, for every $P \in C$ there is $w \in H^0(E)$, $w \neq 0$, with $w(P) = 0$; hence $\deg(A) > 0$. Note that B is spanned by the image of $H^0(E)$; hence if $B \not\cong \mathcal{O}$, we have $h^0(A) \leq 2$. Thus $\deg(A) \leq g + 1$ and if $A \not\cong 0$ then B is a g^1_k; if $t := -s_1 = \deg(A) - \deg(B)$ we have $t \leq g + 1 - k$. Then under this hypothesis a bundle E with $s_1 < 0$ does not exist if $k \geq g + 1$; for $k \leq g + 1$ its existence is the existence of a g_k^1.

Let us note that by one of the definitions of the coboundary in the cohomology of (1), if $B = \mathcal{O}$ the unique section of B lifts to E if and only if (1) splits.

(1.3.) This short subsection is a continuation of the Subsection (1.2.) with a different problem as target, but the same tools. Now we are interested in the smallest positive degree for a spanned rank 2 vector bundle, E, on C which induces a birational map to a Grassmannian.

To avoid trivial cases we suppose $h^0(E) \geq 3$. If $h^0(E) = 3$, then we have $E \cong f^*(T\mathbb{P}^2(-1))$ with $f : C \to \mathbb{P}^2$ a birational morphism on its image. Since $\det(T\mathbb{P}^2(-1)) \cong \mathcal{O}_{\mathbb{P}^2}(1)$, $\deg(E) = \deg(f(C))$, i.e. the smallest degree of such bundles is just given by the smallest integer k such that C has a g_k^2.

For the same reasons if we take a bundle E with $h^0(E) = 4$, but we consider maps to a Grassmannian induced by a proper subspace of $H^0(E)$ of dimension three, then the smallest degree that we have to consider is the smallest integer, m, such that it exists a linear series g_m^2. Then the situation is analogous to the case $h^0(E) = 3$.

Hence from now on, we consider the first non immediate case: $h^0(E) = 4$ and $V = H^0(E)$. If we consider reducible bundles, the smallest $\deg(E)$ is $a + b$, where a and b are the smallest integers such that C has a g_a^1 and a g_b^1 not composed with an involution. But among the vector bundles in $\mathrm{Ext}(A, B)$ there may be indecomposable vector bundles with lower degree (e.g. we refer to remark 1.11). So we have to consider indecomposable bundles.

Let A be a maximal degree subbundle of E. B is spanned. $a := \deg(A)$, $b := \deg(B)$. Let us suppose $\dim J = 2$ i.e. B is a g_b^1. Then A is a g_a^1 and E is spanned iff either A and B are spanned. Assume that A and B are base point free pencils. If $b - a \leq g$ then by Corollary 1.2 of [LN] (if $b - a < 0$ is obvious), there is always a non zero extension (1). It is easy to check that under all our assumptions the induced map $h_E : C \to G(2, 4)$ is birational if and only if A and B are not composed with the same involution. Hence the role of A and B is the same and the sign of the integer $b-a$ gives only the degree stability of E and the dimension of the family of extensions with a fixed choice of A and B.

Since generically E has finitely many maximal degree line subbundles and always it has at most a one dimensional family of maximal degree line subbundles (see [LN], Corollary 1.2 and [L2], Proposition 6.1), this is essentially the dimension of the isomorphism classes of the bundles constructed using A and B. For instance, if C has at least two pencils of minimal degree, this is the solution of our original problem under our restrictive assumptions. Not only this is the case for C with general moduli, but even if C has only one pencil with minimal degree, taking it as A or B is usually the best choice for finding E of low degree. But for very particular curves the situation may be more complicated and interesting. Let us consider the case of hyperelliptic and trigonal curves.

Example 1.11. Let us assume C hyperelliptic. Then at least one between A and B have to be non special and the best choice is if the other one is the g_2^1 (assuming $g \geq 2$); in this way we have $g+3$ has minimal $\deg(E)$ obtained. However if C is hyperelliptic, a better choice with B spanned is obtained by taking $A = \mathcal{O}$ and B a general (hence birational) g_{g+2}^2.

Example 1.12. Assume C trigonal and $g \geq 5$ (hence with a unique g_3^1). We make the assumption that $h^1(A) > 0$ and $h^1(B) > 0$. Since any special spanned line bundle L on C is such that either L or $K \otimes L^*$ is the multiple of the g_3^1 defined as the minimal integer $a+b$ depends on the Maroni invariant of C (see e.g. [Sc] and [CKM]). Such a invariant is defined as the first integer w such that, if R is the g_3^1, $h^0(R^{\otimes w}) > w+1$ (assuming that $R^{\otimes w}$ is special, otherwise, as always, if C is a general trigonal curve, under these assumptions there is no bundle E (see [B2])).

§2.

This section contains a construction of possible *degenerations* of an embedding of a smooth curve C contained in a Grassmannian (see Theorems 3.3 and 3.4). and a following application of Theorem 3.3 to map associated to derived bundles (introduced in [Pe1] and [Pe2]) of a spanned vector bundle on C (see Theorems 3.5 and 3.6).

(2.1.) *Definitions and notations.*

Let C be a smooth curve (usually of genus $g \geq 2$) embedded in any Grassmannian G, say $G = G(r, v)$; let Q be the tautological rank r quotient bundle and S the tautological rank $v-r$ subbundle. Set $E := Q|C$ and let $V \subseteq H^0(C, E)$ be the vector space inducing the given embedding (h_V) of C into G.

Let $\mathrm{Pr}(C, V, E; G)$ (or just $\mathrm{Pr}(C, V; G)$ or $\mathrm{Pr}(C, V)$) be the closure in the Hilbert scheme $\mathrm{Hilb}(G)$ of the family of all $h^*(C)$ with $h \in \mathrm{Aut}^0(G)$. Let us note that $\mathrm{Aut}(G)$ is connected unless $v = 2r$, when it has two connected components corresponding to the duality of $G(r, v)$ and $G(v-r, v)$.

Let $\mathrm{Cl}(C, G)^{**}$ be the closure in $\mathrm{Hilb}(G)$ of all embeddings of the abstract curve C into G and $\mathrm{Cl}(C, V, G)$ be the integral component of $\mathrm{Cl}(C, G)^{**}_{\mathrm{red}}$ containing the given embedding of C into G.

Our aim is to find reducible curves $D \in \mathrm{Pr}(C, V, E; G)$ (see Theorem 3.3) and in $\mathrm{Cl}(C, V, G)$ (see Theorem 3.4). For related results, see [B3].

Definition 2.1. A tree T in a Grassmannian $G := G(r, v)$ is a reduced and connected curve with $p_a(T) = 0$; moreover it has no loop, its irreducible components are smooth and rational and the irreducible components, say $T_1, \ldots, T_s$, of T containing $P \in \mathrm{Sing}(T)$ have tangent cone at P spanning a Zariski tangent space of maximal dimension.

Usually we will use only nodal trees.

We choose an order, say $T_1, \dots, T_h$, for the irreducible components of T; let $a_u := (a_{u1}, \dots, a_{ur})$, $1 \le u \le h$, be an ordered r-ple of non negative integers with $a_{ui} \ge a_{uj}$ if $i \ge j$; let $\tau := (h; a_1, \dots, a_h)$.

Definition 2.2. T has *type* τ (with respect to the chosen order of its components) if for every u the splitting type of $Q|T_u$ is a_u.

A "line" D of G is a smooth rational curve $D \subseteq G$ with $Q|D$ of splitting type $(1, 0, \dots, 0)$, i.e., a line of the projective space $\mathbb{P}(H^0(G, \mathcal{O}_G(1)))$ in which G is embedded by the Plücker embedding.

Let us define the degree of a reduced curve C in G as the degree of C with respect to the Plücker embedding, i.e. it is $\deg(\mathcal{O}_G(1)|C)$. Let E be a rank r vector bundle on C; fix a vector space $V \subseteq H^0(C, E)$ and $P \in C$; V spans E at P if and only if $\dim(V \cap H^0(E(-P))) = \dim(V) - r$; V embeds E at P if and only if it spans E at P and $\dim(V \cap H^0(E(-2P))) < \dim(V) - r$.

A bamboo T in G is a nodal tree in G with "lines" as irreducible components and such that the "lines" form a chain, i.e. such that we may order its irreducible components $T_1, \dots, T_h$, $h := \deg(T)$, in such a way that $T_i \cap T_j$ if and only if $|i - j| \le 1$; note that we have $\mathrm{card}(T_i \cap T_{i+1}) = 1$ for every $i < h$.

(2.3.) Let us fix a smooth genus g curve C and a rank r vector bundle E on C. Take a linear subspace $V \subseteq H^0(C, E)$ spanning E and inducing an embedding $h_V : C \to G(r, v)$ with $v := \dim(V)$. It is easy to check that a general vector space $W \subseteq V$ with $\dim(W) \ge r + 2$ induces an embedding of C. Indeed fix $P \in C$; a general $W'' \subseteq V$ with $\dim(W'') = r + 1$ spans E and induces an embedding $h_{W''}$ of C at P. Hence $h_{W''}$ is an embedding outside finitely many points of C. Furthermore, it is easy to check that for a general W'' the morphism $h_{W''}$ has differential vanishing at most of order 1 at every points of C; hence adding another suitable section to W'' we find W which induces an embedding of C.

Theorem 2.4. *Fix a smooth genus g curve C and a rank r vector bundle E on C. Take a linear subspace $V \subseteq H^0(C, E)$ spanning E and inducing an embedding $h_V : C \to G(r, v)$. Set $v := \dim(V)$; fix an integer $w \ge r + 2$, a vector space $W \subseteq V$ with $\dim(W) = w$, W spanning a rank r subsheaf E' of E and inducing in this way an embedding of C into $G(r, w) \subseteq G(r, v)$, a point $P \in C$ and a bamboo $T \subseteq G$ intersecting $h_W(C)$ only at $h_W(P)$, the intersection being quasi-transversal. Assume $\deg(T) = \deg(E) - \deg(E') \le v - w$. Assume that the scheme E/E' has support at P and that it is a quotient of $\mathcal{O}_C$, i.e. that it is the Cartier divisor $\deg(T)P$ on C. Then $h_W(C) \cup T \in \mathrm{Pr}(C, V, E; G)$.*

Proof. First assume $v - w = 1$. Note that E' is obtained by E as a kernel of a surjection from E to a skykraper sheaf of length 1 supported at P. We take another hyperplane

W'' of V and assume that W'' spans E and induces in this way an embedding of C into G. Set $A := W'' \cap W$. Note that a general hyperplane B of V with $A \subset B$ spans E and induces an embedding of C into G. The result follows from the completeness of Hilb(G) exactly as in the rank 1 case considered in [BE1], (first part of the proof of 2.1 with particular attention to the picture of page 4), [B1], §7, and [BE2], Remark 0.11. The case $v - w > 1$ follows very easily by induction (exactly as in [BE1], proof of Proposition 2.1). □

Theorem 2.5. *Fix a smooth genus g curve C and a rank r vector bundle E on C. Take a linear subspace $V \subseteq H^0(C, E)$ spanning E and inducing an embedding $h_V : C \to G(r, v)$. Set $v := \dim(V)$; fix an integer $w \geq r+2$, a vector space $W \subseteq V$ with $\dim(W) = w$, W spanning a rank r subsheaf E' of E and inducing in this way an embedding of C into $G(r, w) \subseteq G(r, v)$, a point $P \in C$ and a smooth rational curve $D \subset G$ intersecting $h_W(C)$ only at $h_W(P)$, the intersection being quasi-transversal. Assume $\deg(D) = \deg(E) - \deg(E') \leq v - w$. Assume that the scheme E/E' is supported by P and that it is a quotient of $\mathcal{O}_C$. Then $h_W(C) \cup D \in \mathrm{Cl}(C, V, E; G)$.*

Proof. Let R be the abstract reducible nodal curve which is an union of C and a curve $A \cong \mathbb{P}^1$ which transversally intersect C only at P (hence with $p_a(R) = p_a(C)$).

Consider the map π of R into $G := G(r, v)$ which coincides with h_V on C and maps A into the point $h_V(P)$ (hence with $(\pi^*(Q))|A \cong r\mathcal{O}_A$). We claim that π is a smooth point of the scheme of all morphisms from R to G. Note that for a Mayer–Vietoris type exact sequence , $h^1(N_R) = 0$, with N_R the normal bundle of R. Then by the general theory of deformations of maps with fixed source and target it remains to check that the elementary anti-transformation (i.e. increasing the degree by 1) with respect to the one dimensional vector subspace of the fiber over $C \cap A$ of the normal sheaf $N_{\pi|A}$ of $\pi|A$ (which is a rank r vector bundle) has splitting type $(0, \dots, 0, -1)$ and not $(1, 0, \dots, 0, -2)$ (or if $r = 1$ the type is (-1)). Hence the claim is local around A and we may assume R rational.

The case $r = 1$ is trivial because $\deg(N_{\pi|A}) = -1$. Then we may take a suitable quotient to reduce ourselves to the case $r = 1$ and to conclude. □

With Theorem 2.4 instead of the quotations from [B3] (essentially Lemma 1.8 of [B3]), we can modify the proofs of Theorems 0.1 and 0.2 of [B4] and then we can obtain respectively the following results on the derived bundles on spanned vector bundles on C (we refer to [Pe1] and [Pe2] for definitions and results about derived bundles).

Theorem 2.6. *Let C be a smooth curve of genus g and E a rank r vector bundle on C such that for a general $P \in C$ we have $h^1(E(-P)) = 0$. Fix an integer v with $r < v \leq h^0(E)$ and a general vector space $V \subseteq H^0(E)$ with $\dim(V) = v$. Then we have:*

(i) *If $v \geq 2r$ the first derived bundle of (E, V) is 0.*

(ii) *If $v > 2r$ the first torsion sheaf of (E, V) is* 0.

(iii) *If $v = 2r$ the first torsion sheaf of (E, V) has degree* $2\deg(E) - 2r + 2rg$.

(iv) *If $v < 2r$ the first derived bundle of (E, V) has rank $2r - v$.*

Theorem 2.7. *Fix integers r, v and i with $1 < r < v < 2r$, $i \geq 1$. Set $a := 2r - v$, $b := [(r + a - 1)/a]$ and $r' = ab$. Fix a smooth genus g curve C and a rank r vector bundle on C such that for a general $P \in C$ we have $h^1\big(E(-iP)\big) = 0$. Let $V \subseteq H^0(E)$ be a general linear space with $\dim(V) = r$. Let $drk_i := r_{i-1} - r_i$ be the i-th differential rank of (E, V); so if $i \leq b$, then $drk_i = a$, if $i = b + 1$ then $drk_i = r' - r$ and if $i > b + 1$ then $drk_i = 0$.*

§3.

Here we assume $g = 1$ and characteristic 0 (unless otherwise stated). We will consider only the case of the complete embedding induced by the space of global sections, $H^0(E)$, of an indecomposable rank r vector bundle E on the elliptic curve C. First, we will analyze the conditions under which jet bundles of E are spanned (see Proposition 3.1). Since the kernel of the evaluation surjection $H^0(E) \to E$ is again an indecomposable rank $d - r$ vector bundle (see Remark 4.3), by the fundamental Duality Theorem ([Pe1], Theorem II.6.1 and Corollary II.6.4, [Pe2], Theorem 7.1) we reduce at once the case of derived bundles to the case of higher order jets (see Theorem 3.4 for the statements of the results obtained in this way).

Just by Riemann–Roch and the classification and cohomological properties of indecomposable bundles on an elliptic curve ([At]) we have the following result.

Proposition 3.1. *Let E be an indecomposable rank r vector bundle of degree d on the elliptic curve C. Set $d := kr + y$ with $0 \leq y < r$, $k > 0$ and $k > 1$ if $y = 0$. Then:*

(a) *$H^0(E)$ spans the jets of E to order exactly $k - 1$ if $y \neq 0$, of order $k - 2$ if $y = 0$;*

(b) *if $y = 0$, $H^0(E)$ spans the $k - 1$ jets of E outside the $(k-1)^2$ points $P \in C$ with $\mathcal{O}_C\big((k-1)P\big) \cong \det(E)$.*

Remark 3.2. If characteristic $p > 0$ and p divides $k - 1$ one has to change the statement and the proof of Proposition 3.1 only in part (b); in this case $H^0(E)$ spans the $k - 1$ jets of E outside the points of C of order dividing $k - 1$, i.e., if $k - 1 = p^e m$ with $(p, m) = 1$, take $p^e m^2$ for C not supersingular and m^2 for C supersingular.

Remark 3.3. Let E be an indecomposable rank r vector bundle of degree d. Set $d := kr + y$ with k, y integers and $0 \leq y < r$. Assume $k \geq 1$ and $y \geq 1$, i.e., assume E spanned and stable. Let F be the rank $d - r$ vector bundle of degree $-d$ arising as kernel of the surjective evaluation map $H^0(E) \to E$. Here we will check the well-known fact that F is indecomposable. Assume by contradiction $F = A \oplus B$ with $\mathrm{rank}(A)/\deg(A) \leq \mathrm{rank}(B)/\deg(B)$. Then $H^0(E)/A$ is a subbundle of E with $\mathrm{rank}(H^0(E)/A)/\deg(H^0(E)/A) \geq \mathrm{rank}(E)/\deg(E)$. The existence of such bundle is impossible because E is stable.

Theorem 3.4. *Let E be an indecomposable rank r vector bundle of degree d on the elliptic curve C. Set $d := rk + y$ with $0 \leq y < r$. Assume $k > 0$ and $y > 0$ (i.e., assume E spanned and stable). Then for every $i \geq 1$ the i-th differential rank $r_{i-1} - r_i$ of $\big(E, H^0(E)\big)$ is $\min\{d - r, r - (i-1)(d-r)\}$.*

§4.

Here we introduce a very important problem (even for the irreducibility of other moduli problems of higher dimensional varieties, of stable sheaves on higher dimensional varieties and of the Hilbert scheme of curves in $\mathbb{P}^n$). Up to now we have only easy (but useful) results on this problem, but we hope to obtain much more.

Jointly with results of the previous papers to which we refer in the introduction, we got a lot of informations on certain subsets (or "strata") of a moduli scheme of rank r stable vector bundles on C and on all the set of vector bundles of rank r on C. It is very important from many point of views to know when a stratum is in the closure of another stratum and when the closure of a stratum intersects another stratum.

Consider again the extension (1) of sheaves on an integral complete variety Y. Assume that both A and B vary on some integral parameter space T. On the open subset U of T in which the corresponding Ext^1 has minimal dimension by [BPS] the relative Ext^1 is a vector bundle and hence an integral variety. Assume $U \neq T$. We try to find conditions which imply that the reduction of the total space of the relative Ext^1 is an integral variety when A and B vary on all the parameter space T.

The building block for our results (see Corollary 4.5) is the rank 2 case proved in [LN], Proposition 3.1, and Proposition 4.2 below, i.e. the stratification of the moduli space $M(2, d)$ of rank 2 stable bundles of degree d according to the degree of the maximal subbundle.

For the proof of 4.2 we need the following lemma.

Lemma 4.1. *Fix an ordered set $(M_1, \ldots, M_r)$ with $M_i \in \mathrm{Pic}(C)$, and set $F := \bigoplus M_i$, $1 \leq i \leq r$. Assume the existence of a filtration $\{E_i\}$, $0 \leq i \leq r$, of the rank r vector bundle E with $E_0 = \{0\}$, $\mathrm{rank}(E_i) = i$ if $i \geq 1$ and $E_i/E_{i-1} \cong M_i$ for every $i \geq 1$. Then F is the flat limit of a family of bundles all isomorphic to E.*

Proof. Let e_i, $(1 \leq i < r)$, be the class of the extension

$$0 \to E_i \to E_{i+1} \to M_{i+1} \to 0 \tag{3}$$

We find by induction on r the flat family taking as extension classes te_i, $t \in \mathbb{K}$, $t \neq 0$, t going to 0. □

We believe that the following result is well known, but we have not found a published proof.

Proposition 4.2. *Fix an integer s with $0 < s \leq g-1$. Let $M(2, d; L)$ be the moduli scheme of rank 2 stable vector bundles on C with determinant L. Then the stratum with $s_1 = s$ is in the closure of the stratum with $s_1 = s+2$.*

Proof. Let us observe that the proposition is trivially valid in the cases $s \geq g-3$. Infact in these cases the integer $s+2$ reaches the highest value as possible ($s+2 = g-1$, *if* $d \equiv g-1$, $s+2 = g$ if $d \equiv g$); therefore since s_1 is lower semicontinuos its associated stratum is open. Then let us suppose that $s \leq g-4$. Let π be the canonical projection of $P(E)$ on C.

Now let us suppose that $P(E)$ has only one minimal section, σ, (i.e., E has only one maximal line subbundle). We can identify σ with C and a selected point P in $\operatorname{supp} \sigma$ with a precise point of C called again P. Since a fiber of π and σ has intersection 1, we can choose a general point Q of the fibre through P, distinct from P, such that σ does not pass through Q; by Lemma 4.3 of [LN] the vector bundle E', obtained by an elementary transformation of $P(E)$ in Q, has invariant $s_1(E') = s_1(E) + 1$.

If E' has a finite number of maximal subbundles then $E' \otimes \mathcal{O}(P)$ has a finite number of maximal line subbundles; hence with an elementary transformation of $E' \otimes \mathcal{O}(P)$ with respect to a point Q' in the fiber on P of $P(E' \otimes \mathcal{O}(P))$, we can find a vector bundle G with invariant $s_1(G) = s_1(E) + 2$. As Q' varies in the fibre we find a flat family of vector bundles in the stratum with $s_1 = s_1(E) + 2$ parameterized by $\boldsymbol{k}$ whose flat limit is E.

It remains to show that we can choose E' and E with a finite number of maximal line subbundles. Let us recall that by Proposition 3.3 of [LN], for every $0 \leq s \leq g-2$ there is an open dense subset, V_s, of the stable vector bundles of rank 2 with $s_1(E) = s$ such that every vector bundle E of V_s has exactly one maximal subbundle. Therefore we proceed in two steps; first we restrict ourselves to consider the extensions of torsion sheaves with support in P and elements in the open subset, V_{s+1}, of stable vector bundles of rank 2 and degree $d-1$ with a finite number of maximal line subbundles; then we restrict the space of these extensions to the open subspace of vector bundles with a finite number of maximal line subbundles (again by Proposition 3.3 of [LN]). Finally let us choose E in this space. □

For 4.4 we need the following remark.

Remark 4.3. With no restriction on $\operatorname{rank}(A)$ and $\operatorname{rank}(B)$ in (1), assume that E is simple. This implies $h^0(A \otimes B^*) = 0$. Hence by Riemann–Roch and semicontinuity the set of extensions (1) with E simple is an open subset of a vector space of dimension $h^1(A \otimes B^*) = \chi(A \otimes B^*) = \deg(A)\operatorname{rank}(B) - \deg(B)\operatorname{rank}(A) + \operatorname{rank}(A) \cdot$

rank$(B)(1-g)$ (hence it depends only on $\deg(A)$, $\deg(B)$, rank(A) and rank(B)).

Remark 4.4. Let us define a vector bundle Q a *general extension* of other two, say M and P with respect to L, if it belongs to the open subset of such extensions Q' of the same pair, such that $\mathrm{Ext}^1(Q', L)$ has minimal dimension.

Let P, P', M, M', L vector bundles on C. Let S (resp. S') the set of all vector bundles on C which are extensions by L of a general extension of M by P (resp. of M' by P'). Assume that both S and S' contain simple bundles and that the set of general extensions of M' by P' is in the closure of the set of general extensions of M' by P'. Set u (resp. u') the minimal dimension of Ext^1-group with respect to L, and suppose that $u = u'$. Then we claim that S' is in the closure of S. To check the claim use [BPS] and the assumption on the existence of simple bundles to apply Remark 4.3. In an analogous way we prove:

Corollary 4.5. *Fix line bundles P, P', M, M', L on C with* $\deg(P) + \deg(M) = \deg(P') + \deg(M')$, $\deg(P) \leq \deg(P') < \deg(M')$, $\deg(L) < \deg(M')$. *Let S (resp. S') the set of all vector bundles on C which are stable extensions by L of a stable vector bundle which is an extension of M by P (resp. of M' by P'). Then S' is in the closure of S.*

Proof. Let U (resp. U') be the set of stable vector bundles on C which are extensions of M by P (resp. of M' by P'). By [LN], Proposition 1.2, we have $U' \neq \emptyset$. Since $s_1 = \deg(M) - \deg(P) < \deg(M') - \deg(P') = s_1{}'$, by Proposition 4.2 U' is in the closure of U. By the assumption on $\deg(L)$ we have $S' \neq \emptyset$, $S \neq \emptyset$. Hence we conclude by Remark 4.4. □

Remark 4.6. If the reader is interested in analogous results in families of vector bundles with fixed determinant, he just has to assume in 4.4 and 4.5 the hypothesis $\det(P) \otimes \det(M) \cong \det(P') \otimes \det(M')$.

References

[At] M. Atiyah, Vector bundles over an elliptic curve, Proc. London Math. Soc. 7 (1957), 414–452, reprinted in: M. Atiyah, Collected Works, Vol. I, pp. 103–143, Oxford Science Publications, Claredon Press, Oxford, 1988.

[B1] E. Ballico, On the postulation of canonical curves in $\mathbb{P}^3$, Ark. Math. 22 (1984), 139–151.

[B2] E. Ballico, A remark on linear series on general k-gonal curves, Boll. UMI (7), 3-A (1989), 195–197.

[B3] E. Ballico, Curves in Grassmannians and flag varieties, Boll. U.M.I. (7), 9-A (1995), 313–327.

[B4] E. Ballico, Weierstrass loci for vector bundles on curves, preprint.

[B5] E. Ballico, Vector bundles on curves and Brill–Noether–Gieseker theory, III, preprint n. 455 Dept. of Math. University of Trento.

[BE1] E. Ballico, Ph. Ellia, On degeneration of projective curves, in: Algebraic Geometry — Open Problems, pp. 1–15, Lecture Notes in Math. 997, Springer-Verlag, 1983.

[BE2] E. Ballico, Ph. Ellia, On the postulation of a general projection of a curve in $\mathbb{P}^N$, $N \geq 4$, Ann. Mat. Pura Appl. 147 (1987), 267–301.

[BR1] E. Ballico, B. Russo, Vector bundles on curves and Brill–Noether–Gieseker theory, I, preprint n. 450 Dept. of Math., University of Trento.

[BR2] E. Ballico, B. Russo, Vector bundles on curves and Brill–Noether–Gieseker theory, II, preprint n. 452 Dept. of Math., University of Trento.

[BPS] C. Banica, M. Putinar, G. Schumacher, Variation der globalen Ext in Deformationen kompakter komplexer Raüme, Math. Ann. 250 (1980), 135–155.

[Ba] W. Barth, Transplanting cohomology class in complex projective space, Amer. J. Math. 92 (1970), 951–961.

[CKM] M. Coppens, C. Keem, G. Martens, Primitive linear series on curves, Manuscripta Math. 77 (1992), 237–264.

[Ha] R. Hartshorne, Varieties of low codimension in projective space, Bull. Amer. Math. Soc. 80 (1974), 1017–1032.

[Hi1] A. Hirschowitz, Rank techniques and jump stratifications, in: Vector bundles on algebraic varieties, Proceedings Bombay 1984, pp. 159–205, Oxford University Press, 1987.

[Hi2] A. Hirschowitz, Problemes de Brill–Noether en rang superieur, preprint.

[Hi3] A. Hirschowitz, Problemes de Brill–Noether en rang superieur, C. R. Acad. Sci. Paris Sér. I 304 (1988), 153–156.

[L1] H. Lange, Zur Klassification von Regelmannigfaltigkeit, Math. Ann. 262 (1983), 447–459.

[2] H. Lange, Höhere Secantenvarietäten und Vectorbündel auf Kurven, Manuscripta Math. 52 (1985), 63–80.

[L3] H. Lange, Universal Families of Extensions, J. Algebra 83 (1983), 101–112.

[LN] H. Lange, M. S. Narasimhan, Maximal subbundles of rank two vector bundles on curves, Math. Ann. 266 (1983), 55–72.

[N] Vector Bundles on Algebraic Curves, Problem list — Europroj. Workshop Version, Liverpool, 6–10 April 1994, edited and distributed by P. E. Newstead.

[Pe1] D. Perkinson, Jet bundles and curves in Grassmannians, Ph. D. thesis, Chicago, 1990.

[Pe2] D. Perkinson, Curves in Grassmannians, Trans. Amer. Math. Soc., to appear.

[Sc] F. O.Schreyer, Sygyzies of canonical curve and special linear series, Math. Ann. 275 (1986), 105–137.

[Ta] X.-J. Tan, Some results on the existence of rank two special stable vector bundles, Manuscripta Math. 75 (1992), 365–373.

[Te1] M. Teixidor, Brill–Noether theory for vector bundles of rank 2, Tôhoku Math. J. 43 (1991), 123–126.

[Te2] M. Teixidor, On the Gieseker–Petri map for rank 2 vector bundles, Manuscripta Math. 75 (1992), 375–382.

On Calabi–Yau complete intersections in toric varieties

Victor V. Batyrev and Lev A. Borisov*

Abstract. We investigate Hodge-theoretic properties of Calabi–Yau complete intersections V of r semi-ample divisors in d-dimensional toric Fano varieties having at most Gorenstein singularities. Our main purpose is to show that the combinatorial duality proposed by second author agrees with the duality for Hodge numbers predicted by mirror symmetry. It is expected that the complete verification of mirror symmetry predictions for singular Calabi–Yau varieties V in arbitrary dimension demands considerations of so called *string-theoretic Hodge numbers* $h_{\rm st}^{p,q}(V)$. We restrict ourselves to the string-theoretic Hodge numbers $h_{\rm st}^{0,q}(V)$ and $h_{\rm st}^{1,q}(V)$ $(0 \leq q \leq d-r)$ which coincide with the usual Hodge numbers $h^{0,q}(\widehat{V})$ and $h^{1,q}(\widehat{V})$ of a *MPCP*-desingularization $\widehat{V}$ of V. We prove the duality for $(0,q)$-Hodge numbers and for the alternating sum of $(1,q)$-Hodge numbers. The complete duality for $(1,q)$-Hodge numbers will be checked only for the case of Calabi–Yau complete intersections in projective spaces.

1. Introduction

It was conjectured in [1] that the polar duality for reflexive polyhedra induces the mirror involution for Calabi–Yau hypersurfaces in Gorenstein toric Fano varieties. The second author has proposed a more general duality which conjecturally induces the mirror involution also for Calabi–Yau *complete intersections* in Gorenstein toric Fano varieties. The verfication of predictions for Gromov–Witten invariants of Calabi–Yau complete intersections in ordinary and weighted projective spaces [15, 14] as well as in some toric varieties [2] (see also [5, 8, 9, 11]) can be considered as partial confirmations of this generalized mirror construction.

If two *smooth* n-dimensional Calabi–Yau manifolds V and W form a mirror pair, then their Hodge numbers must satisfy the relation

$$h^{p,q}(V) = h^{n-p,q}(W), \tag{1.1}$$

for all $0 \leq p, q \leq n$.

*Supported by DFG.

However, the combinatorial involutions which were constructed in [1, 6] relate families of *singular* Calabi–Yau varieties. If V is a Calabi–Yau complete intersection of semi-ample divisors in a Gorenstein toric Fano variety $\mathbf{P}$, then there exists always a partial desingularization $\pi : \widehat{V} \to V$ (*MPCP*-desingularization of V) such that:

- $\widehat{V}$ is again a Calabi–Yau complete intersection of semi-ample divisors in a projective toric variety $\widehat{\mathbf{P}}$;
- $\widehat{V}$ and $\widehat{\mathbf{P}}$ have only Gorenstein terminal abelian quotient singularities.

If $\widehat{V}$ is smooth (this is always the case for $n \leq 3$), we can use $\widehat{V}$ instead of V for verification of the duality (1.1). In general, we have to change the cohomology theory and to consider the so called *string-theoretic Hodge numbers* $h_{\text{st}}^{p,q}(V)$ for singular V. For mirror pair of *singular* n-dimensional Calabi–Yau varieties V and W, we must have the duality for the string-theoretic Hodge numbers:

$$h_{\text{st}}^{p,q}(V) = h_{\text{st}}^{n-p,q}(W), \tag{1.2}$$

for all $0 \leq p, q \leq n$.

The main properties of the string-theoretic Hodge numbers were considered in [4]. These numbers satisfy the Poincaré duality and we have $h_{\text{st}}^{p,q}(V) = h_{\text{st}}^{p,q}(\widehat{V})$ for all $0 \leq p, q \leq n$. Moreover, one has the following properties:

Proposition 1.1. *Let $h^{p,q}(\widehat{V})$ denote the usual (p,q)-Hodge number of $\widehat{V}$. Then*

(i) $h_{\text{st}}^{p,q}(V) = h^{p,q}(\widehat{V})$ *for all* $p = 0, 1$ *and* $0 \leq q \leq n$;

(ii) $h_{\text{st}}^{p,q}(V) = h^{p,q}(\widehat{V})$ *for all* $0 \leq p, q \leq n$ *if* $\widehat{V}$ *is smooth.*

The main purpose of this paper is to verify the duality (1.2) for string-theoretic $(0, q)$ and $(1, q)$-Hodge numbers of Calabi–Yau complete intersections V and its mirror partner W predicted by the construction in [6]. According to 1.1(i), it is sufficient to check the analgous duality for the usual Hodge numbers of the corresponding *MPCP*-desingularizations $\widehat{V}$ and $\widehat{W}$. We prove the duality for $(0, q)$-Hodge numbers and for the alternating sum of $(1, q)$-Hodge numbers. The complete duality for $(1, q)$-Hodge numbers will be checked only for the case of Calabi–Yau complete intersections in projective spaces.

In Section 2 we remind necessary facts from the theory of toric varieties. Section 3 is devoted to basic properties of complete intersections in toric varieties. In Section 4 we explain the relation between Calabi–Yau complete intersection in Gorenstein toric Fano varieties and nef-partitions of reflexive polyhedra. In Section 5 we prove the duality (1.2) for $(0, q)$-Hodge numbers and give explicit formulas for them. It turned out to be not so easy to derive general formulas for $(1, q)$-Hodge numbers and to prove the duality (1.2) for $(1, q)$-Hodge numbers in full generality. In Sections 6 and 7 we give explicit formulas and prove the duality only for the alternative sum of $(1, q)$-Hodge numbers; i.e., for the Euler characteristics of the sheaves of 1-forms on $\widehat{V}$ and

$\widehat{W}$. In Section 8, we derive explicit formulas for $(1, q)$-Hodge numbers of Calabi–Yau complete intersections of *ample* divisors. Finally, in Section 9 we establish the duality (1.2) for all $(1, q)$-Hodge numbers of Calabi–Yau complete intersections in projective spaces and their mirror partners.

2. Basic notations and statements

Let M and $N = \mathrm{Hom}(M, \mathbb{Z})$ be dual free abelian groups of rank d, $M_{\mathbb{R}}$ and $N_{\mathbb{R}}$ the real scalar extensions of M and N, $\langle *, * \rangle \; : \; M_{\mathbb{R}} \times N_{\mathbb{R}} \to \mathbb{R}$ the canonical pairing. We consider M (resp. N) as the maximal lattice in $M_{\mathbb{R}}$ (resp. in $N_{\mathbb{R}}$). We denote by $\mathbf{T}$ the *affine algebraic torus* over $\mathbb{C}$:

$$\mathbf{T} := \mathrm{Spec}\, \mathbb{C}[M] \cong \mathbb{C}[X_1^{\pm 1}, \dots, X_d^{\pm 1}].$$

By a *lattice polyhedron* in $M_{\mathbb{R}}$ (resp. in $N_{\mathbb{R}}$) we always mean a convex polyhedron of dimension $\leq d$ whose vertices belong to M (resp. in N). The *relative interior* of Δ is the set of interior points of Δ which is considered as a subset of the minimal $\mathbb{R}$-linear affine subspace containing Δ. For any lattice polyhedron Δ, we denote by $l^*(\Delta)$ the number of lattice points in the relative interior of Δ. We set $b(\Delta) = (-1)^{\dim \Delta} l^*(\Delta)$ and denote by $l(\Delta)$ the number of lattice points in Δ.

A lattice polyhedron Δ defines the *projective toric variety* over $\mathbb{C}$:

$$\mathbf{P}_{\Delta} = \mathrm{Proj}\, S_{\Delta},$$

where S_{Δ} is the monomial subalgebra in the polynomial ring

$$\mathbb{C}[X_0, X_1^{\pm 1}, \dots, X_d^{\pm 1}]$$

spanned as $\mathbb{C}$-linear space by monomials $X_0^k X_1^{m_1} \dots X_d^{m_d}$ such that the corresponding lattice point $(m_1, \dots, m_d) \in M$ belongs to $k\Delta$. We remark that the dimension of $\mathbf{P}_{\Delta}$ equals $\dim \Delta$. If $\dim \Delta = d$, then $\mathbf{P}_{\Delta}$ can be considered as a projective compactification of $\mathbf{T}$. In the latter case, the irreducible components $\mathbf{D}_1, \dots, \mathbf{D}_n$ of $\mathbf{P}_{\Delta} \setminus \mathbf{T}$ one-to-one correspond to $(d-1)$-dimensional faces $\Theta_1, \dots, \Theta_n$ of Δ. We denote by $\mathbf{e}_1, \dots, \mathbf{e}_n$ the primitive lattice points in N which define the linear equations for the affine hyperplanes containing $\Theta_1, \dots, \Theta_n$ (in other words, $\mathbf{e}_1, \dots, \mathbf{e}_n$ are primitive integral interior normal vectors to faces $\Theta_1, \dots, \Theta_n$).

The is another well-known definition of toric varieties $\mathbf{P}_{\Delta}$ via the normal fan $\Sigma = \{\sigma_B\}$ consisting of all rational polyhedral cones

$$\sigma_B = \mathbb{R}_{\geq 0}\mathbf{e}_{i_1} + \dots + \mathbb{R}_{\geq 0}\mathbf{e}_{i_s} \subset N_{\mathbb{R}}$$

corresponding to those subsets $B = \{i_1, \dots, i_s\} \subset \{1, \dots, n\}$ for which the intersection $\Theta_{i_1} \cap \dots \cap \Theta_{i_s}$ is not empty. In this situation, we also use the notation $\mathbf{P}_{\Sigma}$ for toric varieties associated with Σ.

It is known that every invertible sheaf $\mathcal{L}$ on any toric variety $\mathbf{P}$ admits a $\mathbf{T}$-linearization. By this reason, we shall consider in this paper only $\mathbf{T}$-linearized invertible sheaves on toric varieties. A $\mathbf{T}$-linearization of $\mathcal{L}$ induces the M-grading of the cohomology spaces

$$H^i(\mathbf{P}, \mathcal{L}) = \bigoplus_{m \in M} H^i(\mathbf{P}, \mathcal{L})(m).$$

The convex hull $\Delta(\mathcal{L})$ of all lattice points $m \in M$ for which $H^0(\mathbf{P}, \mathcal{L})(m) \neq 0$ will be called the *supporting polyhedron for global sections of* $\mathcal{L}$.

Recall the following well-known statement [7, 16]:

Theorem 2.1. *There is one-to-one correspondence*

$$\mathcal{L} \cong \mathcal{O}_{\mathbf{P}}(a_1\mathbf{D}_1 + \cdots + a_n\mathbf{D}_n) \leftrightarrow \varphi, \quad a_i = \varphi(\mathbf{e}_i), \ i = 1, \ldots, k,$$

between $\mathbf{T}$*-linearized invertible sheaves* $\mathcal{L}$ *on* $\mathbf{P} = \mathbf{P}_\Sigma$ *and continious functions* $\varphi : N_\mathbb{R} \to \mathbb{R}$ *which are integral (i.e.,* $\varphi(N) \subset \mathbb{Z}$*) and* $\mathbb{R}$*-linear on every cone* $\sigma \in \Sigma$*. Moreover, the supporting polyhedron for the global sections of* $\mathcal{L} = \mathcal{L}(\varphi)$ *equals*

$$\Delta(\mathcal{L}) = \{x \in M_\mathbb{R} | \langle x, y \rangle \geq -\varphi(y) \text{ for all } y \in N_\mathbb{R}\}.$$

By *semi-ample invertible sheaf* $\mathcal{L}$ on a projective toric variety $\mathbf{P}_\Delta$ we always mean an invertible sheaf $\mathcal{L}$ generated by global sections. We have the following [7, 16]:

Proposition 2.2. $\mathcal{L}$ *is semi-ample if and only if the corresponding* Σ*-piecewise linear* φ *is upper convex. Moreover, any semi-ample invertible sheaf* $\mathcal{L}$ *(together with a* $\mathbf{T}$*-linearization) on a toric variety* $\mathbf{P}$ *is uniquely determined by its supporting polyhedron* $\Delta(\mathcal{L})$*:*

$$\mathcal{L} \cong \mathcal{O}_{\mathbf{P}}(a_1\mathbf{D}_1 + \cdots + a_n\mathbf{D}_n), \text{ where } a_i = -\min_{x \in \Delta(\mathcal{L})} \langle x, \mathbf{e}_i \rangle.$$

Definition 2.3. Let Δ and Δ' be two lattice polyhedra. Then we call a polyhedron Δ' a *Minkowski summand* of Δ if there exist a positive integer μ and a lattice polyhedron Δ'' such that $\mu\Delta = \Delta' + \Delta''$.

Using 2.2, one easily obtains:

Proposition 2.4. *A lattice polyhedron* Δ' *is the supporting polyhedron for global sections of a* $\mathbf{T}$*-linearized semi-ample invertible sheaf on* $\mathbf{P}_\Delta$ *if and only if* Δ' *is a Minkowski summand of* Δ.

In the sequel, we shall use many times the following statement:

Theorem 2.5. *Let D be a nef-Cartier divisor (or, equivalently, $\mathcal{O}_{\mathbf{P}}(D)$ is a semi-ample invertible sheaf) on a projective toric variety $\mathbf{P} = \mathbf{P}_\Delta$, Δ' the lattice polyhedron supporting the global sections $\mathcal{O}_{\mathbf{P}}(D)$. Then*

$$H^i(\mathbf{P}, \mathcal{O}_{\mathbf{P}}(-D)) = 0, \ \textit{if } i \neq \dim \Delta'$$

$$H^i(\mathbf{P}, \mathcal{O}_{\mathbf{P}}(-D)) = l^*(\Delta'), \ \textit{if } i = \dim \Delta' .$$

In particular the Euler characteristic $\chi(\mathcal{O}_{\mathbf{P}}(-D))$ equals $b(\Delta')$.

Proof. Let $k = \dim \Delta'$. Then the invertible sheaf $\mathcal{O}_{\mathbf{P}}(D)$ defines the canonical morphism

$$\pi_D : \mathbf{P} \to \mathbf{V} = \operatorname{Proj} \oplus_{n \geq 0} H^0(\mathbf{P}, \mathcal{O}_{\mathbf{P}}(nD)),$$

where $\mathbf{V} \cong \mathbf{P}_{\Delta'}$ is a k-dimensional projective toric variety, and $\mathcal{O}_{\mathbf{P}}(D) \cong \pi_D^* \mathcal{O}_{\mathbf{V}}(1)$. Therefore,

$$H^i(\mathbf{P}, \mathcal{O}_{\mathbf{P}}(-D)) \cong H^i(\mathbf{V}, \mathcal{O}_{\mathbf{V}}(-1)).$$

Hence $H^i(\mathbf{P}, \mathcal{O}_{\mathbf{P}}(-D)) = 0$ for $i > k$. On the other hand, $H^i(\mathbf{V}, \mathcal{O}_{\mathbf{V}}(-1)) = 0$ for $i < k$, and $H^k(\mathbf{V}, \mathcal{O}_{\mathbf{V}}(-1)) = l^*(\Delta')$ (see [7, 12]). □

A complex d-dimensional algebraic variety W is said to have only *toroidal singularities* if the m-adic completion of the local ring (R, m) corresponding to any point $p \in W$ is isomorphic to the m_σ-completion of a semi-group ring (S_σ, m_σ), where $S_\sigma = \mathbb{C}[\sigma \cap M]$ for some d-dimensional rational convex polyhedral cone $\sigma \subset M_{\mathbb{R}}$ with vertex at $0 \in M$, and the maximal ideal $m_\sigma \subset S_\sigma$ is generated by all non-constant monomials. We formulate without proof the following technical statement which is a generalization of the classical Bertini's theorem:

Theorem 2.6. *Let W be a complex projective algebraic variety with only toroidal singularities, $\mathcal{L}$ a semi-ample invertible sheaf on W, $D \subset W$ the set of zeros of a generic global section of $\mathcal{L}$. Then D again has only toroidal singularities. In particular, D is irreducible if* $\dim D > 0$.

3. Complete intersections

Let $\Delta_1, \ldots, \Delta_r$ be lattice polyhedra in $M_{\mathbb{R}}$. In this section $\mathbf{P}$ denotes the d-dimensional toric variety $\mathbf{P}_\Delta$, where $\Delta = \Delta_1 + \cdots + \Delta_r$ (without loss of generality, we assume $\dim \Delta = \dim M_{\mathbb{R}} = d$). By 2.4, the lattice polyhedron Δ_i is the support polyhedron for global sections of some semi-ample invertible sheaf $\mathcal{L}_i$ on $\mathbf{P}$ ($i = 1, \ldots, r$). We identify $H^0(\mathbf{P}, \mathcal{L}_i)$ with the space of all Laurent polynomials $f_i \in \mathbb{C}[M]$ having Δ_i as the Newton polyhedron.

Definition 3.1. Lattice polyhedra $\Delta_1, \ldots, \Delta_r$ are called k-dependent if there exist $n > 0$ and n-element subset

$$\{\Delta_{i_1}, \ldots, \Delta_{i_n}\} \subset \{\Delta_1, \ldots, \Delta_r\},$$

such that

$$\dim(\Delta_{i_1} + \cdots + \Delta_{i_n}) < n + k - 1.$$

Lattice polyhedra which are not k-dependent will be called k-independent.

Remark 3.2. It follows immediately from definition that the k-independence of lattice polyhedra $\Delta_1, \ldots, \Delta_r$ implies the l-independence for $1 \leq l \leq k$.

Theorem 3.3. *Let $Z_f \subset \mathbf{T}$ be a complete intersection of r affine hypersurfaces $Z_{f_1}, \ldots, Z_{f_r}$ defined by a general system of the equations $f_1 = \cdots = f_r = 0$ where f_i is a general Laurent polynomial with the Newton polyhedra Δ_i ($i = 1, \ldots, r$). Denote by Z_i the closure of Z_{f_i} in $\mathbf{P}$ ($i = 1, \ldots, r$). Let*

$$V = Z_1 \cap \cdots \cap Z_r.$$

Then the following statements hold:

(i) *V is non-empty if and only if $\Delta_1, \ldots, \Delta_r$ are 1-independent;*

(ii) *if $\Delta_1, \ldots, \Delta_r$ are 2-independent, then V is irreducible;*

(iii) *if $\Delta_1, \ldots, \Delta_r$ are k-independent ($k \geq 3$), then $h^1(\mathcal{O}_V) = \cdots = h^{k-2}(\mathcal{O}_V) = 0$.*

Proof. Denote by $\mathcal{K}^*$ the Koszul complex

$$\mathcal{O}_{\mathbf{P}}(-Z_1 - \cdots - Z_r) \to \cdots \to \sum_{i<j} \mathcal{O}_{\mathbf{P}}(-Z_i - Z_j) \to \sum_i \mathcal{O}_{\mathbf{P}}(-Z_i) \to \mathcal{O}_{\mathbf{P}}.$$

There are two spectral sequences $'E$ and $''E$ converging to the hypercohomology $\mathbf{H}^*(\mathbf{P}, \mathcal{K}^*)$ (cf. [10]):

$$'E_2^{p,q} = H^p(\mathbf{P}, \mathcal{H}^q(\mathcal{K}^*)),$$
$$''E_2^{p,q} = H^q(\mathbf{P}, H^p(\mathcal{K}^*)).$$

Since $\mathcal{K}^*$ is an acyclic resolution of $\mathcal{O}_V$, we have $\mathcal{H}^q(\mathcal{K}^*) = 0$ for $q \neq r$, and $\mathcal{H}^r(\mathcal{K}^*)) = \mathcal{O}_V$, the first spectral sequence degenerates and we get the isomorphisms

$$\mathbf{H}^{r+p}(\mathbf{P}, \mathcal{K}^*) \cong H^p(\mathbf{P}, \mathcal{O}_V) \cong H^p(V, \mathcal{O}_V).$$

On the other hand, the second spectral sequence does not degenerate in general. The term $''E_2^{p,q}$ is the cohomology of the bicomplex

$$''E_1^{p,q} = \bigoplus_{\{i_1, \ldots, i_{r-p}\}} H^q(\mathbf{P}, \mathcal{O}_{\mathbf{P}}(-Z_{i_1} - \cdots - Z_{i_{r-p}})).$$

We can compute $''E_1^{p,q}$ using Proposition 2.5. The statements (i)–(iii) will follow from the consideration of $''E$.

(i) It is clear that V is empty if and only if Z_f is empty. Assume that $\Delta_1, \ldots, \Delta_r$ are 1-dependent. Then there exists $n \geq 1$ such that $d' = \dim(\Delta_{i_1} + \cdots + \Delta_{i_n}) < n$. This means that we can choose the coordinates $X_1, \ldots, X_d$ on $\mathbf{T}$ in such a way that the polynomials $f_{i_1}, \ldots, f_{i_n}$ depends only on some d' of them. Therefore, we obtain an overdetermined system $f_{i_1} = \cdots = f_{i_n} = 0$; i.e., V is empty.

Assume now that V is empty, i.e., $\mathcal{K}^*$ is acyclic. If some Δ_i is 0-dimensional, then $\Delta_1, \ldots, \Delta_r$ are 1-dependent, and everything is proved. Otherwise, one has the non-zero cycle in $''E_1^{r,0} \cong {}''E_2^{r,0} \cong H^0(\mathbf{P}, \mathcal{O}_{\mathbf{P}}) \cong \mathbb{C}$ which must be killed by some next non-zero differential

$$d_l \,:\, ''E_l^{r-l,l-1} \to E_l^{r,0} \; (l \geq 2).$$

Therefore $''E_l^{r-l,l-1} \neq 0$ for some $2 \leq l \leq r$. This implies that there exists an l-element subset $\{i_1, \ldots, i_l\} \subset \{1, \ldots, r\}$ such that

$$H^{l-1}(\mathbf{P}, \mathcal{O}_{\mathbf{P}}(-Z_{i_1} - \cdots - Z_{i_l})) \neq 0.$$

Applying 2.5, we see that there exists an l-element subset

$$\{\Delta_{i_1}, \ldots, \Delta_{i_l}\} \subset \{\Delta_1, \ldots, \Delta_r\}$$

such that

$$\dim(\Delta_{i_1} + \cdots + \Delta_{i_l}) = l - 1,$$

i.e., $\Delta_1, \ldots, \Delta_r$ are 1-dependent.

(ii) Assume that $\Delta_1, \ldots, \Delta_r$ are 2-independent. By 2.5, one has

$$''E_1^{r-s,s-1} = {}''E_1^{r-s,s} = 0 \;\text{ for }\; 1 \leq s \leq r.$$

Hence

$$''E_l^{r-s,s-1} = {}''E_l^{r-s,s} = 0 \;\text{ for } 1 \leq s \leq r, l \geq 1.$$

So $\mathbf{H}^r(\mathbf{P}, \mathcal{K}^*) \cong \mathbb{C} \cong H^0(V, \mathcal{O}_V)$. Therefore V is connected. By 2.6, V is irreducible.

(iii) Assume that $\Delta_1, \ldots, \Delta_r$ are k-independent ($k \geq 3$). By the same arguments using 2.5, we obtain $\mathbf{H}^r(\mathbf{P}, \mathcal{K}^*) \cong \mathbb{C}$, and $\mathbf{H}^{r+1}(\mathbf{P}, \mathcal{K}^*) = \cdots = \mathbf{H}^{r+k-2}(\mathbf{P}, \mathcal{K}^*) = 0$. This implies $h^1(\mathcal{O}_V) = \cdots = h^{k-2}(\mathcal{O}_V) = 0$. □

Remark 3.4. Khovansky announced ([13] p. 41) that the statement 3.3(i) was first discovered and proved by D. Bernshtein using properties of mixed volumes in the following equivalent form:

The affine variety Z_f is empty if and only if there exists an l-dimensional affine subspace of $M_{\mathbb{R}}$ containing affine translates of some $l + 1$ polyhedra from $\{\Delta_1, \ldots, \Delta_r\}$.

Corollary 3.5. *Assume that all lattice polyhedra $\Delta_1, \ldots, \Delta_r$ have positive dimension, $l^*(\Delta_1 + \cdots + \Delta_r) = 1$, $d = \dim(\Delta_1 + \cdots + \Delta_r)$, and for any proper subset*

$\{k_1, \ldots, k_s\} \subset \{1, \ldots, r\}$ *one has* $l^*(\Delta_{k_1} + \cdots + \Delta_{k_s}) = 0$. *Then the following statements hold:*

(i) V *is empty if and only if* $d = r - 1$;

(ii) V *consists of* 2 *distinct points if and only if* $d = r$;

(iii) V *is a smooth irreducible curve of genus* 1 *if and only if* $d = r + 1$.

(iv) V *is a nonempty irreducible variety of dimension* $d - r \geq 2$ *having the property* $h^1(\mathcal{O}_V) = \cdots = h^{d-r-1}(\mathcal{O}_V) = 0$ *and* $h^0(\mathcal{O}_V) = h^{d-r}(\mathcal{O}_V) = 1$ *if and only if* $d \geq r + 2$.

Proof. It follows from Proposition 2.5 and our assumptions that $''E_1^{r,0}$ and $''E_1^{0,d}$ have dimension 1 and all remaining spaces $''E_1^{p,q}$ are zero.

(i) If $r = d + 1$, then Z_f is empty by dimension arguments. On the other hand, if Z_f is empty, then $''E_l^{p,q}$ becomes acyclic for $l \gg 0$. Note that the only nontrivial one-dimensional spaces $''E_1^{r,0}$ and $''E_1^{0,d}$ can kill each other only via the non-zero differential

$$d_l : {}''E_l^{0,d} \to {}''E_l^{r,0}$$

where $r = d + 1$ and $l = r$, i.e., Z_f is empty if and only if $r = d + 1$.

Assume $r > d + 1$. Then

$$\mathbb{C} \cong {}''E_1^{0,d} \cong {}''E_\infty^{0,d} \cong \mathbf{H}^d(\mathbf{P}, \mathcal{K}^*).$$

On the other hand, the isomorphism $\mathbf{H}^{r+p}(\mathbf{P}, \mathcal{K}^*) \cong H^p(C, \mathcal{O}_V)$ implies $\mathbf{H}^i(\mathbf{P}, \mathcal{K}^*) = 0$ for $i < r$. Contradiction.

(ii) If $r = d$, then

$$\mathbb{C}^2 \cong {}''E_1^{0,d} \oplus {}''E_1^{r,0} \cong {}''E_\infty^{0,d} \oplus {}''E_\infty^{r,0} \cong H^0(V, \mathcal{O}_V).$$

Since V is nonempty, one has $\dim V = 0$, i.e., V consists of 2 distinct points.

(iii)–(iv) For $r \leq d - 1$, we have isomorphisms

$$\mathbb{C} \cong {}''E_1^{r,0} \cong H^0(V, \mathcal{O}_V),$$

$$\mathbb{C} \cong {}''E_1^{0,d} \cong H^{d-r}(V, \mathcal{O}_V),$$

and

$$0 \cong {}''E_1^{p,q} \cong H^{d-r}(V, \mathcal{O}_V) \text{ if } p + q \neq r, d.$$

This proves (iii)–(iv). □

4. Calabi–Yau varieties and nef-partitions

Definition 4.1. A lattice polyhedron $\Delta \subset M_{\mathbb{R}}$ is called *reflexive* if $\mathbf{P}_\Delta$ has only Gorenstein singularities and $\mathcal{O}_{\mathbf{P}}(1)$ is isomorphic to the anticanonical sheaf which is considered together with its natural **T**-linearization. In this case, we call $\mathbf{P}_\Delta$ a *Gorenstein toric Fano variety*.

Remark 4.2. Since the **T**-linearized anticanonical sheaf on $\mathbf{P}_\Delta$ is isomorphic to $\mathcal{O}_{\mathbf{P}_\Delta}(\mathbf{D}_1 + \cdots + \mathbf{D}_n)$, it follows from the above definition that any reflexive polyhedron Δ has $0 \in M$ as the unique interior lattice point. Moreover,

$$\Delta = \{x \in M_{\mathbb{R}} | \langle x, \mathbf{e}_i \rangle \geq -1, \; i = 1, \ldots, n\},$$

where $\mathbf{e}_1, \ldots, \mathbf{e}_n$ are the primitive integral interior normal vectors to codimension-1 faces $\Theta_1, \ldots, \Theta_n$ of Δ. These properties of reflexive polyhedra were used in another their definition [1].

Theorem 4.3. [1] *Let Δ be a reflexive polyhedron as above. Then the convex hull $\Delta^* \subset N_{\mathbb{R}}$ of the lattice vectors $\mathbf{e}_1, \ldots, \mathbf{e}_n$ is again a reflexive polyhedron. Moreover, $(\Delta^*)^* = \Delta$.*

Definition 4.4. The lattice polyhedron Δ^* is called *dual* to Δ.

Using the adjunction formula, 2.2 and 2.4, we immediately obtain:

Proposition 4.5. *Assume that a d-dimensional Gorenstein toric Fano variety $\mathbf{P}_\Delta$ contains a $(d-r)$-dimensional complete intersection V $(r < d)$ of r semi-ample Cartier divisors $Z_1, \ldots, Z_r$ such that the canonical (or, equivalently, dualizing) sheaf of V is trivial. Then there exist lattice polyhedra $\Delta_1, \ldots, \Delta_r$ such that*

$$\Delta = \Delta_1 + \cdots + \Delta_r.$$

Definition 4.6. Let $\Sigma \subset N_{\mathbb{R}}$ be the normal fan defining a Gorenstein toric Fano variety $\mathbf{P}_\Delta$, $\varphi : N_{\mathbb{R}} \to \mathbb{R}$ the integral upper convex Σ-piecewise linear function corresponding to the **T**-linearized anticanonical sheaf (by 2.1, $\varphi(\mathbf{e}_i) = 1, i = 1, \ldots, n$), $\Delta = \Delta_1 + \cdots + \Delta_r$ a decomposition of Δ into a Minkowski sum of r lattice polyhedra Δ_j $(j = 1, \ldots, r)$, $\varphi = \varphi_1 + \cdots + \varphi_r$ the induced decomposition of φ into the sum of integral upper convex Σ-piecewise linear functions φ_j (see 2.2). Then $\Pi(\Delta) = \{\Delta_1, \ldots, \Delta_r\}$ is called a *nef-partition of* Δ if $\varphi_j(\mathbf{e}_i) \in \{0, 1\}$ for $i = 1, \ldots, n$, $j = 1, \ldots, r$.

Remark 4.7. There are reflexive polyhedra Δ which admit decompositions into Minkowski sum of two lattice polyhedra, but do not admit any nef-partition. For example, let $\Delta \subset \mathbb{R}^2$ be the convex hull of 4 lattice points: $(1, 0)$, $(-1, 0)$, $(0, 1)$ and $(0, -1)$. Then Δ is a 2-dimensional reflexive polyhedron which admits a decomposition into Minkowski sum of two 1-dimensional polyhedra Δ_1 and Δ_2, where Δ_1 is the convex hull of $(-1, 0)$ and $(0, -1)$, and Δ_2 is the convex hull of $(0, 0)$ and $(1, 1)$. However, Δ does not admit any nef-partition.

Remark 4.8. The notation in 4.6 are a little bit different from definitions and notations in [3, 6], but they are essentially equivalent.

Definition 4.9. Let $\Pi(\Delta) = \{\Delta_1, \dots, \Delta_r\}$ be a nef-partition of a reflexive polyhedron Δ. We define the lattice polyhedron $\nabla_j \subset N_{\mathbb{R}}$ $(j = 1, \dots, r)$ as the convex hull of $0 \in N$ and all lattice vectors $\mathbf{e}_i \in \{\mathbf{e}_1, \dots, \mathbf{e}_n\}$ such that $\varphi_j(\mathbf{e}_i) = 1$.

Theorem 4.10. [6] *The Minkowski sum* $\nabla = \nabla_1 + \cdots + \nabla_r$ *is a reflexive polyhedron. Moreover,* $\Pi(\nabla) = \{\nabla_1, \dots, \nabla_r\}$ *is a nef-partition of* ∇, *and one has also*

$$\nabla^* = \mathrm{Conv}\{\Delta_1, \dots, \Delta_r\},$$

$$\Delta^* = \mathrm{Conv}\{\nabla_1, \dots, \nabla_r\}.$$

Definition 4.11. The nef-partition $\Pi(\nabla)$ is called the *dual nef-partition.*

Example 4.12. Let $\Delta_j \subset M_{\mathbb{R}}^{(j)}$ is a reflexive polyhedron $(j = 1, \dots, r)$, $\nabla_j := \Delta_j^* \subset N_{\mathbb{R}}^{(j)}$ the corresponding dual reflexive polyhedron $(j = 1, \dots, r)$. We set $M_{\mathbb{R}} := M_{\mathbb{R}}^{(1)} \oplus \cdots \oplus M_{\mathbb{R}}^{(r)}$, $N_{\mathbb{R}} := N_{\mathbb{R}}^{(1)} \oplus \cdots \oplus N_{\mathbb{R}}^{(r)}$. Then $\Pi(\Delta) = \{\Delta_1, \dots, \Delta_r\}$ is a nef-partition of the reflexive polyhedron $\Delta = \Delta_1 + \cdots + \Delta_r \subset M_{\mathbb{R}}$, and $\Pi(\nabla) = \{\nabla_1, \dots, \nabla_r\}$ is the dual nef-partition of the reflexive polyhedron $\nabla = \nabla_1 + \cdots + \nabla_r \subset N_{\mathbb{R}}$.

By [1], a maximal projective triangulation $\mathcal{T}$ of $\Delta^* \subset N_{\mathbb{R}}$ defines a *MPCP*-desingularization $\pi_{\mathcal{T}} : \widehat{\mathbf{P}}_\Delta \to \mathbf{P}_\Delta$. The $\mathbf{T}$-invariant divisors on $\widehat{\mathbf{P}}_\Delta$ one-to-one correspond to lattice points on the boundary $\partial\Delta^*$.

Definition 4.13. Denote by $\mathcal{V}(\Delta^*)$ the set $\partial\Delta^* \cap N$. If v is a lattice point in $\mathcal{V}(\Delta^*)$, then the corresponding toric divisor on $\widehat{\mathbf{P}}_\Delta$ will be denoted by $\mathbf{D}(v)$.

Since the anticanonical sheaf on $\widehat{\mathbf{P}}_\Delta$ is semi-ample, $\Delta_1, \dots, \Delta_r$ are supporting polyhedra for global sections of some semi-ample invertible sheaves $\widehat{\mathcal{L}}_1, \dots, \widehat{\mathcal{L}}_r$ on $\widehat{\mathbf{P}}_\Delta$.

Definition 4.14. Let $\Pi(\Delta)$ be a nef-partition, $Z_f \subset \mathbf{T}$ a complete intersection of r general affine hypersurfaces $Z_{f_1}, \dots, Z_{f_r}$ defined by a general system of the polynomial equations $f_1 = \cdots = f_r = 0$ where f_i is a general Laurent polynomial with the Newton polyhedra Δ_i $(i = 1, \dots, r)$. Denote by $\widehat{Z}_i$ the closure of Z_{f_i} in $\widehat{\mathbf{P}}_\Delta$ $(i = 1, \dots, r)$. Define V (resp. $\widehat{V}$) as the closure of Z_f in $\mathbf{P}_\Delta$ (resp. in $\widehat{\mathbf{P}}_\Delta$; i.e., $\widehat{V} = \widehat{Z}_1 \cap \cdots \cap \widehat{Z}_r$). If $\Pi(\nabla)$ is the dual nef-partition, then the corresponding general complete intersection in $\mathbf{P}_\nabla$ (resp. in $\widehat{\mathbf{P}}_\nabla$) will be denoted by W (resp. $\widehat{W}$).

By the adjunction formula and 2.6, one immediately obtains:

Proposition 4.15. *In the above notations, assume that V is nonempty and irreducible. Then $\widehat{V}$ is an irreducible $(d - r)$-dimensional projective algebraic variety with at most Gorenstein terminal abelian quotient singularities and trivial canonical class. In particular, $\widehat{V}$ is smooth if $d - r \leq 3$.*

In the next section we prove the following:

Theorem 4.16. *Let $\Pi(\Delta)$ be a nef-partition, $\Pi(\nabla)$ the dual nef-partition, V (resp. W) corresponding to $\Pi(\Delta)$ (resp. to $\Pi(\nabla)$) general complete intersections in $\mathbf{P}_\Delta$ (resp. in $\mathbf{P}_\nabla$). Then the following statements hold:*

(i) *V is nonempty if and only if W is nonempty;*

(ii) *V is irreducible if and only if W is irreducible;*

(iii) *$h^i(\mathcal{O}_V) = h^i(\mathcal{O}_W)$ for $0 \leq i \leq d - r$.*

Corollary 4.17. *Let $\Pi(\Delta)$ be a nef-partition, $\Pi(\nabla)$ the dual nef-partition, $\widehat{V}$ (resp. $\widehat{W}$) corresponding to $\Pi(\Delta)$ (resp. $\Pi(\nabla)$) general complete intersections in $\widehat{\mathbf{P}}_\Delta$ (resp. in $\widehat{\mathbf{P}}_\nabla$). Then $\widehat{V}$ is an irreducible $(d - r)$-dimensional projective algebraic variety with at most Gorenstein terminal abelian quotient singularities and trivial canonical class if and only if $\widehat{W}$ has the same properties.*

5. Semi-simplicity principle for nef-partitions

Let Δ be a d-dimensional reflexive polyhedron in $M_{\mathbb{R}}$, $\Pi(\Delta) = \{\Delta_1, \ldots, \Delta_r\}$ a nef-partition of Δ.

Definition 5.1. We say that $\Pi(\Delta)$ *splits into a direct sum*

$$\Pi(\Delta) = \Pi(\Delta^{(1)}) \oplus \cdots \oplus \Pi(\Delta^{(k)})$$

if there exist convex lattice polyhedra $\Delta^{(1)}, \ldots, \Delta^{(k)} \subset \Delta$ satisfying the conditions

(i) $d = \dim \Delta^{(1)} + \cdots + \dim \Delta^{(k)}$ and

$$\Delta = \Delta^{(1)} + \cdots + \Delta^{(k)};$$

(ii) for $1 \leq i \leq k$, the lattice point 0 is cointained in the relative interior of $\Delta^{(i)}$, $\Delta^{(i)}$ is reflexive, and

$$\Pi(\Delta^{(i)}) = \{\Delta_j \subset \Delta \mid \Delta_j \subset \Delta^{(i)}\}$$

is a nef-partition of $\Delta^{(i)}$.

Definition 5.2. Assume that $\Pi(\Delta)$ splits into a direct sum

$$\Pi(\Delta) = \Pi(\Delta^{(1)}) \oplus \cdots \oplus \Pi(\Delta^{(k)}).$$

Denote by $M_{\mathbb{R}}^{(i)}$ the minimal $\mathbb{R}$-linear subspace of $M_{\mathbb{R}}$ containing $\Delta^{(i)}$ ($i = 1, \ldots, k$). We set also $M^{(i)} = M \cap M_{\mathbb{R}}^{(i)}$ ($i = 1, \ldots, k$). We say that $\Pi(\Delta)$ splits into the direct sum *over* $\mathbb{Z}$ if

$$M = M^{(1)} \oplus \cdots \oplus M^{(k)}.$$

It is easy to see the following:

Proposition 5.3. *Assume that a nef-partition $\Pi(\Delta)$ splits over $\mathbb{Z}$ into a direct sum $\Pi(\Delta^{(1)}) \oplus \cdots \oplus \Pi(\Delta^{(k)})$. Then the Gorenstein toric Fano variety $\mathbf{P}_\Delta$ is the product of the Gorenstein toric Fano varieties $\mathbf{P}_{\Delta^{(i)}}$ $(i = 1, \ldots, k)$. Moreover,*

$$V = V^{(1)} \times \cdots \times V^{(k)},$$

$$\widehat{V} = \widehat{V}^{(1)} \times \cdots \times \widehat{V}^{(k)},$$

where $V^{(i)}$ (resp. $\widehat{V}^{(i)}$) is the Calabi–Yau complete intersections defined by the nef-partition $\Pi(\Delta^{(i)})$ in $\mathbf{P}_{\Delta^{(i)}}$ (resp. in $\widehat{\mathbf{P}}_{\Delta^{(i)}}$), $i = 1, \ldots, k$.

Remark 5.4. In 4.12 we gave a simplest example of a nef-partition $\Pi(\Delta)$ which splits into a direct sum $\{\Delta_1\} \oplus \cdots \oplus \{\Delta_r\}$ over $\mathbb{Z}$. In general situation, if $\Pi(\Delta)$ splits into a direct sum $\Pi(\Delta) = \Pi(\Delta^{(1)}) \oplus \cdots \oplus \Pi(\Delta^{(k)})$, then $M^{(1)} \oplus \cdots \oplus M^{(k)}$ is a sublattice of finite index in M, but not necessarily the lattice M itself.

Example 5.5. Let $\Delta = \Delta_1 + \Delta_2 \subset \mathbb{R}^4$ where

$$\Delta_1 = \text{Conv}\{(1,0,0,0), (0,1,0,0), (-1,0,0,0), (0,-1,0,0)\};$$

$$\Delta_2 = \text{Conv}\{(0,0,1,0), (0,0,0,1), (0,0,-1,0), (0,0,0,-1)\}.$$

We define the lattice $M \subset \mathbb{R}^4$ as

$$M = \left(\frac{1}{2}, \frac{1}{2}, \frac{1}{2}, \frac{1}{2}\right) + \mathbb{Z}^4.$$

Then Δ splits into direct sum $\Pi(\Delta) = \{\Delta_1\} \oplus \{\Delta_2\}$. But it is not a splitting over $\mathbb{Z}$, because $M_1 \oplus M_2$ is a sublattice of index 2 in M.

Definition 5.6. We call a nef-partition $\Pi(\Delta)$ *reducible* if there exists a subset

$$\{k_1, \ldots, k_s\} \subset \{1, \ldots, r\} \ (0 < s < r)$$

such that $\Delta_{k_1} + \cdots + \Delta_{k_s}$ contains 0 in its relative interior. Nef-partitions which are not reducible are called *irreducible*.

Remark 5.7. (i) Notice that $\Delta_{k_1} + \cdots + \Delta_{k_s}$ contains 0 in its relative interior if and only if

$$l^*(\Delta_{k_1} + \cdots + \Delta_{k_s}) = 1.$$

(ii) It is clear that an irreducible nef-partition has no nontrivial splitting into a direct sum.

Theorem 5.8. *Any nef-partition $\Pi(\Delta)$ has a unique decomposition into direct sum of irreducible nef-partitions.*

Proof. Assume that $\Pi(\Delta)$ is reducible. Choose lattice polyhedra $\Delta_{k_1}, \dots, \Delta_{k_s}$ such that $\Delta' = \Delta_{k_1} + \cdots + \Delta_{k_s}$ contains 0 in its relative interior. Denote

$$\Lambda' = \mathrm{Conv}(\Delta_{k_1}, \dots, \Delta_{k_s}).$$

It is clear that Λ' also contains 0 in its relative interior and has the same dimension as $\Delta_{k_1} + \cdots + \Delta_{k_s}$. Denote by $M'_{\mathbb{R}}$ the minimal linear subspace in $M_{\mathbb{R}}$ containing Δ'. Denote by $\psi_1, \dots, \psi_r$ the piecewise linear functions on $M_{\mathbb{R}}$ which determine the dual nef-partition $\Pi(\nabla)$ (see 4.6). We set $\psi' = \psi_{k_1} + \cdots + \psi_{k_s}$, $\psi'' = \psi_1 + \cdots + \psi_r - \psi' = \psi_{j_1} + \cdots + \psi_{j_{r-s}}$. Then

$$\Lambda' = \{x \in M'_{\mathbb{R}} \mid \psi'(x) \leq 1\}.$$

Since ψ' is a non-negative integral convex piecewise linear function having value 1 at the relative boundary $\partial\Lambda' \subset M'_{\mathbb{R}}$, the polyhedron Λ' is reflexive.

Since ψ' is convex and non-negative

$$C' = \{x \in M_{\mathbb{R}} \mid \psi'(x) = 0\}$$

is a convex subset of $M_{\mathbb{R}}$ containing all non-negative linear combinations of vertices of $\Delta_{j_1}, \dots, \Delta_{j_{r-s}}$. One has $M'_{\mathbb{R}} \cap C' = 0$. Assume that 0 is not contained in the relative interior of C'. Then, by separateness theorem for convex sets, there exists a non-zero element $y' \in N_{\mathbb{R}}$ such that $\langle M'_{\mathbb{R}}, y' \rangle = 0$, and $\langle C', y' \rangle \geq 0$. Therefore $\langle v, y' \rangle \geq 0$ for all vertices v of $\Delta_1, \dots, \Delta_r$. Contradiction. Thus 0 is contained in the relative interior of C'. Since $\mathbb{R}_{\geq 0}C' \subset C'$, C' is a linear subspace and $M_{\mathbb{R}} = M'_{\mathbb{R}} \oplus C'$. We put $M''_{\mathbb{R}} = C'$.

Define

$$\Lambda'' = \{x \in M''_{\mathbb{R}} \mid \psi''(x) \leq 1\}.$$

Then $\Lambda'' = \mathrm{Conv}(\Delta_{j_1}, \dots, \Delta_{j_{r-s}})$, and $\Delta'' := \Delta_{j_1} \oplus \cdots \oplus \Delta_{j_{r-s}}$ contains 0 in its relative interior. Therefore Λ'' is also reflexive.

Thus we obtain $\nabla^* = \mathrm{Conv}(\Lambda', \Lambda'')$, $\dim \Lambda' + \dim \Lambda'' = d$, and

$$\Pi(\Delta') = \{\Delta_{k_1}, \dots, \Delta_{k_s}\},$$

$$\Pi(\Delta'') = \{\Delta_{j_1}, \dots, \Delta_{j_{r-s}}\}$$

are nef-partitions of the reflexive polyhedra Δ' and Δ'' corresponding to restricitions of $\psi_{k_1}, \dots, \psi_{k_s}$ (resp. of $\psi_{j_1}, \dots, \psi_{j_{r-s}}$) on $M'_{\mathbb{R}}$ (resp. on $M''_{\mathbb{R}}$). Now the statement of Theorem 5.8 follows by induction. □

Proof of Theorem 4.16. Note that V is nonempty if and only if $h^0(V, \mathcal{O}_V) \neq 0$. By 2.6, V is irreducible if and only if $h^0(V, \mathcal{O}_V) = 1$. Therefore, (i) and (ii) follow from (iii). By 3.5 and 5.7(i), the spectral sequence $''E^{p,q}$ which computes the cohomology of $\mathcal{O}_V$ degenerates at $''E_2^{p,q}$ for all irreducible nef-partitions. By 5.3, we obtain the degeneration of $''E^{p,q}$ at $''E_2^{p,q}$ for all nef-partitions $\Pi(\Delta)$ which split over $\mathbb{Z}$ into a sum of irreducible nef-partitions.

According to Theorem 5.8, in general situation we have a finite covering

$$\pi \ : \ \mathbf{P}_\Delta \to \mathbf{P}_{\Delta^{(1)}} \times \cdots \times \mathbf{P}_{\Delta^{(k)}}$$

where the degree of π equals the index of $M^{(1)} \oplus \cdots \oplus M^{(k)}$ in M, and $\Pi(\Delta^{(1)}), \ldots, \Pi(\Delta^{(k)})$ are irreducible nef-partitions. Let

$$\tilde{\pi} \ ; \ V \to \tilde{V} := V^{(1)} \times \cdots \times V^{(k)}$$

the finite covering induced by π. Then we obtain the canonical homomorphism of the spectral sequences:

$$\tilde{\pi}^* \ : \ {}''\tilde{E}_l^{p,q} \to {}''E_l^{p,q}.$$

By 5.7(i), we have the canonical isomorphism

$${}''\tilde{E}_1^{p,q} \cong {}''E_1^{p,q}.$$

Since ${}''\tilde{E}^{p,q}$ degenerates at ${}''\tilde{E}_1^{p,q}$ (see the above arguments), ${}''E^{p,q}$ also degenerates at ${}''E_2^{p,q}$. Therefore, $h^i(\mathcal{O}_V) = h^i(\mathcal{O}_{\tilde{V}})$ $(i = 1, \ldots, d-r)$. Analogously, $h^i(\mathcal{O}_W) = h^i(\mathcal{O}_{\tilde{W}})$ $(i = 1, \ldots, d-r)$. Thus, we have reduced (iii) to already known case. □

Corollary 5.9. *Let $I = \{1, \ldots, r\}$. Denote by $|J|$ the cardinality of a subset $J \subset I$. Define*

$$E(\Delta, t) = \sum_{i=0}^{d-r} h^i(\mathcal{O}_V)t^i.$$

Then

$$E(\Delta, t) = \sum_{J \subset I} l^*(\sum_{j \in J} \Delta_j) t^{\dim\left(\sum_{j\in J}\Delta_j\right) - |J|}.$$

By Serre duality, and using the natural isomorphisms $H^i(V, \mathcal{O}_V) \cong H^i(\widehat{V}, \mathcal{O}_{\widehat{V}})$ $(i = 1, \ldots, d-r)$, we also obtain:

Corollary 5.10. *Assume that $\widehat{V}$ is nonempty and irreducible. Then*

$$h^i(\widehat{V}, \mathcal{O}_{\widehat{V}}) = h^{d-r-i}(\widehat{W}, \mathcal{O}_{\widehat{W}}).$$

6. $\chi(\Omega^1)$ for complete intersections

Now we want to calculate the Euler characteristic of $\Omega^1_{\widehat{V}}$ of a Calabi–Yau complete intersection $\widehat{V} = \widehat{Z}_1 \cap \cdots \cap \widehat{Z}_r$ in a *MPCP*-desingularization $\widehat{\mathbf{P}} := \widehat{\mathbf{P}}_\Delta$ of a Gorenstein toric Fano variety $\mathbf{P} := \mathbf{P}_\Delta$.

One has the standard exact sequence for a complete intersection:

$$0 \to \mathcal{O}_{\widehat{V}}(-\widehat{Z}_1) \oplus \cdots \oplus \mathcal{O}_{\widehat{V}}(-\widehat{Z}_r) \to \Omega^1_{\widehat{\mathbf{P}}}|_{\widehat{V}} \to \Omega^1_{\widehat{V}} \to 0.$$

On the other hand, for toric varieties with only quotient singularities there exists the exact sequence [7, 16]:

$$0 \to \Omega^1_{\widehat{\mathbf{P}}} \to \mathcal{O}^d_{\widehat{\mathbf{P}}} \to \bigoplus_{v \in \mathcal{V}(\Delta^*)} \mathcal{O}_{\mathbf{D}(v)} \to 0.$$

By transversality of $\widehat{Z}_1, \ldots, \widehat{Z}_r$ to all strata on $\widehat{\mathbf{P}}$ we can restrict the last exact sequence on $\widehat{V}$ without obtaining additional Tor-sheaves:

$$0 \to \Omega^1_{\widehat{\mathbf{P}}}|_{\widehat{V}} \to \mathcal{O}^d_{\widehat{V}} \to \bigoplus_{v \in \mathcal{V}(\Delta^*)} \mathcal{O}_{\mathbf{D}(v)\cap\widehat{V}} \to 0.$$

Consequently, we obtain:

Proposition 6.1.

$$\chi(\Omega^1_{\widehat{V}}) = \chi(\mathcal{O}^d_{\widehat{V}}) - \sum_{j=1}^{r} \chi(\mathcal{O}_{\widehat{V}}(-\widehat{Z}_j)) - \sum_{v \in \mathcal{V}(\Delta^*)} \chi(\mathcal{O}_{\mathbf{D}(v)\cap\widehat{V}}).$$

In order to compute $\chi(\mathcal{O}_{\widehat{V}}(-\widehat{Z}_j))$, we consider the Koszul resolution

$$0 \to \mathcal{O}_{\widehat{\mathbf{P}}}(-\widehat{Z}_1 - \cdots - \widehat{Z}_r) \to \cdots$$

$$\cdots \to \sum_{j<k} \mathcal{O}_{\widehat{\mathbf{P}}}(-\widehat{Z}_j - \widehat{Z}_k) \to \sum_{j} \mathcal{O}_{\widehat{\mathbf{P}}}(-\widehat{Z}_j) \to \mathcal{O}_{\widehat{\mathbf{P}}} \to \mathcal{O}_{\widehat{V}} \to 0.$$

Tensoring it by $\mathcal{O}_{\widehat{\mathbf{P}}}(-\widehat{Z}_i)$ and using 2.5, we get

Proposition 6.2.

$$\chi(\mathcal{O}_{\widehat{V}}(-\widehat{Z}_i)) = -\sum_{j} b(\Delta_i + \Delta_j) + \sum_{j<k} b(\Delta_i + \Delta_j + \Delta_k) - \cdots$$

$$\cdots + (-1)^r b(\Delta_i + \Delta_1 + \cdots + \Delta_r).$$

For the computation of $\chi(\mathcal{O}_{\mathbf{D}(v)\cap\widehat{V}})$ we need the following:

Proposition 6.3. *Let v an arbitrary lattice point in $\mathcal{V}(\Delta^*)$, $\Gamma(v)$ the minimal face of Δ^* containing v. Then $\Gamma(v)$ is a face of a polyhedron ∇_i for some i $(1 \leq i \leq r)$.*

Proof. If v is a vertex of Δ^*, then the statement is evident, because $\Delta^* = \mathrm{Conv}\{\nabla_1, \ldots, \nabla_r\}$.

Assume that $\Gamma(v)$ is a convex hull of $k > 1$ faces of polyhedra ∇_i. Let $v_1, \ldots, v_m$ be vertices of $\Gamma(v)$. By assumption, $m \geq k$. Without loss of generality, we assume that $v_1 \in \nabla_1, \ldots, v_k \in \nabla_k$. Since v belongs to relative interior of $\Gamma(v)$, there exist positive numbers $\lambda_1, \ldots, \lambda_m$ such that $\lambda_1 + \cdots + \lambda_m = 1$ and $\lambda_1 v_1 + \cdots + \lambda_m v_m = v$. Choose an arbitrary vertex w of the dual face $\Gamma^*(v) \subset \Delta$. Since $\Delta = \Delta_1 + \cdots + \Delta_r$, there exist $w_i \in \Delta_i$ such that $w = w_1 + \cdots + w_r$. It follows from definition of dual nef-partitions (cf. [6]) that

$$\langle w_i, v_i \rangle \geq -1, \quad 1 \leq i \leq k$$

and

$$\langle w_i, v_j \rangle \geq 0, \quad i \neq j, \ 1 \leq i \leq k, \ 1 \leq i \leq k.$$

On the other hand,

$$\langle w, v \rangle = -1 = \sum_{i=1}^{m} \sum_{j=1}^{r} \lambda_i \langle w_j, v_i \rangle.$$

Since $\lambda_i > 0$, the last equality is possible only if all vertices $v_1, \ldots, v_m$ belong to *the same* polyhedron ∇_i for some i $(1 \leq i \leq r)$. □

Corollary 6.4.

$$l(\Delta^*) = l(\nabla_1) + \cdots + l(\nabla_r) - r + 1.$$

Definition 6.5. Denote by $\Delta_j(v)$ the face of Δ_j which defines the Newton polyhedron for the equation of $\widehat{Z}_j$ in the toric divisor $\mathbf{D}(v) \subset \widehat{\mathbf{P}}_\Delta$.

It is easy to prove the following:

Proposition 6.6. *Assume that the face $\Gamma(v) \subset \Delta^*$ is also a face of ∇_i. Let $\Gamma^*(v)$ be the dual face of Δ. Then*

$$\Delta_i(v) = \{x \in \Delta_i \mid \langle x, v \rangle = -1\},$$

$$\Delta_j(v) = \{x \in \Delta_j \mid \langle x, v \rangle = 0 \ \textit{if}\ j \neq i\},$$

and

$$\Delta_1(v) + \cdots + \Delta_r(v) = \Gamma^*(v).$$

Tensoring the Koszul resolution of $\mathcal{O}_{\widehat{V}}$ by $\mathcal{O}_{\mathbf{D}(v)}$, we obtain:

Proposition 6.7.

$$\chi(\mathcal{O}_{\mathbf{D}(v) \cap C}) = -\sum_j b(\Delta_j(v)) + \sum_{j<k} b(\Delta_j(v) + \Delta_k(v)) - \cdots$$
$$\cdots + (-1)^r b(\Delta_1(v) + \cdots + \Delta_r(v)).$$

Let $\nabla_i^0 = \nabla \cap \mathcal{V}(\Delta^*)$ $(i = 1, \ldots, r)$, and $I := \{1, \ldots, r\}$. If we rewrite

$$\sum_{v \in \mathcal{V}(\Delta^*)} \chi(\mathcal{O}_{\mathbf{D}(v) \cap C}) = \sum_{i=1}^{r} \sum_{v \in \nabla_i^0} \sum_{J \subset I} (-1)^{|J|+1} b(\sum_{j \in J} \Delta_j(v)) =$$

$$= \sum_{i=1}^{r} \sum_{v \in \nabla_i^0} \sum_{i \notin J \subset I} (-1)^{|J|+1} b(\sum_{j \in J} \Delta_j(v)) + \sum_{i=1}^{r} \sum_{v \in \nabla_i^0} \sum_{i \in J \subset I} (-1)^{|J|+1} b(\sum_{j \in J} \Delta_j(v)),$$

then 6.1 and 6.2 imply:

Theorem 6.8.

$$\begin{aligned} \chi(\Omega^1_{\widehat{V}}) = \chi(\mathcal{O}^d_{\widehat{V}}) &+ \sum_{i=1}^{r} \sum_{J \subset I} (-1)^{|J|+1} b(\Delta_i + \sum_{j \in J} \Delta_j) \\ &+ \sum_{i=1}^{r} \sum_{v \in \nabla_i^0} \sum_{i \notin J \subset I} (-1)^{|J|+1} b(\sum_{j \in J} \Delta_j(v)) \\ &+ \sum_{i=1}^{r} \sum_{v \in \nabla_i^0} \sum_{i \in J \subset I} (-1)^{|J|+1} b(\sum_{j \in J} \Delta_j(v)). \end{aligned}$$

7. Mirror duality for $\chi(\Omega^1)$

We prove that the involution $\Pi(\Delta) \leftrightarrow \Pi(\nabla)$ chahge the sign of $\chi(\Omega^1)$ by $(-1)^{d-r}$ (as it would follow from the expected duality $h^{1,q}(\widehat{V}) = h^{1,d-r-q}(\widehat{W})$, $0 \leq q \leq d-r$):

Theorem 7.1.

$$\chi(\Omega^1_{\widehat{V}}) = (-1)^{d-r} \chi(\Omega^1_{\widehat{W}}).$$

For the proof we need some preliminary statements.

Proposition 7.2. *Let $i \in J \subset I$, $\Delta_i^0 = \Delta_i \cap \mathcal{V}(\nabla^*)$. Then a nonzero lattice point w belongs to the relative interior of the polyhedron $\Lambda = \Delta_i + \sum_{j \in J} \Delta_j$ if and only if $w \in \Delta_i^0$ and the zero point $0 \in N$ belongs to the relative interior of $\sum_{j \notin J} \nabla_j(w)$. Moreover, if this happens, then*

$$\dim \left(\Delta_i + \sum_{j \in J} \Delta_j \right) + \dim \left(\sum_{j \notin J} \nabla_j(w) \right) = d.$$

Proof. First of all, let's check that the interior lattice point w must belong to Δ_i.

This means checking $\langle w, \nabla_k\rangle \geq 0$ for $k \neq i$ and $\langle w, \nabla_i\rangle \geq -1$, which follows easily from the fact that w is a lattice point that belongs to $(1-\epsilon)\Lambda$ for some small positive ϵ.

The polyhedron $\Lambda - w$ is defined by the inequalities

$$\langle x, v\rangle \geq -2 - \langle w, v\rangle, \quad v \in \nabla_i,$$

$$\langle x, v\rangle \geq -1 - \langle w, v\rangle, \quad v \in \nabla_j, \; j \in J, \; j \neq i,$$

$$\langle x, v\rangle \geq 0 - \langle w, v\rangle, \quad v \in \nabla_j, \; j \notin J.$$

Since $\langle w, v\rangle \geq -1$ for $v \in \nabla_i$, and $\langle w, v\rangle \geq 0$ for $v \notin \nabla_i$, only the inequalities $\langle x, v\rangle \geq 0$ for $v \in \nabla_j$, $\langle w, v\rangle = 0$ $(j \notin J)$ give rise to nontrivial restrictions for the intersection of a small neighbourhood of $0 \in M_{\mathbb{R}}$ with $\Lambda - w$. Therefore, it remains to consider the halfspaces defined by the inequalities $\langle x, v\rangle \geq 0$ where $v \in \nabla_j(w)$ $(j \notin J)$. Denote by L_w the convex cone in $M_{\mathbb{R}}$ defined by the inequalities

$$\langle x, v\rangle \geq 0, \quad \text{for all } v \in \nabla_j(w) \text{ and } j \notin J.$$

Then 0 lies in the relative interior of $\Lambda - w$ if and only if L_w is a linear subspace of $M_{\mathbb{R}}$. On the other hand, the cone

$$C_w = \sum_{j \notin J} \mathbb{R}_{\geq 0} \nabla_j(w)$$

is dual to L_w. Moreover, C_w is a linear subspace in $N_{\mathbb{R}}$ if and only if 0 is contained in the relative interior of $\sum_{j\notin J} \nabla_j(w)$. It remains to note that a convex cone is a linear subspaces if and only if the dual cone is a linear subspace. In the latter case, $\dim L_w + \dim C_w = d$, i.e., $\dim\left(\Delta_i + \sum_{j\in J}\Delta_j\right) + \dim\left(\sum_{j\notin J}\nabla_j(w)\right) = d$. □

Corollary 7.3.

$$\sum_{i=1}^{r}\sum_{J\subset I}(-1)^{|J|+1} b(\Delta_i + \sum_{j\in J}\Delta_j) = (-1)^{d-r}\left(\sum_{i=1}^{r}\sum_{w\in\Delta_i^0}\sum_{i\notin J\subset I}(-1)^{|J|+1} b(\sum_{j\in J}\nabla_j(w))\right).$$

Proof. Denote by $(-1)^{\dim\Theta} b^0(\Theta)$ the number of *nonzero* lattice points in the relative interior of a lattice polyhedron Θ. Let $i \in J \subset I$, $J' = J \setminus \{i\}$. Note that 0 is in the relative interior of $\Delta_i + \sum_{j\in J'}\Delta_j$ if and only if 0 is in the relative interior of $\Delta_i + \sum_{j\in J}\Delta_j$. Since $|J| = |J'| + 1$ and

$$\dim\left(\Delta_i + \sum_{j\in J'}\Delta_j\right) = \dim\left(\Delta_i + \sum_{j\in J}\Delta_j\right),$$

we can have

$$\sum_{i=1}^{r}\sum_{J\subset I}(-1)^{|J|+1}b(\Delta_i+\sum_{j\in J}\Delta_j)=\sum_{i=1}^{r}\sum_{i\in J\subset I}(-1)^{|J|+1}b^0(\Delta_i+\sum_{j\in J}\Delta_j).$$

It remains to apply Proposition 7.2 and the property $|J|+|I\setminus J|=r$. □

Proposition 7.4. *Let $i\in J\subset I$, and $v\in\nabla_i^0$ is a lattice point. Then a lattice point w belongs to the relative interior of the polyhedron $\Lambda=\sum_{j\in J}\Delta_j(v)$ if and only $w\in\Delta_i^0$ and the lattice point v belongs to the relative interior of $\nabla_i(w)+\sum_{j\notin J}\nabla_j(w)$. Moreover, if this happens, then*

$$\dim\left(\sum_{j\in J}\Delta_j(v)\right)+\dim\left(\nabla_i(w)+\sum_{j\notin J}\nabla_j(w)\right)=d-1.$$

Proof. We only need to prove the implication in one direction and the formula for the dimensions.

By 6.6, $\langle\Delta_j(v),v\rangle=0$ for $j\in J$, $j\neq i$ and $\langle\Delta_i(v),v\rangle=-1$. Therefore, $\langle w,v\rangle=-1$, in particular, $w\neq 0$. Now we would like to prove that $w\in\Delta_i$. Because of $\Lambda\subset\Delta$, this amounts to checking $\langle w,\nabla_k\rangle\geq 0$ for $k\neq i$. Suppose there exists a vertex v' of ∇_k such that $\langle w,v'\rangle=-1$. Because w lies in the relative interior of Λ, we get $\langle\Lambda,v'\rangle=-1$. However, this is impossible, because $\Delta_j(v)$ contain zero if $j\neq i$, which leads to $\Delta_i(v)\subseteq\Lambda$.

The polyhedron $\Lambda-w$ is defined by the conditions

$$\langle x,v'\rangle\geq -1-\langle w,v'\rangle,\ v'\in\nabla_j,\ j\in J,$$

$$\langle x,v'\rangle\geq 0-\langle w,v'\rangle,\ v'\in\nabla_j,\ j\notin J,$$

$$\langle x,v\rangle=-1-\langle w,v\rangle=0.$$

If we are interested in the neighbourhood of $0\in M$, we are left with the inequalities $\langle x,v'\rangle\geq 0$, where v' is either a vertex of $\nabla_j(w)$ ($j\notin J$), or v' is the vertex of $\nabla_i(w)$, or $v'=-v$. The zero point is in the interior of $\Lambda-w$, hence the convex cone L_w defined by these inequalities is a linear subspace. As a result, the dual to L_w cone

$$C_w=\mathbb{R}_{\geq 0}\nabla_i(w)-\mathbb{R}_{\geq 0}v+\sum_{j\notin J}\mathbb{R}_{\geq 0}\nabla_j(w).$$

is also a linear subspace.

Now we use 6.6 and the equality $\langle w,v\rangle=-1$ to conclude that $\{y\in C_w|\langle w,y\rangle=0\}$ is a linear subspace of dimension one less which equals

$$\mathbb{R}_{\geq 0}\left(\nabla_i(w)-v\right)+\sum_{j\notin J}\mathbb{R}_{\geq 0}\nabla_j(w).$$

Because $(\nabla_i(w) - v)$ and $\nabla_j(w)$ contain zero, this implies that their sum has the same dimension and contains zero in its interior, which proves the first statement of the proposition.

The formula for the dimensions follows from the duality of L_w and C_w. □

Corollary 7.5.

$$\sum_{i=1}^{r} \sum_{v \in \nabla_i^0} \sum_{i \in J \subset I} (-1)^{|J|+1} b(\sum_{j \in J} \Delta_j(v)) =$$

$$= (-1)^{d-r} \left(\sum_{i=1}^{r} \sum_{w \in \Delta_i^0} \sum_{i \in J' \subset I} (-1)^{|J'|+1} b(\sum_{j \in J'} \nabla_j(w)) \right).$$

Proof. We set $J' = \{I \setminus J\} \cup \{i\}$. It remains to apply Proposition 7.4 and the property $|J| + |J'| = r + 1$. □

Proof of Theorem 7.1. It remains to combine the statements 5.10, 7.3 and 7.5 with the formula in 6.8. □

8. Complete intersections of ample divisors

Our next purpose is to give explicit formulas for $(*, 1)$-Hodge numbers for *MPCP*-resolution $\widehat{V}$ of a Calabi–Yau complete intersection of ample divisors. Notice that in this case nef-partition is always irreducible.

First, we glue together the following two exact sequences

$$0 \to \mathcal{O}_{\widehat{V}}(-\widehat{Z}_1) \oplus \cdots \oplus \mathcal{O}_{\widehat{V}}(-\widehat{Z}_r) \to \Omega^1_{\widehat{\mathbf{P}}}|_{\widehat{V}} \to \Omega^1_{\widehat{V}} \to 0,$$

$$0 \to \Omega^1_{\widehat{\mathbf{P}}}|_{\widehat{V}} \to \mathcal{O}^d_{\widehat{V}} \to \bigoplus_{v \in \mathcal{V}(\Delta^*)} \mathcal{O}_{\mathbf{D}(v) \cap \widehat{V}} \to 0$$

and obtain the complex

$$\mathcal{Q}^* \ : \ 0 \to \mathcal{O}_{\widehat{V}}(-\widehat{Z}_1) \oplus \cdots \oplus \mathcal{O}_{\widehat{V}}(-\widehat{Z}_r) \to \mathcal{O}^d_{\widehat{V}} \to \bigoplus_{v \in \mathcal{V}(\Delta^*)} \mathcal{O}_{\mathbf{D}(v) \cap \widehat{V}} \to 0$$

whose cohomology $\mathcal{H}^i$ can be nontrivial only if $i = 1$, and

$$\mathcal{H}^1(\mathcal{Q}^*) \cong \Omega^1_{\widehat{V}},$$

i.e., the hypercohomology of $\mathcal{Q}^*$ coincides with the cohomology of $\Omega^1_{\widehat{V}}[1]$.

Proposition 8.1. *Assume that* $\widehat{Z}_1, \ldots, \widehat{Z}_r$ *are nef- and big-divisors; i.e.,* Δ_i *is a Minkowski summand of* Δ *and* $\dim \Delta_i = \dim \Delta = d$ $(i = 1, \ldots, r)$*. Then*

$$H^k(\mathcal{O}_{\widehat{V}}(-\widehat{Z}_i)) = 0 \ \textit{for}\ i \neq d - r$$

and

$$h^{d-r}(\mathcal{O}_{\widehat{V}}(-\widehat{Z}_i)) = \sum_{J \subset I} (-1)^{r-|J|} l^*(\Delta_i + \sum_{j \in J} \Delta_j).$$

Proof. Consider the Koszul resolution of $\mathcal{O}_{\widehat{V}}(-\widehat{Z}_i)$. Then the corresponding second spectral sequence degenerates in $''E_2$, because our assumptions on $\widehat{Z}_1, \ldots, \widehat{Z}_r$ imply $''E_1^{p,q} = 0$ for $q \neq d$. The latter immediately gives all statements. □

Corollary 8.2. *Let* $\widehat{V}$ *be a complete intersection such that* $d - r \geq 3$*. Then, under assumptions in* 8.1*, the second spectral sequence corresponding to the complex* Q^* *degenerates in* $''E_2^*$*, and one obtains the following relations*

$$h^{d-r-1}(\Omega^1_{\widehat{V}}) = \sum_{i=1}^{r} h^{d-r}(\mathcal{O}_{\widehat{V}}(-\widehat{Z}_i)) - d -$$

$$- \sum_{v \in \mathcal{V}(\Delta^*)} \left(h^{d-r-1}(\mathcal{O}_{\mathbf{D}(v) \cap \widehat{V}}) - h^{d-r-2}(\mathcal{O}_{\mathbf{D}(v) \cap \widehat{V}}) \right),$$

$$h^k(\Omega^1_{\widehat{V}}) = \sum_{v \in \mathcal{V}(\Delta^*)} h^{k-2}(\mathcal{O}_{\mathbf{D}(v) \cap \widehat{V}}) \quad \textit{for}\ 2 \leq k \leq d - r - 2,$$

$$h^1(\Omega^1_{\widehat{V}}) = \sum_{v \in \mathcal{V}(\Delta^*)} h^0(\mathcal{O}_{\mathbf{D}(v) \cap \widehat{V}}) - d,$$

$$h^0(\Omega^1_{\widehat{V}}) = h^{d-r}(\Omega^1_{\widehat{V}}) = 0.$$

Proposition 8.3. *Assume that* $\widehat{Z}_1, \ldots, \widehat{Z}_r$ *are proper pullbacks of ample divisors on* $\mathbf{P}_\Delta$*; i.e.,* Δ_i *and* Δ *are Minkowski summand of each other* $(i = 1, \ldots, r)$*. Choose an element of* $v \in \mathcal{V}(\Delta^*)$*. Let* s *be dimension of the minimal face* $\Theta \subset \Delta^*$ *containing* v*. Then the faces* $\Delta_1(v), \ldots, \Delta_r(v)$ *depend only on* Θ *(we denote these faces by* $\Theta_1^*, \ldots, \Theta_r^*$ *), and the following statements hold:*

(i) *If* $d - r - s - 1 > 0$*, then*

$$h^{d-r-s-1}(\mathcal{O}_{\mathbf{D}(v) \cap \widehat{V}}) = \sum_{J \subset I} (-1)^{r-|J|} l^*(\sum_{j \in J} \Theta_j^*),$$

$$h^0(\mathcal{O}_{\mathbf{D}(v) \cap \widehat{V}}) = 1,$$

and

$$h^i(\mathcal{O}_{\mathbf{D}(v) \cap \widehat{V}}) = 0 \quad \textit{for all}\ i \neq 0, d - r - s - 1.$$

(ii) *If* $d - r - s - 1 = 0$, *then*

$$h^0(\mathcal{O}_{\mathbf{D}(v)\cap\widehat{V}}) = 1 + \sum_{J\subset I}(-1)^{r-|J|}l^*(\sum_{j\in J}\Theta_j^*),$$

and

$$h^i(\mathcal{O}_{\mathbf{D}(v)\cap\widehat{V}}) = 0 \quad \textit{for all } i \neq 0.$$

(iii) *If* $d - r - s - 1 < 0$, *then* $\mathbf{D}(v) \cap \widehat{V}$ *is empty.*

Proof. Since $\widehat{Z}_1, \ldots, \widehat{Z}_r$ are proper pulbacks of ample divisors, the polyhedron Δ^* is combinatorially dual to each of $r+1$ polyhedra $\Delta, \Delta_1, \ldots, \Delta_r$. By this combinatorial duality, the faces $\Delta_i(v) \subset \Delta_i$ are dual to the face Θ and their arbitary sums have the same dimension $d - s - 1$. In the sequel, we denote $\Delta_i(v)$ simply by Θ_i^*.

Consider the Koszul resolution of $\mathcal{O}_{\mathbf{D}(v)\cap\widehat{V}}$. Then the corresponding second spectral sequence degenerates in $''E_2$, because $''E_1^{p,q} = 0$ for $q \neq d-s-1, 0$, $''E_1^{0,p} = 0$ for $p \neq r$, and $''E_1^{0,r} \cong \mathbb{C}$. Now the statements (i)–(iii) are obvious. □

Corollary 8.4. *Let* $\widehat{V}$ *be a complete intersection such that* $d - r \geq 3$. *Then, under the assumptions in* 8.3, *one obtains*

$$h^{d-r-1}(\Omega^1_{\widehat{V}}) = \sum_{i=1}^{r}\left(\sum_{J\subset I}(-1)^{r-|J|}l^*(\Delta_i + \sum_{j\in J}\Delta_j)\right) - d$$

$$- \sum_{\substack{\dim\Theta=0\\ \Theta\subset\Delta^*}}\left(\sum_{J\subset I}(-1)^{r-|J|}l^*(\sum_{j\in J}\Theta_j^*)\right) + \sum_{\substack{\dim\Theta=1\\ \Theta\subset\Delta^*}} l^*(\Theta)\cdot\left(\sum_{J\subset I}(-1)^{r-|J|}l^*(\sum_{j\in J}\Theta_j^*)\right),$$

$$h^k(\Omega^1_{\widehat{V}}) = \sum_{\substack{\dim\Theta=d-r-k-1\\ \Theta\subset\Delta^*}} l^*(\Theta)\cdot\left(\sum_{J\subset I}(-1)^{r-|J|}l^*(\sum_{j\in J}\Theta_j^*)\right)$$

for $2 \leq k \leq d - r - 2$;

$$h^1(\Omega^1_{\widehat{V}}) = \text{Card}\{\text{ lattice points in faces of dimension} \leq d-r-1\} - d$$

$$+ \sum_{\substack{\dim\Theta=d-r-1\\ \Theta\subset\Delta^*}} l^*(\Theta)\cdot\left(\sum_{J\subset I}(-1)^{r-|J|}l^*(\sum_{j\in J}\Theta_j^*)\right),$$

$$h^0(\Omega^1_{\widehat{V}}) = h^{d-r}(\Omega^1_{\widehat{V}}) = 0.$$

Corollary 8.5. *Assume that $r = 1$ and $d \geq 4$. Then the Hodge numbers $h^{p,1}(\widehat{V})$ have the following values*

$$h^{0,1}(\widehat{V}) = h^{d-1,1}(\widehat{V}) = 0,$$

$$h^{1,1}(\widehat{V}) = l(\Delta^*) - d - 1 - \left(\sum_{\substack{\Theta^* \subset \Delta^* \\ \operatorname{codim} \Theta^* = 1}} l^*(\Theta^*) \right) + \left(\sum_{\substack{\Theta^* \subset \Delta^* \\ \operatorname{codim} \Theta^* = 2}} l^*(\Theta^*) \cdot l^*(\Theta) \right),$$

$$h^{d-2,1}(\widehat{V}) = l(\Delta) - d - 1 - \left(\sum_{\substack{\Theta^* \subset \Delta^* \\ \operatorname{codim} \Theta^* = d}} l^*(\Theta) \right) + \left(\sum_{\substack{\Theta^* \subset \Delta^* \\ \operatorname{codim} \Theta^* = d-1}} l^*(\Theta^*) \cdot l^*(\Theta) \right),$$

$$h^{p,1}(\widehat{V}) = \sum_{\substack{\Theta^* \subset \Delta^* \\ \operatorname{codim} \Theta^* = p+1}} l^*(\Theta^*) \cdot l^*(\Theta); \ \textit{for } 2 \leq p \leq d-3 .$$

Proposition 8.6. *Let $\widehat{V}$ be a complete intersection of ample divisors $\widehat{Z}_1, \ldots, \widehat{Z}_r$ such that $d - r \geq 3$. Then one obtains*

$$h^{d-r-1}(\Omega^1_{\widehat{V}}) = \sum_{i=1}^{r} \left(\sum_{J \subset I} (-1)^{r-|J|} l^*(\Delta_i + \sum_{j \in J} \Delta_j) \right) - d$$
$$- \sum_{\substack{\dim \Theta = 0 \\ \Theta \subset \Delta^*}} \left(\sum_{J \subset I} (-1)^{r-|J|} l^*(\sum_{j \in J} \Theta_j^*) \right),$$

$$h^k(\Omega^1_{\widehat{V}}) = 0 \ \textit{for } k \neq 1, d-r-1 ,$$

$$h^1(\Omega^1_{\widehat{V}}) = \text{Card}\{ \text{ lattice points in faces of dimension} \leq d - r - 1\} - d.$$

Proof. If $\widehat{Z}_1, \ldots, \widehat{Z}_r$ are ample, then $\mathbf{P}_\Delta$ has at most terminal singularities; i.e., $l^*(\Theta) = 0$ for all faces of Δ^* of positive dimension. This immediately implies the formulas. □

Example 8.7. *Complete intersection V_{d_1,d_2} of two hypersurfaces in $\mathbf{P}^5$.*

For two cases below, we have

$$\sum_{\substack{\text{codim }\Xi=1\\ \Xi\subset\Delta}} l^*(\Xi) = 30.$$

Case 1. $d_1 = d_2 = 3$. Then $l(\Delta_1) = l(\Delta_2) = 56$, $l^*(2\Delta_1) = l^*(2\Delta_2) = 1$. Therefore,

$$h^{2,1}(V_{3,3}) = (112 - 7) - (30 + 1 + 1) + 0 = 73.$$

Case 2. $d_1 = 2$, $d_2 = 4$. Then $l(\Delta_1) = 21$, $l(\Delta_2) = 126$, $l^*(2\Delta_1) = 0$, $l^*(2\Delta_2) = 21$. Therefore,

$$h^{2,1}(V_{2,4}) = (147 - 7) - (30 + 21 + 0) + 0 = 89.$$

Example 8.8. *Complete intersection* $V_{2,2,3}$ *in* $\mathbf{P}^6$.

$$h^{2,1}(V_{2,2,3}) = (28 + 28 + 84 - 9) - (7 \cdot 6 + 1 + 1 + 7 + 7) + 0 = 73.$$

Example 8.9. *Complete intersection* $V_{2,2,2,2}$ *in* $\mathbf{P}^7$.

$$h^{2,1}(V_{2,2,2,2}) = (4 \cdot 36 - 11) - (8 \cdot 7 + 3 \cdot 4) + 0 = 65.$$

9. Complete intersections in $\mathbf{P}^d$

Consider the case when a nef-partition $\Pi(\Delta) = \{\Delta_1, \ldots, \Delta_r\}$ defines a Calabi–Yau complete intersection $V = V_{d_1,\ldots,d_r}$ in projective space $\mathbf{P}^d$. This means that the polyhedra $\Delta_i = d_i\Lambda$ $(i = 1, \ldots, r)$ are d_i-multiples of a regular d-dimensional simplex Λ and $d_1 + \cdots + d_r = d$. Our purpose is to compute the $(*, 1)$-Hodge numbers of the mirror Calabi–Yau complete intersection $\widehat{W} = \widehat{Y}_1 \cap \cdots \cap \widehat{Y}_r$ defined by the dual nef-partition $\Pi(\nabla) = \{\nabla_1, \ldots, \nabla_r\}$. In this case, ∇_i $(i = 1, \ldots, r)$ is a regular d_i-dimensional simplex. If for some i we have $d_i = 1$, then the Calabi–Yau complete intersection in $\mathbf{P}^d$ reduces to a Calabi–Yau complete intersection in $\mathbf{P}^{d-1}$. So we can assume that $d_i \geq 2$ for all $i = 1, \ldots r$.

Proposition 9.1. *In the above situation, one has*

$$h^k(\mathcal{O}(-\widehat{Y}_i - \sum_{j\in J} \widehat{Y}_j)) = 0 \quad \textit{for all } k$$

unless $J = I$, *or* $J \cup \{i\} = I$. *Moreover*

$$h^k(\mathcal{O}(-\widehat{Y}_i - \sum_{j\in I} \widehat{Y}_j)) = 0 \quad \textit{for } k \neq d\ ,$$

$$h^d(\mathcal{O}(-\widehat{Y}_i - \sum_{j\in I} \widehat{Y}_j)) = l(\nabla_i);$$

and

$$h^k(\mathcal{O}(-\sum_{j\in I} \widehat{Y}_j)) = 0 \quad \textit{for } k \neq d\ ,$$

$$h^d(\mathcal{O}(-\sum_{j\in I} \widehat{Y}_j)) = 1.$$

Proof. Note that the polyhedra ∇_i, ∇_j $(j \in J)$ are regular simplices spanning *linearly independent* subspaces unless $i \in J$, $J = I$, or $J \cup \{i\} = I$. Therefore, $\nabla_i + \sum_{j\in J} \nabla_j$ is a regular simplex having no lattice points in its relative interior unless $i \in J$, $J = I$, or $J \cup \{i\} = I$. If $i \in J$, then $\nabla_i + \sum_{j\in J} \nabla_j$ has no lattice points in its relative interior since $l^*(2\nabla_i) = 0$ for all $i \in I$ $(d_i = \dim \nabla_l \geq 2)$. Hence the first statement follows from 2.5.

If $J = I$, or $j \cup \{i\} = I$, then $\dim(\nabla_i + \sum_{j\in J} \nabla_j) = d$. By 2.5, $h^k(\mathcal{O}(-\widehat{Y}_i - \sum_{j\in I} \widehat{Y}_j)) = 0$ for $k \neq d$. The remaining statements follow from $l^*(\nabla_i + \nabla_1 + \cdots + \nabla_r) = l(\nabla_i)$ and $l^*(\nabla_1 + \cdots + \nabla_r) = 1$. □

Corollary 9.2.

$$h^k(\mathcal{O}_{\widehat{W}}(-\widehat{Y}_i)) = 0 \ \textit{for } k \neq d - r$$

and

$$h^{d-r}(\mathcal{O}_{\widehat{W}}(-\widehat{Y}_i)) = l(\nabla_i) - 1.$$

Proof. Tensoring the Koszul resolution $\mathcal{K}^*$ of $\mathcal{O}_{\widehat{W}}$ by $\mathcal{O}(-\widehat{Y}_i)$, we obtain the degenerated spectral sequence "$E^{p,q}$ from which one immediately obtains the statement. □

Proposition 9.3. *For any vertex w of ∇^* and any subset $J \subset I$, one has*

$$l^*(\sum_{j\in J} \nabla_j(w)) = 0,$$

unless $\dim \sum_{j\in J} \nabla_j(w) = 0$.

Proof. We consider two cases:

Case 1. $J \neq I$. We know that $\sum_{j\in J} \nabla_j(w)$ is always a face of $\sum_{j\in J} \nabla_j$. On the other hand, the linear subspaces spanned by ∇_j are linearly independent. Since all ∇_i are regular simplices, there is no face of $\sum_{j\in J} \nabla_i$ of positive dimension containing a lattice point in its relative interior.

Case 2. $J = I$. If there exists $i \in I$ such that $\dim \nabla_i(w) = 0$, then setting $J' = J \setminus \{i\}$ we reduce all to Case 1.

The polyhedron ∇ has $d+2$ lattice points, but only those lattice points $v \in \nabla$ which satisfy the condition $\langle w, v\rangle \in \{0, -1\}$ might appear in $\nabla_j(w)$. Hence there exists a vertex of ∇ which does not belong to any of $\nabla_i(w)$. Therefore linear subspaces spanned by $\nabla_i(w)$ are linearly independent. We again obtain the same statement, since all $\nabla_i(w)$ are regular simplices. □

Using the Koszul resolution, we obtain

Corollary 9.4.

$$h^k(\mathcal{O}_{\mathbf{D}(w)\cap\widehat{W}}) = 0 \quad for\ k > 0$$

Therefore the second spectral sequence of the 3-term complex Q^* degenerates in $''E_2$ and we have

Proposition 9.5. *Assume that $d - r \geq 3$. Then*

$$h^k(\Omega^1_{\widehat{W}}) = 0 \ \ for \ \ 2 \leq d - r - 2 \,,$$

and

$$h^{d-r-1}(\Omega^1_{\widehat{W}}) = -d + \sum_{i=1}^{r}(l(\nabla_i) - 1) = 1 = h^1(\Omega^1_V).$$

Using the duality for the Euler characteristic from Theorem 7.1

$$-h^1(\Omega^1_{\widehat{W}}) + (-1)^{d-r-1}h^{d-r-1}(\Omega^1_{\widehat{W}})$$

$$= (-1)^{d-r}\left(-h^1(\Omega^1_V) + (-1)^{d-r-1}h^{d-r-1}(\Omega^1_V)\right),$$

we obtain

$$h^1(\Omega^1_{\widehat{W}}) = h^{d-r-1}(\Omega^1_V).$$

Therefore, we get the complete duality for $(1, q)$-Hodge numbers:

Theorem 9.6. *Let V be a Calabi–Yau complete intersection of r hypersurfaces in $\mathbf{P}^d$ and $d-r \geq 3$, $\widehat{W}$ a MPCP-desingularization of the Calabi–Yau complete intersection $W \subset \mathbf{P}_\nabla$ Then*

$$h^q(\Omega^1_{\widehat{W}}) = h^{d-r-q}(\Omega^1_V) \quad for \ \ 0 \leq q \leq d - r.$$

References

[1] V. V. Batyrev, Dual polyhedra and mirror symmetry for Calabi–Yau hypersurfaces in toric varieties, J. Algebraic Geom. 3 (1994) 493–535.

[2] V. Batyrev, D. van Straten, Generalized Hypergeometric Functions and Rational Curves on Calabi–Yau Complete Intersections in Toric Varieties, Commun. Math. Phys. 168 (1995),493–533.

[3] V. V. Batyrev, L. A. Borisov, Dual Cones and Mirror Symmetry for Generalized Calabi–Yau Manifolds, in: Mirror Symmetry II (ed. S.-T. Yau), 1995, 65–80.

[4] V. V. Batyrev, D. I. Dais, Strong McKay Corrrespondence, String-theoretic Hodge Numbers and Mirror Symmetry, to appear in Topology.

[5] P. Berglund, T. Hübsch, On a Residue Representation of Deformations, Koszul and Chiral Rings, IASSNS-HEP-94/97, hep-th/9411131.

[6] L. A. Borisov, Towards the Mirror Symmetry for Calabi–Yau Complete Intersections in Gorenstein Toric Fano Varieties, Preprint 1993, alg-geom/9310001.

[7] V. I. Danilov, Geometry of toric varieties, Russ. Math. Surv., 33 (1978), 97–154.

[8] G. Ellingsrud, S. A. Strømme, Counting twisted cubic curves on general complete intersections, alg-geom/9409006.

[9] G. Ellingsrud, S. A. Strømme, Bott's Formula and Enumerative Geometry, Preprint, November 1994, alg-geom/9411005.

[10] P. Griffiths, J. Harris, Principles of Algebraic Geometry, 1978.

[11] S. Hosono, A. Klemm, S. Theisen and S. T. Yau, Mirror symmetry, mirror map and application to complete intersection Calabi–Yau spaces, Nuclear Phys. B 433 (1995), 501–552.

[12] A. G. Khovansky, Newton polyhedra and toric varieties, Functional. Anal. Appl. 11 (1977), 56–67.

[13] A. G. Khovansky, Newton polyhedra and the genus of complete intersections, Functional. Anal. Appl. 12 (1978), 38–46.

[14] A. Klemm, S. Theisen, Mirror Maps and Instanton Sums for Complete Intersections in Weighted Projective Space, Modern Phys. Lett. A 9 (1994), 1807.

[15] A. Libgober and J. Teitelbaum, Lines on Calabi–Yau complete intersections, mirror symmetry, and Picard-Fuchs equations, Duke Math. J. 69 (1) 29–39, Internat. Math. Res. Notices (1993).

[16] T. Oda, Convex Bodies and Algebraic Geometry - An Introduction to the Theory of Toric Varieties, Ergeb. Math. Grenzgeb. (3) 15, Springer-Verlag, Berlin–Heidelberg–New York–London–Paris–Tokyo 1988.

Asymptotic results for hermitian line bundles over complex manifolds: the heat kernel approach

Thierry Bouche

Abstract. This paper is intended to be an introduction to a heat kernel approach to some problems in complex geometry. We recall and generalize many results from a previous paper. We demonstrate their strength by giving very simple proofs of nontrivial results of global analysis on complex manifolds. Some of the new results here will be used in a joint work with A. Abbes in order to give a simple direct proof in the case of varieties over number fields of the arithmetic Hilbert–Samuel theorem due to Gillet and Soulé.

Introduction

This paper is a continuation of [B1] where we computed the zeroth order asymptotic expansion of the heat kernel associated to high tensor powers of a hermitian line bundle over a complex manifold. Our aim is to indicate an entirely new method of constructing holomorphic (or more generally harmonic) sections of vector bundles over complex manifolds by using heat kernel estimates. This will range from bounds on cohomology groups (as it is known, these can imply lower bounds on the dimension of the space of holomorphic sections under suitable hypothesis), vanishing theorems, to an explicit construction of sections of some vector bundle twisted by high powers of a positive line bundle such that their norm converges to a Dirac mass at some point on the manifold. This construction produces holomorphic sections satisfying arbitrary conditions at some point. As an application, we will derive very simple proofs of well known theorems such as Kodaira vanishing or Kodaira embedding. Our hope is to convince complex geometers that these techniques lead to a partial alternative to Hörmander's L^2 estimates. We also include some generalizations of previous results in [B1], with applications to arithmetic geometry in mind (c.f. [A-B]).

Now, we introduce our notations: X is a compact complex analytic manifold of dimension n, endowed with a hermitian metric ω and associated volume element $dV =$

AMS Classification scheme numbers: 32L05; 53P10, 51N15

Key words: heat kernel, positive line bundle, Kodaira theorems, distortion function.

$\omega^n/n!$. E (*resp.* L) is a hermitian holomorphic vector bundle of rank r (*resp.* 1). For a global C^∞ $(0,q)$-form σ of $E(k) = E \otimes L^{\otimes k}$, $|\sigma(x)|^2$ denotes the pointwise length induced by the given metrics and $\|\sigma\|_2 = \left(\int_X |\sigma|^2 dV\right)^{1/2}$ (*resp.* $\|\sigma\|_\infty = \sup_X |\sigma|$) the L^2 norm (*resp.* the sup norm). We will be primarily concerned with the spectral distribution of the operator $\overline{\Box}_k^q = (\bar\partial^* + \bar\partial)^2$ which is the $\bar\partial$-laplacian (with respect to the given metrics) acting on $(0,q)$-forms with values in $E(k)$ (the exponent q will be omitted when we deal with sections (i.e. $q=0$)). As $\overline{\Box}_k^q$ is elliptic and X is compact it has a discrete spectrum which can be recovered from the *heat kernel* $K_k^q(t,x,y)$ which is the smooth kernel of the operator $e^{-\frac{2t}{k}\overline{\Box}_k^q}$. It enjoys the following expansion: for $j = 0, 1, \ldots$, let μ_j^k be the eigenvalues of $\overline{\Box}_k^q$ (counted with multiplicities), and $(\psi_j)_j$ be an orthonormal L^2 basis of eigen-forms associated to the μ_j^k's, then

$$K_k^q(t,x,y) = \sum_{j\ge 0} e^{-\frac{2t}{k}\mu_j^k}\psi_j(x) \otimes \psi_j^*(y),$$

and is characterized by the following properties:

(1i) $K_k^q \in C^\infty\big(]0,+\infty[\times X \times X, \mathrm{Hom}(\bigwedge^{0,q} T^*X \otimes E(k), \bigwedge^{0,q} T^*X \otimes E(k))\big)$;

(1ii) $\left(\frac{\partial}{\partial t} + \overline{\Box}_k^q\right) K_k^q = 0$ where $\overline{\Box}_k^q$ acts on the first variable;

(1iii) $K_k^q(t,x,y) \to \delta_y$ (Dirac δ-function at point y) if $t \to 0$;

(1iv) $K_k^q(t,x,y) = (K_k^q)^*(t,y,x)$.

If Ω is an open set of X, and $\overline{\Box}_{k,\Omega}^q$ is the operator defined by $\overline{\Box}_k^q$ on Ω, with Dirichlet condition at the boundary, we get in the same way the associated heat kernel $K_{k,\Omega}^q$ satisfying:

(2i) $K_{k,\Omega}^q \in C^\infty\big(]0,+\infty[\times\overline{\Omega} \times \overline{\Omega}, \mathrm{Hom}(\bigwedge^{0,q} T^*X \otimes E(k), \bigwedge^{0,q} T^*X \otimes E(k))\big)$;

(2ii) $\left(\frac{\partial}{\partial t} + \overline{\Box}_k^q\right) K_{k,\Omega}^q = 0$ inside Ω;

(2iii) $K_{k,\Omega}^q(t,\partial\Omega,y) = \{0\}$;

(2iv) $K_{k,\Omega}^q \underset{t\to 0}{\longrightarrow} \delta_y$.

For the sake of convenience, we also introduce the following notations: $e_k^q(t,x)$ (resp. $e_{k,\Omega}^q(t,x)$) is the fiberwise trace of $K_k^q(t,x,x)$ (resp. $K_{k,\Omega}^q(t,x,x)$) considered as an endomorphism of the fiber $E(k)_x$. We shall often call abusively e_k^q the heat kernel of $\overline{\Box}_k^q$. From the formula above we get:

$$e_k^q(t,x) = \sum_{j\ge 0} e^{-\frac{2t}{k}\mu_j^k}|\psi_j(x)|^2. \qquad (*)$$

Our point here is to illustrate how this "heat kernel" technique works out in a number of classical theorems of complex geometry, and even allows to refine them by a more local control on the holomorphic (*resp.* harmonic) objects they deal with. The present paper is thus both of an expository and research nature, the new results discussed in sections 2 or 3 being slight extensions of known results; however, the point of view, and most of our proofs, are new. Further developments of the theory are to be found in the papers [A-B] and [B4], the underlying Riemannian situation is made explicit in [B5]. In the first section of this paper, we state the main theorem from [B1] giving an asymptotic estimation for e_k^q when $k \to +\infty$ and show how it yields fine control on the cohomology groups of $E(k)$. The second section is devoted to estimates and applications of the so-called envelope and distortion functions which give some taste of how the global sections of $E(k)$ behave at a given point of X. Most of the results there are new for $r \geq 2$, $q > 0$. The third section illustrates our strategy of producing holomorphic sections with heat kernel estimates, and includes a self-contained proof of the Kodaira embedding theorem. Throughout the paper, the letter C will denote any constant not depending on x, t nor k.

1. Asymptotic bounds on cohomology

We denote by $\alpha_1, \dots, \alpha_n$ the eigenvalues of the curvature of L with respect to the metric ω. For a multi-index J, we put $\bar{\alpha}_J = \sum_{j \notin J} \alpha_j - \sum_{j \in J} \alpha_j$. We set also the function $\frac{\alpha}{\sinh \alpha t}$ to be $\frac{1}{t}$ when $\alpha = 0$. Let us call

$$e_\infty^q(t,x) = r(4\pi)^{-n}\Big(\sum_{|J|=q} e^{t\bar{\alpha}_J}\Big)\prod_{j=1}^{n} \frac{\alpha_j(x)}{\sinh \alpha_j(x) t}.$$

Then we get from [B1] the

Theorem 1.1.

(1a) *When $k \to +\infty$, the function $k^{-n} e_k^q(t,x)$ converges to $e_\infty^q(t,x)$ uniformly with respect to $x \in X$ and $t \in]0, k^\varepsilon]$ for a given $\varepsilon > 0$, not depending on k.*

(1b) *More precisely, if we fix $x_0 \in X$ and some $\eta \in]0, \frac{1}{6}[$, for any sequence of real numbers r_k such that $r_k k^{-\frac{1}{2}+\eta}$ is bounded above and below by positive constants, and for a sequence of open subsets Ω_k of X containing for all k the geodesic ball of center x_0 and radius r_k, the following convergence occurs*

$$k^{-n} e_{k,\Omega_k}^q(t, x_0) \to e_\infty^q(t, x_0)$$

uniformly with respect to $t \in]0, k^\varepsilon]$ for any $\varepsilon < \eta$.

Remarks. Notice that the rank r of E was omitted in Theorems 1 and 2 of [B1], which was a typo. The precise sense of uniformity here is stated as Theorem 2 in [B1]: it means that the function denoted by the Landau symbol $o(k^n)$ is bounded independently of x and t in the given domain. (1a) is thus only the reformulation of Theorems 1 and 2 (using the remark before Corollary 1). Although (1b) does not appear there, it is also a direct consequence of the whole method of the paper, because (1a) is proven by localizing the estimates of e_k^q at x_0 to those of e_{k,B_k}^q where B_k is a geodesic ball of center x_0 and whose radius satisfies conditions weaker than those on r_k. Since this paper was written, his author has improved some of the results it presents by using Eq. (1b) in full details (see [B4]).

Theorem 1.1 (1a) has first been proven by J.-M. Bismut [Bi] by using probabilistic methods, then generalized by E. Getzler to Heisenberg manifolds. There exists also a simple analytic approach of Demailly [D2] (all these without uniformity with respect to t outside bounded intervals). The original motivation for this theorem is that it implies Demailly's holomorphic strong Morse inequalities [D1]. In fact, it can be viewed as a *local* version of them (by analogy to the expression "local index theorems": it is a pointwise asymptotic estimation of the functions that give, after integration, the inequalities of Demailly). We shall be a little more precise: it is immediate from the infinite sum expansion of the heat kernel (∗) given in the introduction (and the Hodge identification between cohomology and harmonic forms) that

$$\dim H^i(X, E(k)) \leq \int_X e_k^i(k^\varepsilon, x)\, dV.$$

According to theorem 1.1 Eq. (1a), e_∞^q is the uniform limit on X of $k^{-n}e_k^q(t, x)$. If we divide the inequality above by k^n and let k tend to $+\infty$, we obtain as limit of the right hand side the integral of e_∞^q. As the left hand side does not depend on t, we can bound it by taking the right hand side's limit when t goes to infinity. It is then only a matter of simple computations to get the "weak Morse inequalities" of Demailly [D1] (see [B2] for the detailed derivation of the strong form):

Theorem 1.2. *For any* $q = 0, \ldots, n$

$$\dim H^q(X, E(k)) \leq (-1)^q r \frac{k^n}{n!} \int_{X(q)} \left(\frac{i}{2\pi} c(L)\right)^n + o(k^n).$$

Unfortunately, these inequalities fail to be sharp when the order of growth of the cohomology groups is not maximum (a polynomial expansion would be much better although it seems quite out of reach by now: we only have been able to bound the heat kernel to a lower order in k in the case of degenerate curvature, under some additional assumptions, c.f. section 3 of [B3]). The point we want to stress now is that Theorem 1.1 can be much sharper in such a situation. In fact, we are able to prove the

Theorem 1.3 (generalized Kodaira vanishing). *If the curvature of L has at least $n - q + 1$ positive eigenvalues everywhere on X, then*

$$H^i(X, E(k)) = 0 \quad \text{for} \quad i \geq q$$

as soon as k is sufficiently large.

Proof. In our setting, the strategy of the proof is quite simple: we are going to show that the integral over X of the heat kernel $e_k^i(k^\varepsilon, x)$ tends to zero. Therefore it must be less than 1 for k large, but this integral bounds the dimension of $H^i(X, E(k))$, and we are done. We shall now suppose that the eigenvalues of $ic(L)$ are ordered (i.e. $\alpha_1 \leq \ldots \leq \alpha_n$) and denote $\alpha_0 = \inf_X \alpha_q$ which is positive by our hypothesis on L and X. With our notations, we have, for $|J| = i$, $\bar{\alpha}_J \leq -\alpha_1 - \cdots - \alpha_i + \alpha_{i+1} + \cdots + \alpha_n$. For any j, according to the discussion which led to theorem 1.2, $\alpha_j e^{k^\varepsilon \alpha_j} / \sinh k^\varepsilon \alpha_j$ is a bounded function whatever the sign of α_j is. Especially, if $\alpha_j > 0$, $\alpha_j e^{-k^\varepsilon \alpha_j} / \sinh k^\varepsilon \alpha_j \leq Ce^{-2k^\varepsilon \alpha_j}$ for some constant C. But the point is that, if $i \geq q$, $\alpha_i \geq \alpha_0 > 0$; thus $e^{\bar{\alpha}_J t} / \prod_{j=1}^n \sinh \alpha_j t \leq Ce^{-k^\varepsilon \alpha_0}$. Now we use the strong uniformity of (1a) to see that $e_k^i(k^\varepsilon, x) \leq Ck^n e_\infty^i(k^\varepsilon, x)$. Integrating this last inequality over X yields $\dim H^i(X, E(k)) \leq C \operatorname{Vol}(X) k^n e^{-k^\varepsilon \alpha_0}$ thanks to the previous one. This proves the theorem.

Remarks. The note [B4] presents an enhanced version of this result, which does not require strict positivity. The same proof yields the vanishing in degree less or equal to q if $ic(L)$ has at least $q + 1$ negative eigenvalues, which also follows from Serre duality. This remark will be used in theorem 2.1. The Bochner technique proof of theorem 1.3 involves a careful construction of an adapted metric ω for which the control of the curvature eigenvalues is possible. What is remarkable in our heat equation proof is that we just don't have to care about ω.

Before ending this section, we would like to ask:

Question. Can one prove *precise* vanishing theorems by a heat equation method?

For instance, it is clear that one should be able to prove Andreotti–Grauert vanishing for vector bundles on q-complete manifolds. See [B4] for another answer to this question, which is however not specific to the heat equation approach.

2. The envelope and distortion functions

We now turn our attention to generalizations and applications of the "main theorem" in [B1]. We first make the following general observation: the spaces $H^q(X, E(k))$ are finite dimensional vector spaces, thus any two norms on it must be equivalent. When identified with the space $\mathcal{H}^q(X, E(k))$ of harmonic $(0, q)$-forms, it is naturally endowed with the L^2 and sup norms. The first inequality between these two norms: $\|\sigma\|_2^2 \leq \mathrm{Vol}(X)\|\sigma\|_\infty^2$ is obvious and does not depend on k. The second one is more subtle, and must depend on k, for reasons which will soon be clear. Notice that it is quite unnatural to bound the sup norm by an integrated one, except when, for instance, the function $|\sigma|^2$ is plurisubharmonic, but this would mean that L has negative curvature and so no cohomology (at least for $q < n$)! Anyway, one can look for the constant C_k giving the other inequality, one can even be more optimistic and seek a description of what we shall call the *envelope function* for the L^2 norm.

Definition. Let the family $h_1, \ldots, h_m$ be an L^2 orthonormal basis of $\mathcal{H}^q(X, E(k))$, then the distortion function is defined by:

$$b_k^q(x) = \sum_{j=1}^{m} |h_j(x)|^2.$$

The envelope function is:

$$s_k^q(x) = \sup_{\|h\|_2=1} |h(x)|^2.$$

It is clear that the constant C_k is simply the supremum of the envelope function whereas the distortion function is exactly the zero-eigenvalue part of the expansion of the heat kernel (notice also that the distortion function does not depend on the choice of the orthonormal basis (h_j)). The terminology for b_k^0 has the following geometrical interpretation generalizing the one given in [B1] following G. Kempf, when E is the trivial line bundle. When k is large, the global sections of the bundle $E(k)$ define an embedding in a Grassmannian such that $E(k)$ is the pull-back of the universal quotient bundle Q. As Q is a quotient bundle of a trivial one, it has a natural metric whose pull-back on $E(k)$ is proportional to the initial metric. The function b_k^0 is precisely the pointwise ratio (or "distortion") between the initial metric and the pulled-back one. The connection between these two functions is made precise in the

Proposition 2.1. *We have the two inequalities*

$$b_k^q(x) \leq e_k^q(t, x) \quad \textit{for any} \quad t > 0 \tag{2a}$$

$$s_k^q \leq b_k^q \leq M(x) s_k^q \tag{2b}$$

where $M(x)$ is the maximum rank of the subspace of the fiber $\bigwedge^{0,q} T^*X \otimes E(k) \restriction_x$ generated by the global harmonic $(0, q)$-forms.

Proof. (2a) needs no proof. The first inequality in (2b) follows simply from the fact that for any global form h of unit L^2 norm you can find an orthonormal basis containing it, then compute b_k^q with this basis. To prove the second inequality, we construct a L^2 orthonormal basis of $\mathcal{H}^q(X, E(k))$ $h_1, \ldots, h_m$ such that $h_1(x), \ldots, h_M(x)$ form an orthogonal basis of the fiber $\bigwedge^{0,q} T^*X \otimes E(k)) \restriction_x$, and $h_j(x) = 0$ for $j > M(x)$.

This is easily done by induction: we first define the h_j's for $j > M$ as an orthonormal basis of the space of global harmonic forms vanishing at x, then we pick one unit norm element h_M of the orthogonal (which is precisely of dimension M), then choose h_{M-1} in the same space such that its value at x is orthogonal (in the fiber) to that of h_M, and so on. Now, if we take any h of unit norm, it can be written $h = \sum_{j=1}^{m} \gamma_j h_j$ with $\sum |\gamma_j|^2 = 1$, thus $|h(x)|^2 = \sum_{j=1}^{M} |\gamma_j|^2 |h_j(x)|^2 \leq \max_{1\leq j\leq M} |h_j(x)|^2$. This proves that, up to some reordering, $s_k^q(x) = |h_1(x)|^2$. The proof is complete.

Proposition (2b) is completely satisfactory in case $r = 1$, $q = 0$ because it asserts that the two functions are the same. Nevertheless, the case $q = 0$ only has important geometric sense. In fact, the second inequality in (2b) should be an equality in many cases (e. g. if $\mathrm{Aut}(E)$ is transitive on the fibers): we shall show that it is the limiting case when L is positive (for $q = 0$). We now turn to answer (asymptotically) the question we raised at the beginning of this section.

Corollary 2.1. *Let* $C_k = \sup_{\|h\|_2=1} \|h\|_\infty^2$ *for* $h \in \mathcal{H}^q(X, E(k))$ *then*

$$C_k \leq r \sup_X \left|\det_\omega \frac{i}{2\pi} c(L)\right| k^n + o(k^n).$$

Proof. In fact, using (2a) and the estimates of e_∞^q, we get:

$$s_k^q(x) \leq e_k^q(k^\varepsilon, x) \leq r\left|\det_\omega \frac{i}{2\pi} c(L)_x\right| k^n + o(k^n).$$

This proves that, if the determinant is nonzero somewhere on X, we can choose a point on X and k sufficiently large in order that the corresponding term of order k^n dominates e_k^q everywhere on X, whence the result. Here, however, $o(k^n)$ is definitely not under control.

Corollary 2.1 is an improvement of an earlier result of Gromov, Gillet and Soulé [G-S] which only asserted the order of growth of C_k for $q = 0$ under the unnecessary hypothesis that L should be positive. If we require some positivity, we can obtain a much more precise statement generalizing the main theorem of [B1]:

Theorem 2.1. *If the curvature of L has constant signature q (i.e. $X(q) = X$) then the following asymptotic estimation holds:*

$$b_k^q(x) \sim r(-1)^q \det_\omega \frac{i}{2\pi} c(L)_x (k^n + o(k^n))$$

uniformly over X.

Remarks. Notice that we know by theorem 1.3 and Serre duality that the cohomology in all other degrees vanish. Of course, by integrating the above relation, one obtains an estimation for the dimension of the only nonvanishing group, which already follows from the Riemann-Roch theorem in this context. The case $q = 0$ (L positive) is specially interesting because it allows to construct many holomorphic sections of $E(k)$, as will be seen in the next section. The theorem of [B1] treats the case $q = 0$, and E trivial. In this case, there is another approach by Tian [T] based on L^2 estimates. We also point out that, according to proposition 2.1, corollary 2.1 is sharp in degree 0 when L is positive and $r = 1$.

Proof. Under our hypothesis, the estimate of theorem 2.1 is the limit of the one for e_k^q when k goes to infinity. The only thing we have to check is that the part of the expansion of $e_k^q(k^\varepsilon, x)$ corresponding to nonzero eigenvalues converges uniformly to zero when $k \to +\infty$. This is done by observing that this part injects *via* $\bar\partial$ (*resp.* $\bar\partial^*$) in the corresponding part of e_k^{q+1} (*resp.* e_k^{q-1}). But, after the proof of theorem 1.3, these two heat kernels tend uniformly to zero at (k^ε, x, x) when $k \to +\infty$. In fact, if $\alpha_0 = \inf_X(|\alpha_q|, \alpha_{q+1})$, we have $e_k^{q+1}(k^\varepsilon, x) \leq Ck^n e^{-2\alpha_0 t}$ and the same bound applies for e_k^{q-1}. On the other hand, the eigen-forms in degree q are controlled by the convergence of their heat kernel, namely: $|\psi_j(x)|^2 \leq e^{\frac{2t}{k}\mu_j^k} e_k^q(t, x) \leq Ck^n e^{\frac{2t}{k}\mu_j^k}$. Using this last inequality for a t strictly smaller than k^ε and combining it with the previous one finishes the proof (which is entirely similar to the one of [B1] except for invoking $\bar\partial^*$).

If X is a submanifold in a manifold Y of dimension m, there is another norm comparison similar to the one we outlined at the beginning of this section: suppose L is defined and positive on the whole of Y, then you would like to compare the L^2 norm on $H^0(X, E(k))$ with the quotient L^2 norm inherited from $H^0(Y, E(k))$ when k is sufficiently large (which yields the surjectivity of the restriction map). When X is the zero locus of a section of a vector bundle on Y, such results are well-known [O], [M], but do not lead to optimal constants (with respect to the k-dependency). However, once again, there must be some constant for linear algebraic reasons in *any* case, and we can hope to have the following asymptotic answer:

Conjecture 2.1. Let Y (resp. X) be a complex-analytic manifold (resp. submanifold of Y) of dimension m (resp. n), E and L as above (L positive). Let $s \in H^0(X, E(k))$, and $\|s\|_{2,X}$ (resp. $\|s\|_{2,Y}$) denote the L^2 norm (resp. quotient L^2 norm) of s. Put $c_k^1 = \inf_{\|s\|_{2,Y}=1} \|s\|_{2,X}^2$ and $c_k^2 = \sup_{\|s\|_{2,Y}=1} \|s\|_{2,X}^2$, then there exists positive constants c_1, c_2 such that

$$c_1 k^{n-m} \leq c_k^1 \leq c_k^2 \leq c_2 k^{n-m}.$$

More precisely, we conjecture that $k^{m-n} c_k^i$ converge to the same limit for $i = 1, 2$, and that this limit only depends on the metric on Y and the curvature of L.

A solution to this conjecture could lead to a drastic improvement in the paper [A-B]. Some evidence for it can be found in the study of the case $Y = \mathbb{P}^m$ and $L = \mathcal{O}(1)$ where the expected limiting constant is the degree of the projective manifold X (i.e. its volume with respect to the $ic(L)^n$ volume element). The results of section 3 also lead to its complete verification in the case $n = 0$ (X a point).

3. Construction of peak sections when L is positive

We conclude this paper by results of existence of sections of $E(k)$ using the main theorems of sections 1 and 2. We suppose from now on that L is positive and X is endowed with the Kähler metric $\omega = ic(L)$.

Theorem 3.1. *Fix $x_0 \in X$. Then, for any $k \geq 0$ there exists a global section σ_k of $E(k)$ of unit L^2 norm such that for any geodesic ball of radius r_k such that $r_k\sqrt{k} \to +\infty$, the following bound holds*

$$\int_{B(x_0,r_k)} |\sigma_k|^2 dV \geq 1 - o(1). \tag{3}$$

Where $o(1)$ means some quantity tending to zero when $k \to +\infty$. Moreover, $|\sigma_k(x)|^2 = o(k^n)$ uniformly on compact subsets of $X \setminus \{x_0\}$.

Remarks. Theorem 3.1 generalizes slightly Lemma 2.1 of [T] which was proven by using L^2 methods. In fact, his lemma produces sections of L^k with prescribed Taylor expansion of order p and mass almost concentrated at x_0, called *peak* sections. This is the same as producing sections of a power of the line bundle over a blown-up manifold we shall construct in the proof of the Kodaira embedding theorem, where we will see that it is positive. The first proof of theorem 3.1 we shall give is in some sense the converse to the one of the lemma of Tian. The reason is that Tian obtains as a consequence of his lemma a special case of our theorem 2.1 (see lemma 3.2 of [T]). Here, we use theorem 2.1 to prove theorem 3.1. In both cases, the heart of the proof is the computation of the same integral.

Proof. Let U be a small open subset of X containing x_0 on which there exist holomorphic coordinates $(z_1, \dots, z_n)$ centered at x_0, and a local frame ℓ (*resp.* $(e_1, \dots, e_r)$) of L (*resp.* E) such that

$$\begin{aligned} < e_\lambda(z), e_\mu(z) > &= \delta_{\lambda\mu} + O(|z|) \\ |\ell(z)|^2 &= 1 - |z|^2 + O(|z|^4) \\ \omega_z &= i\partial\bar\partial|z|^2 + O(|z|^2) \end{aligned}$$

where $|z|$ denotes the Euclidean norm in $\mathbb{C}^n$. Now, we define σ_k to be one unit norm generator of the orthogonal to the space of global sections vanishing at x_0 such that $|\sigma_k(x_0)|^2 = s_k(x_0)$ (this was shown to be possible during the proof of proposition 2.1). Then we know that $|\sigma_k(x_0)|^2 = s_k(x_0) = \left(\frac{k}{2\pi}\right)^n + o(k^n)$. Define $f_k = (f_k^1, \dots, f_k^r)$ to be the holomorphic function on U representing σ_k, namely $\sigma_k = \sum_\lambda f_k^\lambda e_\lambda \otimes \ell^k$. Because of the mean inequality for holomorphic functions in $\mathbb{C}^n$ we have:

$$|f_k(0)|^2 \leq \frac{1}{\mathrm{Vol}(S(0,t))} \int_{S(0,t)} |f_k(z)|^2 d\sigma(z) \tag{4}$$

where $d\sigma$ is the canonical volume element of the Euclidean sphere of radius t in $\mathbb{C}^n$. We denote $a(n) = \mathrm{Vol}(S(0,1)) = 2\pi^n/(n-1)!$. Multiplying (4) by $(1-t^2)^k$ and integrating with respect to t over $]0, r_k]$ we get

$$\begin{aligned} |f_k(0)|^2 a(n) \int_0^{r_k} t^{2n-1}(1-t^2)^k dt &\leq \int_0^{r_k} dt \int_{S(0,t)} |f_k(z)|^2 (1-|z|^2)^k d\sigma(z) \\ &= \int_{B(0,r_k)} |f_k(z)|^2 (1-|z|^2)^k d\lambda(z). \end{aligned} \tag{5}$$

Here $d\lambda$ is the Lebesgue measure on $\mathbb{C}^n$ hence $dV = \frac{\omega^n}{n!} = 2^n d\lambda + O(|z|^2)$. The real integral on the left of (5) is easily computed: $\int_0^{r_k} t^{2n-1}(1-t^2)^k dt \sim \frac{(n-1)!k!}{2(n+k)!}$ because the assumption $r_k k^{1/2} \to +\infty$ insures us that terms of the form $(1-r_k^2)^k = e^{k\log(1-r_k^2)} \leq e^{-kr_k^2}$ tend to zero. This implies that the left hand side of (5) has limit 2^{-n} when $k \to +\infty$. But this concludes the argument because the right hand side is asymptotically equal to the integral of $|\sigma_k|^2$ over $B(x_0, r_k)$ up to the constant factor 2^{-n} as soon as we choose $r_k \to 0$.

From now on, we omit q in the notation as it is always zero.

Lemma 3.1. *If L is positive*

$$s_k(x) \sim \left(\frac{k}{2\pi}\right)^n \tag{6}$$

uniformly over X.

Proof. Theorem 2.1 and the inequality $rs_k \geq b_k$ due to (2b) in proposition 2.1 yield the lower bound. Now, suppose we have at some point x_0 $\limsup k^{-n} s_k(x_0) > (2\pi)^{-n}$. For any k we can choose again σ_k to be a unit norm element of the L^2 orthogonal to the space of sections of $E(k)$ vanishing at x_0 such that $|\sigma_k(x_0)|^2 = s_k(x_0)$. By the argument above this would imply for a k large enough:

$$\int_{B(x_0,r_k)} |\sigma_k|^2 dV > 1,$$

a contradiction. The uniformity comes from the fact that all estimates which lead to the contradiction are uniform.

The last statement of theorem 3.1 comes from the same analysis. If

$$\limsup k^{-n} \sigma_k(x) > 0$$

there would be a non-negligible contribution of neighborhoods of x to the global norm of σ_k, which contradicts the previous statement. Now, we derive the

Corollary 3.1 (Kodaira embedding theorem). *If L is positive, the map $\Phi_{kL} : X \to \mathbb{P}\big(H^0(X, L^k)^*\big)$ defined by the linear system $|kL|$ is a projective embedding for k sufficiently large.*

Proof. The first observation is that theorem 2.1 implies that $|kL|$ is base point free when k is large because the nonzero limit of b_k at any point forces some section to be nonzero at this point. Thus Φ_{kL} is a well-defined map on X. Moreover, it is injective because, if two points were not separated by sections of L^k, the orthogonal of the space of sections vanishing at both points would be one dimensional. But then, the proof of theorem 3.1 shows that the unit norm generator of this space should have its mass concentrated near each one of the two points, thus a total norm greater than $2-\varepsilon$. What remains to check is that the sections of L^k separate infinitely near points. The derivation of this result from a precise vanishing theorem goes back to Kodaira himself who proved his embedding theorem as a consequence of his precise vanishing for positive line bundles together with the "quadratic transform" technique (blow-up of a point). If we restrict ourselves to the heat kernel method, thus to Theorem 1.3, the same technique applies and leads in turn to prove m-jet ampleness rather than very ampleness. We sketch hereafter this standard argument. Let us choose a point x on X and call $\widetilde{X} \xrightarrow{\pi} X$ the blow-up of X at x. Let us also call D the exceptional divisor, and $\mathcal{O}(D)$ the associated line bundle, whose restriction to $D \simeq \mathbb{P}(T_x^* X)$ is isomorphic to $\mathcal{O}(-1)$, thus negative. Using the fact that $\pi^* L$ is positive on any tangent vector to $\widetilde{X}$ except those tangent to D, it is easy to check that the line bundle $\widetilde{L} = \pi^* L^{k_0} \otimes \mathcal{O}(-D)$ is positive for some $k_0 \geq 0$. Theorem 1.3 yields $H^1(\widetilde{X}, \widetilde{L}^m) = 0$ for some large $m \geq 2$. If we call $\mathcal{I}_x$ the maximal ideal sheaf associated to x, the space of m-jets of L^{mk_0} at x is $H^0(X, L^{mk_0} \otimes \mathcal{O}_X/\mathcal{I}_x^m)$. Using the exact sequence

$$0 \to \mathcal{I}_x^m \to \mathcal{O}_X \to \mathcal{O}_X/\mathcal{I}_x^m \to 0,$$

and blowing it up at x, we obtain the diagram

$$\begin{array}{ccccc} H^0(\widetilde{X}, \pi^* L^{mk_0}) & \to & H^0(D, \pi^* L^{mk_0} \otimes \mathcal{O}_{\widetilde{X}}/\mathcal{O}_{\widetilde{X}}(-mD)) & \to & H^1(\widetilde{X}, \widetilde{L}^m) = 0 \\ \pi^* \uparrow & & \pi^* \uparrow & & \\ H^0(X, L^{mk_0}) & \to & H^0(X, L^{mk_0} \otimes \mathcal{O}_X/\mathcal{I}_x^m) & & \end{array}$$

where vertical arrows are isomorphisms. This yields the result. □

Remarks. After corollary 3.1 we know that X is an algebraic projective manifold. Thus, any coherent sheaf on X admits a global syzygy. A simple induction on the length of the syzygy shows that the vanishing theorem 1.3 is valid for any coherent sheaf E. This is known as theorem A (see for instance [G-H]). Our results of the previous section also lead to an easy direct proof of the so-called theorem B in the case of vector bundles (twisted by ideal sheaves of zero-dimensional subschemes):

Corollary 3.2. *If L is positive and E is any vector bundle of rank r, let $Z \subset X$ be a finite set, and $x \in X \setminus Z$. Then the fiber $E(k)_x$ is generated by global sections of $E(k)$ vanishing on Z for k sufficiently large (k uniform with respect to x in compact subsets of $X \setminus Z$).*

Proof. We treat the case $Z = \{x_0\}$ for simplicity, the general case being similar. The pointwise norm at $x \neq x_0$ of the peak section σ_k associated to x_0 is negligible relatively to $k^{\frac{n}{2}}$, and thus does not alter the estimate for b_k nor s_k at x (This statement is uniform for x outside any ball B of center x_0). Therefore, if we call $b'_k(x) = b_k(x) - |\sigma_k(x)|^2$ we have $b'_k \sim b_k$ uniformly outside B. But b'_k is the sum of the squares of the pointwise norms of an orthonormal basis of the space of sections of $E(k)$ vanishing at x_0 (because σ_k is orthogonal to this space). Denote by $M'(x)$ the dimension of the linear subspace of $E(k)_x$ generated by global section vanishing at x_0. The above remark, lemma 3.1 and (an obvious adaptation of) proposition 2.1 imply: $rs_k \sim b'_k \leq M'(x)s_k \leq rs_k$. As $M'(x)$ and r are integers, it is clear that they must be equal for k large, which precisely means the announced property.

An alternative proof of Theorem 3.1. We would like to finish this paper with a second proof of theorem 3.1 which illustrates a possible use of (1b) in theorem 1.1. It can also be seen as an introduction to the delicate localization procedure we use in our joint paper [A-B]. Let us first introduce some notations: given an open subset Ω of X and some $\mu > 0$, we shall denote $\mathcal{H}_k(\Omega, \mu)$ the direct sum of the eigen-spaces of $\overline{\Box}_{k,\Omega}$ corresponding to eigenvalues $\mu_j^k(\Omega) \leq k\mu$. Now, if we denote (somewhat abusively) μ_1^k the first nonzero eigenvalue of $\overline{\Box}_k$ on X, we have the well known

Lemma 3.2. *If α_0 is the infimum of the curvature eigenvalues on X, we have the lower bound:*

$$\liminf k^{-1}\mu_1^k \geq \alpha_0.$$

Proof. This fact follows immediately from the proof of theorem 2.1 because we showed there that

$$\sum_{j\geq 1} e^{-\frac{2\mu_j^k}{k}t} \leq \int_X e_k^1(t,x)dV \leq Ck^n e^{-2\alpha_0 t}$$

if $t \leq k^\varepsilon$. $e^{-\frac{2\mu_1^k}{k}t}$ has thus the same upper bound. Take the Log of this inequality and let t tend slowly to infinity. □

Now, we claim that

Lemma 3.3. *For $\mu < \alpha_0$ and $k \gg 0$, the map*

$$\Psi_{k,\mu} : \mathcal{H}_k(\Omega, \mu) \to H^0(X, L^k)$$

which is simply the orthogonal projection is injective. Moreover, it is not far from being an isometry: For $u \in \mathcal{H}_k(\Omega, \mu)$ one has the estimate

$$\|\Psi_{k,\mu}(u) - u\|_2^2 \leq (k\mu/\mu_1^k)\|u\|_2^2.$$

Proof. Define $H_\Omega(u) = \int_\Omega < \frac{1}{k}\overline{\Box}_k u, u > dV$ (*resp.* $H(u) = \int_X < \frac{1}{k}\overline{\Box}_k u, u > dV$). For $u \in \mathcal{H}_k(\Omega, \mu)$ we have $H(u) = H_\Omega(u) \leq \mu\|u\|_2^2$ thus, if $u \neq 0$ $\Psi(u)$ cannot be zero because this would imply $H(u) \geq (\mu_1^k/k)\|u\|_2^2 \geq (\alpha_0 - o(1))\|u\|_2^2 > \mu\|u\|_2^2$. This proves the first assertion while the second boils down to the Pythagorean theorem: if $u = \Psi(u) + u_1$ is the orthogonal decomposition, we have $\|u\|_2^2 = \|\Psi(u)\|_2^2 + \|u_1\|_2^2$ with $H(u_1) = H(u) \leq \mu\|u\|_2^2$ and $H(u_1) \geq (\mu_1^k/k)\|u_1\|_2^2$. □

Now, our construction goes like this: for any $\varepsilon > 0$ choose a $\mu < \alpha_0\varepsilon$ and any unit norm section τ_k of $\mathcal{H}_k(B_k, \mu)$ where $B_k = B(x_0, r_k)$. This is possible because theorem 1.1 (1b) implies that the first eigenvalue of $\frac{1}{k}\overline{\Box}_{k,B_k}$ tends to zero. In fact, if it was not the case, one can find a subsequence k_ν such that $e_{k_\nu, B_{k_\nu}}(k_\nu^\varepsilon, x_0, x_0) \to 0$, which contradicts (1b) (the estimates are the same as for the proof of theorem 2.1). Thus, for any $\mu > 0$ the space $\mathcal{H}_k(B_k, \mu)$ is not trivial when k is large. We put $\hat{\sigma}_k = \Psi(\tau_k)$ and we obtain $\hat{\sigma}_k + \tilde{\sigma}_k = \tau_k$ with $\|\tilde{\sigma}_k\|_2^2 \leq \varepsilon$. Whence $\|\hat{\sigma}_k\|_2^2 = \|\tau_k\|_2^2 - \|\tilde{\sigma}_k\|_2^2 \geq 1 - \varepsilon$ and $\varepsilon \geq \int_X |\hat{\sigma}_k - \tau_k|^2 dV \geq \int_{X \setminus B_k} |\sigma_k|^2 dV$. So the section $\sigma_k = \frac{\hat{\sigma}_k}{\|\hat{\sigma}_k\|_2^2}$ is the global holomorphic section we are looking for.

Concluding remarks. It has been shown that many important results in analytic geometry can be derived from Eq. (1a). One can even improve some of these results by giving them some flavor of what they should mean at a point or, to say it in another manner, what is their local analogue. Anyway, the global results we have been able to prove in this paper are somewhat outdated. This is because strict positivity was a necessary condition in order to get rid of the non-harmonic part in the heat kernel (because strict positivity forces the first eigenvalue of $\overline{\Box}_k$ to go to infinity with k). There are thus two possible ways of development for this theory. One should be able, following a conjecture of Siu [S], to find estimates for the first eigenvalue of $\overline{\Box}_k$ under weaker conditions than positivity. (We shall discuss in the forthcoming paper [B4] what we can do in this direction with the tools developed here: this is quite far from what could be expected!) For instance, if L has semi-positive curvature everywhere, and positive definite at some point, many of the results we discussed have weaker analogues (e.g. Grauert–Riemenschneider theorems). More generally (and we come to the second point), the same question holds for nef and big line bundles: the overlap of L^2 methods in this case is due to the use of singular metrics. It would be quite marvelous to obtain Shokurov's base point freeness *via* an analogue of theorem 2.1. The other point is thus: is it possible to develop these devices for singular metrics?

The last remark we would like to write is due to Y.-T. Siu. Since our main interest is in holomorphic (or harmonic) objects, why should we work with the whole spectrum of $\overline{\Box}_k^q$, and then try to get rid of its greater part rather than work directly with holomorphic objects like the Bergman kernel? For instance, if you develop in a power series along the fibers the Bergman kernel of the unit ball in the total space L^*, the value along the diagonal of the coefficient of the kth power should look much like our distortion function b_k. The (technical) reason for which we worked with the heat kernel is that it is a very supple function, which allows lots of manipulations (among which the localization properties as in (1b) is not the least): it is less rigid than the Bergman kernel. Moreover, results like theorem 1.2 or the second proof of theorem 3.1 show that a little walk outside holomorphy can lead to powerful controls of holomorphic objects. Anyway, a vector-bundle approach to the Bergman kernel is certainly a worthy direction for future research.

References

[A-B] Abbes, A., Bouche, Th., Sur le théorème de Hilbert–Samuel (arithmétique), C. R. Acad. Sci. Paris Sér. I Math. 317 (1993), 589–591, and Ann. Inst. Fourier (Grenoble) 45 (1995), 375–401.

[Bi] Bismut, J.-M., Demailly's asymptotic inequalities: a heat equation proof, J. Functional. Anal. 72 (1987), 263–278.

[B1] Bouche, Th., Convergence de la métrique de Fubini-Study d'un fibré linéaire positif, Ann. Inst. Fourier (Grenoble) 40 (1990), 117–130.

[B2] Bouche, Th., Distortion function and the heat kernel of a positive line bundle, in: Complex Analysis, Proceedings of the international workshop held in Wuppertal (K. Diederich ed.), February 1990 (1991), 60–63.

[B3] Bouche, Th., Noyau de la chaleur et inégalités de Morse holomorphes pour un fibré en droites à courbure dégénérée, Bull. Sci. Math. 2° Sér. 116 (1992) 167–183.

[B4] Bouche, Th., Two vanishing theorems for holomorphic vector bundles of mixed sign, Prépublication de l'Institut Fourier n° 256, Grenoble , Math. Z. 218 (1995), 519–526.

[B5] Bouche, Th., Sur le noyau de la chaleur associé aux puissances tensorielles d'un fibré en droites complexes, Séminaire de théorie spectrale et géométrie (Grenoble) 12 (1993–1994), 41–49.

[D1] Demailly, J.-P., Champs magnétiques et inégalités de Morse pour la d''-cohomologie, Ann. Inst. Fourier (Grenoble) 35 (1985), 189–229.

[D2] Demailly, J.-P., Holomorphic Morse Inequalities, in: Several Complex Variables and Complex Geometry, Part. 2 (S. G. Krantz ed.), Providence, A.M.S. Proceedings of Symposia in Pure Math. 52 (1991), 93–114.

[G-H] Griffiths, P., Harris, J., Principles of algebraic geometry, J. Wiley & Sons, New York 1978.

[G-S] Gillet, H., Soulé, C., Amplitude arithmétique, C. R. Acad. Sci. Paris Sér. I Math. 309 (1989), 929–932.

[Ma] Manivel, L., Un théorème de prolongement L^2 de sections holomorphes d'un fibré hermitien, Math. Z. 212 (1993), 107–122.

[O] Ohsawa, T., On the extension of L^2 holomorphic functions II, Publ. Res. Inst. Math. Sci. 24 (1988), 265–275.

[S] Siu, Y.-T., A vanishing theorem for semipositive line bundles over non-Kähler manifolds, J. Differential Geom. 19 (1984), 431–452.

[T] Tian, G., On a set of polarized Kähler metrics on algebraic manifolds, J. Differential Geom. 32 (1990) 99–130.

Effective bounds for Nori's connectivity theorem

Robert Braun and Stefan Müller-Stach

Abstract. The paper consists of two parts: First we give explicit bounds for a theorem of M. Nori in the case of complete intersections using the reformulation given by K. Paranjape. It implies in particular: Let Y be a general complete intersection of type $(2, a, a)$ in $\mathbb{P}^7$ with $a \geq 9$, then Y carries a nontrivial 1-cycle modulo algebraic equivalence which is Abel–Jacobi equivalent to zero (Nori's example). The discussion of sharpness of these results leads to the second part: Let $X \subset \mathbb{P}^n_{\mathbb{C}}$ be an arbitrary subscheme defined by r equations of degrees $d_1, \dots, d_r$. We give a modern proof of Predonzan's theorem, that there exists a sharp bound $n(d_1, \dots, d_r; m)$ such that X contains a linear $\mathbb{P}^m_{\mathbb{C}}$ whenever $n \geq n(d_1, \dots, d_r; m)$.

Key words: algebraic cycle, complete intersection, linear subvariety.

1. Introduction

We recall Nori's connectivity theorem:

Theorem [Nori 93]. *Let $(X, \mathcal{O}_X(1))$ be a smooth, projective, polarized variety with $\mathcal{O}_X(1)$ ample. Let $f : \mathcal{Y} \to S$ be the universal family of complete intersections of type $(d_1, \dots, d_r)$ on X with base space $S \subset \prod \mathbb{P}H^0(\mathcal{O}_X(d_i))$ Zariski open. If $g : T \to S$ is any smooth base change, $\mathcal{Y}_T = \mathcal{Y} \times_S T \subset X \times T$ and* $\min(d_i)$ *sufficiently large, then*

$$H^i(\mathcal{Y}_T, \mathbb{Q}) \cong H^i(X \times T; \mathbb{Q})$$

for $i \leq 2(\dim X - r) - 1$.

Here *sufficiently large* means, that there is a number $N = N(X, \mathcal{O}_X(1), r)$ depending on X, $\mathcal{O}_X(1)$ and r only, such that the theorem holds for $\min(d_i) \geq N$. From this Nori obtains the following

Corollary [Nori 93]. *Let $(X, \mathcal{O}_X(1))$ and $d_1, \dots, d_r$ be as above and* $\min(d_i) \geq N(X, \mathcal{O}_X(1), r)$. *Assume that there is a codimension p-cycle ξ on X with $p <$* $\dim X - r$ *and which is homologous to zero on each hyperplane section. Then if*

$[\xi] \neq 0$ in $H^{2p}_{\mathrm{pr}}(X, \mathbb{Q})$, $\xi|_Y$ is not algebraically equivalent to zero in $CH^p(Y)$, where $Y \subset X$ is a general complete intersection of type $d_1, \ldots, d_r$ on X.

Griffiths first discovered nontrivial cycles modulo algebraic equivalence in codimension two on a quintic threefold using Abel–Jacobi invariants. With the corollary above Nori was able to produce the first known example of an interesting cycle in the kernel of the Abel–Jacobi map:

Example. Let $X = Q^{2p}$ be the $2p$-dimensional smooth quadric in $\mathbb{P}^{2p+1}$ and $Y \subset Q^{2p} a$ $(p-1)$-fold complete intersection of type $(d, \ldots, d)$. Take ξ = difference of two p-dimensional linear subspaces from different rulings. The corollary then applies for d sufficiently large.

Other applications can be found in [Green 94]. Nori did not give any explicit bounds in his paper. Later M. S. Ravi gave the bound $N(Q^6, \mathcal{O}(1), 2) \leq 72$ for $p = 3$ in [Ravi 93]. In this paper we are able to improve this bound to $N \leq 9$ for $p = 3$ by using a method from [Paranjape 94]. Our first result is:

Theorem 1. *$N(Q^{2p}, \mathcal{O}(1), p-1) \leq 2p+3$ for $p \geq 1$. More generally if $X \subset \mathbb{P}^{n+s}$ is itself a complete intersection of type $(a_1, \ldots, a_s)$ with $2 \leq a_1 \leq a_2 \leq \cdots \leq a_s$, then*

$$N(X, \mathcal{O}_X(1), r) \leq \begin{cases} n \cdot a_s + \sum_{i=1}^{s}(a_i - 1) - 2r - 1, & \text{if } a_s \geq 3 \\ s + 2n - 2r & \text{if all } a_i = 2 \end{cases}$$

$(n = \dim X)$.

It was a very nice observation by K. Paranjape ([Paranjape 92]) that the sharpness of any bound can be estimated by looking at the linear subspaces contained in a general complete intersection Y of type $(d_1, \ldots, d_r)$ on X.

We were thus led to our second problem: Fix multidegrees $2 \leq d_1 \leq \cdots \leq d_r$ and look at the set of all subschemes X defined by r equations of degrees $d_1, \ldots, d_r$. Let $n(d_1, \ldots, d_r; m; \mathbb{C})$ be the infimum of all n, such that every X as above which is contained in $\mathbb{P}^n_{\mathbb{C}}$, contains a linear $\mathbb{P}^m_{\mathbb{C}}$ itself. Similarly let $l(d_1, \ldots, d_r; m; \mathbb{C})$ be the infimum of all n such that in the definition of $n(d_1, \ldots, d_r; m; \mathbb{C})$ above we have additionally that

$$CH_m(X) \otimes \mathbb{Q} \cong \mathbb{Q} \cdot [\mathbb{P}^m_{\mathbb{C}}].$$

We give a proof of the theorem of A. Predonzan ([Predonzan48]) with methods of modern enumerative geometry to avoid the general position arguments in *op. cit.*:

Theorem 2.

$$n(d_1, \ldots, d_r; m; \mathbb{C}) = \frac{1}{m+1} \sum_{i=1}^{r} \binom{d_i + m}{m} + m.$$

Acknowledgements. We thank the following organizations for supporting our research project: the organizers of the Trento conference, DFG Forschungsschwerpunkt and the Universities Bayreuth and Essen.

2. Effective bounds for $N(X, \mathcal{O}_X(1))$

Let $X \subset \mathbb{P}^{n+s}$ be a smooth complete intersection of type $(a_1, \dots, a_s)$, where we assume $2 \le a_1 \le \dots \le a_s$ without loss of generality. Recall that a sheaf F on X is called *m-regular* if $H^i(X, F(m-i)) = 0$ for all $i \ge 1$. We are interested in the regularity of Ω_X^k for all $0 \le k \le n$.

Lemma 1. *Ω_X^k is $k(a_s - 1) + \sum_{i=1}^s (a_i - 1)$ regular if $a_s \ge 3$. If all $a_i = 2$, then Ω_X^k is $(k + s + 1)$-regular.*

Proof. Recall that by Bott's formulas $\mathcal{O}_{\mathbb{P}^n}$ is 0-regular and $\Omega_{\mathbb{P}^n}^k$ is $(k+1)$-regular for $k \ge 1$. If F is any sheaf on $\mathbb{P}^{n+s}$ which is m-regular, then $F|_X$ is $(m + \sum_{i=1}^s (a_i - 1))$-regular by induction on the restriction sequences. To prove Lemma 1, consider the exact sequence

$$0 \longrightarrow N_X^* \longrightarrow \Omega_{\mathbb{P}^{n+s}}^1|_X \longrightarrow \Omega_X^1 \longrightarrow 0$$

where $N_X^* = \bigoplus \mathcal{O}_X(-a_i)$ is the conormal bundle of X. Therefore Ω_X^k has a resolution

$$(*) \quad 0 \longrightarrow F_{k+1} \longrightarrow F_k \longrightarrow \cdots \longrightarrow F_1 \longrightarrow \Omega_X^k \longrightarrow 0$$

where $F_j = S^{j-1} N_X^* \otimes \Omega_{\mathbb{P}^{n+s}}^{k-j+1}|_X$, $1 \le j \le k+1$. The regularity of a tensor product is the sum of the individual regularities, hence F_j is m_j-regular with

$$m_j = (j-1)a_s + k - j + 2 + \sum_{i=1}^{s} (a_i - 1)$$

for $1 \le j \le k$ and F_{k+1} is m_{k+1}-regular where $m_{k+1} = ka_s + \sum_{i=1}^s (a_i - 1)$. Now we need the following general fact: If in any resolution like $(*)$ we have $m_{k+1} \ge m_j + k - j$ for $1 \le j \le k$, then Ω_X^k is $m_{k+1} - k$ regular. (This follows by splitting $(*)$ into short exact sequences and induction). Now if $a_s \ge 3$ we have $m_{k+1} \ge m_j + k - j$ and hence Ω_X^k is $(m_{k+1} - k = k(a_s - 1) + \sum_{i=1}^s (a_i - 1))$-regular. If all $a_i = 2$ by a similar argument one gets that Ω_X^k is $(k + s + 1)$-regular. □

Let M_k be the regularity of Ω_X^k.

Definition. $M_X := \max\{M_k - k - 1 \mid 0 \le k \le n\}$.

By [Paranjape 92], formula 2.8. we have the estimate

$$N = N(X, \mathcal{O}_X(1), r) \leq 2(\dim X - r) + M_X$$

for the bound needed in Nori's connectivity theorem. For the reader of the original papers: we take $e = n - r - 1$ in the notation of Nori–Paranjape. In our special situation, $X \subset \mathbb{P}^{n+s}$ a complete intersection of type $(a_1, \ldots, a_s)$, we get together with Lemma 1:

Theorem 1. *Let* $n = \dim X$, X *as above, then*

$$N(X, \mathcal{O}_X(1), r) \leq \begin{cases} n \cdot a_s + \sum_{i=1}^{s}(a_i - 1) - 2r - 1, & \text{if } a_s \geq 3 \\ s + 2n - 2r & \text{if } a_s = 2. \end{cases}$$

Examples. (a) Let $X = \mathbb{P}^n$, $r = 1$. We have $M_{\mathbb{P}^n} = 0$ by Bott's formulas. Hence $N(\mathbb{P}^n, \mathcal{O}(1), 1) = 2n - 2$ by the result of Paranjape. This is "sharp" by the observation that a hypersurface Y of degree $d \leq 2n-3$ contains skew lines and therefore contradicts the conjectural implications of Nori's theorem. See the discussions in [Paranjape 92] and [Schoen 93].

(b) Let $X = Q^{2p} \subset \mathbb{P}^{2p+1}$ a smooth quadric, $s = 1$ and $r = p - 1$ as in Nori's example [Nori 93]. We get by Theorem 1:

$$N(X, \mathcal{O}_X(1), p - 1) \leq 2p + 3.$$

Hence for $p = 3$ we regain $N \leq 9$ thus improving [Ravi 93].

3. Precise bounds for $n(d_1, \ldots, d_r; m; \mathbb{C})$

As we already mentioned in the introduction the results in § 1 are related to the following problem:

Fix multidegrees $d_1 \leq d_2 \leq \cdots \leq d_r$ and look at the infimum of all integers n, such that every subscheme $X \subset \mathbb{P}^n$ defined by r equations of degrees $d_1, \ldots, d_r$ contains a linear $\mathbb{P}^m_{\mathbb{C}}$. This number is denoted by $n(d_1, \ldots, d_r; m; \mathbb{C})$. Similarly let $l(d_1, \ldots, d_r; m; \mathbb{C})$ be the infimum over the smaller set where we additionally assume that $CH_m(X) \otimes \mathbb{Q}$ is generated by $\mathbb{P}^m_{\mathbb{C}} \subset X$ over $\mathbb{Q}$. For $k = \mathbb{C}$ these are finite numbers by [Paranjape92]. This implies in particular some results about the representability of Chow groups for $X \subset \mathbb{P}^n_{\mathbb{C}}$ with n sufficiently large. It is therefore of great interest to determine $n(d_1, \ldots, d_r; m; \mathbb{C})$. Paranjape has also given an algorithm to compute the $l(d_1, \ldots, d_r; m; \mathbb{C})$ from $n(d_1, \ldots, d_r; m; \mathbb{C})$ in his paper.

Arno Predonzan proved the following result in 1948 by "counting constants". We give a modern proof avoiding the restriction to smooth varieties.

Theorem 2.

$$n(d_1, \dots, d_r; m; \mathbb{C}) = \frac{1}{m+1} \sum \binom{d_i + m}{m} + m.$$

Taking this for granted, we first check whether Theorem 1 has any chances to be sharp:

Examples. (a) Taking $r = 1, d_1 = d, m = 1$ we obtain $n(d; 1; \mathbb{C}) = \frac{d+1}{2} + 1$. This is the same as to say that a hypersurface of degree $d \leq 2n - 3$ in $\mathbb{P}^n$ contains a line, whether it is smooth or not. See § 2, Example (a).

(b) Let $r = 3, d_1 = 2, d_2, d_3 \geq 2$ and $m = 1$. We obtain

$$n(2, d_2, d_3; 1; \mathbb{C}) = \frac{1}{2}(3 + (d_2 + 1) + (d_3 + 1)) + 1 = \frac{d_2 + d_3 + 7}{2}.$$

This is to say that if $d_2 + d_3 \leq 2n - 7$, any subscheme defined by equations of degrees $2 \leq d_2 \leq d_3$ in $\mathbb{P}^n$ will contain a line. Taking $n = 7$ we obtain $d_2 + d_3 \leq 7$. On the other hand to verify Theorem 1 with X a 6-dimensional quadric we proved $N(Q^6, \mathcal{O}(1), 2) \leq 9$. This would mean $\min(d_2, d_3) \geq 9$. We see that there is a gap and Theorem 1 may not be sharp.

Proof of Theorem 2. Consider the incidence variety

$$I \subset Gr(m, n) \times \prod_{i=1}^{r} \mathbb{P}H^0(\mathbb{P}^n, \mathcal{O}(d_i))$$

given by all (L, F_i) with $L(\cong \mathbb{P}^m_{\mathbb{C}}) \subset V(F_1, \dots, F_r)$. It has two projections

$$\begin{array}{ccc} & I & \\ {}^{p}\swarrow & & \searrow^{q} \\ Gr(m, n) & & \prod \mathbb{P}H^0(\mathbb{P}^n, \mathcal{O}(d_i)) \end{array}$$

q will be generically surjective, if and only if we show that the generic fibre dimension of q is nonnegative. To do this consider the incidence variety

$$Z \subset Gr(m, n) \times \prod_{i=1}^{r} \mathbb{P}^n$$

given by all $(L, x_1, \dots, x_r)$ with $x_i \in L$. We have projections

$$\begin{array}{ccc} & Z & \\ {}^{\alpha}\swarrow & & \searrow^{\beta} \\ Gr = Gr(m, n) & & \prod \mathbb{P}^n \end{array}$$

Let E be the vector bundle

$$E = \alpha_* \beta^* \mathcal{O}_{\mathbb{P}^n}(d_1) \boxtimes \cdots \boxtimes \mathcal{O}_{\mathbb{P}^n}(d_r)$$

on $Gr(m, n)$. One has a surjection

$$S : \prod H^0(\mathbb{P}^n, \mathcal{O}(d_i)) \otimes \mathcal{O}_{Gr} \longrightarrow E$$

i.e. E is globally generated. Now if we choose $F = (F_1, \ldots, F_r) \in \prod \mathbb{P} H^0(\mathcal{O}(d_i))$ generically, this defines a section $S_F \in H^0(Gr, E)$. If $X = V(F_1, \ldots, F_r)$ S_F will also be denoted by S_X. Now

$$S_X \text{ vanishes at } L \in Gr \Leftrightarrow L \subset X.$$

But since $rk(E) = \sum_{i=1}^r \binom{m+d_i}{d_i}$ and $h^0(E) > rk(E)$, S_X will vanish in the right codimension which is $rk(E)$ for F chosen generically. The fibres of q are exactly the loci $V(S_X)$, hence they have the dimension

$$(m+1)(n-m) - \sum \binom{d_i + m}{d_i}$$

generically if this number is ≥ 0 and they are empty otherwise. The final observation is, that since I is projective we may drop the word "generically": q is even surjective if this number is ≥ 0. Theorem 2 now follows. □

References

[Green94] M. Green, Infinitesimal methods in Hodge theory, in: Lecture Notes in Math. 1594, Springer-Verlag, 1994.

[Nori 93] M. Nori, Algebraic cycles and Hodge theoretic connectivity, Invent. Math. 111 (2) (1993), 349–373.

[Paranjape 92] K. Paranjape, Cohomological and cycle theoretic connectivity, Ann. of Math. 140 (1994), 641–660.

[Predonzan 48] A. Predonzan, Intorno agli S_k giacenti sulla varieta intersezione completa di piu forme, Rend. Sc. Acad. Lincei 5 (1948), 238–242.

[Ravi 93] M. S. Ravi, An effective version of Nori's theorem, Math. Z. 214 (1993), 1–8.

[Schoen 93] Ch. Schoen, On Hodge structures and non-representability of Chowgroups, Compositio Math. 88 (1993), 285–316.

Kodaira dimension and fundamental group of compact Kähler manifolds

Frédéric Campana

1. Introduction

Let $c : \tilde{X} \to X$ be the universal covering of the compact connected Kähler manifold X, and $\pi_1(X)$ its fundamental group, which is finitely presented. We shall say that a finitely presented group Γ is *Kähler* if $\Gamma \cong \pi_1(X)$ for some X. We shall consider the following questions.

A. Classify the groups Γ which are Kähler

Not all finitely presented groups Γ are indeed Kähler. The question has thus two aspects:

A1. Construct examples of Kähler groups.

We shall in Section 2 survey some constructions, based on Lie groups and Morse theory.

A2. Find obstructions for a finitely presented Γ to be Kähler.

We shall survey in Section 3 some obstructions derived from Hodge theory and L^2-cohomology theory. In Section 4 we shall study in more detail the obstructions on the completion of the nilpotent part of Γ arising from the study of the Albanese map and rational homotopy theory.

Observe that large classes of finitely presented groups are not known to be either Kähler or non-Kähler, even among the simplest ones (nilpotent, for example).

B. Describe the structure of $\tilde{X}$

The guiding principle is here Shafarevitch's conjecture, which asserts that $\tilde{X}$ should be holomorphically convex (i.e.: proper over a Stein base). We shall survey in Sections 5, 6 some partial results in this direction. The methods are geometric, but L^2-cohomology should be essential for future developments.

C. Relate $\pi_1(X)$ and the Kodaira dimension of X

In riemannian geometry, positivity assumptions on Ricci curvature restrict the complexity of $\pi_1(M)$, M a riemannian manifold. Just in the same way, $\pi_1(X)$ simplifies when $\kappa(X)$ (or some variants measuring the geometric positivity of Ω^1_X) becomes smaller. Again, the methods used here rest on L^2-cohomology.

Remark. All known compact Kähler manifolds are deformation of some projective one: in particular they are diffeomorphic to it, and hence have the same π_1. It might thus be possible that Kähler groups are projective (i.e.: isomorphic to $\pi_1(X)$ for some projective X). This is true for surfaces and threefolds (see 6.5). We shall also see (4.5.4) some other results which points in this direction.

Many interesting topics can not be discussed here. Let us mention the theory of harmonic maps to symmetric spaces or Bruhat–Tits buildings, which deal with the finite dimensional semi-simple representations of Kähler groups (see: [G-S], [Mok], [Si], [Zuo]).

These notes are the written version of a series of lectures given during the Trento school in June 1994.

I would like to thank the organizers, and especially M. Andreatta and G. Bolondi for their kindness and efficiency, as well as the participants for their attention. Several improvements in the exposition are due to observations of L. Bonavero and J. Winkelmann.

2. Construction of examples (survey)

We shall survey here some constructions based essentially on Morse theory. This survey follows closely D. Arapura's [A1].

2.1. Introduction

Recall that any compact differentiable manifold M has a fundamental group Γ which is *finitely presented* (i.e.: of the form $\Gamma = F_r/R$, where F_r is the free group on a finite set of generators $\{x_1, \dots, x_r\}$). Thus F_r consist of all words $W = x_{i_1}^{\alpha_1} \cdot x_{i_2}^{\alpha_2} \dots x_{i_N}^{\alpha_N}$, where $N \geq 0$ (if $N = 0$, $W = e$ the unit element), $\alpha_j \in \mathbb{Z}$ and $i_j \neq i_{j+1}$. The product is defined by concatenation (and obvious *simplification* rules: $x_i \cdot x_i^{-1} = e$). Moreover, R is the *normal* subgroup of F_r generated by a set of words $(W_\lambda)_{\lambda \in \Lambda}$ with Λ *finite* (if we drop this assumption, Γ is just said to be *finitely generated*). (The fact that $\pi_1(M)$ is finitely presented follows easily from Van Kampen's theorem).

Conversely, any finitely presented group Γ is isomorphic to $\pi_1(M)$ for some 4-

dimensional smooth manifold. A construction (described to me by M. Katz in Nancy) goes like this:

If $\Gamma = F_r$, just take $M = \sharp_r(S^1 \times S^3)$ (by Van Kampen's theorem[1]; the sign $\sharp_r$ stands for the connected sum of r copies). To create one additional relation W_λ in $\pi_1(M)$, realize this W_λ by a smooth S^1 in M, and take a small tubular neighborhood N of S^1. Thus $N \cong S^1 \times S^3$, and $\partial N \cong S^1 \times S^2 \cong \partial(S^2 \times S^2)$. Now delete N and replace it by $S^2 \times S^2$, identifying the boundaries as above. From Van Kampen's theorem we deduce that we have "simplified" $\pi_1(M)$ by (W_λ), but not more. Remark that the possible $\pi_1(M)$ are classified if $\dim_{\mathbb{R}} M = 1, 2$, but not 3.

It has been shown by Taubes ([Ta], see also [Ga]) that if M is any compact smooth 4-dimensional manifold, there exists $k_M \in \mathbb{Z}^+$ such that $M' = M\sharp_r\mathbb{P}_2(\mathbb{C})$ carries an anti-self dual riemannian metric, to which a twistor space $\tau : Z(M') \to M'$ is attached by R. Penrose's construction. Thus: $Z(M') = Z$ is a 3-dimensional *complex* compact manifold, and τ a smooth (non-holomorphic!) submersion with holomorphic fibers $\mathbb{P}_1(\mathbb{C}) \cong S^2$. We thus get: $\pi_1(Z) \cong \pi_1(M') \cong \pi_1(M)$ and the (amazing) result that any finitely presented group Γ is $\pi_1(Z)$ for some 3-dimensional compact complex manifold.

This construction however never produces interesting Kähler groups, because if $Z(M') = Z$ is Kähler (resp. bimeromorphic to Kähler), then M' is S^4 or $\mathbb{P}_2(\mathbb{C})$ ([Hit]) (resp. homeomorphic to $\sharp_r\mathbb{P}_2(\mathbb{C})$ ([Ca1])), so that $\pi_1(Z) = \{1\}$.

We now survey some constructions of finitely presented Kähler groups Γ.

2.2. Finite groups are Kähler

We give here a construction due to R.I. Shafarevitch ([Sha], p. 402) (see also [Se1], p. 527 for another similar construction). Let G be a finite group, realized as subgroup of σ_m, the symmetric group on m letters, for m sufficiently large.

Let $P = (\mathbb{P}_3(\mathbb{C}))^m$, on which σ_m acts by permuting the factors. Let $P' = P/\sigma_m$ be the natural normal quotient projective variety, embedded in $\mathbb{P}_N(\mathbb{C})$ using a basis of σ_m-invariant homogeneous polynomial of a sufficiently high degree. Then $S' = \mathrm{Sing}(P')$ has codimension at least 3 in P', since the natural map $\sigma : P \to P'$ is unramified over $P^* := \{(p_1, \ldots, p_m) \mid P_i \neq P_j \ (i \neq j)\}$. Let $P'^* = \sigma(P^*)$, so that $\sigma^* : P^* \to P'^*$ is unramified and Galois with group σ_m. Now cut P'^* by a general $(3m-2)$-plane of $\mathbb{P}_N(\mathbb{C})$ to get X', which is a smooth projective surface, with $X := \sigma^{-1}(X')$ smooth and connected (by Bertini). We thus get an unramified Galois covering $\sigma : X \to X'$ of group σ_m. But X is a complete intersection of $(3m-2)$-hypersurfaces in P, and so is simply-connected by Lefschetz's hyperplane section theorem. Thus $\pi_1(X') = \sigma_m$.

[1]Recall that it asserts, under mild restrictions, that if S is a topological space covered by two open subsets $S_i (i = 1, 2)$ such that $S_1 \cap S_2 = S_0$ is connected, then: $G \cong G_1 *_{G_0} G_2$ with $G = \pi_1(S)$, $G_i = \pi_1(S_i)$ $(i = 0, 1, 2)$.

To construct a projective surface X'' with $\pi_1(X'') = G$, just take the intermediate covering $X'' = X/G$.

2.3. Product of Kähler groups

Products of Kähler groups and finite index subgroups of Kähler groups are again Kähler.

2.4. Free abelian groups

Free abelian groups of even rank are Kähler. Indeed: $\mathbb{Z}^{\oplus 2r} = \pi_1(E^r)$ if E is any elliptic curve. More generally, $\mathbb{Z}^{\oplus 2r} \cong \pi_1(A)$ if A is any r-dimensional abelian manifold, or any of its complete intersections of dimension at least two.

2.5. Heisenberg groups of even rank

We shall say that Γ is a *Heisenberg group* of rank $r \geq 2$ if

1) $[\Gamma, \Gamma] \cong \mathbb{Z}$ (where $[\Gamma, \Gamma]$ is the commutator subgroup of Γ, that is: the subgroup of Γ generated by all commutators $[x, y] = xyx^{-1}y^{-1}$ of elements x, y of Γ). Remark that $[\Gamma, \Gamma] = e$ means that Γ is abelian, and that $[\Gamma, \Gamma]$ is a normal (and even characteristic) subgroup of Γ.
2) $[\Gamma, [\Gamma, \Gamma]] = \{e\}$ (or equivalently, $[\Gamma, \Gamma]$ is contained in the center $Z(\Gamma)$ of Γ).
3) $\Gamma/[\Gamma, \Gamma]$ is abelian, free of rank r.

These groups are just the most commutative ones beyond the abelian groups (among torsion free groups).

Proposition ([Ca2]). *For any $r \geq 4$, there exist Heisenberg groups Γ_{2r} of even rank $2r \geq 8$, which are Kähler.*

We just describe the construction when $r = 2m \geq 4$ with m even (it can easily be adapted to the case m odd by taking factors of unequal dimension):

Let A be an abelian variety of dimension $2m$, with *finite* regular map

$$f : A \to \mathbb{P}_m(\mathbb{C}) \times \mathbb{P}_m(\mathbb{C}) = P.$$

(Such maps are easily constructed by taking $A = A_1 \times A_1$).

Let $\Delta = \Delta_g \subset P$ be the graph of a *generic* automorphism $g \in \mathbb{P}\mathrm{Gl}(m+1, \mathbb{C})$ of $\mathbb{P}_m(\mathbb{C})$. (By composing with g^{-1} the identification of the second factor of P, we can and shall assume that Δ is the diagonal of P). Take $X := f^{-1}(\Delta)$. Then X is smooth, connected and one has an exact sequence:

$$1 \to \mathbb{Z} \to \pi_1(X) \to \pi_1(A) \to 1 \qquad (2.1)$$

such that $\mathbb{Z} = [\Gamma, \Gamma]$.

We shall explain briefly why this is so, following [F-L], Section 10: Let $H_i (i = 1, 2)$ be two disjoint m-dimensional projective linear subspaces of $\mathbb{P}_{2m+1}(\mathbb{C})$, let $\tilde{P} := \mathbb{P}_{2m+1}(\mathbb{C}) - \{H_1 \cup H_2\}$, and let $\pi : \tilde{P} \to P \left(= (\mathbb{P}_m(\mathbb{C}))^2\right)$ be the natural projection defined in homogeneous coordinate by:

$$\pi([z_o : z_1 : \ldots : z_m : \zeta_0 : \ldots : \zeta_m]) = ([z], [\zeta]),$$

where $[z] = [z_o : z_1 : \ldots : z_m]$ and similarly for $[\zeta]$ (of course we take H_1 (resp. H_2) defined by the equations: $[\zeta] = 0$ (resp. $[z] = 0$).

Notice that $\pi : \tilde{P} \to P$ is naturally a $\mathbb{C}^*$-bundle, the fiber above $([z], [\zeta])$ being $[\lambda z : \zeta]$ for $\lambda \in \mathbb{C}^*$. Let $\tilde{\Delta} \subset \tilde{P}$ be the m-dimensional linear projective subspace of $\mathbb{P}_{2m+1}(\mathbb{C})$ defined by the equation: $[z : \zeta] \in \tilde{\Delta}$ iff $z = \zeta$. The restriction $\pi : \tilde{\Delta} \to \Delta$ is then an isomorphism.

Let $\tilde{A} := A \times_P \tilde{P}$; $\tilde{f} : \tilde{A} \to \tilde{P}$ be deduced from $f : A \to P$ by the base change $\pi : \tilde{P} \to P$: the natural map $\alpha : \tilde{A} \to A$ is thus a $\mathbb{C}^*$-bundle, too. Let $\tilde{X} := \alpha^{-1}(X) = \tilde{f}^{-1}(\tilde{\Delta})$, and notice that $\alpha : \tilde{X} \to X$ is an isomorphism.

The situation can be summarized in the following commutative diagram:

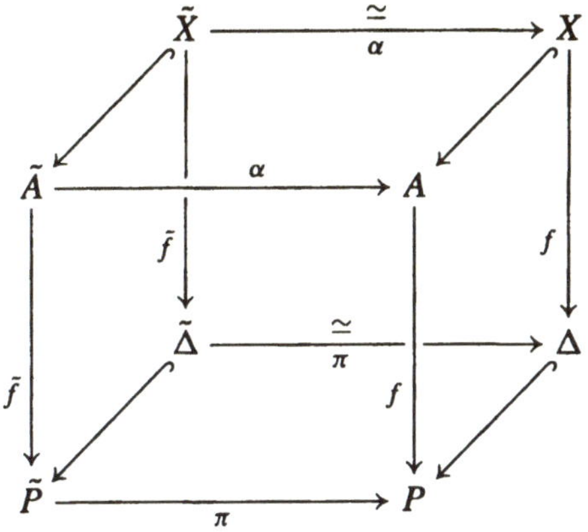

From this diagram, we now get at the π_1-level the following commutative diagram:

$$\begin{array}{ccccccccc} & & & & \pi_1(\tilde{X}) & \overset{\cong}{\to} & \pi_1(X) & \to & 1 \\ & & & & \tilde{\chi}_* \downarrow & & \downarrow \chi_* & & \\ 1 & \to & \pi_1(\mathbb{C}^*) & \to & \pi_1(\tilde{A}) & \to & \pi_1(A) & \to & 1 \end{array}$$

in which the bottom line is derived from the long homotopy sequence of the $\mathbb{C}^*$-fibration $\alpha : \tilde{A} \to A$, using the fact that $\pi_2(A) = \pi_2(\mathbb{C}^{2m}) = 0$.

The essential fact is now that the map $\tilde{\chi}* : \pi_1(\tilde{X}) \to \pi_1(\tilde{A})$ is an isomorphism. This is easily derived from the following stratified Morse-theoretic result which extends in several directions the classical theorem of Lefschetz on hyperplane sections.

Theorem ([Go-Ma], p. 195). *Let $g : V \to \mathbb{P}_N(\mathbb{C})$ be an algebraic finite dominant map from the quasi-projective variety V. Let L be the d-dimensional linear projective subspace such that $W := g^{-1}(L)$ is smooth compact. Then $\pi_i(V, W) = 0$ if $i \leq d$; in particular $\pi_1(V) \cong \pi_1(W)$ as soon as $d \geq 2$.*

A proof of similar results, which avoids stratified Morse theory may be found in [S-V].

If we now apply the preceding result to $(g, V, N, L) = (\tilde{f}, \tilde{A}, 2m+1, \tilde{\Delta})$, we indeed obtain the natural isomorphism $\tilde{\chi}_* = \pi_1(\tilde{X}) \overset{\cong}{\to} \pi_1(\tilde{A})$, which gives us an exact sequence:

$$1 \to \mathbb{Z}_0 = \pi_1(\mathbb{C}^*) \to \Gamma := \pi_1(X) \to \mathbb{Z}^{\oplus 4m} = \pi_1(A) \to 1.$$

2.5.2. It remains to check that: $\mathbb{Z} \cong [\Gamma, \Gamma]$; but obviously $[\Gamma, \Gamma] \subset \mathbb{Z}_0$ since $\pi_1(A)$ is abelian. If we had $[\Gamma, \Gamma] = e$, Γ were commutative and thus isomorphic to $\mathbb{Z}^{\oplus 4m+1}$, which contradicts Hodge theory ($b_1(X)$ is even: see below). Thus $[\Gamma, \Gamma]$ is cyclic of finite index δ, say in $\mathbb{Z}_0$. That $[\Gamma, [\Gamma, \Gamma]] = \{e\}$ follows from the fact that if we let Γ act by conjugation on $\mathbb{Z}_0$ (which is normal), this action descends to a representation $\rho : \pi_1(A) \to \mathrm{Aut}(\mathbb{Z}_0)$; this last group is however reduced to $\mathbb{Z}_2$ (the exchange of generators). In our situation this representation is trivial, since $\tilde{A} = A \times_P \tilde{P}$ has two "ends" (or equivalently, the S^1-bundle over A associated to $\tilde{A}$ is orientable).

2.5.3. Remark. The result would immediately follow if we could prove that $f_* : \pi_2(A, X) \to \pi_2(P, \Delta) \cong \mathbb{Z}$ is an isomorphism. Observe also that the last equality shows that Δ is not a complete intersection of ample divisors (if $m \geq 2$). (This last result can also be derived from an easy computation using intersection numbers).

2.5.4. Remark. A closer look at the proof above shows (see [C-T2] for these remarks, and (1.4.8) for more elementary details on Heisenberg groups) that the extension class $\lambda \in \mathrm{Hom}(\bigwedge^2 \mathbb{Z}^{\oplus 4m}, \mathbb{Z}) = H^2(A, \mathbb{Z})$ is nothing but $f^*(\omega, -\omega)$, where $\omega \in H^2(\mathbb{P}_m, \mathbb{Z})$ is the cohomology class of $\mathcal{O}(1)$. In particular, it is non degenerate, of type $(1, 1)$.

Because λ cannot be a principal polarisation, $\pi_1(X)$ is never $H_{4m,1}$ in the above construction. Let $e = (e_1 \leq \ldots \leq e_4)$ be the elementary divisors of λ ($m = 2$). The following e's are achieved([C-T2] Section 6):

$$e = (2, 2, 2, 2);\ e = (1, 1, 5, 5);\ e = (1, 1, 3, 3).$$

2.5.5. Remark. The statement of the result of [Go-Ma] above is not the original one. We simplified it to adapt to our situation. The reduction to this result is taken essentially from [F-L], Section 10. An other similar example was already given in [S-V] which in fact shows that some Γ_{12} is Kähler. Note that this unnoticed example in

[S-V] existed before (and disproved) a conjectured made by J. Carlson and D. Toledo ([C-T1]) that a nilpotent Kähler group should have a subgroup of finite index which is abelian.

2.5.6. Heisenberg groups. Exact sequences $1 \to \mathbb{Z} \to \Gamma \to G \to 1$ where $G \cong \mathbb{Z}^{\oplus r}$ is abelian, and $\mathbb{Z} \subset Z(\Gamma)$ correspond to elements $\lambda \in \mathrm{Hom}_{\mathbb{Z}}(\bigwedge^2 G, \mathbb{Z}) = H^2(G, \mathbb{Z})$ as follows:

1) If this extension is given, $\lambda : \bigwedge^2 G \to \mathbb{Z}$ is defined by $\lambda(a, b) = [a, b]$, which is skew-symmetric and bilinear (by the commutator formula):

$$[ac, b] = [a, [c, b]] \cdot [c, b] \cdot [a, b]].$$

2) If λ is given, define Γ to be generated by $(a_1, \dots a_r), \zeta$ subject to the relations: $[a_i, a_j] = \zeta^{\lambda}(a_1, a_j)$.

Exercise. Let Γ' be a finite index subgroup of Γ, Heisenberg of rank r. Show that Γ' is itself Heisenberg of rank r (indeed : $[\Gamma', \Gamma']$ is of finite index in $[\Gamma, \Gamma]$).□

Let now $r = 2m$ and $\lambda : \bigwedge^2 G \to \mathbb{Z}$ be non-degenerate, as above; by an appropriate choice of a $\mathbb{Z}$-basis of G, λ takes the form: $\sum_{\alpha=1}^{m} e_\alpha(z_{2\alpha-1} \wedge z_{2\alpha})$, where $e = (e_1 \leq e_2 \leq \dots \leq e_m)$ are positive integers.

Thus Γ is nothing but the subgroup $H_{2m,e}$ of index $(e_1 \dots e_n)$ of $H_{2m,1}$ consisting of all integer-valued $(n+2) \times (n+2)$-matrices $g_{a,b,c}$ of the form:

$$g_{a,b,c} = \begin{pmatrix} 1 & a & c \\ 0 & I_n & b \\ 0 & 0 & 1 \end{pmatrix}$$

where I_n is the $n \times n$ unit matrix, $c \in \mathbb{Z}$, $a \in \mathbb{Z}^n$, ${}^t b \in \mathbb{Z}^n$.

Finally, if $E \to A$ is an orientable $\mathbb{C}^*$-bundle over the Abelian variety A, one has a central group extension $1 \to \pi_1(\mathbb{C}^*) \cong \mathbb{Z} \to \pi_1(E) \to \pi_1(A) \to 1$ which shows that $\pi_1(E)$ is a Heisenberg group of rank $r = 2 \dim A$). This extension is characterized as above by the class $\lambda \in H^2(A, \mathbb{Z}) = \mathrm{Hom}_{\mathbb{Z}}(\bigwedge^2 G, \mathbb{Z})$ if $G = \pi_1(A)$.

This situation arises above in the proof of (2.1), where $\alpha : \tilde{A} \to A$ is the f-lift from the $\mathbb{C}^*$-bundle $\tilde{P} \to P$. We thus get (from the homology spectral sequence of this fibration) a natural exact sequence:

$$0 \to H_2(\tilde{P}, \mathbb{Z}) \cong \mathbb{Z} \to (P, \mathbb{Z}) \cong \mathbb{Z}^{\oplus 2} \to H_1(\mathbb{C}^*, \mathbb{Z}) \cong \mathbb{Z} \to 0$$

which determines the class $(\omega, -\omega) \in H^2(P, \mathbb{Z})$ of this $\mathbb{C}^*$-bundle. Thus: $\lambda = f^*(\omega, -\omega)$ as stated above in (2.5.3).

2.5.7. Remark. Let X be an m-dimensional manifold with $\pi_1(X)$ a Heisenberg group H_{2m} of rank $r = 2s \geq 8$ as constructed in 2.5 above. By cutting X with the inverse image of a generic linear subspace of appropriate dimension, we get a surface $S = S_s$ with $\pi_1(S) = \pi_1(X) \cong H_{2S}$.

We shall check that $c_1^2(S_m)/c_2(S) = 2(1 - 1/s)$ $(s \geq 4)$. (Notice that for the example S' given in [S-V], one has $c_1^2(S')/c_2(S') = 5/3$, which corresponds to the value $s = 6$ above). The smallest interval containing ratios $c_1^2(S)/c_2(S)$ for S surfaces with $\pi_1(S)$ a Heisenberg group thus contains $[3/2, 2]$.

To compute the value $c_1^2(S)/c_2(S)$, it is sufficient to observe first that, by smoothness: $N_{S|A} \cong f^* N_{L|P}$ if $S = f^{-1}(L)$, if $N_{Y|Z}$ is the normal bundle of Y in Z. This equality has been communicated to me by A. Sommese.

The rest of the computation is straightforward, since:

$$1 = c(A) = c(S) \cdot c(N_{S|A}) = c(S) \cdot f^* c(N_{L|P})$$

if $c(V)$ is the total Chern class of V.

An easy computation shows that: $c(N_{L|P}) = (1 + H)^{s-1}$ if H is the class of $\mathcal{O}(1)$. Thus $(c_1^2/c_2)(S_m) = 2\frac{(s-1)^2}{(s-1)s} = 2(1 - \frac{1}{s})$ as claimed. (The first values are: 3/2; 8/5; 5/3; 12/7; 7/4; 16/9 corresponding respectively to $s = 4, 5, 6, 7, 8, 9$).

2.6. Fundamental groups of curves ($g \geq 1$)

2.6.1. Let $C_g = C$ be a (compact connected) curve of genus $g \geq 1$. Then $\Gamma_g := \pi_1(C_g)$ admits the following finite presentation: $\Gamma_G = F_{2g}/R$, where F_{2g} is the free group on $2g$ generators (a_j, b_j) $(1 \leq j \leq g)$, and R is the normal subgroup generated by the single "relation" $\prod_{j=1}^{g}[a_j, b_j]$. This is usually written: $\Gamma_g = \langle a_1, \ldots, a_g, b_1, \ldots, b_g \mid \prod_{j=1}^{g}[a_j, b_j]\rangle$. We thus get for $g = 1 : \Gamma_1 \cong \mathbb{Z}^{\oplus 2} = \langle a, b \mid ab = ba\rangle$. Observe that Γ_g contains F_{2g-1} (and hence F_2), as soon as $g \geq 2$.

2.6.2. Remarks. Let[2] G_g be a curve of genus $g \geq 1$, and $C_{g,p}$ be the curve with $p \geq 1$ distinct points deleted. Then:

(1) $C_{g,p}$ is homotopic to a wedge W_r of $r = (2g + p - 1)$ circles (obvious, it follows also from Morse theory). The drawing for $g = p = 2$ is:

[2] I thank M. Katz for pointing this out to me.

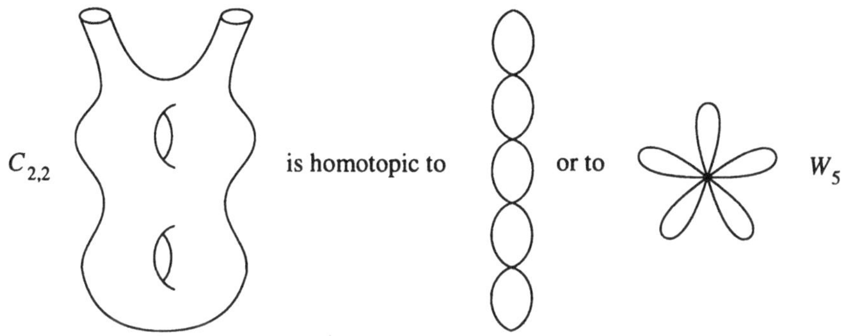

(2) $\pi_1(C_{g,p}) \cong \pi_1(W_{2g+p-1}) \cong F_{2g+p-1}$.

The last assertion is by induction on r from Van Kampen's theorem.

We shall now use these facts to show the famous:

2.6.3. Theorem (Nielsen–Schreier). *Let Γ be a subgroup of finite index d in F_r. Then Γ is isomorphic to $F_{\tilde{r}}$, with $\tilde{r} = 1 + d(r-1)$.*

2.6.4. Remarks. When Γ is of infinite index in F_r, it is still true that Γ is a free group (on countably many generators). This too can be proven as below. Although a very simple and more elementary proof of the above result is given in [Se2] §3, it seems worth to give a proof based on the familiar Riemann–Hurwitz formula.

Proof. Let $C_{g,p}$ be such that $2g+p-1 = r$ (we assume $r \geq 2$, otherwise the assertion is trivial). Let $f^* : \tilde{C}^* \to C_{g,p}$ be the finite étale covering of $C_{g,p}$ of fundamental group Γ: then $\tilde{C}^*$ can be compactified to $\tilde{C}$ by adding say $\tilde{p}$ points, so that f^* extends to a holomorphic d-sheeted (possibly ramified) cover $f : \tilde{C} \mapsto C_g$. Let $\tilde{g}$ be the genus of $\tilde{C}$; we thus have $\pi_1(\tilde{C}^*) \cong F_{\tilde{r}}$ is free on $\tilde{r} = 2\tilde{g} + \tilde{p} - 1$ generators. We compute $\tilde{r}$.

From the Riemann–Hurwitz formula, we get:

$$2(\tilde{g}-1) = 2d(g-1) + R, \text{ where } R = \sum_{\tilde{x} \in \tilde{C}} (e_{\tilde{x}}(f) - 1) \tag{2.2}$$

is the degree of the ramification divisor $D = \sum (e_{\tilde{x}}(f) - 1) \cdot \tilde{x}$ of f.

Let $C_g, p = C_g - P$, with $P = \{x_1, \dots, x_p\}$, and $\tilde{P} = f^{-1}(P)$; we have: $d.p = \sum_{\tilde{x} \in \tilde{P}} e_{\tilde{x}}(f) - 1) + \sharp \tilde{P} = R + \tilde{p}$. Thus:

$$d \cdot p = R + \tilde{p}. \tag{2.3}$$

From (2.2) we get $2(\tilde{g}-1) = 2d(g-1)\tilde{p} + dp$, or: $2\tilde{g} + \tilde{p} - 2 = d(2g+p-1)$, which is nothing but 2.6.3.

2.6.7. Remark. The proof does not need the precise form of $\pi_1(C_g)$ given in 2.6.1, and only rests on a topological (not analytical) interpretation of the genus g.

2.7. Cocompact lattices of SU$(n, 1)$

Let $G = \mathrm{SU}(n, 1)$ be the group of unimodular $(n+1)\times(n+1)$-matrices with complex coefficients which preserve the quadratic form

$$q(z, \zeta) = z_1\bar{\zeta}_1 + \cdots + z_n\bar{\zeta}_n - \sqrt{2}z_{n+1}\bar{\zeta}_{n+1}$$

on $\mathbb{C}^{n+1}$. Thus $G = \{g \in \mathrm{Sl}(n+1, \mathbb{C}) \mid {}^t\bar{g}Qg = Q\}$ where Q is the diagonal matrix with coefficients $(1, \ldots, 1, -\sqrt{2})$.

We let G act on $\mathbb{P}_n(\mathbb{C})$ in the natural way: it is easy to check that, for this action there are exactly three orbits:

1. the $(n-1)$-dimensional quadric $Q_{n-1} := \{[z] \mid Q(z, z) = 0\}$;
2. the ball $B_n = \{[z] \mid q(z, z) < 0\}$, identified with the ball of radius $\sqrt[4]{2}$ in $\mathbb{C}^n$;
3. its complement $C_n = \{[z] \mid q(z, z) > 0\}$.

For the above action, the stabilizer of $[0 : \ldots : 0 : 1] = p_0$ is the maximal compact subgroup $K \cong U(n) = \left\{\begin{pmatrix} B & 0 \\ 0 & \delta \end{pmatrix} \mid B \in U(n) \text{ and } \delta = (\det B)^{-1}\right\}$, so that G/K is naturally identified with B_n, by mapping $g \in G$ to $g \cdot p_0$.

Let now Γ_0 be an (arithmetic) discrete cocompact subgroup of $G = \mathrm{SU}(n, 1)$: it always admit a torsion free subgroup Γ of finite index. Such a subgroup Γ now acts freely and properly discontinuously on $B_n = G/K$. The quotient $X = \Gamma\backslash B_n = \Gamma\backslash G/K$ is thus a compact complex manifold with $\pi_1(X) \cong \Gamma$. And X is Kähler, because B_n carries a Kähler form which is $\mathrm{SU}(n, 1)$-(and hence Γ)-invariant.

It follows from a theorem of Kodaira (or from 7.2.2 below) that X is projective, because the holomorphic n-form $\omega(z) = dz_1 \wedge \ldots \wedge dz_n$ is L^2 with respect to the natural $\mathrm{SU}(n, 1)$-invariant metric on B_n.

Observe that every curve of genus $g \geq 2$ is covered by B_1, and thus obtained by the preceeding construction.

The delicate point is the construction of Γ_0. An explicit example in the situation above is $\Gamma = G(\mathbb{Z}[i, \sqrt{2}])$ the subgroup of G consisting of all g with coefficients in the ring $\mathbb{Z}[i, \sqrt{2}]$ ([Bo1,2]).

2.7.1. Remarks. The above construction can be generalized in two ways:

1. One can replace G/K above by any irreducible hermitian symmetric space of non-compact type (i.e.: the bounded symmetric domains, see [He], pp. 518, 527 for

their classification).

The existence theorem for Γ still applies in this more general situation.

2. One can take (see [To1]) for Γ any torsion-free discrete arithmetic lattice in $G \cong SU(n, 1)$ ($\Gamma\backslash G$ is only required to have finite volume), provided G is realized as an algebraic group defined over $\mathbb{Q}$ in such a way that the $\mathbb{Q}$-rank of G is one. In this case, $(\Gamma\backslash G/K) = Y^*$ can be compactified as a singular projective variety by adding only finitely many points. A Morse-theoretic argument now shows that cutting Y^* with a projective linear space of appropriate dimension, one gets a projective smooth surface X such that $\pi_1(X) \cong \pi_1(Y^*) \cong \Gamma$.

A concrete example is obtained by taking: $G = \{g \in \mathrm{Sl}(n+1, \mathbb{C}) \mid g \text{ preserves } \bar{q}\}$, where $\bar{q}$ is the quadratic form $\bar{q}(z, \zeta) = z_1\bar{\zeta}_1 + \cdots + z_n\bar{\zeta}_n - z_{n+1} \cdot \bar{\zeta}_{n+1}$, and Γ torsion-free of finite index in $G \cap \mathrm{Sl}(n+1, \mathbb{Z}[i])$.

See [To1] for more details on this construction, where it was introduced.

2.8. Non-arithmetic cocompact lattices in SU(2, 1)

Recall that a group Γ is called *arithmetic* if it can be embedded as an *arithmetic subgroup* of some algebraic subgroup G defined over $\mathbb{Q}$ of $\mathrm{Gl}(n, \mathbb{C})$ for some n. (For such $G \subset \Gamma$, Γ is arithmetic if $\Gamma \cap \mathrm{Gl}(n, \mathbb{Z})$ is of finite index both in Γ and in $G_{\mathbb{Z}} := G \cap \mathrm{Gl}(n, \mathbb{Z})$).

See [Se3], [Bo2] for properties of these groups. Remark that the Kähler groups constructed by the procedure of 2.7 are arithmetic. Moreover, a deep theorem of G. Margulis ([Ma]) asserts in particular that lattices in simple groups of rank at least 2 are arithmetic.

There actually exist non-arithmetic cocompact lattices in SU(2, 1) (which has of course rank 1), or equivalently, there exists projective surfaces X which are:

1. covered by the ball $B_2 \subset \mathbb{C}^2$;
2. of non-arithmetic π_1 in SU(2, 1).

Remark that the condition 1 above can be replaced by 1′ below, due to S.T. Yau's solution of the Calabi conjecture:

1′. $c_1^2(X) = 3c_2(X)$. (Recall that by Bogomolov–Miyaoka–Yau inequality, one has $c_1^2 \leq 3c_2$ as soon as $\kappa(X) \geq 0$).

On the other hand, such extremal surfaces have been constructed in particular by F. Hirzebruch ([Hir]), providing both arithmetic and non-arithmetic Kähler groups in SU(2, 1). We briefly describe a construction in the latter category.

Take X to be minimal surface with function field:

$\mathbb{C}(z_1/z_3; z_2/z_3; (\ell_2/\ell_1)^{1/5}; \ldots; (\ell_9/\ell_1)^{1/5})$ where $[z_1 : z_2 : z_3]$ are homogeneous coordinates on $\mathbb{P}_2(\mathbb{C})$ and the ℓ_k's are defined by: $(z_i^3 = z_j^3)$ $(1 \leq i \leq j \leq 3)$.

Thus X appears as a smooth model of a Galois cover of $\mathbb{P}_2(\mathbb{C})$ branched along 9 lines, with group $\mathbb{Z}_5^{\oplus 8}$.

It is shown in [Hir] that $c_1^2(X) = 3c_2(X)$, and in [Y-Y] that $\Gamma = \pi_1(X)$ is commensurable with the monodromy group Γ_1 of a certain hypergeometric differential equation introduced by Appel. Further, it is shown by G. Mostow ([Mo]) that Γ_1 is not arithmetic.

2.9. Non-residually finite Kähler groups

The preceeding examples of Kähler groups are linear, or realized as subgroups of some linear groups.

Examples of Kähler groups which are not linear, because not even residually finite, were first given in [To2].

Recall first that a group Γ is said to be *residually finite* if $\Gamma_F = \{1\}$, were Γ_F is by definition the intersection of all subgroups of finite index in Γ: it is a normal (even characteristic) subgroup of Γ. Let $\Gamma/\Gamma_F := \Gamma^{\text{alg}}(X)$ the quotient group. When $\Gamma = \pi_1(X)$, $\Gamma^{\text{alg}} = \pi_1^{\text{alg}}(X)$ is called here the *algebraic fundamental group of* X, although it is not the projective limit of Galois groups of finite Galois étale covers of X.

A result of A. Malcěv ([Mal1]) asserts that a linear group is residually finite.

The construction of D. Toledo in [To2] provides a Kähler group Γ for which Γ_F is a free group on infinitely many generators: in particular Γ is not linear. The fact that Γ is not residually finite is derived from a difficult group-theoretic result of M. Ragunathan [Ra], which asserts that any lattice Λ in the universal cover $\widetilde{\text{spin}}(2, m)$ of spin$(2, m)$, (m odd) contains $8 \cdot \mathbb{Z}$, where $\mathbb{Z} = \ker(\widetilde{\text{spin}}(2, m) \to (2, m))$.

Toledo's construction can briefly be described as follows: let $n \geq 2$ be even, and let X_n be one component of the symmetric space for the group $SO(2, n)$: it is the open subset of the Grassmannian of unoriented 2-planes in $\mathbb{R}^{n+2}$ on which $q(x, x) = \sqrt{2}(x_1^2 + x_2^2) - (x_3^2 + \cdots + x_{n+2}^2)$ is definite positive. Thus X_n is hermitian symmetric, and appears as the open subset of the quadric $Q_n \subset \mathbb{P}_{n+1}(\mathbb{C})$ defined by: $b(z, z) = 0$ and $b(z, \bar{z}) > 0$, where b is the complex linear extension of q. Let $v = (0, \ldots, 0, 1)$, so that $q(v, v) < 0$, and let $v^\perp$ the b-orthogonal complement of v: restricted to $v^\perp$, q has signature $(2, n-1)$. The subgroup $SO^+(2, n-1)$ of $SO^+(2, n)$ fixing v is isomorphic to $SO^+(2, n-1)$, and leaves $v^\perp$ invariant. Thus X_{n-1} naturally embeds in X_n as its subset of 2-planes contained in $v^\perp$.

Let now Γ be torsion-free discrete and cocompact in $SO^+(2, n)$, for example: Γ of finite index in $(SO^+(2, n) \cap Sl(n + 2, \mathbb{Z}'[\sqrt{2}]))$, where $\mathbb{Z}'[\sqrt{2}]$ is the ring of integers of $\mathbb{Q}[\sqrt{2}]$.

Then Γ acts effectively on X_n. Let $M := \Gamma \backslash X_n$: it is a projective manifold with $\pi_1(M) \cong \Gamma$.

Let $\Gamma_0 := \Gamma \cap SO^+(2, n-1)$: it is cocompact in $SO^+(2, n-1)$ and is the subgroup of Γ that leaves X_{n-1} invariant. Moreover, $D := \Gamma_0 \backslash X_{n-1}$ embeds in M as a divisor with negative normal bundle. It is then shown in [To2] that if the contraction that contracts D to a point p is $q : M \to V$, then:

1. V is projective with one singular point p.
2. $\pi_1(M-D) = \pi_1(V-\{p\}) = \Phi$ is the fundamental group of a projective manifold X (the argument is Morse-theoretic, analogous to the one used in 2.7.1 and in [To1]).
3. If $\Phi_0 = \pi_1(\partial N)$, where N is a tubular neighborhood of D in M, the natural map $\varphi : \Phi_0 \to \Phi$ induced from the inclusion $\partial N \hookrightarrow M - D$ is injective. Observe that Φ_0 is a central extension of Γ_0 given by:
$$1 \to Z \cong \mathbb{Z} \to \Phi_0 \to \Gamma_0 \to 1.$$
4. Φ_0 is a cocompact lattice in $\widetilde{\mathrm{SO}}(2, m)$, the covering group of $\mathrm{SO}^+(2, m)$ corresponding to its infinite cyclic factor in $\pi_1(\mathrm{SO}^+(2, m)) = \pi_1(\mathrm{SO}(2) \times \mathrm{SO}(m)) \cong \mathbb{Z} \times \mathbb{Z}_2$.

   The result of Ragunathan can now easily be applied, since $\widetilde{\mathrm{spin}}(2, m)$ is the universal cover of $\widetilde{\mathrm{SO}}(2, m)$, to see that Φ_0 (and hence Φ) is not residually finite.

Remark. Although the above group Φ is far from being residually finite, it admits still subgroups of arbitrary large finite index. A question asked by Serre ([Se2]) is: is the Higman 4-group Γ the fundamental group of a projective manifold ?

This group Γ is finitely presented, with the 4 generators a_j $(1 \leq j \leq 4)$ subject to the 4 relations: $[a_{j+1}, a_j] = a_j$ $(1 \leq j \leq 4; 4 + 1 = 1)$.

This group is infinite, but has no non-trivial subgroup of finite index ([Se2], p. 18); it is not Kähler by [G-S], 9.1, and [Se2], p. 19.

Exercise. Show the last assertion (if $\psi : \Gamma \to \Delta$ is a surjective homomorphism of groups with Δ finite, let p be the smallest prime number dividing one of the orders n_j of $\bar{a}_j = \psi(a_j)$ in Δ.

Assume that $p|n_1$. Since $a_1^{2^{n_2}} = a_2^{n_2} a_1 a_2^{-n_2}$, we get: $2^{n_2} \equiv 1(n_1) \equiv 1(p)$. Thus $p \neq 2$, and $2 \not\equiv 1(p)$. Consider now one prime factor q of the order $N \neq 1$ of the image of 2 in $(\mathbb{Z}_p)^*$: q divides n_2. But $1 < N < p$).

Remark. Presently, no Kähler infinite group without non-trivial subgroup of finite index is known.

3. Obstructions (partial survey)

Let Γ be a finitely generated group, and $[\Gamma, \Gamma]$ its commutator subgroup. Let $H_1(\Gamma) := \Gamma/[\Gamma, \Gamma]$: it is a finitely generated abelian group. We denote its rank with $b_1(\Gamma) := \mathrm{rank}_{\mathbb{Z}}(H_1(\Gamma))$. Again, this survey is inspired by [A1], from which most of the examples are taken.

3.1. The first Betti number of a Kähler group

3.1.1. The first Betti number $b_1(\Gamma)$ of a Kähler group is even. This follows from Hodge theory: let $\Gamma = \pi_1(X)$ for X a Kähler manifold. Then $H_1(\Gamma) = H_1(X)$ (by Hurewicz's theorem); so $b_1(\Gamma) = b_1(X)$. Now by Hodge decomposition: $H^1(X, \mathbb{C}) \cong H^0(X, \Omega^1) \oplus H^1(X, \mathcal{O}) \cong H^0(X, \Omega^1) \oplus \overline{H^0(X, \Omega^1)}$ and $b_1(X) = 2h^0(X, \Omega^1)$.

3.1.2. Examples.

1. F_r (the free group on r generators) is not Kähler if r is odd. In particular, $\mathbb{Z}$ is not Kähler (although $\mathbb{Z} \cong \pi_1(S)$ if S is a Hopf surface). Indeed
$$b_1(F_r) = r.$$
2. Let $\Gamma = \mathbb{Z}^{\oplus 2} \ltimes \mathbb{Z}_2$ be the semi-direct product, where $\mathbb{Z}^2$ acts on $\mathbb{Z}^{\oplus 2}$ by the involution $\alpha : (a, b) \to (a, -b)$. One has thus an exact sequence (of groups):
$$1 \to \mathbb{Z}^{\oplus 2} \to \Gamma \xrightarrow{\pi} \mathbb{Z}_2 \to 1,$$
and $\Gamma = \{(a, b, \zeta) \in \mathbb{Z} \times \mathbb{Z} \times \mathbb{Z}_2\}$, the product being defined by:
$$(a, b, \zeta) \cdot (a', b', \zeta') = (a + a', b + (-1)^{\zeta} b', \zeta + \zeta').$$
It is easy to check that $[\Gamma, \Gamma] = \{(0, b, 0) \mid b \in \mathbb{Z}\}$. Thus $H_1(\Gamma) \cong \mathbb{Z} \oplus \mathbb{Z}_2$ and $b_1(\Gamma) = 1$. So Γ is not Kähler (although $\ker(\pi) = \mathbb{Z}^{\oplus 2}$ is of index 2 and Kähler, so that it is not sufficient for Γ to be Kähler to have a subgroup of finite index which is Kähler).
3. An even simpler description of this fact is $\Gamma = \langle a, b \mid aba^{-1} = b^{-1} \rangle$, which is $\pi_1(B)$, B the Klein bottle. Indeed, $b_1(\Gamma) = 1$, but Γ has an index two subgroup isomorphic to $\mathbb{Z}^{\oplus 2} = \pi_1(S^1 \times S^1)$.
4. Let Γ be the finitely presented group: $\langle a, b \mid [a, b] = b \rangle$. Then Γ is a solvable group with $b_1(\Gamma) = 1$, so that Γ is not Kähler.

We check these assertions:

Let $\sigma : \Gamma \to \mathbb{Z}$ be the surjective group homomorphism which maps $\gamma = a^{\alpha_1} b^{\beta_1} \dots a^{\alpha_k} b^{\beta_k}$ to $(\alpha_1 + \dots + \alpha_k)$. This is well-defined (because the sum of exponents of a in the single relation defining Γ is zero). Let Δ be the kernel of σ: it is isomorphic to the additive group of the subring $\mathbb{Z}[1/2]$ of $\mathbb{Q}$. Thus, it is abelian (but not finitely generated, so that Γ is solvable, but not polycyclic). The isomorphism $\Delta \cong \mathbb{Z}[1/2]$ is obtained by sending $a^p b a^{-p}$ to $\left(\frac{1}{2}\right)^p$ for $p \in \mathbb{Z}$, noticing that their elements commute and generate Δ.

It remains to check that $[\Gamma, \Gamma] = \Delta$, which is obvious, since $[\Gamma, \Gamma]$ contains $b = [a, b]$ as well as $a^{-1}ba$ and these two elements generate Δ.

3.1.3. Remark. It is very interesting to know whether there exist solvable, but non polycyclic, groups of finite presentation which are Kähler. Recall that a solvable group Γ is polycyclic iff it satisfies one of the following equivalent conditions:

1. Any subgroup of Γ is finitely generated;
2. Γ has a finite composition series all quotients of which are cyclic;
3. Γ can be embedded as a discrete subgroup of some solvable real Lie group.

3.2. The first Betti number of a subgroup of finite index is even

If Γ is Kähler, $b_1(\Gamma')$ is even for any Γ' of finite index in Γ (Indeed: Γ' is Kähler too, so 3.1 applies to it as well).

3.2.1. Examples.

1. F_r is not Kähler (for any r, even even). Indeed: if r is odd this follows from 3.1. Otherwise, assume $r \geq 2$ and let $\alpha : F_r \to \mathbb{Z}^{\oplus r} = H_1(F_r)$ be the natural quotient homomorphism. Let $\Gamma' := \alpha^{-1}(2\mathbb{Z} \oplus \mathbb{Z}^{r-1})$: then Γ' has index 2 in F_r. It is thus isomorphic to $F_{\tilde{r}}$ with $\tilde{r} = 1 + 2(r-1)$ by Nielsen–Schreier's theorem (see 2.6 above), but $\tilde{r}$ being odd, Γ' and hence F_r is not Kähler.

   Notice that if $(a_1, \ldots, a_r)$ are the generators of f_r, then Γ' is the *normal* subgroup of F_r generated by $S = (a_1^2, a_2, \ldots, a_r)$, but that the subgroup Γ'' of F_r generated by S is of infinite index in F_r.

   Exercise. Show that Γ' is generated by

   $$S' = (a_1^2, a_2, \ldots, a_r, [a_1, a_2], [a_1, a_3], \ldots, [a_1, a_r])$$

   which has indeed $\tilde{r}$ elements.

2. Let H_2 be the 3-dimensional Heisenberg group of rank 2: it is the group of 3×3 upper triangular matrices with integral coefficients and 1's on the diagonal. It is easy to check that it admits an exact sequence:

   $$1 \to \mathbb{Z} \to H_2 \to \mathbb{Z}^{\oplus 2} \to 1$$

   with $\mathbb{Z} = [H_2, H_2] = Z(H_2)$ (the center of H_2).

   From this one easily deduces that for any subgroup H of finite index in H_2, $[H, H]$ is of finite index in $\mathbb{Z}$ (exercise) so that $b_1(H) = 2$ for any such H. Nevertheless, H_2 is not Kähler. This follows from the following obstruction, due to Johnson–Rees ([J-R]).

3.3. Johnson–Rees's obstruction

Theorem ([J-R]). *If* Γ *is Kähler and if* $b_1(\Gamma) > 0$, *there exists* $\lambda : H^2(\Gamma, \mathbb{R}) \to \mathbb{R}$ *linear such that* $\lambda \circ s : \bigwedge^2 H^1(\Gamma, \mathbb{R}) \to \mathbb{R}$ *is a non-degenerate bilinear form where* $s : \bigwedge^2 H^1(\Gamma, \mathbb{R}) \mapsto H^2(\Gamma, \mathbb{R})$ *is the natural linear map.*

(We define $H^i(\Gamma, \mathbb{R})$ to be $H^i(K, \mathbb{R})$ if K is any $K(\Gamma, 1)$, that is any CW-complex with *contractible* universal cover such that $\pi_1(K) \cong \Gamma$ (this does not depend on K which is well-defined up to homotopy. See §5 below for more details on group cohomology).

Proof. Recall that, if X is any finite CW-complex and if $\Gamma = \pi_1(X)$, such a K can be constructed by adding to X finitely many cells of dimension 3 to kill first $\pi_2(X)$, then adding cells of dimension 4 to kill $\pi_3(X')$, and so on.

The resulting CW-complex K has thus $\pi_1(K) = \pi_1(X) = \Gamma$, and $\pi_i(K) = 0$ ($i \geq 2$): its universal cover is thus contractible (see [G-M], p. 35). Let $h : X \hookrightarrow K$ be the above embedding, and $h^i : H^i(K, \mathbb{R}) \to H^i(X, \mathbb{R})$ the resulting maps. We know that h^1 is isomorphic, that h^2 is injective (since $H^2(K, X; \mathbb{R}) = 0$, because only cells of dimensions at least 3 have been added) and we have a commutative diagram

$$\begin{array}{ccc} H^2(\Gamma, \mathbb{R}) & \mapsto & H^2(X, \mathbb{R}) \\ \uparrow & & \uparrow s \\ \bigwedge^2 H^1(\Gamma, \mathbb{R}) & \overset{\cong}{\to} & \bigwedge^2 H^1(X, \mathbb{R}) \end{array} \tag{3.4}$$

Return to the case X compact and Kähler. One defines $\Lambda : \bigwedge^2 H^1(\Gamma, \mathbb{R}) \to \mathbb{R}$ by: $\Lambda(\alpha, \beta) = \int_X h^1(\alpha) \wedge h^1(\beta) \wedge [\omega]^{n-1}$, where $[\omega]$ is a Kähler class on X. By Lefschetz hard theorem, this bilinear form is non degenerate.

Moreover, $\wedge$ factors through s, which shows the result.

3.3.2. Example. Let us consider $H_2 = \Gamma$ introduced in 3.2 above. We shall see that $b_1(H_2) = 2 > 0$, but that $s : \bigwedge^2 H^1(\Gamma, \mathbb{R}) \to H^2(\Gamma, \mathbb{R})$ is zero, which certainly contradicts the preceding criterion.

First a $K(\Gamma, 1)$ is easily constructed: it is just the quotient $N = \Gamma \backslash \tilde{N}$ of the nilpotent Lie group $\tilde{N} = \left\{ \begin{pmatrix} 1 & a & c \\ 0 & 1 & b \\ 0 & 0 & 1 \end{pmatrix} \mid (a, b, c) \in \mathbb{R}^3 \right\}$ modulo the subgroup

$$\Gamma = \left\{ \begin{pmatrix} 1 & a & c \\ 0 & 1 & b \\ 0 & 0 & 1 \end{pmatrix} \mid (a, b, c) \in \mathbb{Z}^3 \right\}.$$

So that $H^1(N, \mathbb{R}) = \mathbb{R}[\mathrm{d}a] \oplus \mathbb{R}[\mathrm{d}b]$.

Nevertheless $\mathrm{d}a \wedge \mathrm{d}b = \mathrm{d}u$, with $u = \mathrm{d}c + a\ \mathrm{d}b$, which descends to N.

This shows that $s = 0$.

3.3.3. Corollary ([J-R]). *Let Γ_1, Γ_2, K be finitely generated groups such that Γ_1 and Γ_2 admit non-trivial subgroups Γ_1' and Γ_2' of finite index. Then: $\Gamma := (\Gamma_1 * \Gamma_2) \times K$ is not Kähler.*

Recall that $(\Gamma_1 * \Gamma_2)$ is the *free product* of Γ_1 and Γ_2, consisting of words, the letters of which are alternatively in Γ_1 and Γ_2, without any further relations. For example: $F_r * F_s \cong F_{r+s}$. If Γ_1 (resp. Γ_2) has the presentation $\langle a_1, \dots, a_r \mid w_1, \dots, w_\rho\rangle$ (resp. $\langle b_1, \dots, b_s \mid w_1', \dots, w_\sigma'\rangle$), then $G * H$ has the presentation $\langle a_1, \dots, a_r, b_1, \dots, b_s \mid w_1, \dots, w_\rho, w_1', \dots, w_\sigma'\rangle$.

For example: $\mathbb{Z}_2 * \mathbb{Z}_3 = \langle a, b \mid a^2 = b^3 = 1\rangle$ is not Kähler (but it is isomorphic to $\pi_1(\mathbb{P}_2(\mathbb{C}) - C)$ where C is the curve of affine equation $(x^2+y^2)^3 + (y^3+1)^2 = 0$, a classical example of Zariski). In particular $\mathbb{Z}_2 * \mathbb{Z}_3$ is infinite.

3.3.4. Exercise. Show that $H^1(\Gamma_1 * \Gamma_2, \mathbb{R}) = H^1(\Gamma_1, \mathbb{R}) \oplus H^1(\Gamma_2, \mathbb{R})$, that $H^2(\Gamma_1 * F_r, \mathbb{R}) = H^2(\Gamma_1, \mathbb{R})$ and that $H^2(F_r, \mathbb{R}) = 0$ if Γ_i are any f.g. groups.

3.3.5. *Proof of* 3.3.3 *(in the simpler case where $K = (1)$).* We follow the proof of [J-R].

Case 1. $\Gamma_1 = F_r$ $(r > 0)$; then $H^1(\Gamma, \mathbb{R}) = H^1(\Gamma_1) \oplus H^1(\Gamma_2) = H^1(F_r) \oplus H^1(\Gamma_1)$ (where $H^i(F_j) = H^i(\Gamma_j, \mathbb{R})$). Then for any $\lambda : H^2(\Gamma) = H^2(F_r) \oplus H^2(\Gamma_2) = H^2(\Gamma_2) \to \mathbb{R}$, $\lambda \circ s : \bigwedge^2(H^1(\Gamma) \to \mathbb{R}$ is degenerate, since $H^1(\Gamma_1)$ is isotropic for $\lambda \circ s$ and orthogonal to $H^1(\Gamma_2)$.

Case 2. Let $\Psi_i : \Gamma_i \to (\Gamma_i/\Gamma_i') = \Delta_i$ (we can assume that Γ_i' is normal in Γ_i), and $\Psi : \Gamma \to \Delta_1 \times \Delta_2$ the morphism which extends it. By Kurosh's theorem (a generalization to free products of the theorem of Nielsen–Schreier, see [Se2], p. 77), if (m_i) is the order of Δ_i $(i = 1, 2)$, then $\ker \Psi \cong F_n * (\Gamma_1'' * \Gamma_2'')$, where $n = (m_1 - 1)(m_2 - 1) \geq 1$, and Γ_1'' is a free product of m_j copies of $\ker(\Psi_i)$ (where $1 \leq i \neq j \leq 2$). Now $\ker \Psi$ has finite index in Γ and is not Kähler by Case 1, and thus so is Γ.

3.4. Nilpotent groups and nilpotent completion of Kähler groups

This topic will be considered with more details in §4 below. We shall only state one of the known obstructions:

3.4.1. Proposition ([Ca3]). *Let Γ be a nilpotent Kähler group. Then either $\Gamma/$*Torsion *is abelian, or $b_1(\Gamma) \geq 8$.*

For example: the fundamental group of the Iwasawa manifold is not Kähler (see §4 below).

The proof of [C-T1] uses rational homotopy theory to show 3.4.1 in the special case where Γ is Heisenberg; the proof of [Ca3] avoids it and rests on more elementary arguments.

As a consequence, the Heisenberg groups H_r are Kähler iff $r \geq 8$ is even. Let us remark that these Heisenberg groups are presently the only nilpotent groups (without abelian subgroups of finite index) known to be Kähler.

The case of polycyclic groups should reduce to that of nilpotent groups due to a result announced by D. Arapura and M. Nori: if $\Gamma = \pi_1(X)$ for some projective manifold X, then Γ should contain a nilpotent group of finite index, provided Γ is polycyclic.

The proof should rest on characteristic p methods.

This result should probably extend to the case of Kähler groups.

3.4.2. Example ([A1]). Let $\Gamma := \mathbb{Z}^2 \ltimes \mathbb{Z}^2$ where $\mathbb{Z}^2$ acts on $\mathbb{Z}^2$ by $T \in \mathrm{SL}(2, \mathbb{Z})$. Then Γ is polycyclic, since it is an extension of $\mathbb{Z}^2$ by $\mathbb{Z}^2$. By the above result, Γ should be Kähler only if T is of finite order in $\mathrm{Sl}(2, \mathbb{Z})$, because this is precisely the condition for Γ to have a nilpotent subgroup of finite index. Conversely, if T has finite order, $\Gamma = \pi_1(S)$ for some bielliptic surface ($c_1(S) = 0$, $b_1(S) = 2$), classified in [Be1] p. 113.

3.5. Ends of Kähler groups

3.5.1. The number of ends of a finitely generated group Γ.

3.5.1.1. Let first $\tilde{Z}$ be a locally compact topological space which is countable at infinity. We shall define the *number of ends* of $\tilde{Z}$, denoted by $e(Z) \in \{0, 1, 2, \ldots, \infty\}$ as $e(Z) = \lim_{n \to \infty} N(\tilde{Z}, n)$, where $N(\tilde{Z}, n)$ is the number of non-compact connected components of $\tilde{Z} - V_n$, where $(V_n)_{n \in \mathbb{N}}$ is an increasing sequence of relatively compact open subsets of $\tilde{Z} = \bigcup_{n \geq 0} V_n$.

Thus $e(\tilde{Z}) = 0$ iff $\tilde{Z}$ is compact. If $\tilde{Z}$ is $\mathbb{R}^n$, then $e(\tilde{Z}) = 2$ if $n = 1$, but $e(\tilde{Z}) = 1$ if $n \geq 2$. Finally

$$e(\mathbb{C}^2 - \{0\}) = 2.$$

3.5.1.2. Let now Γ be any finitely generated group, and Z be any *compact* CW-complex such that $\pi_1(Z) = \Gamma$.

We define $e(\Gamma) := e(\tilde{Z})$, where $\tilde{Z}$ is the universal covering of Z: this number does not depend on the choice of Z (see below).

3.5.1.3. Examples.

1. $e(\mathbb{Z}^n) = 1$ if $n \geq 2$, but $e(\mathbb{Z}) = 2$. To see this, just take $Z = (S^1)^n$.
2. $e(\Gamma) = 0$ iff Γ is finite.
3. $e(\Gamma) = e(\Gamma')$ if Γ and Γ' are *commensurable* (i.e.: if they have subgroups of finite index which are isomorphic).
4. If Γ is nilpotent and not commensurable with $\mathbb{Z}$, then $e(\Gamma) = 1$. We can assume that Γ is torsion free (by dividing out its torsion subgroup, which is finite since Γ is finitely generated). It can thus be embedded in a nilpotent Lie group $\tilde{N}$ (which is homeomorphic to $\mathbb{R}^n$ for some $n \geq 2$) by a theorem of Malčev ([Mal2]) in such a way that it is discrete and cocompact there. Then just take for Z the nilmanifold $Z = \Gamma\backslash\tilde{N}$.
5. If $F_r = \Gamma$ is the free group of r generator, then $e(F_r) = \infty$ if $r \geq 2$. Indeed: just take for Z a wedge of r circles. Its universal covering is the *Cayley graph* of F_r described as follows:
6. Let Γ be any finitely generated group with generators $(a_1, \ldots, a_r)$. The *Cayley graph* $C(\Gamma)$ of Γ is the graph whose vertices are the elements of Γ and with two edges g and h connected iff $gh^{-1} = a_i^{\pm 1}$ for some $i \in \{1, \ldots, r\}$. This graph is connected because the a_i generate Γ.

   Notice also that any relation $a_{i_1}^{\alpha_1} \cdot \cdots \cdot a_{i_k}^{\alpha_k} = 1$ ($\alpha_j \in \mathbb{Z}$) means that the path in $C(\Gamma)$ starting from 1 and going successively to

   $$a_{i_k}^{\varepsilon(k)}; a_{i_k}^{2\varepsilon(k)}, \ldots, a_{i_k}^{\alpha_k}, a_{i_{k-1}}^{\varepsilon(k-1)} a_{i_k}^{\alpha_k}; \ldots; a_{i_1}^{(|\alpha_k|-1)\cdot\varepsilon(1)} \cdot a_{i_2}^{\alpha_2} \cdot \cdots \cdot a_{i_k}^{\alpha_k}$$

   is a loop. Here we have assumed that $\alpha_j \neq 0$ $(1 \leq j \leq k)$ and defined $\varepsilon(j) = \alpha_j/|\alpha_j| = \operatorname{sgn}(\alpha_j)$. Thus $C(\Gamma)$ is a tree iff Γ is the free group on the generators $(a_1, \ldots, a_r)$.

   It is now (relatively) easy to check that $e(\Gamma)$ as defined above is independent on the choice of Z: the question is largely reduced to comparing $e(\tilde{Z})$ and $e(\tilde{Z}')$ where $\tilde{Z}$ and $\tilde{Z}'$ are the Cayley graph of Γ associated to different sets of generators. (see [S-W] for more details).

Let us describe $C(\Gamma)$ for $\Gamma = \mathbb{Z}^{\oplus 2}$ and $\Gamma = F_2$ (with the obvious generators): $C(\mathbb{Z}^{\oplus 2})$ is the infinite grid in $\mathbb{R}^2$, while $C(F_2)$ is the infinite tree with 4 edges emanating from each vertex.

3.5.2. Properties of $e(\Gamma)$. Let Γ be a finitely generated group. Then:

1. $e(\Gamma) \in \{0, 1, 2, \infty\}$
2. $e(\Gamma) = 2$ iff Γ admits an infinite cyclic subgroup of finite index.
3. If $e(\Gamma) = \infty$, Γ contains a subgroup isomorphic to F_2.

For the proof, we refer to [S-W].

3.5.3. Remark. It follows from the first two statements that if Γ is Kähler, then $e(\Gamma)$ is either 0, 1 or ∞.

Actually, the last possibility can be excluded, too:

3.5.4. Theorem ([A-B-R]; see also [Gr1]). *Let* $1 \to K \to \Gamma \to H \to 1$ *be an exact sequence of groups, with* Γ *Kähler and* K *finitely generated. Then* $e(H) \leq 2$.

3.5.5. Corollary. *Let* Γ *be Kähler. Then* $e(\Gamma) \leq 1$ *(and* Γ *is thus not a free product amalgamated over a finite group: we shall not define this notion (see* [S-W] *for this), but this conclusion is stronger than* 3.3.3*).*

3.5.6. Example [A1]. The Braid group B_n is not Kähler. This group is the fundamental group of the set of n unordered distinct points of $\mathbb{C}$.

The proof of 3.5.4 given in [A-B-R] is based on L_2 cohomology theory, and may be divided in 4 steps:

Step 1. $e(H) = \infty \Rightarrow H$ contains an $F_2 \Rightarrow H$ is not amenable and $e(H) \geq 2 \Rightarrow H^1EL_2(H) \neq 0$. We shall not define this last notion because:

Step 2. From [Do], $H^1EL_2(H) \cong H^1EL_2(\tilde{X})$ if $\pi : \tilde{X} \to X$ is the Galois cover with group Γ of some compact riemannian manifold X with $\pi_1(X) \cong H$. Here $\tilde{X}$ is equipped with the metric of X lifted by π, and $H^1EL_2(\tilde{X}) = \{\omega \in H^0_{(2)}(\mathcal{E}^1(\tilde{X})) \mid \omega = dg \text{ and } \Delta g = 0\}$ is thus the vector space of smooth 1-forms ω on $\tilde{X}$ which are L_2 and of the form $\omega = dg$ for some harmonic function g on $\tilde{X}$. (The definition of $H^1EL_2(H)$ is similar, but combinatorial). We of course choose X to be compact Kähler and its metric to be Kähler from now on, here.

Step 3. ([Gr]) If $H^1EL_2(\tilde{X}) \neq 0$, there exists a *proper* connected H-equivariant holomorphic map $\Phi : \tilde{X} \to \tilde{Y}$ where $\tilde{Y}$ is a hyperbolic Riemann surface (i.e.: covered by the unit disk).

This step is the one where Kähler geometry plays a role. It was originally established by M. Gromov in [Gr1]. We shall show it later in 5.1.19 by a slightly different argument.

Step 4. In the situation $\Phi : \tilde{X} \to \tilde{Y}$ of Step 3, we have:

1) $\pi_1(\tilde{Y}) = \{1\}$ so that $\tilde{Y}$ is the unit disk and $e(H) = 1$ as asserted. More precisely:

2) H is commensurable with $\pi_1(C)$ where C is a compact curve of genus $g \geq 2$.

This step rests essentially on Riemann surface theory.

Indeed: K surjects on $\pi_1(\tilde{Y})$: by hypothesis, K is f.g., thus so is $\pi_1(\tilde{Y})$. It follows that $\pi_1(\tilde{Y}) \cong F_r$ for some $r \geq 0$.

We have $r \geq 2$ if $r \neq 0$, because otherwise $H^1EL_2(H) = 0$. If $r \geq 2$, the automorphism group of $\tilde{Y}$ is finite, contradicting the fact that H is infinite since $H^1EL_2(H) \neq 0$.

It thus follows that H has a finite index subgroup which is discrete cocompact and torsion free in $SU(2, 1)$. Hence 2) follows.

3.5.7. Other obstructions. We shall not survey here other obstructions dealing essentially with semi-simple representation of Kähler groups and based either on the theory of harmonic maps ([C-T3], [G-M], [Mok], [Siu]) or representation varieties ([G-M], [Si], [Zuo]).

Other obstructions may be deduced also from [Be2]; [G-L].

4. Nilpotent completion of Kähler groups

4.1. Introduction

The first obstruction for a finitely presented group Γ to be Kähler is that $b_1(\Gamma) = \mathrm{rk}_{\mathbb{Z}}(\Gamma/\Gamma_2)$. It is natural, because $b_1(\Gamma) = \mathrm{rk}_{\mathbb{Z}}(\Gamma/\Gamma_2)$, with $\Gamma_2 := [\Gamma, \Gamma]$ the second commutator groups of Γ, to extend this study to the so-called *lower central series* of Γ, consisting of higher-order commutator subgroups $\Gamma_n := [\Gamma, \Gamma_{n-1}]$ of Γ. We denote by $\Gamma^{\mathrm{nilp}} := \Gamma/\Gamma'_\infty$, where Γ'_∞ is the intersection of the (torsion free) lower central series of Γ. We call it the *nilpotent limit of* Γ.

The Albanese map $\alpha : X \to \mathrm{Alb}(X)$, when X is compact Kähler, plays the central role in this study. Let indeed $\Gamma := \pi_1(X)$. Then:

$$\alpha_* : H_1(\Gamma, \mathbb{Q}) = H_1(X, \mathbb{Q}) \to H_1(\mathrm{Alb}(X), \mathbb{Q})$$

is an isomorphism, by construction. We shall extend this to Γ^{nilp} as follows: if Y is a smooth model of $\alpha(X) := Y_0 \subset \mathrm{Alb}(X)$, we can lift α to Y, and still write $\alpha : X \to Y$.

Let $\Delta := \pi_1(Y)$; then $\alpha_* : \Gamma^{\mathrm{nilp}} \to \Delta^{\mathrm{nilp}}$ is (up to torsion) injective and has an image of finite index. In other words: the information carried by the Albanese map contains all of the lower central series of Γ, not only its first term, although only this one is used in the construction of the Albanese map.

The proof is elementary (it does not even use Hodge theory), but uses cohomology of groups. The first two sections are devoted to introducing the needed tools.

Immediate consequences of this result are that:

1) $b_1(\Gamma) \geq 8$ if Γ is nilpotent and Kähler, but Γ/Torsion is not abelian

2) $\pi_1(X)^{\mathrm{nilp}}$ is, up to an abelian factor, $\pi_1(A)^{\mathrm{nilp}}$ if A is an algebraic reduction of X.

Rational homotopy theory gives deeper information, when used together with Hodge theory: the main result of [D-G-M-S] implies in particular that $\pi_1(X)^{\mathrm{nilp}} \otimes_{\mathbb{Z}} \mathbb{R}$ is determined by the kernel of the natural map $\bigwedge^2 H^1(X, \mathbb{R}) \to H^2(X, \mathbb{R})$.

This fact is exploited in [C-T1] to show that certain nilpotent groups are not Kähler, and to recover 1) above, when $\pi_1(X)$ is a Heisenberg group.

4.2. Commutator calculus

Let Γ be any group. We define for any two subgroups A, B of Γ by $[A, B]$ the subgroup of Γ generated by all commutators $[a, b]$ with $(a, b) \in A \times B$: if A and B are normal (resp. characteristic) in Γ, so is $[A, B]$.

The *lower central series* of Γ is the decreasing sequence of characteristic subgroups $(\Gamma_n)_{n\geq 1}$ of Γ inductively defined by: $\Gamma_1 = \Gamma$; $[\Gamma, \Gamma_n] = \Gamma_{n+1}$ $(n \geq 1)$.

We shall denote by $\Gamma_\infty := (\cap_{n\geq 1}\Gamma_n)$ the intersection of all terms of the lower central series of Γ.

We say that Γ is *step-n nilpotent* if $\Gamma_n \neq (1)$ but $\Gamma_{n+1} = (1)$, and that Γ is *nilpotent* if Γ is step-n nilpotent for some n. In particular: Γ is step-1 nilpotent means that Γ is abelian, but $\Gamma \neq (1)$.

We say that Γ is *residually nilpotent* if $\Gamma_\infty = (1)$.

4.2.1. Exercise. Check that if $1 \to K \to G \to H \to 1$ is an exact sequence of groups, then:

1. If G is nilpotent, so are K and H.
2. If K and H are nilpotent, and if this extension is central (i.e.: $K \subset C(G)$, the center of G) so that K is actually abelian, then G is nilpotent too. Estimate its nilpotency class from that of H.
3. If G is nilpotent, show that $C(G) \neq (1)$, and that:
4. Any subgroup of G is finitely generated if so is G.
5. Let $\Gamma = \pi_1(B)$, B the Klein bottle, so that $\Gamma = \langle a, b \mid aba^{-1} = b^{-1}\rangle$. Show that the subgroup N of Γ generated by b is normal and gives an exact sequence $1 \to N \to \Gamma \to \mathbb{Z} \to 1$. Compute the lower central series of Γ; show in particular that Γ is not nilpotent, but that $\cap_{n\geq 1}\Gamma_n = 1$.

Let us now consider the quotients Γ/Γ_n of the lower central series of Γ: it's easy to check that this group is step-m $(m \leq n)$ nilpotent for each n, that the exact sequence

$$1 \to (\Gamma_n/\Gamma_{n+1}) \to (\Gamma/\Gamma_{n+1}) \to (\Gamma/\Gamma_n) \to 1$$

is central, and that moreover, if : $\Gamma \to \Delta \to 1$ is a morphism of groups with Δ step-n nilpotent, then $\varphi = \varphi_n \circ \gamma_n$ for some morphism $\varphi_n : \Gamma/\Gamma_{n+1} \to \Delta \to 1$ where $\gamma_n : \Gamma \to \Gamma/\Gamma_n$ is the natural quotient.

For each n, we also have a natural exact sequence:

$$1 \to \Gamma/\Gamma_\infty \to \Gamma/\Gamma_\infty \to \Gamma/\Gamma_n \to 1$$

so that (Γ/Γ_∞) appears as the limit of the (Γ/Γ_n).

Finally, all these constructions are functorial:

If $\alpha : \Gamma \to \Delta$ is a group morphism, we shall denote with $\alpha_n : \Gamma/\Gamma_n \to \Delta/\Delta_n$ the map it induces (for $1 \leq n \leq \infty$).

4.2.2. Examples.

1. (Γ/Γ_2) is nothing but $H_1(\Gamma)$, the abelianization of Γ.

2. Let $U_{n,p} = \left\{ \begin{pmatrix} 1 & 0 & \cdots & 0 & * & * \\ & \ddots & \ddots & \ddots & \ddots & * \\ & & \ddots & \ddots & \ddots & 0 \\ & 0 & & \ddots & \ddots & \vdots \\ & & & & \ddots & 0 \\ & & & & & 1 \end{pmatrix} \right\}$

   be the subgroup of $(n+1) \times (n+1)$ unipotent complex matrices with the first $(p-1)$ diagonals above the principal diagonal being 0, for $1 \leq p \leq n$. We notice: $\mathcal{U}_{n,1} = \mathcal{U}_n$ is the group of unipotent matrices, and $\mathcal{U}_{n,n} = (1)$. Remark that $[\mathcal{U}_{n,p}, \mathcal{U}_{n,q}] = \mathcal{U}_{n,p+q}$, which shows that $(\mathcal{U}_n)_p = \mathcal{U}_{n,p}$ and that $\mathcal{U}_n$ is step-n nilpotent. The same properties hold for $\mathcal{U}_n(\mathbb{Z})$, its subgroup consisting of matrices with integer coefficients.

3. Let $F = F_r$ be the free group on $r \geq 2$ generators $(a_1, a_2, \ldots, a_r)$. This groups is not nilpotent, since it contains F_2 which maps surjectively to $\Gamma = \langle a_1, a_2 \mid [a_1, a_2] = a_2 \rangle$, which is not nilpotent (see 3.1.2.3 for its structure: $[\Gamma, \Gamma] = \Delta \cong \mathbb{Z}[1/2]$); Hence the lower central series of F is infinite.

   It can be shown that for any n, F_n/F_{n+1} is a free abelian group of rank $d(n, r)$ where: $d(n, r) = \frac{1}{n} \sum_{d|n} \mu(d) \cdot r^{n/d}$ (Witt's Formula; see [Se4] elementary proof).

   Here $\mu : \mathbb{N}^+ \to \{-1, 0, +1\}$ is the Möbius function, defined by:

$$\mu(d) = \begin{cases} 0 & \text{if } d \text{ is divisible by } p^2 \text{ for some prime number } p \\ (-1)^k & \text{otherwise, where } k \text{ stands for the number} \\ & \text{of primes dividing } d \end{cases}$$

   Moreover: $F_\infty = (1)$ (i.e. $\cap_{n\geq 1} F_n = (1)$), a result due to Magnus, which also means that $F = F_r$ is residually nilpotent. Let us show this, which is easy: let A_r be the associative algebra consisting of formal power series in the non-commutative indeterminates $X_1, \ldots, X_r$.

   Let $\Phi : F \to A^*$ be the group morphism from F to the group A^* of units of A defined by:

$$\Phi(a_{i_1}^{\alpha_1}, \ldots, a_{i_k}^{\alpha_k}) = (1 + X_{i_1})^{\alpha_1} \ldots (1 + X_{i_k})^{\alpha_k}$$

   for $\alpha_j \in \mathbb{Z}$, where $(1 + X_i)^{-1} = \sum_{m\geq 0} (-1)^m X_i^m$.

   We have the following properties:

   1. Φ is injective.
   2. $\Phi(F_n) \subset 1 + A_n$, where A_n is the two sided ideal of A consisting of positive

series with no monomial of degree less than n. (A more precise statement is true, which is more difficult to get, but not needed here, namely $\Phi(F_n) = (1 + A_n) \cap \Phi(F)$).

3. From these two properties the assertion $F_\infty = (1)$ follows. Let us check them: for $a \in A$, we put $\nu(a) = \infty$ if $a = 0$, and $\nu(a) = \delta$, the smallest degree of a monomial with nonzero coefficient in a, otherwise. An easy check now shows that $\Phi([g, h]) \in 1 + A_{m+n}$ if g (resp. h) belongs to G is such that $\Phi(g)$ (resp. $\Phi(h)$) belongs to $1 + A_m$ (resp. $1 + A_n$). Because Φ is injective, one gets the statement.
4. All the properties above still hold if $\Gamma = \pi_1(C_g)$ is the fundamental group of a compact Riemann surface of genus $g \geq 2$; here $\Gamma_\infty = (1)$ ([Ba], unfortunately the proof is not as easy as in the example 3); (Γ_n/Γ_{n+1}) is still a free group of rank $d(n, g)$, which are determined from the equality (due to J. Labute): $(1 - gt + t^2) = \prod_{n\geq 1}(1 - t^n)^{d(n,g)}$.
5. On the other hand, there are also residually finite Kähler groups Γ for which $b_1(\Gamma') = 0$ for any subgroup Γ' of finite index in Γ. Just take Γ torsion free and cocompact discrete in a simple Lie group of hermitian type. For such a Kähler group, $\Gamma'^{\text{nilp}} = (1)$ for any such Γ'; the nilpotent completion does not give any information.

4.2.3. We now want to consider torsion free nilpotent groups. Recall ([Hirs]) that if G is a finitely generated nilpotent group, its torsion elements form a finite characteristic subgroup Tors(G). Thus: $G/\text{Tors}(G)$ is also a finitely generated nilpotent group, moreover torsion free.

For any finitely generated Γ, we shall denote for $n \geq 1$ by Γ'_n the kernel of the composite map: $\Gamma \to \Gamma^n := \Gamma/\Gamma_n \to \Gamma^n/\text{Tors}(\Gamma^n) := \Gamma'^n$

Thus: Γ'_n is a characteristic subgroup of Γ, and Γ'^n enjoys with respect to torsion free nilpotent groups the same universal property as Γ^n does with respect to nilpotent groups.

Again the construction of Γ'_n and Γ'^n is functorial in Γ. Let $\Gamma'_\infty := \cap_{n\geq 1}\Gamma'_n$ and $\Gamma'^{\text{nilp}} := (\Gamma/\Gamma'_\infty)$; if $\varphi : \Gamma \to \Delta$ is morphism of group,

$$\varphi'_\infty : \Gamma'^{\text{nilp}} \to \Delta'^{\text{nilp}}$$

is the corresponding morphism.

Observe that Γ'^{nilp} is torsion free.

4.2.4. Let us give examples of finitely presented groups Γ which are torsion free, but such that the natural epimorphism of groups has infinite kernel (answering a question raised by L. Bonavero):

1. Let $\Gamma = \langle a, b \mid [a, b] = b^{-2}\rangle$ be the fundamental group of the Klein bottle. Let

N be the subgroup generated by b; it is normal and gives us an exact sequence:

$$1 \to N \cong \mathbb{Z} \to \Gamma \to \mathbb{Z}_2 \to 1,$$

where $\mathbb{Z}_2 = (\pm 1)$ acts on b by: $(-1) \cdot b = b^{-1}$.

We then have: $\Gamma_2 := [\Gamma, \Gamma] = [\Gamma, N] = 2N$ (additive notation, here!), generated by b^2. Inductively one gets: $\Gamma_{n+1} = 2^n \cdot N$ and thus: $\Gamma_\infty = 1$.

On the other hand, the exact sequence:

$$1 \to N/_{2^n \cdot N} \to \Gamma/_{\Gamma_{n+1}} = \Gamma/_{2^n \cdot N} \to \Gamma/N \to 1$$

shows that (Γ/Γ_{n+1}) is torsion for every $n \geq 1$.

Thus: $\Gamma/\Gamma'_{n+1} = 1$ and $\Gamma'_{n+1} = \Gamma = \Gamma'_\infty$.

Observe that Γ is solvable (but not nilpotent), a similar example has been communicated by J. Winkelmann:

2. Let $\Gamma = \mathbb{Z} \ltimes N$, where the generator $(+1)$ of $\mathbb{Z}$ acts on $N = \mathbb{Z}^2$ by the matrix $A = \begin{pmatrix} 5 & 2 \\ 2 & 1 \end{pmatrix}$. Again we have $\Gamma_2 := [\Gamma, \Gamma] = [\Gamma, N] = k \cdot N$ with $k = 2$, and $\Gamma_\infty = (1)$ with $\Gamma'_\infty = \Gamma$. (Notice that any matrix $A = \begin{pmatrix} ab+1 & b \\ a & 1 \end{pmatrix}$ with $k = \gcd(a, b) \geq 2$ would work).

4.3. Cohomology of groups — Stallings' criterion

Our basic tool will be the following:

4.3.1. Theorem (J. Stallings [St]). *Let $R = \mathbb{Z}$ or $\mathbb{Q}$; let $\phi : \Gamma \to \Delta$ be a morphism of groups. Assume that $\varphi_{*,i} : H_i(\Gamma, R) \to H_i(\Delta, R)$ is isomorphic for $i = 1$, and surjective for $i = 2$. Then:*

(1) *if $R = \mathbb{Z}$, $\varphi_n : \Gamma/\Gamma_n \to \Delta/\Delta_n$ is isomorphic ($1 \leq n < \infty$) and injective if $n = \infty$. If φ is onto, so is φ_∞;*

(2) *if $R = \mathbb{Q}$, $\varphi'_n : \Gamma/\Gamma'_n \to \Delta/\Delta'_n$ is injective ($1 \leq n < \infty$) and its image has finite index in Δ/Δ'_n if $n < \infty$.*

Remark. If φ is onto, or of *finite index* (i.e. if so is its image in Δ), so are φ_∞ and φ'_∞. This is easy to see.

We shall here only sketch the proof of 4.3.1(1), the case where $R = \mathbb{Q}$ being analogous.

Before doing this, we shall recall or introduce some basic notion of homology of groups. Although there are several approaches to this, we restrict to the topological one, which is more geometric.

4.3.2. Definition.

(1) Let be a finitely generated group. A $K(\Gamma, 1)$ is a CW-complex K such that its universal cover $\tilde{K}$ is contractible, and $\pi_1(K) \cong \Gamma$; K is well-defined up to homotopy.

(2) $H_i(\Gamma, R) := H_i(K(\Gamma, 1), R)$ and $H^i(\Gamma, R) := H^i(K(\Gamma, 1), R)$ if $R = \mathbb{Z}, \mathbb{Q}, \mathbb{R}, \mathbb{C}$. By (1), this doesn't depend on K.

4.3.3. Examples.

1. $\Gamma = F_r$; one can take $K(F_r, 1) = W_r$, a wedge of r-circles (by Van Kampen's theorem), since $\tilde{W}_b$ is a tree, hence contractible. One gets:

$$H_i(F_r, \mathbb{Z}) = \begin{cases} 0 & \text{if } i \geq 2 \\ \mathbb{Z}^{\oplus r} & \text{if } i = 1 \\ \mathbb{Z} & \text{if } i = 0. \end{cases}$$

2. $\Gamma = \pi_1(C_g)$; take $K(\Gamma, 1) = C_g$ (since C_g is uniformized by the unit disk, which is contractible).
One gets:

$$H_i(\pi_1(C_g), \mathbb{Z}) = \begin{cases} 0 & \text{if } i \geq 3 \\ \mathbb{Z} & \text{if } i = 0, 2 \\ \mathbb{Z}^{\oplus 2g} & \text{if } i = 1. \end{cases}$$

3. $\Gamma = \mathbb{Z}^{\oplus r}$; take $K(\Gamma, 1) = (S^1)^r$. One gets:

$$H_i(\mathbb{Z}^{\oplus r}, \mathbb{Z}) = \begin{cases} 0 & \text{if } i \geq r+1 \\ \binom{r}{i} & \text{if } 0 \leq i \leq r \end{cases}$$

4. If Γ is a torsion free nilpotent group, it can be realized as a cocompact discrete subgroup of some nilpotent simply-connected Lie group G, with Lie algebra $\mathcal{G}$. This is a result of A. Malčev ([Mal2]; see also Bourbaki, Groupes et Algèbres de Lie; III §9, ex. 33). Thus $\exp : \mathcal{G} \to G$ is an algebraic isomorphism, and G is contractible.

   Thus $M = G/\Gamma$ is a $K(\Gamma, 1)$, which allows one to compute $H_1(\Gamma, \mathbb{Z})$.
The computation of $H^i(\Gamma, \mathbb{R})$ can be reduced to computations in the Lie Algebra $\mathcal{G}$ of G. More precisely, let $(\bigwedge^* \mathcal{G}^*, d)$ be the exterior algebra of *right-invariant* differential forms on G; this is a finite-dimensional graded-commutative differential algebra. One has a natural injection $\rho : A = (\bigwedge^* \mathcal{G}^*, \mathrm{d}) \to (\mathcal{E}^*(M), \mathrm{d}) = B$ in the De Rham complex of real forms on M, which is a map of graded-commutative differential algebra. At the cohomological level ρ is an isomorphism (i.e. so is $\rho^* : H^*(A) \to H^*(B)$). We shall sketch the proof of this result, originally due to K. Nomizu ([No]). One proceeds by induction on the dimension n of G: if $n(G) = 1$, G is abelian, M is a real torus and any d-closed p-form α is cohomologous to unique translation invariant p-form $\tilde{\alpha}$ obtained by averaging α over the

compact group M.

In the general case, there exists a (positive dimensional) connected *central* closed Lie subgroup K of G such that $\Delta := K.\Gamma/K$ and $K \cap \Gamma$ are discrete and cocompact in $G/K := H$ and K, respectively. (This is due to Y. Matsushima).

The exact sequence of groups: $1 \to K \to G \to H \to 1$ gives rise to a *principal* fibration $f : M \to N := H/\Delta$ with fiber the torus $F = K/K \cap \Gamma$.

The same F-averaging argument as before now shows that if (C, d) is the graded-commutative differential algebra of F-invariant forms on M, and $\sigma : (C, d) \to B$ (resp. $\tau : A \to (C, d)$) the natural inclusion, then $\sigma^* : H^*(C, d) \leftrightarrow H^*(B)$ is an isomorphism.

We are thus now reduced to show only that $\tau^* : H^*(A) \to H^*(C, d)$ is isomorphic, too, since $\rho = \sigma \circ \tau$. Let $\mathcal{H}$ be the Lie Algebra of $H = G/K$, $A' := (\bigwedge^* \mathcal{H}^*, d)$ be the algebra of right-invariant forms on H, and $B' = (\mathcal{E}^*(N), d)$ be the De Rham algebra of N. Let $\rho' : A' \to B'$ be the natural inclusion. We also have natural inclusions $f^* : A' \to A$ and $f^* : B' \to B$.

We also have a natural fibration of $(\mathcal{E}^*(M), d) = B$ relative to f, namely: $B_0 = B \supset B_1 \supset \cdots \supset B_r \supset \cdots$, where $B_r := \{\alpha \in B \mid i_{X_1} \cdot i_{X_2} \ldots i_{X_{p-r+1}} \alpha = 0$ for any $X_1, \ldots, X_{p-r+1}$ in $\mathcal{K}$, the Lie algebra of K, if α is a p-form$\}$.

Here i_X is the interior product with the vector field X. More concretely, the forms in B_r are those which cannot be written locally with less than r basic 1-form; i.e. $\alpha(f^*\mathrm{d}n_1 \wedge \cdots \wedge f^*\mathrm{d}n_r) \wedge \beta$, where the $\mathrm{d}n_i$'s are local coordinate on N.

Of course, this filtration induces also a filtration on $A = (\bigwedge^* \mathcal{G}^*, d)$.

It is now easy to check that:

$$B_r/B_{r-1} \cong (f^*B') \otimes \bigwedge^r \mathcal{K}^*$$

$$A_r/A_{r-1} \cong (f^*A') \otimes \bigwedge^r \mathcal{K}^*.$$

For the E_2-terms of the spectral sequences defined by the above filtrations on A and B and abutting to $H^*(A)$ and $H^*(B)$, one gets in a similar way:

$$E_2(B) = H^*(f^*B') \otimes \bigwedge^r \mathcal{K}^*$$

$$E_2(A) = H^*(f^*A') \otimes \bigwedge^r \mathcal{K}^*$$

By induction: $H^*(f^*A') \cong H^*(f^*B')$, and the natural map

$$\rho^* : H^*(A) \to H^*(B)$$

is bijective, as desired.

5. In general, cyclic finite groups Γ have periodic homology; they don't admit $K(\Gamma, 1)$ which are finite CW-complexes.

4.3.4. Remarks.

1. From Hurewicz's theorem, one gets: $H_1(\Gamma, R) = (\Gamma/[\Gamma, \Gamma]) \otimes_{\mathbb{Z}} R$.
2. A description of $H_2(\Gamma, \mathbb{Z})$ has been given by H. Hopf:

$$H_2(\Gamma, \mathbb{Z}) \cong N \cap [F, F]/[N, F] \text{ if } \Gamma \cong F/N$$

with F a free group, is any presentation of Γ.

3. Let $1 \to M \to \Gamma \to \Delta \to 1$ be an exact sequence of group. From H. Hopf's description of $H_2(\Gamma, \mathbb{Z})$, one gets the following exact sequence (with $R = \mathbb{Z}$):
[*] $H_2(\Gamma) \to H_2(\Delta) \to M/[\Gamma, M] \to H_1(\Gamma) \to H_1(\Delta) \to 0$
4. Conversely, this exact sequence gives H. Hopf's description of $H_2(\Gamma)$, when applied to: $1 \to N \to F \to \Gamma \to 1$, recalling that $H_2(F) = 0$.
We shall prove 4.3.4.3 [*] directly, using topological considerations:

First step. There exist a fibration $\varphi : B\Gamma \to B\Delta$ with fiber BM (here $B\Gamma$ denotes a $K(\Gamma, 1)$, and $\widetilde{B\Gamma}$ its universal cover). Just take $\varphi : (\widetilde{B\Delta} \times \widetilde{B\Gamma})/\Gamma = (\widetilde{B\Gamma} \times \widetilde{B\Delta}/M)/\Delta \cong \widetilde{BM} \times \widetilde{B\Delta}/\Delta := E \to (\widetilde{B\Delta}/\Delta) = B\Delta := B$. The point is that $\widetilde{B\Gamma}$ is also a $\widetilde{BM}$.

Second step. If $\varphi : E \to B$ is any fibration with fiber F, the associated homology spectral sequence $E^r_{p,q} = H_p(E, H_q(F, \mathbb{Z})) \to H_{p+q}(E, \mathbb{Z})$ provides a 5-term exact sequence:

$$H_2(E, \mathbb{Z}) \to H_2(B, \mathbb{Z}) = E^2_{2,0} \to H_1(F, \mathbb{Z})/\pi_1(B) =$$

$$= E^2_{0,1} \to H_1(E, \mathbb{Z}) \to H_1(B, \mathbb{Z}) = E^2_{1,0} \to 0.$$

(In this case it can be checked directly without using spectral sequences).

Using the definition of $H_i(\Gamma, \mathbb{Z})$, we get the exact sequence 4.3.4.3 [*], because $F = BM$ here, and $\pi_1(B) = \Delta$ acts by conjugation on $H_1(F, \mathbb{Z}) = M/[M, M]$. Thus:

$$H_1(F, \mathbb{Z})/\pi_1(B) = M/[\Gamma, M].$$

The proof of Stalling's criterion can now be completed: we proceed by induction on n. The cases $n = 1$ and 2 are trivial and part of the hypothesis, respectively. We thus assume that $\Gamma/\Gamma_n \to \Delta/\Delta_n$ is isomorphic. From the following commutative diagram:

$$\begin{array}{ccccccccc} 1 & \to & \Gamma_n & \to & \Gamma & \to & \Gamma/\Gamma_n & \to & 1 \\ & & \downarrow\cong & & \downarrow & & \downarrow\cong & & \\ 1 & \to & \Delta_n & \to & \Delta & \to & \Delta/\Delta_n & \to & 1 \end{array}$$

we deduce from 4.3.4.3 the next one, which has exact rows, ($R = \mathbb{Z}$):

$$\begin{array}{ccccccccccc} H_2(\Gamma) & \to & H_2(\Gamma/\Gamma_n) & \to & \Gamma_n/\Gamma_{n+1} & \to & H_1(\Gamma) & \to & H_1(\Gamma/\Gamma_n) & \to & 0 \\ \sigma\downarrow & & \| & & \tau\downarrow & & \downarrow\cong & & \| & & \\ H_2(\Delta) & \to & H_2(\Delta/\Delta_n) & \to & \Delta_n/\Delta_{n+1} & \to & H_1(\Delta) & \to & H_1(\Delta/\Delta_n) & \to & 0 \end{array}$$

The map τ is thus surjective. Its injectivity follows from the surjectivity of σ by an easy diagram chasing. Thus τ is isomorphic, and thus so is

$$\varphi_{n+1} : \Gamma/\Gamma_{n+1} \to \Delta/\Delta_{n+1},$$

by another easy argument.

A topological consequence of J. Stalling's criterion is:

4.3.5. Corollary. *Let $f : K \to L$ be a continuous map between two CW-complexes. Let $R = \mathbb{Z}$ or $\mathbb{Q}$. Assume that $f_* : H_i(K, R) \to H_i(L, R)$ is isomorphic for $i = 1$, and onto for $i = 2$. Put $\phi = f_* : \Gamma \to \Delta$ with $\Gamma = \pi_1(K)$ and $\Delta = \pi_1(L)$. Then:*

1. *if $R = \mathbb{Z}$, $\phi_\infty : \Gamma^{\text{nilp}} \to \Delta^{nilp}$ is isomorphic if ϕ is onto;*
2. *if $R = \mathbb{Q}$, $\phi'_\infty : \Gamma'^{\text{nilp}} \to \Delta'^{\text{nilp}}$ is injective of finite index if $\phi : \Gamma \to \Delta$ has finite index.*

Proof. By Stalling's result, it is sufficient to check that $f_{*,i} : H_i(\Gamma, R) \to H_i(\Delta, R)$ is an isomorphism for $i = 1$, and onto for $i = 2$.

By adding cells of dimension at least 3 to K and L, one gets CW-complexes K' and L' and embeddings: $\chi : K \to K'$ and $\lambda : L \to L'$ such that:

1. $\tilde{K}'$ and $\tilde{L}'$ are contractible. Thus, K' (resp. L') is a $K(\Gamma, 1)$ (resp. $K(\Delta, 1)$).
2. Because $\pi_i(L') = 0\,(i \geq 2)$, there exists $f' : K' \to L'$ which makes the following diagram commutative:

$$\begin{array}{ccc} K & \xrightarrow{f} & L \\ \chi\downarrow & & \downarrow\lambda \\ K' & \xrightarrow{f'} & L'. \end{array}$$

3. Moreover, since only cells of dimension at least 3 have been added,

$$H_i(K', K; R) = H_i(L', L; R) = 0 \quad (i = 1, 2).$$

The following commutative diagram has thus isomorphic arrows if $i = 1$; if $i = 2$, f_* and λ_* are surjective, hence so is f'_*; we thus have the claimed properties.

$$\begin{array}{ccccccc} & & H_i(K, R) & \xrightarrow{f_*} & H_i(L, R) & & \\ & & \chi_*\downarrow & & \downarrow\lambda_* & & \\ H_i(\Gamma, R) & = & H_i(K', R) & \xrightarrow{f'_*} & H_i(L', R) & = & H_i(\Delta, R). \end{array}$$

4.4. Reduction to the Albanese image

Let X be a compact Kähler manifold, and let $\alpha_0 : X \to \mathrm{Alb}(X)$ be its Albanese map. Observe that, by Hodge theory, $(\alpha_0)_* : H_1(X, \mathbb{Q}) \to H_1(\mathrm{Alb}(X), \mathbb{Q})$ is an isomorphism.

(The example of Hopf surfaces S, for which $\pi_1(S) \cong \mathbb{Z}$, but $\mathrm{Alb}(S)$ is a point, shows that this is not the case for any compact complex manifold).

Let $Y_0 := \alpha(X)$, and $\delta : Y \to Y_0$ be a desingularization of Y_0. Modifying X if necessary (this does not change its π_1), we shall assume that α_0 lifts to $\alpha : X \to Y$ in such a way that $\alpha_0 = \delta \circ \alpha$.

Remark that $\alpha_* : \Gamma := \pi_1(X) \to \Delta := \pi_1(Y)$ has finite index σ, bounded from above by the number of connected components in the generic fiber of α (see [Ca1]). Remark also that α_* need not be surjective.

4.4.1. Theorem ([Ca2]). *In the above situation,*

$$(\alpha_*)'_\infty : (\Gamma/\Gamma'_\infty) \to (\Delta/\Delta'_\infty)$$

is injective, and has finite index $\sigma' \leq \sigma$.

Proof. Because of Stalling's criterion, it is sufficient to check that $\alpha_{*,i} : H_i(X, \mathbb{Q}) \to H_i(Y, \mathbb{Q})$ is isomorphic if $i = 1$ and surjective if $i = 2$.

The case $i = 1$ is clear, because $(\alpha_0) : X \to \mathrm{Alb}(X)$ induces an isomorphism at H_1 level, factorizes through α, and $\alpha_* : \pi_1(X) \to \pi_1(Y)$ has finite index.

The case $i = 2$ uses a basic property of Kähler geometry:

4.4.2. Lemma (see [Fu1]). *Let $\alpha : X \to Y$ be any surjective holomorphic map between complex manifolds, with X compact and Kähler. Then:*

$$\alpha^* : H^i(Y, \mathbb{R}) \to H^i(X, \mathbb{R})$$

is injective for any i (the same argument works also for $H^{p,q}$).

Proof. Let $\beta \in H^i(Y, \mathbb{R})$ and $\gamma \in H^{n-i-2r}(Y, \mathbb{R})$ be represented by d-closed forms $\tilde{\beta}$ and $\tilde{\gamma}$ respectively, where $n = \dim X$ and $(n-r) = \dim Y$. Let ω be a Kähler form on X. Let $c := \int_F \omega^r > 0$ for F a generic fiber of α: this values does not depend on F because ω is d-closed.

By Fubini's theorem: $\int_X \alpha^*(\tilde{\beta} \wedge \tilde{\gamma}) \wedge \omega^r = c \cdot \int_Y \tilde{\beta} \wedge \tilde{\gamma}$.

If $\beta \neq 0$, there exists γ such that the right-hand side does not vanish. Thus $\alpha^*(\beta)$ does not vanish either.

4.4.3. Remarks.

1. The above proof shows that the Kähler condition is superfluous when $r = \dim X - \dim Y = 0$.
2. Let $\alpha : S \to \mathbb{P}_1(\mathbb{C})$ be the algebraic reduction of a Hopf surface S of algebraic dimension one. Then $\alpha^* : H^2(\mathbb{P}_1(\mathbb{C}), \mathbb{Z}) \to H^2(S, \mathbb{Z})$ is the zero map, since $b_2(S) = b_2(S^1 \times S^3) = 0$. Of course (as already seen above), S is not Kähler.

   Another typical example where the injectivity at H^2 level is violated, although α^* is isomorphic at H^1 level, is the Iwasawa manifold.
3. For A any subring of $\mathbb{C}$ (with unit), let

$$G_A := \left\{ \begin{pmatrix} 1 & a & c \\ 0 & 1 & b \\ 0 & 0 & 1 \end{pmatrix} \mid (a, b, c) \in A^3 \right\}.$$

This is a subgroup of $G_\mathbb{C}$, itself a nilpotent complex Lie group of class 2, with center $Z(G_\mathbb{C}) = \{(a, b, c) \in A^3 \mid a = b = 0\} = [G_\mathbb{C}, G_\mathbb{C}]$. It is easy to check that $G_{\mathbb{Z}[i]}$ is discrete cocompact in $G_\mathbb{C}$. The quotient manifold $Z := G_\mathbb{C}/G_{\mathbb{Z}[i]}$ is thus compact, homogeneous with $\mathrm{Aut}_\mathcal{O}(Z) = G_\mathbb{C}$ (acting on the left), and $\pi_1(Z) = G_{\mathbb{Z}[i]} := \Gamma$.

Again $Z(\Gamma) = [\Gamma, \Gamma] = \{(a, b, c) \in (\mathbb{Z}[i])^{\oplus 3} \mid a = b = 0\}$. The natural projection $\tilde{\alpha} : \mathbb{C}^3 \to \mathbb{C}^2$ defined by: $\tilde{\alpha}(a, b, c) = (a, b)$ is thus a group homomorphism, which descends to the Albanese map $\alpha_Z : Z \to \mathrm{Alb}(Z) \cong E^{\oplus 2}$, where $E = (\mathbb{C}/\mathbb{Z}[i])$ is the Gauss elliptic curve. Moreover,

$$\alpha_Z^* : H^2(\mathrm{Alb}(Z), \mathbb{C}) = \wedge^2 H^1(Z, \mathbb{C}) \to H^2(Z, \mathbb{C})$$

is not injective, since $\alpha^*(da \wedge db) = du$, where $u = dc + adb$ are the obvious left-invariant holomorphic one-forms on $G_\mathbb{C}$ (which thus descend to Z.)

4.5. Applications

Let again X be compact Kähler, and $\Gamma := \pi_1(X)$.

4.5.1. Corollary. *Assume that $\alpha : X \to \mathrm{Alb}(X)$ is surjective. Then: $\Gamma'_1 = \Gamma'_\infty$ and $\Gamma'^{\mathrm{nilp}} = H_1(X, \mathbb{Z})/\mathrm{Torsion}$.*

Proof. Indeed, α_* is surjective in that case, and $\Delta'^{\mathrm{nilp}} = H_1(X, \mathbb{Z})/\mathrm{Torsion}$, if $\Delta = \pi_1(Y) = \pi_1(\mathrm{Alb}(X))$.

4.5.2. Corollary. *Assume that either $\kappa(X) = 0$ or that $a(X) = 0$. Then: $\Gamma'^{\mathrm{nilp}} = H_1(X, \mathbb{Z})/\mathrm{Torsion}$.*

Recall that $\kappa(X)$ (resp. $a(X)$) is the Kodaira–Iitaka dimension of X (resp. the algebraic dimension of X).

Proof. In both cases α is surjective ([Ka1] and [U] respectively), so that 4.5.1 applies.

4.5.3. Remarks.

1. It will be shown below (6.5.7) that if $a(X) = 0$ and $\chi(\mathcal{O}_X) \neq 0$, then $\pi_1(X)$ is in fact finite of cardinality at most 2^{n-1} if $n = \dim X$.

   The remaining case $a(X) = \chi(\mathcal{O}_X) = 0$ is more difficult, including for example tori and odd-dimensional Calabi–Yau manifolds.
2. Nevertheless, one conjectures that in both cases ($\kappa(X) = 0$ or $a(X) = 0$) $\pi_1(X)$ should have a subgroup of finite index which is abelian. The above result 4.5.2 thus "reduces" to show that $\Gamma'^{\text{nilp}} = \Gamma$ for some Γ of finite index in $\pi_1(X)$.

   One can extend 4.5.2 as follows: let X again be compact and Kähler; modifying X if needed, we can assume that it has an algebraic reduction $\beta : X \to A$ which is holomorphic, with A smooth and projective (see [U], §3 for this notion).

4.5.4. Corollary. *Let $\beta : X \to A$ be any holomorphic algebraic reduction of X as above. Let $d := q(X) - q(A)$ where $q(X) := h^0(X, \Omega^1_X)$ is the irregularity of X. Then, there is a (non-canonically) split exact sequence of groups:*

$$0 \to \mathbb{Z}^{\oplus 2d} \to \pi_1(X)'^{\text{nilp}} \to \pi_1(A)'^{\text{nilp}} \to 0.$$

4.5.5. Remarks. Of course, this result does not say anything when X is Moishezon (i.e. β bimeromorphic) and reduces to 4.5.2 (case $a(X) = 0$) when $a(X) = 0$. It just shows that the only interesting nilpotent completions of Kähler groups appear in the projective case. One may even conjecture that the above statement should remain true for π_1, and not only for $\pi_1'^{\text{nilp}}$.

See 6.5 below for more on this and its reduction to the case $a(X) = 0$.

Proof of 4.5.4. One has a commutative diagram:

$$\begin{array}{ccc} X & \overset{\beta}{\to} & A \\ \alpha_X \downarrow & & \downarrow \alpha_A \\ \text{Alb}(X) & \overset{\beta'}{\to} & \text{Alb}(A). \end{array}$$

Because β is surjective with connected fibers, so is β'. Thus $\text{Alb}(A) = \text{Alb}(X)/T$ where T is some subtorus of $\text{Alb}(X)$; let $Y_0 := \alpha_X(X)$ and $B_0 := \alpha_A(A)$. Denote $\beta'' : Y_0 \to B_0$ the restriction of β' to Y.

4.5.6. Lemma. *We have $Y_0 = \beta'^{-1}(B_0)$, or equivalently $\beta'' : Y_0 \to B_0$ is a principal fiber bundle with fiber T.*

Before giving the proof, we show how it implies 4.5.4. Let $\delta : B \to B_0$ be a smooth model of B_0. We can assume, modifying A and X if necessary, that $\alpha_A : A \to B_0$ lifts to $\alpha : A \to B$ such that $\alpha_A = \delta \circ \alpha$. Now replace Y_0 by $Y = Y_0 \times_{B_0} B$ to get a commutative diagram:

$$\begin{array}{ccc} X & \stackrel{\beta}{\to} & A \\ \tilde{\alpha}_X \downarrow & & \downarrow \tilde{\alpha}_A \\ Y & \stackrel{\tilde{\beta}}{\to} & B \end{array}$$

in which the map $\tilde{\beta} : Y \to B$ is a principal fiber bundle with fiber T.

From this we get a commutative diagram of groups (written multiplicatively) with exact rows (and $\mathbb{Z}$ coefficients):

$$\begin{array}{ccccccccc} & & \pi_1(T) & \to & \pi_1(T) & \to & \pi_1(B) & \to & 1 \\ & & \downarrow\cong & & \downarrow\tau & & \downarrow & & \\ 1 & \to & H_1(T) & \to & H_1(Y) & \to & H_1(B) & \to & 1 \end{array}$$

The exactness at $H_1(T) = H_1(T, \mathbb{Z})$ comes from the fact that $H_1(Y, \mathbb{Q}) = H_1(X, \mathbb{Q})$ $= H_1(\mathrm{Alb}(X), \mathbb{Q}) \cong H_1(\mathrm{Alb}(A), \mathbb{Q}) \oplus H_1(T, \mathbb{Q})$.

The bottom exact sequence is thus non-canonically split, which lifts to a non-canonical splitting of the top exact sequence, by composition with τ.

It now only remains to apply 4.4.1, which shows that:

$$\alpha'_{X,*,\infty} : \pi_1(X)'^{\mathrm{nilp}} \to \pi_1(X)'^{\mathrm{nilp}}$$

and

$$\alpha'_{A,*,\infty} : \pi_1(A)'^{\mathrm{nilp}} \to \pi_1(B)'^{\mathrm{nilp}}$$

are both injective and of finite index, to conclude, since also $\beta_* : \pi_1(X) \to \pi_1(A)$ is surjective.

Proof of 4.5.6. Let $\beta' : Z_0 := \beta'^{-1}(B_0) \to B_0$; this is a principal fiber bundle with fiber T. Let

$$\Psi : X \to (Y_0 \times_{B_0} A) \stackrel{j}{\hookrightarrow} Z_0 \times_{B_0} A,$$

where $\Psi = \alpha_X \times \beta$, and let $X_0 := \Psi(X)$ be the image of X. We have of course: $a(X) \geq a(X_0) \geq a(A)$ since the natural projection from X_0 to A is surjective (see [U]). Because $a(X) = a(A)$, we also have: $a(X_0) = a(A)$. Moreover, $X_0 \hookrightarrow Z_0 \times_{B_0} A$, which is a principal fiber bundle over A with fiber T. The following lemma 4.5.7 thus shows that $X_0 = Z_0 \times_{B_0} A$ (since Y generates $\mathrm{Alb}(X)$, see [U]), which shows in particular that $Y_0 = Z_0$. This is precisely what 4.5.6 claims.

4.5.7. Lemma. *Let $f : U \to A$ be principal analytic fiber bundle with fiber T a complex torus, and U, A compact complex manifolds, with A projective. Let $X_0 \subset U$ be an irreducible compact analytic subset such that:*

1. *$a(X_0) = a(A)$;*
2. *the general fiber $X_{0,a} = X_0 \cap f^{-1}(a)$, $a \in A$, is irreducible and generates (in the sense of [U]) the fiber $U_a = f^{-1}(a) \cong T$.*

Then $X_0 = U$.

Proof. Let $G \subset \mathrm{Aut}^0(T)$ be the unit connected component of the isotropy subgroup of $X_{0,a}$, for a generic in A; it is a subtorus of $\mathrm{Aut}^0(T)$, and hence independent of a generic (these have no moduli). Let

$$X_1 := X_0/G \subset U/G \xrightarrow{\Psi_1} A$$

be the natural quotient for this action. We have again $a(X_1) = a(A)$ since $a(X_0) \geq a(X_1) \geq a(A)$.

Moreover the generic fiber $X_{1,a}$ of X_1 over A generates the torus $(U/G)_a \cong T/G$, by hypothesis.

On the other hand, if dim $X_1 >$ dim A, $X_{1,a}$ is of general type ($\kappa(X_{1,a}) =$ dim $X_{1,a}$) by K.Ueno's decomposition theorem ([U], 10.9).

This implies then that $a(X_1) = a(A) + \dim X_{1,a}$ because $\mathcal{L}_m := (\Psi_1)_*(K_{X_1}^{\otimes m})$ provides, for m sufficiently large, a bimeromorphic embedding $\phi_m : X_1 \to \mathbb{P}(\mathcal{L})$ over A (see again [U], 2.10). But $\mathbb{P}(\mathcal{L}_m)$ is projective, since so is A.

We thus get: dim $X_{1,a} = 0$, and so $G = \mathrm{Aut}^0(T)$, so that $X_0 = U$.

We considered in 4.5.1 the case when $\alpha : X \to \mathrm{Alb}(X)$ is onto; the next step is when the image is a divisor in $\mathrm{Alb}(X)$.

4.5.8. Corollary. *Let $\alpha : X \to \mathrm{Alb}(X)$ be the Albanese map of the compact Kähler (connected) manifold X. If $Y_0 := \alpha(X)$ is a divisor of $\mathrm{Alb}(X)$, then either: $\kappa(Y_0) \geq 2$ and $\pi_1(X)'^{\mathrm{nilp}} = H_1(X, \mathbb{Z})/\mathrm{Torsion}$ or $\kappa(Y_0) = 1$. (Note that $\kappa(Y_0) = 1$ means that Y_0 is a principal fiber bundle over a smooth curve C_0 of genus $g \geq 2$ with fiber a codimension 2 subtorus of $\mathrm{Alb}(X)$). In this last case, $\pi_1(X)'^{\mathrm{nilp}} \cong \pi_1(C_0) \oplus \mathbb{Z}^{\oplus 2(q(X)-1)}$.*

Proof. We apply K. Ueno's decomposition theorem again: let $G \subset \mathrm{Aut}^0(\mathrm{Alb}(X))$ be the unit component of the isotropy group of Y_0. We thus get a commutative diagram:

$$\begin{array}{ccc} \mathrm{Alb}(X) & \xrightarrow{\pi} & \mathrm{Alb}(X)/G := C \\ \tilde{j} \uparrow & & j \uparrow \\ Y_0 & \xrightarrow{\pi} & Z := \pi(Y_0) \end{array}$$

where π is the natural projection j and $\tilde{j}$ the natural injections. Of course $Y_0 = \pi^{-1}(Z_0)$ by construction, and Z_0 is a divisor of the torus C. By the arguments used in the proof of 4.5.4 above, it is sufficient to show that $\pi_1(Z)'^{\mathrm{nilp}} = H_1(Z, \mathbb{Z})/\mathrm{Torsion}$, if $\delta : Z \to Z_0$ is a smooth model of Z, with dim $Z_0 = \kappa(Y_0)$ ([U], 10.9). Moreover in this case Z_0 is an ample divisor of C ([Mu], p. 25 or [We], §6). From Lefschetz hyperplane section theorem ([Mil], p. 41) we get that $j^* : H^2(C, \mathbb{Z}) \to H^2(Z_0, \mathbb{Z})$ is injective, since dim $Z_0 > 1$ if $\kappa(Y_0) \neq 1$. It now follows from [Del], 8.2.7 that $(j \circ \delta)^* : H^2(C, \mathbb{Z}) \to H^2(Z, \mathbb{Z})$ is injective too. The assertion now follows from 4.4.1.

4.5.9. Corollary ([Ca2], [C-T2]). *Let X be compact Kähler, and assume that $\pi_1(X)$ is nilpotent, but not almost-abelian (i.e.: no subgroup of $\pi_1(X)$ of finite index is abelian). Then: $b_1(X) \geq 8$.*

4.5.10. Remark. This bound is sharp, due to the examples given in 2.5, which also show that the case where $\operatorname{codim}(Y_0, \operatorname{Alb}(X)) \geq 2$ is much harder to classify.

Proof. Assume $b_1(X) \leq 4$. Then α_X is surjective, or $\alpha_X(X) = Y_0$ is a curve of genus g at least 2, and $\alpha_X : X \to Y$ surjective shows that $\alpha_* : \pi_1(X) \to \pi_1(Y)$ is surjective too (its fibers are connected). This is not possible, since $\pi_1(Y)$ is not nilpotent (it contains a subgroup isomorphic to F_{2g-2}). Assume now that $b_1(X) = 6$. Let $d :=$ dim Y_0, with $Y_0 = \alpha_X(X)$. The cases $d = 1, 3$ are excluded as above. The case $d = 2$ is excluded too, since Y_0 is either ample, or projects to a curve of genus $g \geq 2$.

4.5.11. Remark. [Ca2] borrowed from [C-T1] the use of P. Deligne's result [Del], 8.2.7 as above in 4.5.8. On the other hand, J. Carlson and D. Toledo used [Ca2], to modify their original proof in [C-T1] so as to remove the additional assumption they made that $\pi_1(X)$ was a Heisenberg group.

4.5.12. Example [Ca2]. Let Γ be the fundamental group of the Iwasawa manifold. Then Γ is not Kähler, since $b_1(\Gamma) = 4$, but $\Gamma \neq H_1(\Gamma, \mathbb{Z})/\text{Torsion} = H_1(\Gamma, \mathbb{Z}) \cong \mathbb{Z}^{\oplus 4}$.

4.6. Malčev completion of a finitely generated group

Let Γ be a finitely generated group, and $\Gamma_n \subset \Gamma$, $(n \geq 1)$ its torsion free lower central series. For each $n \geq 1, \Gamma'^n := (\Gamma/\Gamma'_{n+1})$ is thus a torsion free finitely generated nilpotent group of class n.

4.6.1. A result of A. N. Malčev ([Mal2]) now asserts that such a group determines a unique simply connected nilpotent (of class n) real Lie group $N_n(\Gamma)$ in which Γ'^n can be embedded as a discrete cocompact subgroup. Let $M_n(\Gamma) := N_n(\Gamma)/\Gamma'^n$ be the natural quotient nilmanifold.

4.6.2. Let $\mathcal{L}_n(\Gamma)_{\mathbb{R}}$ be the Lie algebra of $N_n(\Gamma)$; it is defined over $\mathbb{Q}$, because $N_n(\Gamma)$ has a cocompact discrete subgroup. We denote by $\mathcal{L}_n(\Gamma)$ a rational form of $\mathcal{L}_n(\Gamma)_{\mathbb{R}}$ so that $\mathcal{L}_n(\Gamma) \otimes_{\mathbb{Q}} \mathbb{R} \cong \mathcal{L}_n(\Gamma)_{\mathbb{R}}$.

4.6.3. We thus get a tower of nilmanifolds

$$\to M_{n+1}(\Gamma) \to M_n(\Gamma) \to \cdots \to M_1(\Gamma)$$

and a tower of rational Lie algebras:

$$\mathcal{L}_{n+1}(\Gamma) \to \mathcal{L}_n(\Gamma) \to \cdots \to \mathcal{L}_1(\Gamma) \to 0.$$

Let $\mathcal{L}(\Gamma)$ be the projective limit of this tower of Lie algebras. Because for each n, $\mathcal{L}_n(\Gamma)/[\mathcal{L}_n(\Gamma), \mathcal{L}_n(\Gamma)] \cong \mathcal{L}_1(\Gamma)$ and $\mathcal{L}_n(\Gamma)$ is nilpotent, we have also a surjective morphism of Lie algebras: $\Psi_n : F(H_1(\Gamma, \mathbb{Q}) \to \mathcal{L}_n(\Gamma)$, where $F = F\big(H_1(\Gamma, \mathbb{Q})\big)$ is the free Lie algebra generated over $\mathbb{Q}$ by a $\mathbb{Q}$-basis of $H_1(\Gamma, \mathbb{Q}) = H_1(\Gamma, \mathbb{Z}) \otimes_{\mathbb{Z}} \mathbb{Q}$. (this is a very easy assertion; see [Serre: Groupes et Algébres de Lie], or [Bourbaki, Groupes et Algébres de Lie, chap. II] for the notion of a free Lie Algebra). Let I_n be the kernel of Ψ_n, and $I := \cap_{n \geq 1} I_n$. We thus also get the isomorphism:

4.6.4. $\mathcal{L}(\Gamma) = F(H_1(\Gamma, \mathbb{Q}))/I$.

The rational Lie algebra $\mathcal{L}(\Gamma)$ is called the *Malčev completion of* Γ (this also applies usually to the projective limit of the groups $N_n(\Gamma)$).

4.6.5. Example. Let $\Gamma = \pi_1(C_g)$, where C_g is a curve of genus $g \geq 2$. Thus:

$$\Gamma = \langle a_1, b_1, \ldots, a_g, b_g \mid \prod_{j=1}^{g} [a_j, b_j] = 1 \rangle.$$

In this case $\mathcal{L}(\Gamma) = F(a_1, b_1, \ldots, a_g, b_g)/I$, where I is the ideal generated by the single relation $[a_1, b_1] + [a_2, b_2] + \cdots + [a_g, b_g]$. Observe that I is homogeneous, generated by its elements of degree two; this is a general phenomenon in Kähler geometry (see 4.7 below).

4.6.6. Exercise. Let Γ be as above. We construct a graded Lie algebra $\mathcal{L}(\Gamma)$ over $\mathbb{Z}$ as follows:

$$\mathcal{L}(\Gamma) := \oplus_{n \geq 1} L_n(\Gamma),$$

where $L_n(\Gamma) := \Gamma_n / \Gamma_{n+1}$ (an abelian Lie group with finitely many generators).

For $X \in L_p(\Gamma)$ and $Y \in L_q(\Gamma)$, one defines: $[X, Y] = [\bar{X}, \bar{Y}] \pmod{\Gamma_{p+q+1}}$. Check that this is well-defined, and defines actually a Lie algebra structure on $\mathcal{L}(\Gamma)$ (Jacobi identity is difficult, and rests on a formula of Hall. See [Se4] as above).

Let $L(\Gamma)_{\mathbb{Q}} := \bar{L}(\Gamma) \otimes_{\mathbb{Z}} \mathbb{Q}$. Check that $L(\Gamma)_{\mathbb{Q}}$ is the graded Lie algebra associated to a suitable filtration of $\mathcal{L}(\Gamma)$.

4.7. Formality of compact Kähler manifolds

Note. In Rational Homotopy Theory, a space Z is said to be *formal* if its $\pi_*(Z) \otimes \mathbb{Q}$ is computable (by D. Sullivan's procedure) from $H_*(Z, \mathbb{Q})$ (with its ring structure). The basic result is :

4.7.1. Theorem ([D-G-M-S]). *Let X be a compact connected Kähler manifold, let $\Gamma := \pi_1(X)$. Then: $\mathcal{L}(\Gamma) = F(H_1(X, \mathbb{Q}))/I$, where I is the homogeneous ideal of $F = F(H_1(X, \mathbb{Q}))$ generated by*

$$I_2 := \operatorname{Im}(H_2(X, \mathbb{Q}) \xrightarrow{\lambda} \bigwedge^2 H_1(X, \mathbb{Q}),$$

the map dual to $\bigwedge^2 H^1(X, \mathbb{Q}) \xrightarrow{\lambda^*} H^2(X, \mathbb{Q})$ (Note that I_2 is a $\mathbb{Q}$-vector subspace of $F_2 = \bigwedge^2 H_1(X, \mathbb{Q})$, the space of 2-commutators of F).

4.7.2. Remark. In [D-G-M-S], the above result (and others) are proved with $\mathbb{R}$-coefficients, and not stated in the above precise form. The above formulation is that of [C-T2], deduced from results of J. Morgan. The improvement to $\mathbb{Q}$-coefficients is due to D. Sullivan.

4.7.3. Example. In the situation of $X = C_g$ (see 4.6.5), one gets the statement of 4.6.5, even without knowing a presentation of $\pi_1(X)$. The result is especially sharp in that case, since $\pi_1(X) = \pi_1(X)^{\text{nilp}}$.

4.7.4. Remarks.

1. $\lambda : H_2(X, \mathbb{Q}) \to \bigwedge^2 H_1(X, \mathbb{Q}) = H_2(\text{Alb}(X), \mathbb{Q})$ is nothing but $(\alpha_X)_*$, the direct image by the Albanese map. This provides the link with the approach described in 4.4.
2. It is now a simple matter to deduce 4.4.1 from 4.7.1.
3. Let us say that a Lie algebra $\mathcal{L}$ is *quadratic* if it has a *quadratic presentation*; this means that L is isomorphic to F/I, where F is a free Lie Algebra on finitely many generators, and I is a homogeneous ideal of F generated by elements of degree two (i.e.: linear combinations of 2-commutators). The above theorem then says that $\mathcal{L}(\pi_1(X))$ is quadratic if X is compact Kähler. This eliminates from the Kähler class any finitely generated group Γ such that $\mathcal{L}(\Gamma)$ is not quadratic (see example 4.7.5 below). However, as soon as $\mathcal{L}(\Gamma)$ is quadratic, 4.4.1 applies just as well as 4.7.1.

4.7.5. Example ([C-T2]). Let L_n be the Lie Algebra of strictly upper-triangular $(n+1)\times(n+1)$ matrices. Then L_n is not quadratic for $n \geq 2$. This is a nilpotent Lie Algebra of class n. It is easy to see that it can be presented as: $L_n = F/I$, where $F = F(X_1, \ldots, X_n)$ and $I_2 := I \cap F_2$ is generated by $[X_i, X_j], n \geq j \geq i+2 \geq 1$. One can now check that if I were generated by I_2, then the free Lie Algebra $F(X_1, X_2)$ would inject in $\mathcal{L}$, contradicting its nilpotency.

4.7.6. Example ([C-T2]). Nilpotent quadratic Lie algebras of arbitrarily large nilpotency class have been constructed by S. Chen. Here is an example of class 3 (of course, an example of class 2 is provided by the Heisenberg Lie Algebra $H_{2r}, r \geq 2 : H_{2r} = F(X_1, Y_1, \ldots, X_r, Y_r)/I$, where I is generated by

$$[X_i, X_j], [Y_i, Y_j]\ (i < j); \quad [X_i, Y_i] - [X_j, Y_j]\ (i < j).$$

Let $k \geq 2$ and

$$\mathcal{L}(k) := F(X_1, \ldots, X_k; Y_{ij}(i \leq j))/I$$

where I is generated by

$$[X_i, X_j], [Y_{ij}, Y_{lm}], [X_i, Y_{lj}]\ (i \neq j, l),\ [X_j, Y_{ij}] - [X_l, Y_{il}]\ (i < l,\ \forall i),$$

where $Y_{ij} = Y_{ji}$.

The dimension of $\mathcal{L}(k)$ is $(k^2+5k+2)/2$, its abelianization has dimension $k(k+3)/2$, and its center 1.

4.7.7. Remarks. Presently, no restriction except for quadratic presentation and even-dimensionality of the abelianization is known on the Malčev Lie Algebra of Kähler groups. When the group is nilpotent, the only further restriction is the rank of its abelianization (at least 8, and even). However besides abelian and Heisenberg groups no Kähler examples are known. In particular, no nilpotent torsion free groups of class 3 or more are known to be Kähler. It would be interesting to fill in the huge gap left open, either by examples or by further restrictions.

The basic ingredient in the proof of 4.7.1 is the $\partial\bar{\partial}$-Lemma of Kähler geometry. We shall now briefly explain why $\mathcal{L}(\pi_1(X))$ can be computed from differential forms, but not prove 4.7.1 for which we refer to [D-G-M-S], although a simpler proof should exist, since Postnikov towers are then explicitly given as in 4.6.1.

4.8. $\mathcal{L}(\pi_1(K))$, K a finite CW-complex

Let K be a finite CW-complex, and $\Gamma := \pi_1(K)$. We apply the constructions of 4.6 above to this Γ. Let first embed K in some $K(\Gamma, 1)$, say $\bar{K}$, as in 4.3.5. Because we have natural morphisms of groups $\Gamma \to \Gamma'^n = \Gamma/\Gamma'_n$, and because $M_n(\Gamma)$ is a

$K(\Gamma'_n, 1)$, we also have continuous maps, well defined up to homotopy; which yield a commutative diagram:

$$\begin{array}{ccccc} & & & & M_{n+1}(\Gamma) \\ & & & \overset{\mu_{n+1}}{\nearrow} & \downarrow \gamma_n \\ X & \xrightarrow{j} & \bar{K} & \xrightarrow{\mu_n} & M_n(\Gamma) \\ & & & & \downarrow \\ & & & \mu_1 \searrow & \vdots \\ & & & & \downarrow \gamma_1 \\ & & & & M_1(\Gamma) \end{array} \tag{4.5}$$

Applying now $H_2(\cdot) = H_2(\cdot, \mathbb{Q})$ we get a commutative diagram:

$$\begin{array}{ccccc} & & & & H_2(\Gamma'^{n+1}) \\ & & & \nearrow & \downarrow \gamma_{n*} \\ H_2(X) & \xrightarrow{j_*} & H_2(\bar{K}) = H_2(\Gamma) & \xrightarrow{\mu_{n*}} & H_2(\Gamma'^{n}) \\ & & & & \downarrow \\ & & & \mu_{1*} \searrow & \vdots \\ & & & & \downarrow \gamma_{1*} \\ & & & & H_2(\Gamma'^{1}) \end{array}$$

Recall now that j_* is surjective (4.3.5), and that we have exact sequences:

$$H_2(\Gamma/\Gamma'_{n+1}) \xrightarrow{\gamma_{n*}} H_2(\Gamma/\Gamma'_n) \to (\Gamma'_n/\Gamma'_{n+1}) \otimes \mathbb{Q} \to 0$$

and:

$$H_2(\Gamma) \xrightarrow{\mu_{n*}} H_2(\Gamma/\Gamma'_n) \to (\Gamma'_n/[\Gamma, \Gamma'_n) \otimes \mathbb{Q} \to 0.$$

This shows that $(\mu_n \circ j)_*$ has the same image that γ_{n*}. Thus: $(\Gamma'_n/\Gamma'_{n+1}) \otimes \mathbb{Q}$ is the cokernel of the natural map $(\mu_n \circ j)_* : H_2(X) \to H_2(M_n(\Gamma))$.

Moreover, the central extension

$$0 \mapsto (\Gamma'_n/\Gamma'_{n+1}) \otimes \mathbb{Q} \to N_{n+1}(\Gamma) \to N_n(\Gamma) \to 0$$

is uniquely determined by the map:

$$H_2((\Gamma'_n/\Gamma'_{n+1}), \mathbb{Q}) \to H_2(\Gamma/\Gamma'_n) \to (\Gamma'_n/\Gamma'_{n+1}) \otimes \mathbb{Q} \to 0,$$

which is an element of $H^2(\Gamma/\Gamma'_n, (\Gamma'_n/\Gamma'_{n+1}) \otimes \mathbb{Q})$.

We have thus shown that the knowledge of $H_2(X) \to H_2(M_n(\Gamma))$ determines the next stage $N_{n+1}(X)$ (thus $\mathcal{L}_{n+1}(\pi_1(X))$, but not $M_{n+1}(\Gamma)$).

4.8.2. Example. When $n = 1$, and $K = X$ a compact Kähler manifold, one can take $\mu_1 = \alpha_X$, the Albanese map. For the next stage, one gets a central extension:

$$1 \to (\Gamma'_2/\Gamma'_3) \otimes \mathbb{Q} \to N_2(\pi_1(X)) \to N_1(\pi_1(X)) \to 1$$

where $(\Gamma'_2/\Gamma'_3) \otimes \mathbb{Q} = \text{Cokernel}(H_2(X) \to H_2(\text{Alb}(X)) = \bigwedge^2 H_1(X))$.

We recognize here the degree two term of $\mathcal{L}(\pi_1(X))$, as given in 4.7.1

The computation of $\mathcal{L}(\pi_1(X))$ can be made by dualizing the 1-minimal model of the algebra of $\mathbb{Q}$-differential forms of D. Sullivan's theory.

Using $\mathbb{R}$-coefficients and another approach, we shall briefly explain why one can expect such a result when $K = M$ is a compact smooth manifold:

By K. S. Chen's theory of iterated path integrals (see [D-G-M-S] for a reference), the maps $\mu_n : M \to M_n(\pi_1(M))$ above can be chosen to be smooth.

Now the cohomology of $M_n(\Gamma), \Gamma = \pi_1(M)$, is the same as that of $\Gamma'^m = \pi_1(M_n(\Gamma))$, since $M_n = M_n(\Gamma)$ is a $K(\Gamma'^n, 1)$ as noted above.

But we have seen above 4.3.3(4) that $H^*(M_n) = H^*(\mathcal{E}^*(M_n)) = H^*(\mathcal{A}_n)$, where $\mathcal{A}_n$ is the algebra of N_n-right invariant differential forms on N_n (which descend to M_n).

We get in this way an increasing sequence of algebras:

$$\mathcal{A}_1 \subset \mathcal{A}_2 \subset \cdots \subset \mathcal{A}_n \subset \cdots$$

whose inductive limit is nothing but the 1-minimal model (over $\mathbb{R}$) of the De Rham algebra $\mathcal{E}^*(M)$.

Remark finally that at each stage, the differential algebra $(\mathcal{A}_n, d)$ is dual to the Lie algebra $(\mathcal{L}_n(M), [\ ,\ ])$ in the following sense: $\mathcal{A}_n$ is dual to $\mathcal{L}_n(M)$ (over $\mathbb{Q}$, in fact), and: $du(x \wedge y) = u([x, y])$ for $u \in \mathcal{A}_n$ of degree one, and X, Y in $\mathcal{L}_n(M) = Lie(N_n)$. (see also [G-M] for another description of this 1-minimal model).

5. Structure of the universal cover

5.1. The Shafarevitch conjecture

This is the following statement:

5.1.1. Conjecture ([Sh], p. 407). Let X be a projective manifold, and $c : \tilde{X} \to X$ its universal cover. Is it true that $\tilde{X}$ is holomorphically convex?

We shall consider, more generally, this conjecture when X is compact Kähler. Due to the existence of a Remmert reduction for holomorphically convex analytic spaces, this conjecture splits in two parts:

5.1.2. Conjecture. Let X be a compact Kähler manifold and $c : \tilde{X} \to X$ its universal cover.

S1 Does there exist a *proper* holomorphic map $\tilde{\Phi}_X : \tilde{X} \to \tilde{Y}$ with connected fibers and $\tilde{Y}$ a normal analytic space such that any *compact connected* analytic subset $\tilde{A} \subset \tilde{X}$ is contained in some fiber of $\tilde{\Phi}_X : \tilde{X} \to \tilde{Y}$? (Note that $\tilde{\Phi}_X$ is uniquely defined by these conditions, if it exists.)

S2. If S1 holds for X, is Y Stein (i.e., embeddable as a closed analytic subset of some $\mathbb{C}^N$, here)?

We shall consider here only S1, because nothing seems to be known in general concerning S2 (See however [Lár] for an interesting result in this direction). Notice that the conjecture covers both the cases where $\tilde{X}$ is either compact or Stein. We first consider examples which show that no such conclusion holds if one of the assumptions is dropped.

5.1.3. Example. Let S be a Hopf surface. Then $\tilde{S} = \mathbb{C}^2 - \{(0,0)\}$. Thus (S1) holds for $\tilde{S}$, with $\Phi_S = \mathrm{id}_{\tilde{S}}$. But (S2) fails since $\mathbb{C}^2 - \{0,0\}$ is not Stein.

5.1.4. Example. Let $\tilde{Z} = \mathbb{P}_3(\mathbb{C})/L$, where L is the union of the two lines of homogeneous equations $(z_0 = z_1 = 0)$ and $(z_2 = z_3 = 0)$ in $\mathbb{P}_3(\mathbb{C})$ with homogeneous coordinates $[z_0 : z_1 : z_2 : z_3]$.

Let $Z := \tilde{Z}/\Gamma$ where $\Gamma \doteq \mathbb{Z}$ is the group of analytic automorphisms of $\tilde{Z}$ generated by $\gamma : \mathbb{P}_3(\mathbb{C}) \to \mathbb{P}_3(\mathbb{C})$, with $\gamma([z_0 : z_1 : z_2 : z_3]) = [\alpha z_0 : \alpha z_1 : \beta z_2 : \beta z_3]$, where $0 < |\alpha| < |\beta| < 1$ are two complex numbers such that $\alpha^p \neq \beta^q$ for $(p,q) \in \mathbb{Z}^2 - \{(0,0)\}$. It is easy to check that Γ operates properly discontinuously and freely on $\tilde{Z}$, with compact quotient. Thus $\tilde{Z}$ covers the compact complex threefold Z (which is of course not Kähler, since $\pi_1(\tilde{Z}) = (1)$, so that $\pi_1(Z) \cong \mathbb{Z}$ and $b_1(Z) = 1$). However, $\tilde{Z}$ does not satisfy (S1). Indeed: assume there exists a proper map $\tilde{\Phi}_Z : \tilde{Z} \to \tilde{Y}$ with the property (S1).

Then $\tilde{\Phi}_Z$ had to be constant on the projective lines contained in $\tilde{Z} \subset \mathbb{P}_3(\mathbb{C})$. For $\tilde{a} \in \tilde{Z}$; the lines through a contain a nonempty open subset of $\tilde{Z}$, on which $\tilde{\Phi}_Z$ had to be constant. By analytic continuation, $\tilde{\Phi}_Z$ should be constant, contradicting its properness since $\tilde{Z}$ is not compact.

5.1.5. Example. We now show that (S2) can fail for infinite covers $c' : \bar{X} \to X$ of projective X. Of course, c' is *not* the *universal* cover of X. Presumably (S1) should hold however for any cover of a compact Kähler manifold X, (see 5.1.7 below). A similar example is given in [G-S].

Let $\Gamma \cong \mathbb{Z}$ be the group of automorphisms of $\mathbb{C}^* \times \mathbb{C}^*$ generated by γ with: $\gamma(z_1, z_2) = (\alpha z_1, \beta z_2)$ where $(\alpha, \beta) \in \mathbb{C}^2$ is such that: $0 < |\alpha| \leq |\beta| < 1$, and $\alpha^p \beta^q \neq 1$ for any $(p,q) \in \mathbb{Z}^2 - \{(0,0)\}$.

Then Γ is easily checked to act properly discontinously and freely on $\mathbb{C}^* \times \mathbb{C}^*$ with non-compact quotient $\bar{X} = \mathbb{C}^* \times \mathbb{C}^*/\Gamma$. Actually, Γ is a Hopf surface of algebraic dimension zero minus its two divisors. It is easy to check that any holomorphic function on $\bar{X}$ is constant, but that $\bar{X}$ does not contain any compact holomorphic curve. Thus $\bar{X}$ satisfy (S1) but not (S2) above. However, $\bar{X}$ covers the product $E_\alpha \times E_\beta$ of the elliptic curves $E_\alpha = \mathbb{C}^*/\alpha^{\mathbb{Z}}$ and $E_\beta = \mathbb{C}^*/\beta^{\mathbb{Z}}$.

Notice that by moving α and β slightly, $\bar{X}$ becomes holomorphically convex. This shows that (S2) should be very delicate to check.

We now consider a special case of a question asked by Nori and considered in [G-S], which is a consequence of (S1) above:

5.1.6. Question (Nori). Let S be a projective surface, and C a (possibly singular) irreducible rational curve such that $C^2 > 0$. Is $\pi_1(S)$ finite?

It is shown in [Gu] that the answer is yes if the Shafarevitch conjecture is true (actually, (S1) suffices). Indeed: assume (S1) to be true, and let $\tilde{\Phi}_S : \tilde{S} \to \tilde{Y}$ to be as in (S1). Let $\tilde{C} := c^{-1}(C)$ be the inverse image of C in $\tilde{S}$; it is a union of rational curves.

Let $\tilde{C}_0$ be some connected component of $\tilde{C}$; it has to be mapped to one point by $\tilde{\Phi}_S$, because of (S1). If $d := \dim \tilde{Y} = 0$, $\tilde{S}$ is compact and the answer is yes. Otherwise $\tilde{C}_0$ is compact with $\tilde{C}_0^2 > 0$ which contradicts $\bar{d} > 0$.

Another simple (and similar) argument, not using Shafarevitch's conjecture shows that $q(S) = 0$, which is however far from $\pi_1(S)$ finite. Using Kodaira's classification it has been shown in [G-S] that if $\pi_1(S)$ is not finite, then $\kappa(S) = 2$. We shall show this below by a different argument, without using classification of surfaces.

A counterexample to the finiteness of $\pi_1(S)$ would also contradict Shafarevitch's conjecture. We now state a weak solution to (S1):

5.1.7. Theorem. *Let $c : \bar{X} \to X$ be any connected étale cover of the compact Kähler manifold X.*

Then there exists a proper connected quasi-fibration $\bar{\Phi}_X : \bar{X} \to \bar{Y}$ such that for $\bar{x}$ general in $\bar{X}$, the fiber $\bar{\Phi}_X{}^{-1}(\bar{\Phi}_X(\bar{x}))$ of $\bar{\Phi}_X$ through $\bar{x}$ is the largest connected compact analytic subset of $\bar{X}$ containing $\bar{x}$.

(A meromorphic map $\bar{\Phi} : \bar{X} \to \bar{Y}$ is said to be a *quasi fibration* if its restriction to some open Zariski proper connected dense subset of $\bar{X}$ is analytic and proper.

Recall that $\bar{x} \in \bar{X}$ is said to be "general" if it belongs to some countable intersection of Zariski-dense open subsets of $\bar{X}$).

5.1.8. Remark. When $X = \bar{X}/\Delta$, for Δ a quotient group of $\pi_1(X)$, 5.1.7 geometrically means that if some nonempty open subset of X is covered by an analytic family of irreducible analytic compact subsets Y_s, $s \in S$, such that $(j_s \circ \nu_s)_* : \pi_1(\hat{Y}_s) \to \Delta$

has finite image for s in S, then $\bar{Y}_s$ are contained in fibers of $\bar{\Phi}_X$ for s generic in S. Here $j_s : Y_s \to X$ is the inclusion, $\delta_s : \hat{Y}_s \to Y_s$ the normalization, and $\bar{Y}_s$ any connected component of $c^{-1}(Y_s)$.

5.1.9. Canonical representative. First we shall describe a canonical representative of the meromorphic map $\bar{\Phi}_X : \bar{X} \to \bar{Y}$: since $\bar{\Phi}_X^*$ is proper, there exists an analytic map (after possibly shrinking $\bar{Y}^*$) $F^* : \bar{Y}^* \to \mathcal{C}(\bar{X})$, where $\mathcal{C}(\bar{X})$ denotes the Chow Scheme (or cycle space, see [Ba]) of $\bar{X}$ which sends $y \in \bar{Y}^*$ to the reduced fiber of $\bar{\Phi}_X{}^*$ over y. Because $\bar{\Phi}_X$ is proper meromorphic, F^* extends to $\bar{Y}$ as a proper, meromorphic map, say $F : \bar{Y} \to \mathcal{C}(\bar{X})$. Moreover $F : \bar{Y} \to F(\bar{Y}) \subset \mathcal{C}(\bar{X})$ by construction.

We now replace $\bar{Y}$ by $\bar{Y}^+ := F(\bar{Y}) \subset \mathcal{C}(\bar{X})$. Now $\bar{Y}^+ \subset \mathcal{C}(\bar{X})$ parametrizes a universal family of compact analytic cycles of $\bar{X}$. Let $\bar{X}^+ \subset \bar{Y}^+ \times \bar{X}$ be the incidence relation graph of this universal family, with $\bar{\Phi}_X^+ : \bar{X}^+ \to \bar{Y}^+$ and $\bar{\mu}_X^+ : \bar{X}^+ \to \bar{X}$ the restrictions of the natural projections. Observe that $\bar{\mu}_X^+$ is a proper modification, by construction.

We shall call $\bar{\Phi}_X^+ : \bar{X}^+ \to \bar{Y}$ *the canonical S-reduction* of $\bar{c} : \bar{X}^+ \to X$. (Actually if $\bar{Y}^+$ is not normal, we shall first make base change via the normalization of $\bar{Y}^+$).

5.1.10. Γ-reduction. Assume now that $\bar{c} : \bar{X} \to X$ is Galois with group $\Delta = \pi_1(X)/G$, so that Δ is a quotient of $\pi_1(X)$ by a normal subgroup G, and Δ acts properly discontinously and freely on $\bar{X}$, with $X = \bar{X}/\Delta$. The above S-reduction of $\bar{X}$ will then be called the *$\bar{\Gamma}$-reduction of* (X, Δ).

When the above cover is the universal cover $c : \tilde{X} \to X$ of X, we just call it the *$\tilde{\Gamma}$-reduction of X*.

Now if $\bar{c} : \bar{X} \to X = \bar{X}/\Delta$ is Galois of group Δ, then Δ acts also on $\mathcal{C}(\bar{X})$ in the natural way, and fixes $\bar{Y} \subset \mathcal{C}(\bar{X})$, as well as $\bar{X}^+$, on which it acts properly discontinously and freely, equivariantly for the map $\bar{\Phi}_X^+ : \bar{X}^+ \to \bar{Y}$. We thus get a commutative diagram:

$$\begin{array}{ccccccc} \bar{X}^+ & \overset{c^+}{\to} & X^+ & = & X^+/\Delta & \overset{\mu}{\to} & X \\ \bar{\Phi}_X^+ \downarrow & & \downarrow \Phi_X & & & \swarrow \varphi_\Delta & \\ \bar{Y} & \overset{\delta^+}{\to} & Y & = & \bar{Y}/\Delta\,, & & \end{array}$$

in which μ is a proper modification. The meromorphic connected map $\varphi_\Delta : X \to Y$ will be called the *Γ-reduction of* (X, Δ).

When $\Delta = \pi_1(X)$, so that c is the universal cover of X, we just call it the *Γ-reduction of X*.

Observe that, when X is projective, this map has been constructed by different methods by J. Kollár (in [Ko]), who gives it the name of *Δ-Shafarevitch map*. His approach is based on 5.1.8.

5.1.11. Example [Ko]. Let $\Delta := \pi_1(X)/[\pi_1(X), \pi_1(X)] = H_1(X, \mathbb{Z})$. The above Γ-reduction of (X, Δ) is then just the Remmert–Stein factorization of the Albanese map of X (but $\mathrm{Alb}(X)$ is not constructed in the above process, unless the Albanese map is surjective. Note that φ_Δ is holomorphic in that case).

5.1.12. Example [Ko]. When $X = \bar{X}/\Delta$. it may happen that Δ acts non-freely on $\bar{Y}^+$, although it does act freely on $\bar{X}^+$, even in the special case where $\Delta = \pi_1(X)$.

For example, let $\sigma : S \to S$ be a free involution of some $K3$ surface S. Let $\gamma : C \to C$ be the $\pm$ involution of some elliptic curve C, so that $C/\gamma \cong \mathbb{P}_1(\mathbb{C})$. Let $X := S \times C/\tau$, where $\tau = \sigma \times \gamma$. One easily checks that $\varphi_{\pi_1(X)}$ is nothing but the natural projection $\varphi : X \to C/\gamma = \mathbb{P}_1(\mathbb{C})$.

This situation is discussed in detail in [Ko].

5.1.13. Remark. The problem in the above situation is thus that $\Delta = \pi_1(X)/G$ may act with non-trivial (but finite) isotropy groups on $\bar{Y}_\Delta$ if

$$\bar{\Phi}_\Delta : \bar{X} = \bar{X}/\Delta \to \bar{Y}_\Delta$$

is the $\bar{\Gamma}$-reduction of (X, Δ). This also means that the following sequence is not exact in its middle term: $\pi_1(F) \to \pi_1(X) \to \pi_1(Y_\Delta) \to 1$, where F is a generic fiber of $\Phi_\Delta : X \to \bar{Y}_\Delta/\Delta$, the Γ-reduction of (X, Δ).

Of course, this situation can obviously be remedied by taking a suitable finite étale cover $X_1 \to X$, and working with X_1, in the following two situations:

1. Δ has a subgroup of finite index which is torsion free
2. Δ is residually finite (pick the finite union E of isotropy groups of finitely many points $\bar{y}_i$ of $\bar{Y}_\Delta$ for the action of Δ, and choose $\Delta' \subset \Delta$ normal and of finite index such that $\Delta' \cap E = (1)$. If sufficiently many of the $\bar{y}_i$'s have been chosen, by the compactness of Y_Δ, Δ' will act without fixed points on $\bar{Y}_\Delta$). This fact is also observed in [Ko], §5.

5.1.14. Question. Does there always exist a finite index subgroup of $\Delta = \pi_1(X)/G$ which acts without fixed points on $\bar{Y}_\Delta$ in the above situation?

Equivalently, after replacing X by some suitable finite étale cover, is it true that $\ker(\Phi_\Delta)_*$ becomes finite (or $\gamma d(Y_\Delta = \dim Y_\Delta$ see 5.1.15) where $(\Phi_\Delta)_* : \pi_1(X) \to \pi_1(Y_\Delta)$ is induced from Φ_Δ)?

By Fox's solution to Fenchel's conjecture, this is true if $\dim Y_\Delta = 1$ (see [Fo], [Ch]).

5.1.15. Notion of γ-dimension. Let $c : \bar{X} \to X$ be any connected étale cover, and $\bar{\Phi}^+_X : \bar{X}^+ \to \bar{Y}^+$ its (canonical) S-reduction.

Definition. $\dim \bar{Y}^+$ is called the *γ-dimension* of the cover c. When $X = \bar{X}/\Delta$ (or: c is Galois of group Δ), we call it the *γ-dimension of* (X, Δ).

In the special case where $\Delta = \pi_1(X)$, we we just call it the *γ-dimension of* X, denoted $\gamma d(X)$ (and $\gamma d(X, \Delta)$ in the former case).

We note that $\gamma d(X, \Delta)$ is a birational invariant of X (this makes sense, since so are $\pi_1(X)$ and thus Δ). Moreover, for any Δ, one has:

$$0 \leq \gamma d(X, \Delta) \leq n := \dim X$$

In particular:

$$0 \leq \gamma d(X) \leq n.$$

The extreme case means the following: $\gamma d(X) = 0$ iff $\tilde{X}$, the universal cover of X, is not covered by positive-dimensional compact connected analytic subsets. This is the case, in particular, when X is a complex torus, since $\tilde{X}$ is then Stein.

Observe also that $\gamma d(X) = 1$ iff $\pi_1(X)$ is commensurable to $\pi_1(C)$, for C a curve of genus $g \geq 1$. This is due to Fox's solution of Fenchel's conjecture (see [Ch], [Fo]).

5.1.16. Generic finite covers. Let $f : X_1 \to X_2$ be any generically finite surjective analytic (or even meromorphic) map between complex manifolds. Then: $\gamma d(X_1) \geq \gamma d(X_2)$. Indeed: $f_* : \pi_1(X_1) \to \pi_1(X_2)$ has finite index image ([Ca1]).

5.1.17. Hyperplane sections. Let X_1 be a smooth hyperplane section of the projective manifold X. Then $\gamma d(X_1) = \gamma d(X)$ if $\dim X_1 \geq 2$.

This follows directly from the Lefschetz theorem on hyperplane sections ($\pi_1(X_1) = \pi_1(X)$) which shows that $\tilde{c}^{-1}(X_1) := \tilde{X}_1$ is the universal cover of X_1 if so is $\tilde{c} : \tilde{X} \to X$ for X.

Let now $\tilde{\Phi}^+_X : \tilde{X}^+ \to \tilde{Y}^+$ be the $\tilde{\Gamma}$-reduction of X; the restriction of $\tilde{\Phi}^+_X$ to $\tilde{X}_1$ is then surjective, since X_1 is ample.

(The conclusion actually follows from much weaker assumptions on X_1).

5.1.18. Curves and surfaces. The computation of $\gamma d(X)$ is in general quite difficult. For a curve C, however, one has:

$\gamma d(C) = 0$ iff $C = \mathbb{P}_1(\mathbb{C})$, and $\gamma d(C) = 1$ iff $g(C) \geq 1$ (where $g(C)$ is the genus of C).

For surfaces S not of general type, one can compute $\gamma d(S)$ from $q(S)$ and $\chi(\mathcal{O}_S)$ as follows:

1. For a ruled surface $f : S \to C$ one has $\gamma d(S) = \gamma d(C)$. Thus $\gamma d(S) = 0$ if S is rational, and $\gamma d(S) = 1$ otherwise.
2. For surfaces with $c_1(S) = 0$, one has $\gamma d(S) = 0$ if S is $K3$ or Enriques, and $\gamma d(S) = 2$ otherwise (i.e.: if S is hyperelliptic or a torus).
3. When $f : S \to C$ is relatively minimal elliptic with $\kappa(S) = 1, \gamma d(S) = 0, 1$ or 2. The case $\gamma d(S) = 0$ occurs precisely when $C = \mathbb{P}_1(\mathbb{C})$, and f has at least one singular fiber of type ${}_mI_0$, at most 2 such with relatively prime multiplicities. The second case $\gamma d(S) = 1$ occurs precisely when $g(C) \geq 1$ and $\chi(\mathcal{O}_S) \neq 0$, or

$g(C) = \chi(\mathcal{O}_S) = 0$. The last case $\gamma d(S) = 2$ occurs precisely when $\chi(\mathcal{O}_S) = 0$ and $g(C) \geq 1$, i.e.: when S is covered by an elliptic fiber bundle.

4. When $\kappa(S) = 2$, all three cases $\gamma d(S) = 0, 1, 2$ can occur. To realize $\gamma d(S) = 1$, just take an ample surface S in $\Sigma \times E$ where Σ is a $K3$ surface, and E an elliptic curve.

To summarize: for surfaces S which are *not* of general type, $\gamma d(S)$ can be computed from $q(S)$ and $\chi(\mathcal{O}_S)$ by the formula: $\gamma d(S) = q^-(S) + \chi^+(\mathcal{O}_S)$, where $q^-(S) = \inf\big(q(S), 1\big)$ and

$$\chi^+(\mathcal{O}_S) = \begin{cases} 0 & \text{if } \chi(\mathcal{O}_S) \neq 0; \\ 1 & \text{if } \chi(\mathcal{O}_S) = 0. \end{cases}$$

Finally, observe that $\gamma d(S) \leq \kappa(S)$ provided $\kappa(S) \geq 0$ and $\chi(\mathcal{O}_S) \neq 0$. This is not accidental, as we shall see in §6 below.

5.1.19. Example. There is a special case in which 5.1.7 solves affirmatively Shafarevitch's conjecture: let X be compact Kähler, with $\gamma d(X) \leq 1$. Then the Shafarevitch conjecture is true for X.

Indeed, if $\gamma d(X) = 0$, this is trivial since $\tilde{X}$ is compact. If $\gamma d(X) = 1$, then any map $\Phi_{\tilde{X}} : \tilde{X} \to \tilde{Y}$ as in 5.1.7 is holomorphic (since $\tilde{Y}$ is a smooth non-compact Riemann surface), and proper with $\tilde{Y}$ Stein.

5.1.20. Exercise. Let S be a projective smooth surface containing a rational curve C such that $C^2 > 0$. Show that if $\pi_1(S)$ is infinite, then $\gamma d(S) = 2$ (use 5.1.7, but not the classification of surfaces). Recall that Shafarevitch's conjecture implies that $\pi_1(S)$ is finite. Since $q(S) = 0$, we have $\chi(\mathcal{O})_S > 0$ and 6.3.1 below implies that $\kappa^+(S) = \kappa(S) = 2$.

5.1.21. We shall now give the geometric part of Step 3 of the proof of 3.5.4, namely: if $H^1EL_2(\tilde{X}) \neq 0$, there exists a proper connected H-equivariant holomorphic map $\varphi : \tilde{X} \to \tilde{Y}$, where $\tilde{Y}$ is a hyperbolic Riemann surface.

First step: there exists $h : \tilde{X} \to \mathbb{C}$, holomorphic, such that $\int_{\tilde{X}} \|dh\|^2 d\tilde{v} < \infty$, if $d\tilde{v}$ is the volume form on $\tilde{X}$. In the general case, this rests on the fact that the cup-product is zero on the exact part of the 1-dimensional part of L_2-cohomology ([Gr3]) section 3.3.c).

In the special case $H^1(\tilde{X}, \mathbb{R}) = 0$ (or $H \subset [\Gamma, \Gamma]$), this is not needed: just take $dh = \omega^{1,0}$ if $\omega = dg$, with $\Delta g = 0$ and ω is L_2.

From Fubini and the analyticity of h, we get the coarea formula:

$$\int_{\tilde{X}} \|dh\|^2 d\tilde{v} = \int_{\mathbb{C}} \mathrm{vol}(\tilde{X}_\lambda) d\lambda,$$

where $\tilde{X}_\lambda = h^{-1}(\lambda)$, $\lambda \in \mathbb{C}$.

Almost all level sets of $\tilde{X}_\lambda$ have thus finite volume, and hence are compact (use the same argument as in the proof of 5.4.4 below). Now just take for φ the $\tilde{\Gamma}$-reduction of (X, H); it is easy to check the properties stated.

We now give another simple application of 5.1.7 to rationally connected manifolds.

5.2. Rationally connected manifolds

5.2.1. Definition. Let X be compact Kähler. We say that X is *rationally connected* if any two generic points of X can be joined by some connected compact analytic curve C, whose irreducible components are all rational.

We say that X is *strongly rationally connected* if the above C can be chosen to be irreducible.

Recall also that X is said to be *uniruled* if it is covered by (possibly singular) rational curve.

5.2.2. Remarks.

1. It is shown in [Ca1] that X is projective if X is rationally connected.
2. A deep result ([Ko-Mi-Mo1]) shows that rationally connected is equivalent to strongly rationally connected.
3. It is easy to show that $h^i(X, \mathcal{O}_X) = 0$ for $i > 0$ if X is rationally connected, and so $\chi(\mathcal{O}_X) = 1$ in that case. The strongly rationally connected case is treated in [Ca1].
4. The properties rationally connected (and strongly rationally connected) are obviously birational and invariant under finite étale covers.
5. A curve is rationally connected iff it is $\mathbb{P}_1(\mathbb{C})$; a surface is rationally connected iff it is rational. In dimension 3, it follows from S. Mori's theory that a threefold X is rationally connected iff it is birational to some X' which is either: Fano, a Del Pezzo fibering over $\mathbb{P}_1(\mathbb{C})$, or a conic bundle over a rational surface (but X' might have terminal, $\mathbb{Q}$-factorial singularities).
6. Further examples of rationally connected manifolds X are: unirational and Fano manifolds (i.e. $-K_X$ is ample). See [Ca5] and [Ko-Mi-Mo2] for this last statement. The proof in the Fano case rests on characteristic p methods introduced by S. Mori.

Concerning π_1, we have the following:

5.2.3. Theorem ([Ca5], [Ko-Mi-Mo2]). *Let X be rationally connected. Then we have $\pi_1(X) = (1)$.*

Proof. When X is strongly rationally connected, a simple proof is given in [Ca1]. Because rationally connected implies strongly rationally connected ([Ko-Mi-Mo1]), the result follows. We shall here give a more elementary proof, in the spirit of [Ca1], using 5.1.7 above, which has the advantage of avoiding [Ko-Mi-Mo1]'s deformation theory. Because of 5.1.20 (3) and (4), it is sufficient to show that $\pi_1(X)$ is finite, or that $\gamma d(X) = 0$.

Now let $c : \tilde{X} \to X$ be the universal covering of X. Because any rational curve of X can be lifted to a rational curve of $\tilde{X}$, we see that $\tilde{X}$ also satisfies the condition of definition of rationally connected in 5.1.19. It follows that $\tilde{\Phi}_X : \tilde{X}^+ \to \tilde{Y}^+$ is constant, and hence that $\gamma d(X) = 0$, as desired (the argument is the same as in 5.1.4).

5.2.4. Remarks.

1. The argument above to show that $\pi_1(X)$ is finite applies as well to any compact Kähler in which any two generic points can be joined by some connected compact analytic subset all components of which have a simply connected normalization.
2. The hypothesis that X is Kähler is essential. Indeed the threefold Z of example 5.1.4 is rationally connected (but not Kähler) in the sense of 5.1.19. However, $\pi_1(Z) \cong \mathbb{Z}$. This is because $\tilde{Z}$ does not satisfy 5.1.7.
3. For other results in this direction, see [Ca 4].
4. Remark also that 4.1.20 directly follows from the main result of 6. (see 6.3.1)
5. 5.1.21 together with 5.1.20 (6) implies that any Fano manifold X (i.e. $-K_X$ is ample) is simply connected.

We shall now sketch the proof of 5.1.7. For this, we need some preliminary material.

5.3. Volume of compact analytic cycles

Let V be a connected but not necessarily compact Kähler manifold with Kähler form $\bar{\omega}$, and cohomology class $[\bar{\omega}]$.

Let $Z := \sum_{i \in I} n_i Z_i$ be a compact pure p-dimensional cycle (i.e.: I is finite, n_i is a positive integer and Z_i is a compact irreducible p-dimensional complex analytic subset of V for any $i \in I$). For any Z as above, the *volume of* Z (relative to $\bar{\omega}$) is defined by $v(Z) := \int_Z \bar{\omega}^p = [Z][\bar{\omega}^p]$, where $[Z] \in H_2 p(Z, \mathbb{Z})$ is the homology class of Z. When V is projective, and $\bar{\omega}$ is the Fubini–Study form (suitably normalized), $v(Z)$ is nothing but the *degree* of Z.

The set $\mathcal{C}_p(V)$ of all such Z constitutes the Chow scheme of p- dimensional cycles of V, and carries a natural structure of analytic space. The homology class, and hence the volume, remains constant on connected components of $\mathcal{C}_p(Z)$.

We have the fundamental criterion of compacity:

5.3.1. Theorem ([Li], based on [Bis]). *Let $E \subset \mathcal{C}_p(V)$ be any subset. Then E is relatively compact in $\mathcal{C}_p(V)$ iff:*

1. *the volume is bounded on E (i.e.: for some $c > 0$, one has $v(Z) \leq c$ for any $Z \in E$);*
2. *for some compact subset K of V, one has $Z \subset K$ for any $Z \in E$.*

The proof consists of a reduction to Ascoli's theorem, using Weierstrass parametrization and the equicontinuity of families of uniformly bounded holomorphic functions.

5.3.2. Corollary ([Li]). *Assume V is, moreover, compact. The connected components of $\mathcal{C}_p(V)$ are then compact (because v is constant on these, since $d\bar{\omega} = 0$).*

This property has already considerable consequences for the automorphism group of V, developed in [Li].

Note also that 5.2.1 holds also in the non-Kählerian situation, when $\bar{\omega}$ is only hermitian positive. In this case however, the volume is no longer constant on irreducible components on $\mathcal{C}_p(V)$, which may be non compact, even when V is compact.

5.3.3. We now consider the following situation: $V = \bar{X}$, where $c : \bar{X} \to X$ is a (connected) étale cover of some compact Kähler manifold X. Let ω be a Kähler form on X, and $\bar{\omega} := c^*(\omega)$ be its pull-back on $\bar{X}$.

5.3.4. Proposition ([Ca3]). *Let $c : \bar{X} \to X$, ω and $\bar{\omega}$ be as in 5.3.3. Let $C > 0$ and $\bar{A} \subset \bar{X}$ be given, with $\bar{A}$ compact. There exists then $\bar{K} := \bar{K}(\bar{A}, C) \subset \bar{X}$, with $\bar{K}$ compact, such that for any $Z \in \mathcal{C}_p(\bar{X})$ satisfying*

1. *$v(Z) \leq C$*
2. *Z is connected and meets $\bar{A}$, we also have:*
3. *Z is contained in $\bar{K}$.*

Proof (sketch). Let g be the Kähler metric on X associated to ω, and $\bar{d}$ the distance on $\bar{X}$ defined by $\bar{g} := c^*(g)$: $(\bar{X}, \bar{g})$ is complete since X is compact.

From E. Bishop's lemma, there exist universal constants $\alpha, B = B(n)$ where $n = \dim X$, such that:

$$\operatorname{vol}(Z \cap B(a,r)) := \int_{Z \cap B(a,r)} \omega^p \geq (\alpha r)^{2p} B \quad \text{for any} \quad r > 0$$

if $\bar{a} \in Z$.

Let $\bar{K} = \{\bar{x} \in \bar{X} : \bar{d}(\bar{x}, \bar{A}) \leq Nr\}$, where $N > C[(\alpha r)^{2p} \cdot B]^{-1}$, r fixed and sufficiently small. Then $\bar{K}$ is compact since $(\bar{X}, \bar{d})$ is complete, and if Z satisfied

1.,2. above, but were not in $\bar{K}$, one could find N disjoint balls $B(\bar{a}_i, r)$ with $\bar{a}_i \in Z$. Contradiction.

5.3.5. Example. The above result fails if $(\bar{X}, X)$ is choosen to be $(\bar{Z}, Z)$ as in 5.1.4, if one considers the family of projective lines of $\mathbb{P}_3(\mathbb{C})$ not meeting L, and ω is any hermitian positive metric on Z.

5.3.6. Corollary. *Let $c : \bar{X} \to X$ be a connected étale cover of some compact Kähler manifold. Let $S \subset C_p(\bar{X})$ be any irreducible component of $C_p(\bar{X})$, and $G_S \subset S \times \bar{X}$ be the incidence relation graph of the universal family of p-cycles parametrized by S: it is closed analytic in $S \times \bar{X}$. Assume that the generic member Z_s, $s \in S$ is irreducible reduced. Let $p_s : G_s \to \bar{X}$ and $q_s : G_s \to S$ be the natural projections, q_s is proper, by construction. Then p_s is proper, too.*

Proof. This is just a restatement of 5.3.4, since $p_s^{-1}(\bar{A}) := \{s \in S \mid Z_s \text{ meets } \bar{A}\}$ if $\bar{A} \subset \bar{X}$ is given. Observe that all Z_s are connected, since Z_s is irreducible, reduced for s generic in S. Notice also that G_s is itself irreducible.

5.3.7. The conclusion of 5.1.7 will now follow from the following general result 5.4.2 applied to the maps $\hat{p} : \hat{G}_s \to \bar{X}$ and $\hat{q} : \hat{G}_s \to \hat{S}$ obtained from p_s and q_s by composition with the normalization $n : \hat{G}_s \to G_s$ of G_s, and lifting to the normalization $\hat{S}$ of S.

5.4. Generation of meromorphic equivalence relations

5.4.1. Let G, S, Y be irreducible normal analytic spaces, and $p : G \to Y$, $q : G \to S$ analytic, *proper* and surjective maps. Assume that q has connected fibers. For any s in S, let $Z_s := p(q^{-1}(s))$, it is a compact connected analytic subset of Y.

We say that y, y' in Y are G^n-equivalent (respectively G^∞- equivalent) if there exist $s_1, \dots, s_n$, $n \geq 1$, in S such that $\bigcup_{i=1}^{n} Z_{s_i}$ is connected and contains both y and y' (respectively, if they are G^n-equivalent for some $n \geq 1$). Note that G^∞-equivalence is an equivalence relation. For any $y \in Y$, let $G^\infty(y)$ be the G^∞-equivalence class of y.

5.4.2. Theorem ([Ca3]). *In the situation of 5.4.1, there exists a quasi-fibration $\Psi_G : Y \to B_G$, with B_G normal, such that its generic fiber $Y_G := {\Psi_G}^{-1}(b)$, $b \in B_G$, is $G^\infty(y)$ for any $y \in Y_b$. Observe then that Ψ_G, B_G are well-defined up to bimeromorphic equivalence, and that the fibers of Ψ_G are connected.*

Proof of 5.4.2 (sketch). Let $G^n \subset Y \times Y$ be the increasing sequence of closed, symmetric and Y-proper (for either projection) analytic subsets inductively defined by: $G^0 = \Delta_Y$ (the diagonal), and

$$G^{n+1} = p_1 \times p_2\Big((q_1 \times q_2)^{-1}\big(G_S \times G \cap (p_1 \times p_2)^{-1}(G^n)\big)\Big)$$

if $n \geq 0$. Here, p_i (resp. q_i) denote the (proper) natural projections from $G \times G$ to Y (resp. S) for $i = 1, 2$. Then $G^\infty := \cup_{n \geq 0} G^n$ is the graph of the G^∞-equivalence relation. Thus 5.4.2 follows from the following statements (see [Ca3] for details)

5.4.2.1. There exists a unique irreducible component $\bar{G}^\infty$ of G^∞ which contains Δ_Y, it is symmetric and locally irreducible at the generic point of Δ_Y.

5.4.2.2. For generic $y \in Y$, the fiber of $\bar{G}^\infty$ over y (for either projection) is $\bar{G}^\infty(y)$.

Proof of 5.1.7 *(conclusion).* Let $\bar{\Phi}_0 : \bar{X} \to \bar{Y}_0$ be any proper quasi-fibration with connected fibers. Assume that there exists an $S \subset C_p(\bar{X})$ for some $p > 0$ as in Corollary 5.3.6 which is moreover *covering* (i.e.: $p_s : G_s \to \bar{X}$ is surjective, which means that $\bar{X}$ is covered by the $Z_s, s \in S$). Notice that if no such family exists, we are finished. Otherwise we can apply 5.4.2 to $p : \hat{G}_s \to Y_0$ and $q : \hat{G}_s \to \hat{S}$, where $p = \bar{\Phi}_0 \circ \hat{p}_s$. We obtain a proper quasi-fibration $\Psi : Y_0 \to Y_1$, and taking $\bar{\Phi}_1 := \Psi_0 \bar{\Phi}_0 : X \to Y_1$ we get a new proper quasi-fibration with dim $Y_1 <$ dim Y_0. Just stop when this dimension is minimal.

5.4.3. Example (Rational reduction). Let X be compact Kähler; applying 5.4.2 (see [Ca5] for details), one easily sees that there exists a quasi-fibration $\rho_1 : X \to X_1$ with rationally connected fibers, such that any other quasi-fibration $\sigma : X \to Y$ with this property one has $\rho_1 = \tau \circ \sigma$ for some $\sigma : X \to X_1$. We call ρ_1 the *rational quotient* of X. One has: $\pi_1(X) = \pi_1(X_1)$, which is just a relative version of 5.2.3 (see [Ko]). Iterating it, one gets a quasi fibration $\sigma_X : X \to S(X)$, with $S(X)$ non uniruled and with *rationally generated* fibers, such that $S(X)$ is the biggest non-uniruled image of X. Again $\pi_1(X) = \pi_1(S(X))$.

We shall call $\sigma_X : X \to S(X)$ the *rational reduction* of X: its generic fibers are *rationally generated* (which means that their rational reduction is a constant map) and its image $S(X)$ is not uniruled; these properties characterize σ_X up to bimeromorphic equivalence. Observe that "rationally connected" implies "rationally generated", but the converse is not known: for surfaces both are the same and also equivalent to be rational. For threefolds, it is known that both notions are still equivalent (see [Ko-Mi-Mo1]). In general, the equivalence would follow from a positive answer to the following:

5.4.4. Question. Let $f : X \to \mathbb{P}_1(\mathbb{C})$ be analytic surjective, with rationally connected fibers. Is X itself rationally connected? (or: does X contain a *rational* curve C with

$f(C) = \mathbb{P}_1(\mathbb{C})$?). A positive answer to this would show that the rational quotient of X is the same as its rational reduction.

6. Kodaira–Iitaka dimension and fundamental group

6.1. Introduction

In riemannian geometry, non-negativity assumption on the Ricci curvature of a compact C^∞-manifold M imply strong restrictions on its fundamental group $\pi_1(M)$. More precisely:

If Ricci$(M) > 0$, then $\pi_1(M)$ is finite; if Ricci$(M) = 0$, then $\pi_1(M)$ is almost abelian; and if Ricci$(M) \geq 0$, then $\pi_1(M)$ has "polynomial growth", which implies by a deep theorem of M. Gromov ([Gr1]), that it is almost nilpotent. (Here to say that a group Γ *almost* possesses a property (P) means that some subgroup Γ' of finite index in Γ possesses that property).

Of course, analogues of these results also exist in algebraic geometry, since the sign of the Ricci curvature is closely related to the positivity of $-K_X$, for X compact Kähler connected:

If X is Fano (i.e. $-K_X > 0$ (which means: ample)), then $\pi_1(X) = (1)$ (the proof of 5.2.3(6) is algebro-geometric, but rests on characteristic p methods); if $c_1(X) = 0$, then $\pi_1(X)$ is almost abelian. This last result rests on the solution of Calabi's conjecture by S. T. Yau, and the above result. The case of $-K_X \geq 0$ (in a suitable sense) is studied in [D-P-S], where it is shown that $\pi_1(X)$ has subexponential growth (we don't define here what it means).

Again the proof rests on the existence of metrics with Ricci sufficiently positive curvature.

However, such metrics are usually very difficult to construct, and can only exist in special case. Moreover, the desired conclusions (see open questions §7 below) are bimeromorphically invariant, although such metrics are not.

This approach thus seems to be quite deep in the study of the following questions:

1. Assume $\kappa(X) = 0$ (or that $a(X) = 0$); is then $\pi_1(X)$ almost abelian?
2. More generally, assume one has a surjective morphism of *groups* $\varphi : \pi_1(X) \to G$, where G is a *perfect* group (i.e. $G = [G, G]$). Does φ factorize through a surjective meromorphic map $f : X \to Y$ with Y of general type? (i.e.: does there exist such an f, with a morphism $\psi : \pi_1(Y) \to G$ such that $\varphi = \psi \circ f_*$, where $f_* : \pi_1(X) \to \pi_1(Y)$ is induced by f?)

We shall here present another approach to this type of questions through L^2- theory, which gives restrictions on $\pi_1(X)$ directly from algebro-geometric information on the positivity of Ω^1_X.

More precisely, one has:

6.1.1. Comparison Theorem ([Ca4]). *Let X be compact and Kähler, with $\chi(\mathcal{O}_X) \neq 0$. Then either* 1 *or* 2 *below holds:*

1. $\kappa^+ \geq \gamma d(X)$;
2. $\kappa^+ = -\infty$ *and* $\gamma d(X) = 0$ *(so that $\pi_1(X)$ is finite).*

Here $\kappa^+(X)$ is a variant of the Kodaira–Iitaka dimension $\kappa(X)$ of X, and should conjecturally coincide with it when $\kappa(X) \geq 0$; when $\kappa(X) = -\infty$, $\kappa^+(X)$ should be equal: to $\kappa\big(S(X)\big)$ if X is not rationally connected, and to $-\infty$ if X is rationally connected. Here $S(X)$ is the rational quotient of X (see 6.2.8 for this conjecture).

6.2. Kodaira–Iitaka dimensions

6.2.1. Definition. Let Z be a compact connected complex manifold with $n := \dim Z$. For $\mathcal{F}$ any coherent subsheaf of rank r of Ω^p_X, let $\mathrm{Det}(\mathcal{F})$ be the saturation of $\det(\mathcal{F})$ in $\bigwedge^r \Omega^p_X$. Let

$\kappa^+(X) := \sup\{\kappa(X, \mathrm{Det}(\mathcal{F}) \mid \mathcal{F} \subset \Omega^p_X$ any coherent subsheaf with $p > 0$ arbitrary$\}$,

and let

$$\kappa_+(Z) = \sup\{\kappa(Y) \mid \exists g : Z \to Y \text{ a surjective meromorphic map}\}.$$

We have the following obvious properties:

6.2.2. Proposition. *In the situation of* 6.2.1, *one has:*

1. $\kappa^+(Z)$ *and* $\kappa_+(Z)$ *are bimeromorphic invariants of* X;
2. $\kappa^+(Z) \geq \kappa^+(Z')$ *and* $\kappa_+(Z) \geq \kappa_+(Z')$ *if* $f : Z \to Z'$ *is any surjective analytic map;*
3. $n \geq a(X) \geq \kappa^+(Z) \geq \kappa_+(Z) \geq \kappa(Z) \geq -\infty$.

6.2.3. Examples.

1. Let A any complex torus. Then its tangent bundle T_A is trivial. Thus $\kappa^+(A) = \kappa(A) = 0$.
2. More generally, let $f : A \to X$ be any surjective meromorphic map, with A a complex torus. Then $\kappa(X) \leq 0$. A specific example is:
3. Let X be a non-singular model of $A/\pm 1$, with A an n-dimensional complex torus. If $n = 1$, $X = \mathbb{P}_1(\mathbb{C})$ and X is a Kummer surface if $n = 2$. For $n \geq 2$,

$\kappa^+(X) = \kappa(X) = 0$. One easily shows that $\chi(\mathcal{O}_X) = 2^{n-1}$ (because $h^{0,p}(X) = 0$ if p is odd, and 1 if p is even) and that $\pi_1(X) = 1$ (see [U], §16 for details).

4. Assume that X is rationally generated (see 5.4.3). Then: $\kappa^+(X) = -\infty$ (see [Ca4], §3). A relative version is: $\kappa^+(X) = \kappa^+(S(X))$ if $\sigma_X : X \to S(X)$ is the rational reduction of X, when $\dim S(X) > 0$. This reduces the study of κ^+ to the non-uniruled case. We shall state below some conjectures related to S. Mori's minimal model program, when X is projective.

6.2.4. Étale covers. Let $\bar{c} : \bar{X} \to X$ be a finite connected étale cover of degree d. It should be true, although this is not obvious, that $\kappa^+(\bar{X}) = \kappa^+(X)$. At least the following is easy:

6.2.5. Lemma.

1. *Assume that* $\kappa^+(X) = -\infty$. *Then:* $h^0(\bar{X}, \Omega^p_{\bar{X}}) = 0$ *if* $p > 0$, *and* $\chi(\mathcal{O}_{\bar{X}}) = 1$.
2. *Assume that* $\kappa^+(X) = 0$. *Then:* $h^0(\bar{X}, \bar{\mathcal{F}}) \leq r$ *for any coherent subsheaf* $\bar{\mathcal{F}} \subset \Omega^p_{\bar{X}}$ *of rank* r, *any* $p \geq 0$. *In particular* $|\chi(\mathcal{O}_{\bar{X}})| \leq 2^{n-1}$, *since* $h^0(\bar{X}, \Omega^p_{\bar{X}}) \leq \binom{n}{p}$, *where* $n = \dim X$.

6.2.6. Corollary. *Assume that* $\kappa^+(X) \leq 0$. *Let* $\pi_1^{\mathrm{alg}}(X)$ *be the algebraic fundamental group of* X, *see* 1.7. *Then:*

1. $\pi_1^{\mathrm{alg}}(X) = 1$, *if* $\kappa^+(X) = -\infty$
2. $|\pi_1^{\mathrm{alg}}(X)| \leq 2^{n-1}/|\chi(\mathcal{O}_X)| \leq 2^{n-1}$ *if* $\chi(\mathcal{O}_X) \neq 0$ *(where* $|E|$ *stands for the cardinality of the set* E*).*

Indeed, if $\bar{c} : \bar{X} \to X$ is as in 6.2.4, then $\chi(\mathcal{O}_{\bar{X}}) = d\chi(\mathcal{O}_X)$. The assertion follows from the bounds given in 6.2.7 The comparison theorem stated in 6.1 will allow us to extend these result to the "true" topological π_1, using a deeper multiplicative property of $\chi(\mathcal{O}_X)$.

6.2.7. Example.

1. The bound $|\chi(\mathcal{O}_X)| \leq 2^{n-1}$ for X with $\kappa^+(X) = 0$ is sharp: it is attained by example 6.2.3(3) above.
2. The bound $|\pi_1^{\mathrm{alg}}(X)| \leq 2^{n-1}$ is sharp, up to a logarithmic factor 2: let X be the product of $m = n/2$ Enriques surfaces. Then $\pi_1(X) \cong (\mathbb{Z}_2)^m$ and $|\pi_1(X)| = 2^{n/2}$. It would be interesting to know the exact bound.

We now come back to the relationship between κ^+ and κ, which seems to be important, but quite deep:

6.2.8. Conjectures.

1. $\kappa^+(X) = \kappa(X)$ if $\kappa(X) \geq 0$.
2. $\kappa^+(X) = \kappa(S(X))$ if X is uniruled, but not rationally generated.
3. $\kappa^+(X) = -\infty$ iff X is rationally generated.

These conjectures imply that $\kappa^+(X) = \kappa_+(X)$ for any compact Kähler X, and actually follow from quite standard ones, namely:

6.2.9. Proposition ([Ca4]). *One has the following implications:*

1. $(M) \Rightarrow$ 6.2.8(1) *when X is projective;*
2. $(U) \Rightarrow$ 6.2.8(2).

Here (M) and (U) are defined in 6.2.10 below.

6.2.10. Definition. The property (M) (resp. (U), resp. (M_0)) holds if, for any compact connected Kähler manifold X, one has:

(M): if X is projective and $\kappa(X) = 0$, then X is bimeromorphic to some projective variety Z with only $\mathbb{Q}$-factorial canonical singularities and K_Z numerically trivial;

(U): X is uniruled if $\kappa(X) = -\infty$;

(M_0): $\kappa^+(X) = 0$ if $\kappa(X) = 0$.

6.2.11. Remark. When X is a surface, all these properties, and so also 6.2.8 hold classically. In dimension 3 these properties, as well as 6.2.8, hold when X is projective, due to the existence of Minimal models and the proof of the Abundance Conjecture by Y. Kawamata, Y. Miyaoka and S. Mori.

6.2.12. Proposition. $(M) \Rightarrow (M_0)$ *when X is projective.*

Proof. (See [P], where a similar argument was first used.) Since $\kappa(X) = 0$, X is not uniruled. By Y. Miyaoka's generic semi-positivity theorem ([Miy]), and (M), if C is a sufficiently general smooth hyperplane complete intersection curve contained in the smooth locus Z^* of Z (defined as in 6.2.10(M)), then $\Omega^1_{Z^*|C} = E$ is semi-positive (or numerically effective: any rank one quotient of E has non-negative degree).

But $\Omega^{n-1}_{Z^*} \cong (\Omega^1_{Z^*})^* \otimes K_{Z^*}$. Since K_Z is numerically trivial, $F = \bigwedge^r \Omega^p_{Z^*|C}$ and its dual F^* are both semi-positive for any (r, p). Any rank one subsheaf of F has thus non-positive degree. Hence (M_0) for X.

6.2.13. Remark.

1. It would be interesting to show: $(M_0) \Rightarrow$ 6.2.8(1)
2. It follows in particular from 6.2.12 that $\kappa^+(X) = \kappa(X) = 0$ if $c_1(X) = 0$ and if X is projective. Thus conjecture 6.2.8 is true at least in this case. We shall discuss the non-projective case below.

We present a more geometric form of the conjectures 6.2.8.

6.2.14. Remark. If the conjectures 6.2.8 are true, then $\kappa^+(X) = \kappa_+(X)$ for any X, and $\kappa_+(X) = \kappa(X)$ when $\kappa(X) \geq 0$. Observe then that this equality is nothing but the famous $C'_{n,m}$ conjecture of Iitaka.

6.3. Comparison Theorem

6.3.1. Theorem ([Ca4]). *Let X be a compact Kähler manifold with $\chi(\mathcal{O}_X) \neq 0$. Then either* (1) *or* (2) *holds:*

1. $\kappa^+(X) \geq \gamma d(X)$;
2. $\kappa^+(X) = -\infty$ *and* $\gamma d(X) = 0$ *(i.e. $\pi_1(X)$ is finite).*

6.3.2. Remarks.

1. The condition $\chi(\mathcal{O}_X) \neq 0$ is essential, as shown by the case of complex tori X, for which $\kappa^+(X) = 0$ but $\gamma d(X) = n = \dim X$.
2. The case $\chi(\mathcal{O}_X) \neq 0$, $\kappa^+(X) = -\infty$ and $\gamma d(X) = 0$ occurs for X rationally connected ($\chi(\mathcal{O}_X) = 1$ in that case). Conjecturally (6.2.8(2)), this should even be the only case.
3. Assume that $\chi(\mathcal{O}_X) \neq 0$, and that the universal cover $\tilde{X}$ of X does not contain any positive-dimensional compact analytic subset. Thus $\gamma d(X) = n$. By 6.3.1, $\kappa^+(X) \geq n$. So X is projective (observe again that complex tori with $n \geq 2$ are counter-examples if the condition $\chi(\mathcal{O}_X) \neq 0$ is dropped). This is proved in [Gr2], from which the proof of [Ca4] is inspired (see §7 below).
4. The statement of 6.3.1 can be strengthened: assume $\chi(\mathcal{O}_X) \neq 0$ and $\kappa^+(X) \geq 0$. Then there exists a factorization:

$$X \xrightarrow{g} X_1 \xrightarrow{f} Y$$

   with $\Phi_X = f \circ g$, where $\Phi_X : X \to Y$ is the Γ-reduction of X, and $g : X \to X_1$ is the Iitaka reduction ([U]) of $(X, \operatorname{Det} \mathcal{F})$, where $\mathcal{F} \subset \Omega_X^p$ is coherent such that $\kappa(X, \operatorname{Det} \mathcal{F}) = \kappa^+(X)$.
5. Assume now that conjectures 6.2.8 hold. Then 6.3.1 implies: if $\chi(\mathcal{O}_X) \neq 0$ and $\kappa(X) \geq 0$, then $\kappa(X) \geq \gamma d(X)$.

6. Because the conjectures 6.2.8 are proved if $n = \dim X \leq 3$ (and X is projective) if $n = 3$, we have: if $\chi(\mathcal{O}_X) \neq 0$ we have either $\kappa(X) \geq \gamma d(X)$, or X is rationally connected (rational if $n \leq 2$).
7. We find in particular the result obtained in [Gu-S]: if S is a projective surface containing a rational curve C such that $C^2 > 0$, and if $\pi_1(S)$ is not finite, then $\kappa(S) = 2$. Just apply 5.1.18 and the last remark (6) above.

6.3.3. Example. A compact connected smooth surface S is said to be a K3 *surface* if $q(S) = 0$, $K_S \cong \mathcal{O}_S$. Assume moreover that S is Kähler. Then $\kappa^+(S) = \kappa(S)$ and $\chi(\mathcal{O}_S) = 1 - q + p_g = 2$. From 6.3.1 and 6.2.6 we get that $\pi_1(S) = 1$. A Kähler K3-surface is thus simply-connected.

Of course this result is well-known and can be proved (without the Kähler assumption) by deformation theory. We shall see below in §6 that the present proof also applies with minor modifications to cover the (potential) non-Kähler case as well.

6.4. The cases $\kappa^+(X) = 0$ or $-\infty$

Let X be compact Kähler connected with $n := \dim X$. The comparison theorem allows us to replace the algebraic fundamental group by its topological version in 6.2.6.

6.4.1. Corollary. *Assume that $\kappa^+(X) = -\infty$. Then $\pi_1(X) = 1$.*

Proof. It follows from $\kappa^+(X) = -\infty$ that $h^0(X, \Omega_X^p) = 0$ $(p > 0)$. Thus $\chi(\mathcal{O}_X) = 1 \neq 0$, and $\gamma d(X) = 0$ by 6.3.1. Hence $\pi_1(X) = \pi_1^{\text{alg}}(X) = 1$ (by 6.2.6(1)).

6.4.2. Remarks.

1. Actually, from conjecture 6.2.8, one should have even that X is rationally generated. We get the weaker π_1-statement (by 5.4.3).
2. Because $\kappa^+(X) = -\infty$ if X is rationally generated, we get another proof that $\pi_1(X) = 1$.

6.4.3. Corollary. *Assume that $\kappa^+(X) = 0$ and $\chi(\mathcal{O}_X) \neq 0$. Then $\pi_1(X)$ is finite and $|\pi_1(X)| \leq 2^{n-1}/|\chi(\mathcal{O}_X)| \leq 2^{n-1}$.*

Proof. Apply 6.3.1 to get $0 = \kappa^+(X) \geq \gamma d(X) \geq 0$. Thus: $\gamma d(X) = 0$ and $\pi_1(X) = \pi_1^{\text{alg}}(X)$ is finite. The conclusion now follows from 6.2.6(2).

6.4.4. Example. For X a smooth model of $A/\pm 1$, A any n-dimensional complex torus, one has $\kappa^+(X) = \kappa(X) = 0$, $\chi(\mathcal{O}_X) = 2^{n-1}$ and thus $\pi_1(X) = 1$.

It would be interesting to know what is the bound for $|\pi_1(X)|$ when $\kappa^+(X) = 0$ and $\chi(\mathcal{O}_X) \neq 0$: it is at least $2^{n/2}$ for n even (6.4.6 below).

6.4.5. Corollary. *Assume that $c_1(X) = 0$ and that $\chi(\mathcal{O}_X) \neq 0$. Then $\pi_1(X)$ is finite, of cardinality at most $2^{n-1}/|\chi(\mathcal{O}_X)|$ if X is projective.*

Proof. From 6.2.12, we get that $\kappa^+(X) = 0$. Hence 6.3.1 applies.

6.4.6. Remarks.

1. Of course, the above result is well-known; the projectivity assumption is even unnecessary. The proof given here however does not use the existence of Ricci-flat metrics, and is much more elementary. The Kähler case would also follow if one could prove *without Ricci-flat metrics* that compact Kähler manifolds X with $c_1(X) = 0$ have small deformations which are projective.
2. From the existence of Ricci-flat metrics, one actually sees that if $c_1(X) = 0$, then $\chi(\mathcal{O}_X) = 0$ for n odd, and $|\chi(\mathcal{O}_X)| \leq 2^{n/2}$ if n is even. This bound $2^{n/2}$ is attained by the product of $(n/2)$ K3 surfaces. The example of products of m Enriques surfaces shows that the bound on $|\pi_1(X)|$ cannot be less than $2^{n/2}$ when $c_1(X) = 0$, $\chi(\mathcal{O}_X) \neq 0$.

6.4.7. Corollary. *Assume that $\kappa(X) = 0$, $\chi(\mathcal{O}_X) \neq 0$, and that X is a projective threefold. Then: $|\pi_1(X)| \leq 4$ (thus $\pi_1(X)$ is either trivial, $\mathbb{Z}_2, \mathbb{Z}_3, \mathbb{Z}_4$ or $\mathbb{Z}_2 \times \mathbb{Z}_2$).*

Indeed, in this case $\kappa(X) = \kappa^+(X)$ by the solution of Mori's minimal model program for threefolds.

6.4.8. Remark. Observe that $\kappa^+(X) = 0$ is a natural generalization of the property: $h^0(X, \Omega_X^p) \leq \binom{n}{p}$ $(p \geq 0)$. This last property has been conjectured by K. Ueno when $\kappa(X) = 0$. It has only been proved for $p = 1$ by Y. Kawamata ([K2]). The conjecture 6.2.8(1) appears thus as a strengthening of K. Ueno's one.

We now study the non-projective case.

6.5. The non-projective case

We first start with a brief introduction to non-projective Kähler manifolds.

6.5.1. Let X again be an n-dimensional compact Kähler connected manifold, and $\beta : X \to A$ be any algebraic reduction of X (see [U], §3). We can always assume that β is analytic and A smooth projective; recall that β has connected fibers, that

$\dim A = a(X)$ is called the algebraic dimension of X, and that $\beta^* : \mathcal{M}(A) \to \mathcal{M}(X)$ induces an isomorphism between the field of global meromorphic functions.

In fact, A dominates every projective variety to which X meromorphically maps surjectively. The condition $a(X) = 0$ thus means that every global meromorphic function on X is constant. The "general" complex torus T of dimension $n \geq 2$ has $a(T) = 0$ (see [We]).

6.5.2. Curves. If $n = 1$, X is projective. Thus $\beta = \mathrm{id}_X$.

6.5.3. Surfaces. If $n = 2$ and X is not projective, there are two cases (K. Kodaira): $a(X) = 1$ and $\beta : X \to A$ is an elliptic fibration; or $a(X) = 0$ and X is either a torus or a K-surface.

6.5.4. Threefolds. If $n = 3$, and X is not projective, let X_a be the generic fiber of β. Then ([Fu2]) one of the following statements holds.

1. $a(X) = 2$ and X_a is an elliptic curve.
2. $a(X) = 1$ and X_a is either a torus, the almost homogeneous $\mathbb{P}(F_2)$, or a K3-surface S with $a(S) = 2$. In this last case, $X = S \times A/G$ for some finite group G. In the former, F_2 is the non-split extension $0 \to \mathcal{O}_C \to F_2 \to \mathcal{O}_C \to 0$ over some elliptic curve C (the proof can be considerably simplified and the result generalized by using [Ca6]).
3. $a(X) = 0$. Then X is either *Kummer*, a $\mathbb{P}^1$-fiber space over a Torus or K surface S with $a(S) = 0$, or *simple* and non-Kummer. Recall that X is said to be Kummer if it is bimeromorphic to T/G for some torus T and finite group G.

6.5.5. Definition. We say that X is *simple* if $n \geq 2$ and if it is not covered by its irreducible compact analytic subsets Y with $0 < \dim Y < n$.

6.5.6. Examples.

1. If X is simple, then $a(X) = 0$ and $q(X)$ is either 0 or n (because any surjective meromorphic map $f : X \to Y$ is either constant or generically finite)
2. The general torus T of dimension $n \geq 2$ is simple.
3. If a surface X is simple, it is either a torus, or a K3 surface.
4. The next "simple" example is a general deformation of $X_0 = \widetilde{S \times S}/\tilde{\sigma}$ where S is a K3 surface, $\widetilde{S \times S}$ the blow-up along the diagonal and $\tilde{\sigma}$ the lifting of the involution of $s \times S$ which exchange both factors. It is due to A. Fujiki, and has been extended to an arbitrary number of factors by A. Beauville.
5. No example of a simple non-Kummer threefold X is known. If some such X exists, it cannot be constructed by the standard means; for example, it is not a

deformation of some projective threefold (see [Ca3]). However, for such an X, we have $\pi_1(X) \cong 1, \mathbb{Z}_2$ or $\mathbb{Z}_3$ (see 6.5.8 below).

Our main application of 6.3 in this context is:

6.5.7. Corollary. *Assume that* $a(X) = 0$ *and* $\chi(\mathcal{O}_X) \neq 0$. *Then* $\pi_1(X)$ *is finite, and* $|\pi_1(X)| \leq 2^{n-1}/|\chi(\mathcal{O}_X)| \leq 2^{n-1}$.

Proof. Because $0 = a(X) \geq \kappa^+(X)$, we can apply 6.4.1.3. When $n = 3$, one gets a better result:

6.5.8. Corollary. *Let* $a(X) = q(X) = 0$ *and* $n = 3$. *Then* $\pi_1(X) \cong 1, \mathbb{Z}_2$ *or* $\mathbb{Z}_3$.

Proof. We first show that $\chi(\mathcal{O}_X) \geq 1$. Indeed, since X is not projective: $h^{2,0}(X) > 0$ (by a result of K. Kodaira).

Because $a(X) = 0$, $h^{3,0}(X) \leq 1$. Thus: $\chi(\mathcal{O}_X) = 1 - h^{1,0} + h^{2,0} - h^{3,0} \geq h^{2,0} \geq 1$. We apply now 6.4.1 and 5.3.2 to get: $|\pi_1(X)| \leq 2^{3-1} = 4$. To exclude the case $|\pi_1(X)| = 4$, let $c : \tilde{X} \to X$ be the universal cover of X. If $|\pi_1(X)| = 4$, we have: $\chi(\mathcal{O}_{\tilde{X}}) = 4$, and $\chi(\mathcal{O}_X) = 1$. Thus: $h^{1,0}(\tilde{X}) = 0$, $h^{3,0}(\tilde{X}) = 0 = h^{3,0}(X)$. Contradiction, since then $\chi(\mathcal{O}_X) = 1 + h^{2,0}(X) \geq 2$.

6.5.9. Example. Let S be a K surface with $\mathrm{Pic}(S) = H^1(S, \mathcal{O}_S^*) = 1$. Thus: $a(S) = 0$. Then: $X = \mathbb{P}(T_S)$ has: $a(X) = q(X) = 0$ and $n = 3$. We also have: $\pi_1(X) = 1$, in accordance with 6.5.8.

6.5.10. Remarks.

1. The natural generalization of 6.5.8 is the following: let $\beta : \bar{X} \to \bar{A}$ be the algebraic reduction of $\bar{X}$, some suitable finite étale cover of X. The induced map $\beta_* : \pi_1(\bar{X}) \to \pi_1(\bar{A})$ is surjective, since β has connected fibers. The assertion should be: β_* is bijective. The proof is as above, using the strengthened version 6.3.2(4) of 6.3.1. The difficulty is that $(\Phi_X)_* : \pi_1(\bar{X}) \to \pi_1(\bar{Y})$ has not finite kernel, in general. Other arguments however show that this is plausible.
2. The proof of the above statement only depends on whether $\ker\Phi_*$ is finite, where $\Phi_* : \pi_1(\bar{X}) \to \pi_1(\bar{Y})$ is induced from the Γ-reduction $\Phi : \bar{X} \to \bar{Y}$ of $\bar{X}$. This statement is thus true if $\pi_1(X)$ is either residually finite or has a torsion free subgroup of finite index.
3. If $a(X) = 1$ and $\chi(\mathcal{O}_X) \neq 0$, it is true that $\beta_* : \pi_1(\bar{X}) \to \pi_1(\bar{A})$ is an isomorphism. This extend 6.5.7 to this case. However the degree of $\bar{X} \to X$ is not uniformly bounded in that case.
4. If the above statement is true, $\Gamma = \pi_1(X)$ is an extension

$$1 \to F \to \pi_1(X) = \Gamma \to \pi_1(\bar{A}) = \Delta \to 1$$

with F a finite group. This raises the question: for such an extension (F finite, Δ "projective" and Γ Kähler) is it true that Γ itself is "projective"? (i.e. of the form $\pi_1(B)$ for B smooth projective). The general form of the above statement is:

6.5.11. Conjecture. Let $\bar{\rho} : \bar{X} \to \bar{A}$ be the algebraic reduction of some suitable finite étale cover $\bar{X}$ of X (we don't assume $\chi(\mathcal{O}_X) \neq 0$); let $q = q(\bar{X}) - q(\bar{A})$. Then: Ker $\bar{\rho}_*$ is a (non-canonically) split exact sequence: $1 \to \mathbb{Z}^{\oplus 2a} \to \pi_1(\bar{X}) \to \pi_1(\bar{A}) \to 1$. In particular, if $a(X) = 0$, $\pi_1(X)$ has a finite index subgroup which is abelian.

7. L^2 Cohomology

7.1. Introduction

We shall here sketch a proof of 6.3.1 (in a more general form).

Let $\bar{c} : \bar{X} \to X$ be any connected Galois cover of group $\Delta = \pi_1(X)/N$ of the compact Kähler manifold X; let ω (resp. g) any Kähler form on X (resp. its associated metric); let also $\bar{\omega}$, $\bar{g}$ be their lift under $\bar{c}$ to $\bar{X}$. All metric notions on $\bar{X}$ (such as being L^2) refer to this Δ-invariant metric $\bar{g}$. Notice that $\bar{X}$ is complete, because X is compact.

The proof of 6.3.1 follows an approach initiated by M. Gromov ([Gr2]) and consists of two main parts:

First produce a non-zero L^2 holomorphic p-form $\bar{u}$ on $\bar{X}$ for some $p > 0$. This rests on analytic methods (Atiyah's L^2-index theorem and Hodge theory), which we briefly explain in 7.3 and 7.4.

Next, construct Δ-invariant meromorphic functions on $\bar{X}$ by Δ-summation of $\bar{u}$. This part is complex-geometric, and explained in 7.2.

7.2. Summation of L^2 holomorphic forms on $\bar{X}$

Let the situation be that described in 7.1 (but X does not need to be Kähler here).

7.2.1. Notations. Let $\Phi_\Delta : X \to Y_\Delta$ be the Γ-reduction of (X, Δ) as in 5.1.10.

Let $\mathcal{F} \subset \Omega^p_X$ be a coherent subsheaf of rank r, $\mathrm{Det}(\mathcal{F}) \subset \bigwedge^r \Omega^p_X$ be the saturation of $\mathrm{Det}(\mathcal{F})$.

Assume $\kappa(X, \mathrm{Det}(\mathcal{F})) \geq 0$ and let $\Psi_\mathcal{F} : X \to Z_\mathcal{F}$ be the Iitaka reduction of X relative to $\mathrm{Det}(\mathcal{F})$ (see [U], §5), given by $h^0(X, (\mathrm{Det}(\mathcal{F}))^{\otimes m})$ for m sufficiently large and divisible.

We say that Φ_Δ factorizes through $\Psi_\mathcal{F}$ if there exists some meromorphic map $\eta : Z_\mathcal{F} \to Y_\Delta$ such that $\eta \circ \Psi_\mathcal{F} = \Phi_\Delta$. (This implies that $\kappa(X, \mathrm{Det}(\mathcal{F})) \geq 0$).

7.2.2. Theorem ([Ca4]). *Let $\bar{u}$ be a non-zero holomorphic p-form on X which is L^2 for some $p \geq 0$. There then exists a coherent subsheaf $\mathcal{F}$ of Ω^p_X, $p > 0$ such that $\Phi_\Delta : X \to Y_\Delta$ factorizes through $\Psi_{\mathcal{F}}$.*

In particular: $\gamma d(X, \Delta) \leq \kappa(X, \mathrm{Det}(\mathcal{F})) \leq \kappa^+(X)$. (Notice that $\kappa^+(X) \geq 0$ and that X does not need to be Kähler).

The proof consists of the following lemmas:

7.2.3. Lemma. *Let $\bar{c} : \bar{X} \to X$ be as above; let L be a hermitian line bundle on X, and let $\bar{L}$ be its hermitian lifting on $\bar{X}$. Let $\bar{s} \in H^0(\bar{X}, \bar{L})$ be a holomorphic L^2 section. Then:*

1. *$\bar{s}$ tends to zero at infinity (i.e. for any ε, a compact subset $\bar{K}(\varepsilon)$ of $\bar{X}$ does exist such that $\| \bar{s}(\bar{x}) \| < \varepsilon$ for $\bar{x} \in \bar{X}$ outside $\bar{K}(\varepsilon)$);*
2. *$\bar{s}^{\otimes q}$ is L^1 for any $q \geq 2$;*
3. *the series $\Delta(\bar{s}^{\otimes q}) := \sum_{\delta \in \Delta} \delta^*(\bar{s}^{\otimes q})$ converges to some Δ-invariant holomorphic section of $\bar{L}$ for any $q \geq 2$.*

The assertions (2) and (3) follow easily from (1), which itself follows from the compactness of X, and the standard inequality $\int_{B(r)} |f(z)|^2 dv(z) \geq C(r)$ for any holomorphic function defined on the ball $B(r)$ of radius r centered at 0 in $\mathbb{C}^n$, with its standard volume, if $|f(0)| \geq 1$.

7.2.4. Lemma (Gromov). *Assume that s is L^1 in 7.2.3, and that $\bar{A} \subset \bar{X}$ is some irreducible analytic subset stabilized by some infinite subgroup Δ_0 of Δ (so that $\bar{A}$ is not compact). There exists $q > 0$ such that the Δ-invariant meromorphic function $\Delta(\bar{s}^{\otimes q})/(\Delta(\bar{s}))^{\otimes q}$ is not constant on $\bar{A}$ if $\bar{s}$ does not vanish identically on $\bar{A}$.*

Proof. Let first $(a_\delta)_{\delta \in \Delta}$ and $(b_\delta)_{\delta \in \Delta}$ be two countable L_1 sets of complex numbers indexed by Δ.

Notice that if $A^q (:= \sum_\Delta a^q_\delta) = B^q (:= \sum_\Delta b^q_\delta)$ for every integer q, then $a_\delta = b_\delta$ up to ordering. Thus: $\Delta(\bar{s}^{\otimes q_0})$ is not identically zero on $\bar{A} \subset \bar{X}$, by hypothesis, for some q_0. We assume $q_0 = 1$ by replacing $\bar{s}$ by $\bar{s}^{\otimes q_0}$ if necessary. Let $\mu : \bar{X} \to \mathbb{C}$ be defined by $\mu = \bar{s}/\Delta(\bar{s})$. Thus: $\Delta(\bar{s}^{\otimes q}) = \Delta(\mu^q) \cdot (\Delta(\bar{s}))^{\otimes q}$ if $q > 0$.

If $\Delta(\mu^q)$ were constant on $\bar{A}$ for each $q > 0$, μ itself were constant too, by the remark above. This contradicts the fact that $\bar{s}$ is nonzero on $\bar{A}$, but goes to zero at infinity.

7.2.5. We now conclude the proof of 7.2.2: let $\bar{u} \in H^0(\bar{X}, \Omega^p_{\bar{X}})$ be as in 7.2.2. By 7.2.3 and the maximum modulus principle, $p > 0$.

Let $\bar{\mathcal{F}} \subset \Omega^p_{\bar{X}}$ be the coherent subsheaf generated by all $\delta^*\bar{u}$, $\delta \in \Delta$. Let r be its rank, observe that $\bar{\mathcal{F}}$ is Δ-invariant, and thus: $\bar{\mathcal{F}} = \bar{c}^*\mathcal{F}$ for some coherent $\mathcal{F} \subset \Omega^p_X$.

Let $\delta_1, \dots, \delta_r$ be in Δ such that the $\bar{u}_i := \delta_i^*(\bar{u})$ generate $\bar{\mathcal{F}}$ at some point $\bar{x} \in \bar{X}$ where $\bar{\mathcal{F}}$ is locally free. Let $\bar{s} := \bar{u}_1 \wedge \dots \wedge \bar{u}_r$: it is a nonzero L^2 section of $\det(\bar{\mathcal{F}})$.

Modifying X, we can assume that $det\bar{\mathcal{F}} = \bar{L} = \bar{c}^*(\det(\mathcal{F})) = \bar{c}^*(L)$ is locally free, and apply 7.2.4 to $(\bar{s}, \bar{L})$: the irreducible components of the level sets of $\Delta(\mu^q)$, which is Δ-invariant, are compact.

They are thus contained in fibers of $\bar{\Phi}_\Delta : \bar{X} \to \bar{Y}_\Delta$, the Γ-reduction of (X, Δ). On the other hand, $\Delta(\mu^q)$ is also the quotient of two Δ-invariant sections of $\bar{L}^{\otimes q}$, namely $\Delta(\bar{s}^{\otimes q})$ and $\Delta(\bar{s})^{\otimes q}$.

It is thus induced from some meromorphic function F on $Z_{\mathcal{F}}$, if q is sufficiently large. This establishes the claim, since the generic level sets of $F_0\Psi_{\mathcal{F}}$ is an irreducible union of fibers of $\Psi_{\mathcal{F}}$, and contained in some fiber of Φ_Δ.

7.2.6. Remark. One can restate 6.3.1 as follows: If there exists some non-zero holomorphic p-form on $\tilde{X}$, then $\kappa^+(X) \geq \gamma d(X)$.

7.3. Hodge theory on complete Kähler manifolds

The situation $\bar{c} : \bar{X} \to X$ is that described in 7.1. The Hodge $*$-operator refers to the metric $\bar{g}$; also $\mathrm{d}^* = *\mathrm{d}*$.

We put $\Delta_\mathrm{d} := \mathrm{dd}^* + \mathrm{d}^*\mathrm{d}$, and similarly for $\Delta_{\bar{\partial}}$, Δ_∂. When just Δ is written, the statement applies to any of these three. *We don't assume X to be Kähler, unless explicitly mentioned.*

In 7.3 and 7.4 we shall explain how a non-zero holomorphic p-form $\bar{u}$ for some $p > 0$ is produced from the hypothesis $\chi(\mathcal{O}_X) \neq 0$.

7.3.1. Proposition. *Let u be an L^2 harmonic $(0, p)$-form for the Δ-operator. Then:*

1. $du = \mathrm{d}^*u = 0$ *(if $\Delta = \Delta_\mathrm{d}$; similarly for ∂, $\bar{\partial}$).*
2. *The conjugate $\bar{u}$ of u is holomorphic if $\Delta = \Delta_{\bar{\partial}}$ and if X is Kähler.*
3. *$*u$ is an L^2 holomorphic 2-form on X if X is a surface and $p = 2$ (we don't assume X to be Kähler here), and if $\Delta = \Delta_{\bar{\partial}}$*

The assertion (1), (which we don't use here) is nothing but Lemma 1.1.B of [Gr2] to which we refer for a detailed (and ingenious) proof. What has to be checked is the equality $\langle \Delta u, u\rangle = \langle \mathrm{d}u, \mathrm{d}u\rangle + \langle \mathrm{d}^*u, \mathrm{d}^*u\rangle$. The trouble is that, because $\bar{X}$ is not necessarily compact, integration by parts cannot be directly applied in the form: $\mathrm{d}u \wedge *v - u \wedge *\mathrm{d}v = \pm \mathrm{d}(u \wedge *v)$ hence $\langle \mathrm{d}u, v\rangle = \langle u, \mathrm{d}v\rangle$, since $\int_{\bar{X}} \mathrm{d}(u \wedge *v)$ needs not be zero a priori. One uses instead what is called the Gaffney's cut-off trick in [Gr2], using the fact that $\bar{X}$ is complete to construct cut-off functions whose support exhaust $\bar{X}$.

The assertion (2) follows directly from the standard Kähler identity $2\Delta_{\bar\partial} = 2\Delta_\partial$ when X is Kähler. This shows that: u is not only $\Delta_{\bar\partial}$, but also Δ_∂-harmonic. Thus: $\partial\bar{u} = \overline{\bar\partial u} = 0$ since u is $\Delta_{\bar\partial}$-harmonic. In particular: $\bar\partial * u = 0$.

7.3.2. Remark. Other properties of Hodge L^2-theory are shown to extend to the complete case in [Gr2], such as Hodge decomposition and Lefschetz theorem.

We have now seen from 7.3.1 how to produce some non-zero L^2 holomorphic p-form on $\bar{X}$ from a $\Delta_{\bar\partial}$-harmonic $(0, p)$-form. It remains to produce one if $\chi(\mathcal{O}_X) \neq 0$. This is explained in 7.4. Before, we give an application:

7.3.3. Example ([Ca3]). Let S be a smooth compact connected surface such that: $K_S^{\otimes m} \cong \mathcal{O}_S$ for some $m > 0$, and $\chi(\mathcal{O}_S) \neq 0$. Then $\pi_1(S) = 1$ or $\mathbb{Z}_2$, and S is either a K3 surface ($K_S = \mathcal{O}_S, q(S) = 0$) or an Enriques surface $K_S^{\otimes 2} = \mathcal{O}_S$ ($K_S \neq \mathcal{O}_S, q(S) = 0$). In particular, a K3 surface is simply connected (since $\chi(\mathcal{O}_S) = 2$ and $\pi_1(S) = \mathbb{Z}_2$ would imply $h^0(\tilde{S}, K_{\tilde{S}}) = 3$ for the universal cover $\tilde{S}$ of S).

Proof. One reduces immediately to the case $m = 1$ by a suitable cover. First, if $K_S \cong \mathcal{O}_S$, then $\chi(\mathcal{O}_S) = 2 - q > 0$. Indeed, otherwise $q \geq 2$ and there exist 2 independent holomorphic 1-forms u_1 and u_2 which are linearly independent. If $u_1 \wedge u_2 \neq 0$, S is a torus and $\chi(\mathcal{O}_S) = 0$. Contradiction. Hence $u_1 \wedge u_2 = 0$; u_1 and u_2 are d-closed (since $\dim S = 2$). By Castelnuovo–De Franchi's Lemma, there exist a holomorphic surjective map $\Psi : X \to C$ where C is a curve of genus $g \geq 2$. Its fiber are elliptic, and Kodaira's formula for K_S shows that $\kappa(S) \geq 1$. Contradiction. (Notice this argument is not needed if one already knows that $\chi(\mathcal{O}_S) > 0$ which is the case if $q(S) = 0$). Thus: $q = 0$ or 1, and $\chi(\mathcal{O}_S) = 2$ or 1.

From 7.4 applied to the universal cover $\tilde{S}$ of S we deduce the existence of a non-zero $\Delta_{\bar\partial}$-harmonic L^2 $(0, p)$-form u, p even, since $\chi(\mathcal{O}_S) > 0$. Assume $\tilde{S}$ is not compact. We have $p = 0$, or 2. The first case is excluded by 7.2.3. Thus $p = 2$, and we apply 7.3.1(3): $\bar{u}$ is a holomorphic L^2 2-form on $\tilde{S}$. Contradiction again because $K_S \cong \mathcal{O}_S$. Thus: $\tilde{S}$ is compact and $\pi_1(S)$ is finite. The rest is obvious.

7.3.4. Remark. Most of the argument above is in fact devoted to prove that: $K_S^{\otimes m} \cong \mathcal{O}_S (m > 0)$ and $\chi(\mathcal{O}_S) \neq 0$ imply: $q(S) = 0$.

7.4. Atiyah's L_2 index theorem

7.4.1. This theorem is the main ingredient to produce a non-zero L_2 harmonic $(0, p)$-form on $\bar{X}$, for some $p > 0$, from the information that $\chi(\mathcal{O}_X) \neq 0$.

The general idea is this: if $\bar{c} : \bar{X} \to X$ is a finite Galois cover of group Γ and degree $d = |\Gamma|$, then:

$$\chi(\mathcal{O}_X) = \frac{1}{d}\chi(\mathcal{O}_{\tilde{X}}) = \frac{1}{|\Gamma|}\chi(\mathcal{O}_{\tilde{X}}).$$

This equality still holds when Γ is infinite, provided $\frac{1}{|\Gamma|}\chi(\mathcal{O}_{\tilde{X}})$ is now interpreted as the *normalized* index of $(\bar{\partial} + \bar{\partial}^*)$ acting on L_2 forms of type $(0, *)$ on $\tilde{X}$.

We first introduce the notion of normalized dimension needed to define this normalized index.

7.4.2. Von Neumann (or Γ-) dimension ([Pa2]). Let Γ be a countable group acting by unitary linear bounded transformations on the (separable) Hilbert space $\mathcal{X}$.

Let $H \subset l^2(\Gamma)\hat{\otimes}_{\mathbb{C}}\mathcal{X} := V$ be a Γ-invariant subspace, for the natural action.

One can then define a notion of Γ-dimension of H, denoted by $\dim_\Gamma H$, such that the following properties D0 to D6 hold:

D0. $0 \leq \dim_\Gamma \leq +\infty$ is a real number or $+\infty$.

D1. $\dim_\Gamma H = 0$ iff $H = 0$.

D2. If H is isomorphic to a dense subspace of H', then $\dim_\Gamma H = \dim_\Gamma H'$.

D3. If $H = H_1 \oplus H_2$, the sum being orthogonal, then: $\dim_\Gamma H = \dim_\Gamma H_1 + \dim_\Gamma H_2$.

D4. $\dim_\Gamma(\bigcap_{n\geq 0} H_n) = \lim_{n\to\infty} \dim_\Gamma(H_n)$ if the sequence of subspaces H_n is decreasing.

D5. $\dim_{\Gamma'}(H) = d \cdot \dim_\Gamma(H)$ if $\Gamma' \subset \Gamma$ is of finite index d.

D6. $\dim_\Gamma(l^2(\Gamma)) = 1$.

7.4.3. Examples ([Pa2]).

1. Let $\Gamma = \mathbb{Z}$ act on $L^2(S^1)$ by: $n \to (f(z) \to z^n f(z))$, where $S^1 = \{z \in \mathbb{C} : |z| = 1\}$. (Note that $L^2(S^1) = l^2(\mathbb{Z})$ by Fourier transform). Let $A \subset S^1$ be measurable, and H_A be the subspace of $f \in L^2(S^1)$ consisting of functions vanishing outside A: it is Γ-invariant. From properties above, one gets easily: $\dim_\Gamma(H_A) = \mu(A)$, the Lebesgue measure of A. (Use D5 to show the invariance of the probability measure $A \to \dim_\Gamma(H_A)$ under translations of order $p \in \mathbb{Z}$).
2. If Γ and $\dim H$ are finite, then $\dim_\Gamma H = \frac{1}{|\Gamma|}\dim H$.

Let us give the definition of $\dim_\Gamma$:

Let $P : V \to V$ be a Γ-invariant linear operator, and $\epsilon = (e_i)_{i\in I}$ be an orthonormal basis of $\mathcal{X}$. Then: $\mathrm{Trace}_\Gamma P := \sum_i \langle Pe_i, e_i\rangle$. This does not depend on ϵ if P is hermitian positive (i.e. $\langle P(v), v\rangle \geq 0$ for v in V)

If $H \subset V$ is Γ-invariant, then: $\dim_\Gamma H := \mathrm{Trace}_\Gamma(\pi_H)$, where π_H is the orthogonal projection to H.

7.4.4. Let $\bar{c} : \bar{M} \to M$ be a connected Galois étale cover of group Γ of a compact riemannian manifold M. Let E, F two complex hermitian vector bundles on M, and $D : C^\infty(M, E) \to C^\infty(M, F)$ be a C^∞ *elliptic* differential operator on M and D^* its adjoint. We denote with $\bar{E}, \bar{F}, \bar{D}, \bar{D}^*, \bar{g}$ the lifts of E, F, D, D^*, g (the metric) to $\bar{M}$ by $\bar{c}$.

Let $\mathrm{Ker}_{(2)} \bar{D} \subset L_2(\bar{M}, \bar{E})$ be the space of L_2 solutions to $\bar{D}.s = 0$: it is a Γ-invariant subspace of $V = l^2(\Gamma)\hat{\otimes} L_2(\bar{U}, \bar{E})$ if $\bar{U} \subset \bar{M}$ is an open fundamental domain for Γ, the operation in $\mathcal{X} := L_2(\bar{U}, \bar{E})$ being trivial.

7.4.5. Theorem ([At]). *Let the situation be that of* 7.4.4.

Then: $\delta := \dim_\Gamma(\mathrm{Ker}_{(2)}(\bar{D}))$ *and* $\delta^* := \dim_\Gamma(\mathrm{Ker}_{(2)}(\bar{D}^*))$ *are finite and* $\mathrm{index}_\Gamma(D) := \delta - \delta^* = \mathrm{index}(D)$.

7.4.6. Example. Let $\bar{c} : \bar{X} \to X$ be as above, but with X a compact complex connected manifold. Take:

$$E = \bigoplus_{p \text{ even}} (T_X^{0,p})^*, \quad F = \bigoplus_{p \text{ odd}} (T_X^{0,p})^*$$

$D = \bar{\partial} + \bar{\partial}^*$ so that $D^* = \bar{\partial} + \bar{\partial}^*$, too. We then get: $\mathrm{index}(D) = \chi(\mathcal{O}_X)$, and thus:

7.4.7. Corollary. *In* 7.4.6, *assume that* $\chi(\mathcal{O}_X) \neq 0$. *Then* $\delta - \delta^* = \chi(\mathcal{O}_X) \neq 0$. *In particular, for some* $p \geq 0$, *there exists an* $\bar{L}_2$ *form of type* $(0, p)$ *on* $\bar{X}$ *which is both* $\bar{\partial}$ *and* $\bar{\partial}^*$*-closed.*

7.4.8. Idea of proof. Let us first consider, as in [At], the case where $\bar{c} : \bar{X} \to X$ has finite degree $\mathrm{d} = |\Gamma|$.

Let H_0 (resp. H_1) be the orthogonal projection to Ker D (resp. Ker D^*), and let G be the Green operator for D. Then: $\mathrm{index}(D) = \mathrm{Trace}(H_0) - \mathrm{Trace}(H_1)$, and:

$$GD = 1 - H_1; \quad DG = 1 - H_0.$$

More generally, if Q is any parametrix for D (see [Gi]), then: $GD = 1 - S_1$ and $DG = 1 - S_0$ for some operator S_i with smooth Kernels (also denoted S_i), thus of trace class ([Ro], 6.12).

We now have:

$(1)_X$ $\mathrm{index}(D) = \mathrm{Trace}(S_0) - \mathrm{Trace}(S_1)$.

This formally follows from: $\mathrm{Trace}(AB) = \mathrm{Trace}(BA)$ for A of trace class and B bounded.

$(2)_X$ $\text{Trace}(S_i) = \int_X \text{Trace}(S_i(x,x))dx$ ([Ro], 6.12).

Let now $\bar{S}_i$, $\bar{Q}$ be the lifts of S_i, Q to $\bar{X}$: we still have properties as above. Thus:

$$\text{index}(D) = \int_X (\text{Trace } S_0 - \text{Trace } S_1)(x,x)dx$$

$$= \frac{1}{d}\int_{\bar{X}} (\text{Trace}(\bar{S}_0) - \text{Trace}(\bar{S}_1))(\bar{x},\bar{x})d\bar{x} = \frac{1}{d}\,\text{index}(\bar{D}) = \text{index}_\Gamma(\bar{D}).$$

When Γ is infinite, the proof is similar, except that one has to replace $(i)_X$ by $(i')_{\bar{X}}$ below $i = 1, 2$:

$(1')_{\bar{X}}$ $\text{index}(\bar{D}) = \text{Trace}_\Gamma(\bar{S}_0) - \text{Trace}_\Gamma(\bar{S}_1)$.

$(2')_{\bar{X}}$ $\text{Trace}_\Gamma(\bar{S}_i) = \int_X \text{Trace}(S_i(x,x))dx (= \text{Trace}(S_i))$.

More precisely: M. Atiyah introduces the algebra $\mathcal{A}$ of all the bounded operators on $L^2(\bar{X}, \bar{E})$ which commute with Γ, it can be identified with

$$l^2(\Gamma)\hat{\otimes}L^2(\bar{U}, \text{Hom}(\bar{E}, \bar{E}))$$

for $\bar{U}$ a fundamental domain for Γ.

Now any A in $\mathcal{A}$ has a Schwartz kernel $\bar{A}$ on $\bar{X} \times \bar{X}$ and $\bar{A}$ is Γ-invariant, thus lifted from some $\bar{A}$ on $(\bar{X} \times \bar{X}/\Gamma)$.

Observe that for either projection $(\bar{X} \times \bar{X}/\Gamma)$ is a fiber bundle over $X = \bar{X}/\Gamma$ with fiber $\bar{X}$ as fiber, which has a smooth section σ with image $(\Delta_{\bar{X}}/\Gamma) \cong X$, with $\Delta_{\bar{X}}$ the diagonal of $\bar{X} \times \bar{X}$. We identify the restriction of $\bar{A}$ to $\Delta_{\bar{X}}/\Gamma$ with a section of $\text{Hom}(E, E)$ on X in that way.

One has now results ([At], §4) which are similar to those in the compact case:

7.4.9. [At], 4.16. $A \in \mathcal{A}$ is Γ-Hilbert–Schmidt (resp. of Γ-trace class) provided $\bar{A}$ is L^2 (resp. smooth and either positive self-adjoint, or compactly supported on $(\bar{X} \times \bar{X}/\Gamma)$). If these last cases, one has:

$$\text{Trace}_\Gamma(A) = \int_X \text{Trace } \bar{A}(x)dx$$

7.4.10. If $A, S \in \mathcal{A}$ with S of Γ-trace class, then $\text{Trace}_\Gamma(AS) = \text{Trace}_\Gamma(SA)$ ([At], 4.13).

The proof now goes just as in the compact case; the only essential point is to choose S_i $(i = 1, 2)$ with kernel supported so close to the diagonal of X that their lifts $\bar{S}_i$ to $\bar{X}$ have kernels (seen on $\bar{X} \times \bar{X}/\Gamma$) *compactly supported*, which is possible because of the existence of the section σ above.

One additional technical difficulty is that $\bar{G}$ is no longer bounded, which requires an approximation argument to be able to use the commutation formula 7.4.10 above. See ([At], 5.5-5.13) for it.

8. Problems

Let X be a compact Kähler manifold, $\bar{X}$ some suitable finite étale cover of X, and $\Gamma = \pi_1(X)$. Let $n = \dim X$.

8.1. Nilpotent completions of Kähler groups

8.1.1. Assume $\pi_1(X)^{\text{nilp}}$ is not nilpotent (i.e. $\Gamma'^n \neq (1)$ for all $n \geq 1$).

a) Is then $\pi_1(X)^{\text{nilp}}$ of exponential growth?

b) Does $\pi_1(X)^{\text{nilp}}$ contain a free group of 2 generators?

c) Does there exist a holomorphic surjective map $\bar{\phi} : \bar{X} \to C$, where C is a curve of genus $g \geq 2$? (This seems unlikely in all generality. A counterexample would be interesting.)

8.1.2. Assume X is a smooth submanifold of a complex torus. Is $\pi_1(X)'^{\text{nilp}} \cong \pi_1(X)$? Otherwise, find a counterexample. What happens if X is only a smooth model of its Albanese image $\alpha_X(X) \subset \text{Alb}(X)$?

8.1.3. Find examples of X such that $K = \text{Ker}(\bigwedge^2 H^1(X, \mathbb{C}) \to H^2(X, \mathbb{C}))$ is contained in (resp. meets non-trivially) $H^{2,0}(X) \oplus H^{0,2}(X)$. (We look only for non-trivial examples, that is: such that no finite étale cover $\bar{X}$ of X admits a surjective holomorphic map to some curve C of genus at least 2).

Note that, after [Ca4], $K = K^{2,0} \oplus K^{1,1} \oplus K^{0,2}$ if $K^{i,j} = K \cap H^{i,j}(X)$. (Examples where $\pi_1(X)$ is nilpotent are more interesting).

8.1.4. Find examples of Γ which are Kähler, torsion free and nilpotent of class 3 or more.

8.2. Kodaira dimension and γ-dimension

The comparison theorem states that $\kappa^+(X) \geq \gamma d(X)$ if $\chi(\mathcal{O}_X) \neq 0$ (and conjecturally: $\kappa(X) \geq \gamma d(X)$) provided 6.2.8(1) is true. This inequality fails "maximally" to hold when $\gamma d(X) = n$ and $\kappa^+(X) = 0$. This is the case when X is covered by a torus. Conversely, denoting by $\bar{X} \to X$ some *suitable* finite étale cover of a given X :

8.2.1. Conjecture. Assume $\kappa(X) = 0$ and $\gamma d(X) = n$. Then $\bar{X}$ is bimeromorphic to a torus. A variant is: Assume $\kappa(X) = 0$ and $\tilde{X}$, the universal cover of X, is Stein. Then $\bar{X}$ is a torus. Observe that this conjecture is true if $\pi_1(X)$ is almost abelian when $\kappa(X) = 0$.

8.2.2. Conjecture. Assume that $\kappa(X) = 0$. Let $\bar{q}(X) = \sup\{q(\bar{X})$ if $\bar{X} \to X$ is finite étale $\}$. Then: $0 \leq \bar{q}(X) \leq \gamma d(X) \leq n$. Then: $\bar{q}(X) = \gamma d(X)$. More precisely: $\pi_1(\bar{X}) \cong \pi_1(\mathrm{Alb}(\bar{X}))$ for some $\bar{X}$.

8.2.3. Conjecture. A relative version of 8.2.2 is: assume that $\kappa(X) \geq 0$ and let $\Psi : \bar{X} \to I(\bar{X})$ be the Iitaka reduction of $\bar{X}$. Let $q := q(\bar{X}) - q(I(\bar{X}))$. There is a non-canonically split exact sequence:

$$1 \to \mathbb{Z}^{\oplus 2q} \to \pi_1(\bar{X}) \to \pi_1(I(\bar{X})) \to 1.$$

8.2.4. Remark. There exist compact Kähler manifolds X with $\chi(\mathcal{O}_X) = 0$, but such that all of their étale covers possess non-zero L_2 holomorphic p-forms for some $p > 0$. For example: Calabi–Yau manifolds of odd dimension.

Another interesting problem is the study of the representation of $\pi_1(X)$ on the space of holomorphic p-forms which are L_2 for some $p > 0$ on the universal cover $\tilde{X}$ of X.

Observe indeed that this representation is unitary and *almost faithful* (its kernel is finite), which follows easily from §7 above.

References

[A1] Arapura D., Fundamental groups of smooth projective varieties, Preprint 1993.

[A2] Arapura D., Higgs line bundles, Bull. Amer. Math. Soc. 26 (1992), 310–314.

[A-B-R] Arapura D., Bressler P., Ramachandran M., On the fundamental group of compact Kähler manifolds, Duke Math. J. 68 (1992), 477–488.

[At] Atiyah M., Elliptic operators, discrete groups and Von Neumann Algebras, Astérisque 32-33 (1976), 43–72.

[Ba] Baumslag G., On generalized free products, Math. Z. 78 (1962), 423–438.

[Be1] Beauville A., Surfaces algébriques complexes, Astérisque 54 (1978).

[Be2] Beauville A., Annulation du H^1 pour les fibrés en droites plats, Proc. Bayreuth, Conf. on Alg. Geom. (1950), Lecture Notes in Math. 1507 (1991).

[Bis] Bishop E., Conditions for the analyticity of certain sets, Michigan Math. J. 11 (1964), 289-304.

[Bo1] Borel A., Compact Clifford–Klein forms of symmetric spaces, Topology 2 (1963), 111–122.

[Bo2] Borel A., Introduction aux groupes arithmétiques, Hermann, Paris 1969.

[Ca1] Campana F., On twistor spaces of class C, J. Differential Geom. 33 (1991), 541–549.

[Ca2] Campana F., Remarques sur les groupes de Kähler nilpotents, Ann. Sci. École Norm. Sup. 28 (1995), 307–316, annoncé C.R.A.S. 317 (1993), 777–780.

[Ca3] Campana F., Remarques sur les revétement universel des varietés Kählériennes compacts, Bull. Soc. Math. France. 122 (1994), 255–284 .

[Ca4] Campana F., Fundamental group and positivity of cotangent bundles, J. Algebraic Geom. 4 (1995), 487–505.

[Ca5] Campana F., Connexité rationelle des varietés de Fano, Ann. Sci. École Norm. Sup. 25 (1992), 539–545.

[Ca6] Campana F., Reduction d'Albanese d'un morphisme propre et faiblement Kählérien, Comp. Math. 54 (1985), 378–416.

[C-T1] Carlson J., Toledo D., Notes on nilpotent Kähler groups, unpublished.

[C-T2] Carlson J., Toledo D., Quadratic presentation and Kähler groups, to appear in J. Geom. Anal.

[Ch] Chau T. C., A note concerning Fox's paper on Fenshel's conjecture, Proc. Amer. Math. Soc. 88 (1983), 584–586.

[Del] Deligne P., Théorie de Hodge III, Publ. Math. IHES 44 (1974), 5–77.

[D-G-M-S] Deligne P., Griffiths P., Morgan J., Sullivan D., Real homotopy theory of Kähler manifolds, Invent. Math. 29 (1975), 245–274.

[D-P-S] Demailly J. P., Peternell T., Schneider M., Kähler manifolds with numerically effective Ricci class, Comp. Math. 89 (1993), 217–240.

[Do] Dodziuk J., De Rham–Hodge theory for L^2-cohomology of infinite coverings, Topology 16 (1977), 157–165.

[Fo] Fox R. N., On Fenchel's conjecture about F-groups, Math. Tidsskr. B (1952), 61–65.

[Fu1] Fujiki A., On automorphism groups of compact Kähler manifolds. Invent. Math. 44 (1978), 225–258.

[Fu2] Fujiki A., Semi-simple reductions of compact complex varieties, Publ. Elie Cartan (Nancy) 8 (1983), 79–133.

[F-L] Fulton W., Lazarsfeld R., Connectivity and its applications in Algebraic Geometry, Lecture Notes in Math. 869, 26–92.

[Ga] Gauduchon P., Variétés Riemanniennes autoduales, Sem. Bourbaki. Exp. 767, Mars 1993.

[Go-Ma] Goresky M., MacPherson R., Stratified Morse Theory, Springer 1988.

[Go-Mi] Goldman W., Milsson J., Deformation theory of representations of fundamental groups of compact Kähler manifolds, Publ. Math. IHES 67 (1988), 43–96.

[G-M] Griffiths P., Morgan J., Rational homotopy theory, Birkhäuser (1981).

[Gr-L] Green M., Lazarsfeld R., Deformation theory, generic vanishing theorem, and some conjectures of Enriques, Catanese and Beauville, Invent. Math. 90 (1987), 389–407.

[Gr1] Gromov M., Groups of polynomial growth and expanding maps, Publ. Math. IHES 53 (1981), 53–73.

[Gr2] Gromov M., Kähler hyperbolicity and L_2-theory, J. Differential Geom. 33 (1991), 263–292.

[Gr3] Gromov M., Groupe fondamental des variétés Kählériennes, CRAS 308 (1989), 67–70.

[Gu] Gurjar R., Coverings of Algebraic Varieties, Adv. Stud. Pure Math. 10 (T. Oda, ed.), Kinokuniya 1987, 179–184.

[G-S] Gurjar R., Shastri, Covering space of an elliptic surface, Comp. Math. 54 (1985),95–104.

[G-S] Gromov M., Schoen R., Harmonic maps into singular spaces and p-adic superrigidity for lattices in groups of rank one, Publ. IHES 76 (1992), 165–246.

[He] Helgason S., Differential Geometry, Lie Groups and Symmetric Spaces, Academic Press 1978.

[Hirs] Hirsch K. A., Infinite soluble groups II, Proc. Lond. Math. Soc. 44 (1938), 336–345.

[Hir] Hirzebruch F., Arrangements of lines and algebraic surfaces, Progr. Math. 36 (1983) Birkhäuser, 113–140.

[Hit] Hitchin N. J., Kählerian twistor spaces, Proc. Lond. Math. Soc. 43 (1981), 133–150.

[J-R] Johnson F., Rees E., On the fundamental group of a compact complex algebraic manifold, Bull. London Math. Soc. 19 (1987), 463–466.

[Ka1] Kawamata Y., Characterization of abelian varieties, Comp. Math. 43 (1981), 253–276.

[Ka2] Kawamata Y., Abundance theorem for minimal threefolds, Invent. Math. 108 (1992), 229-246.

[Ka3] Kawamata Y., Matsuda K., Matsuki K., Introduction to the Minimal Model Program, Adv. Stud. Pure Math. 10., T. Oda, ed., Kinokuniya (1987), 283–360.

[Ko] Kollár J., Shafarevitch maps and plurigenera of Algebraic Varieties, Invent. Math. 113 (1993), 177–215.

[Ko-Mi-Mo1] Kollár J., Miyaoka Y., Mori S., Rationally Connected Varieties, J. Alg. Geometry 1 (1992), 429–448.

[Ko-Mi-Mo2] Kollár J., Miyaoka Y., Mori S., Rational connectedness and boundedness for Fano Manifolds, J. Differential Geom. 36 (1992), 765–769.

[Li] Lieberman D., Compactness of the Chow Scheme, LN 670, Séminaire F. Norguet III (1977), 140–186.

[Lár] Lárusson F., An extension theorem for holomorphic functions, Preprint 1993

[Mal1] Malčev A., On the faithful representation of infinite groups by matrices, Amer. Math. Soc. Transl. 45 (1965), 1–18.

[Mal2] Malčev A., On a class of homogeneous spaces, Amer. Math. Soc. Transl. Ser. 1, vol. 9, 276–307.

[Ma] Margulis G., Discrete subgroubs of semi-simple Lie groups, Springer-Verlag 1990.

[Mil] Milnor J., Morse Theory, Ann. Math. Stud. 51, Princeton Univ. Press, 1963.

[Miy] Miyaoka Y., On the Kodaira dimension of minimal threefolds, Math. Ann. 281 (1988), 325–332.

[Mok] Mok N., Factorizations of semi-simple representations of Kähler groups, Invent. Math. 110 (1982), 557–614.

[Mori1] Mori S., Threefolds whose canonical bundles are not numerically effective, Ann. Math. 116 (1982), 133–176.

[Mori2] Mori S., Flip theorem and the existence of minimal models for 3-folds, J. Amer. Math. Soc. 1 (1988), 117–153.

[Mo] Mostow G., Existence of non-arithmetic monodromy groups, Proc. Nat. Acad. Sc. USA 78 (1981), 5948–5950.

[Mu] Mumford D., Abelian Varieties, Tata Inst. Lect. Notes, 1968.

[No] Nomizu K., On the cohomology of compact homogeneous spaces of nilpotent Lie groups, Ann. Math. 59 (1958), 531–538.

[Pa1] Pansu P., Sous-groupes discrets des groupes de Lie: rigidité, arithméticité, Seminaire Bourbaki, exp. 778 (1993–1994).

[Pa2] Pansu P., Introduction to L^2 numbers, preprint 1993.

[P] Peternell T., Minimal Varieties with trivial canonical class, Math. Z. 217 (1994), 377-407.

[Ra] Ragunathan M., Torsion in cocompact lattices in coverings of $Spin(2, n)$, Math. Ann. 266 (1984), 403–419.

[Ro] Roe, J., Elliptic operators, topology and asymptotic methods, Pitman Res. Notes Math. Ser. 179, Longman, New York 1988.

[S-W] Scott P., Wall T., Homological group theory, London Math. Series 36. S. Wall, ed., Cambridge University Press 1979, 137–204.

[Se1] Serre J. P., Collected Works (I), 527–530.

[Se2] Serre J. P., Arbres, Amalgames, SL_2, Asterisque 46 (1977).

[Se3] Serre J. P., Arithmetic groups, in: Homological group theory. (C. T. C. Wall, ed.), 105-136. London Math. Soc. Lecture Notes Ser. 36.

[Se4] Serre J. P., Groupes et Algébres de Lie. Benjamin 1965.

[Sha] Shafarevitch I. R., Basic Algebraic Geometry, Springer Verlag 1977.

[Si] Simpson C., Higgs bundles and local systems, Publ. Math. IHES. 75 (1992), 5–95.

[Siu] Siu Y. T., Complex analyticity of harmonic maps and the strong rigidity of compact Kähler manifolds, Ann. Math. 112 (1980), 73–111.

[St] Stallings J., Homology and central series of groups, J. Algebra 2 (1965), 170–181.

[S-V] Sommese A., Van de Ven A. Homotopy groups of pullbacks of varieties, Nagoya Math. J. 102 (1986), 79–90.

[Ta] Taubes C. H. The existence of anti-self dual conformal structures, J. Differential Geom. 36 (1992), 163–253.

[To1] Toledo D., Examples of fundamental groups of compact Kähler manifolds, Bull. Lond. Math. Soc. 22 (1990), 339–343.

[To2] Toledo D., Existence of non-residually finite Kähler groups, Publ. IHES 77 (1993), 103–119.

[U] Ueno K., Classification Theory of Algebraic Varieties and Compact Complex Spaces, Lecture Notes in Math. 439, Springer 1975.

[Y-Y] Yamazaki T., Yoshida M., On Hirzebruch examples of surfaces with ${c_1}^2 = 3c_2$, Math. Ann. 266 (1984), 421–431.

[We] Weil A., Varietés Kählériennes, Hermann, Paris 1958.

[Zuo] Zuo K., Factorizations of nonrigid Zariski dense representations of π_1 of projective algebraic manifolds, preprint 1992.

Fundamental groups with few relations

*Fabrizio Catanese**

1. Introduction

The purpose of this article is to point out a simple proof of the following two results: the first is due to Siu ([Siu]), was reproved by Beauville ([Be]), and is a particular case of the second one. The second one is due to Gromov ([Gro]) and was reproved by Green and Lazarsfeld ([G-L]).

Theorem 1 (Theorem of the irrational pencil (Siu, Beauville)). *Let X be a compact Kähler manifold, let Γ be the fundamental group $\pi_1(X)$, let Π_g be the fundamental group of a Riemann surface C of genus $g \geq 2$.*

If there exists a surjection $\Gamma \twoheadrightarrow \Pi_g$ then X admits an irrational pencil of genus $g' \geq g$, i.e., a surjective holomorphic map f with connected fibres $f : X \to C$ where C is a Riemann surface of genus g'.

Theorem 2 (Theorem of the few relations (Gromov, Green–Lazarsfeld)). *Let X be a compact Kähler manifold, let Γ be the fundamental group $\pi_1(X)$, and assume there is a surjection $\Gamma \twoheadrightarrow G$ where G is a finitely presented group $G = \langle a_1, \ldots, a_n \mid R_1(a), \ldots, R_m(a)\rangle$ with $n \geq m+2$. Then there is an integer $g \geq 2$ with $2g \geq n-m$ and an irrational pencil f of genus g. If moreover $\Gamma = G$, then the Albanese map is a pencil, i.e.,* $\operatorname{Im} f^* = H^1(X, \mathbb{C})$.

In fact, since clearly Theorem 1 is a particular case of Theorem 2, our purpose here is to show that the above result is a direct consequence of our theorem of the isotropic subspace ([Ca], Theorem 1.10). In turn, this result was a straightforward consequence of the Hodge decomposition, of the Castelnuovo–de Franchis theorem and of the Künneth formula.

---

*Research performed in the realm of the *"A.G.E." project*, contract ERBCHRXCT 940557, of M.U.R.S.T. 40% of G.N.S.A.G.A of C.N.R. The paper was written while the author was "distaccato" at the Accademia dei Lincei.

Theorem 3 (Theorem of the isotropic subspace [Ca]). *Let X be a compact Kähler manifold. Then there is a bijection between the following data:*

i) *irrational pencils $f : X \to C$ of genus $g \geq 2$;*

ii) *subspaces V of $H^1(X, \mathbb{C})$ of the form $V = U \bigoplus \overline{U}$ where U is a subspace of dimension $g \geq 2$ of $H^1(X, \mathbb{C})$ which is a maximal isotropic subspace for the wedge product map $\bigwedge : H^1(X, \mathbb{C}) \times H^1(X, \mathbb{C}) \to H^2(X, \mathbb{C})$*

Under the above correspondence $V = f^* H^1(C, \mathbb{C})$.

Moreover every maximal isotropic subspace U of dimension $g \geq 2$ determines a unique pencil $f : X \to C$ of genus g and is the pullback of a maximal isotropic subspace in $H^1(C, \mathbb{C})$.

Proof of Theorem 2. Let Y be a classifying space $K(G, 1)$ for the group G.

The 2-skeleton of Y is obtained attaching m 2-cells to a wedge of n circles, whence there is an integer $b \leq m$ such that the rank of $H^1(Y, \mathbb{C})$ is $\geq n - m + b$, while the rank of $H^2(Y, \mathbb{C})$ is $\leq b$. Therefore, for $\psi \in H^1(Y, \mathbb{C})$ we have that wedge product with $\psi : H^1(Y, \mathbb{C}) \to H^2(Y, \mathbb{C})$ has a kernel of dimension $\geq n - m$.

Thus, by our assumption, ψ is contained in an isotropic subspace of dimension ≥ 2.

Now, to the surjection $\Gamma \to G$ there corresponds (cf. e.g. [Sp], Theorem 11, p. 428)

i) a continuous map $\phi : X \to Y$ inducing $\Gamma \to G$, and moreover such that

ii) $\phi^* : H^1(Y, \mathbb{C}) \to H^1(X, \mathbb{C})$ is 1-1, (since the dual homomorphism on homology is surjective).

Denote by W the subspace of $H^1(X, \mathbb{C})$ given by the image of ϕ^*. Let U' be a maximal isotropic subspace of W: by what we have seen, necessarily the dimension of U' is at least 2.

By the isotropic subspace theorem, take a maximal isotropic subspace U of $H^1(X, \mathbb{C})$ containing U': U determines a unique irrational pencil f.

Setting $V = f^* H^1(C, \mathbb{C})$ there are two cases:

1) $V \supset W$, and then $2g \geq n - m$, as we wanted, or

2) $\forall f$ constructed as above, $V' = V \cap W$ is a proper subspace of W.

But all these irrational pencils f, e.g. by the isotropic subspace theorem, are completely determined by the injection $f^* : H^1(C, \mathbb{Z}) \to H^1(X, \mathbb{Z})$, hence there is at most a countable number of them.

Since every vector is obviously contained in a maximal isotropic subspace, W is the union of a countable number of proper subspaces V'; since W is a complex vector space, 2) leads to a contradiction and the main statement of Theorem 2 is proved.

Finally, if $\Gamma = G$, our argument shows that $W = H^1(X, \mathbb{C})$ and then clearly $V = W$, i.e., Im $f^* = H^1(X, \mathbb{C})$ as claimed. □

References

[Be] A. Beauville, Appendix to [Ca], Invent. Math. 104 (1991), 289.

[Ca] F. Catanese, Moduli and classification of irregular Kähler manifolds (and algebraic varieties) with Albanese general type fibrations, Invent. Math. 104 (1991), 263–288.

[G-L] M. Green, R. Lazarsfeld, Higher obstructions to deforming cohomology groups of line bundles, J. Amer. Math. Soc. 4 (1991), 87–103.

[Gro] M. Gromov, Sur le groupe fondamental d'une varieté Kählerienne, C. R. Acad. Sci. Paris 308, D.I, (1989), 67–70.

[Siu] Y. T. Siu, Strong rigidity for Kähler manifolds and the construction of bounded holomorphic functions, in: Discrete groups in geometry and analysis (R. Howe, ed.), Birkhäuser 1987, pp. 124–151.

[Sp] E. H. Spanier, Algebraic Topology, Mac Graw-Hill, New York 1966, reprinted by Springer Verlag, 1981.

Some remarks on the obstructedness of cones over curves of low genus

Ciro Ciliberto, Angelo Felice Lopez*, and Rick Miranda***

1991 Mathematics Subject Classification: Primary 14C05; secondary 14J10, 14B07.

1. Introduction

Let C be a smooth irreducible curve of genus g and L a very ample line bundle on C of degree d. Embedding C in $\mathbb{P}^r = \mathbb{P}H^0(L)^*$ we can consider the cone $X_{C,L} \subset \mathbb{P}^{r+1}$ over C with vertex a point in $\mathbb{P}^{r+1} - \mathbb{P}^r$. When $g = 0$ it was shown by Pinkham ([P]) that rational normal cones are obstructed, that is they represent singular points of their Hilbert schemes, as soon as $r \geq 4$. One of the purposes of the present article is to employ the technique of Gaussian maps to study the obstructedness of cones over curves of *positive* genus g and degree $d \gg g$.

For $k \geq 1$ set $R(\omega_C \otimes L^{k-1}, L) = \operatorname{Ker}\{H^0(\omega_C \otimes L^{k-1}) \otimes H^0(L) \to H^0(\omega_C \otimes L^k)\}$ and let

$$\Phi_{\omega_C \otimes L^{k-1}, L} : R(\omega_C \otimes L^{k-1}, L) \to H^0(\omega_C^2 \otimes L^k)$$

be the Gaussian map defined locally by $\Phi_{\omega_C \otimes L^{k-1}, L}(s \otimes t) = s dt - t ds$. Moreover let $\gamma_{C,L}^k = \operatorname{corank} \Phi_{\omega_C \otimes L^{k-1}, L}$, and $\gamma_{C,L} = \operatorname{corank} \Phi_{\omega_C, L}$.

As is well-known now the integers $\gamma_{C,L}^k$ may be used to compute the dimension of the tangent space to the Hilbert scheme at points representing cones $X_{C,L}$ over C (see for example [W1], [St], [T], [CM], [CLM1], [CLM2]).

We will see in Section 2 that there is an interesting difference for the obstructedness of such cones between the nonhyperelliptic case for $g \geq 3$ and the hyperelliptic case $g = 1, 2$. For $g \geq 3$ and $d \gg g$ if C is not hyperelliptic and L is any line bundle on C of degree d, we have $\gamma_{C,L} = 0$ ([W2], [BEL]), and the cone $X_{C,L}$ is unobstructed.

---

* Research partially supported by the MURST national project "Geometria Algebrica"; the authors are members of GNSAGA of CNR.

** Research partially supported by an NSA grant.

On the other hand if C is hyperelliptic and $d \gg g$ then $\gamma_{C,L} > 0$ ([St]); in particular if $g = 1, 2$ and $d \geq 3g+7$, we will show in Section 4, that the cone $X_{C,L}$ is obstructed and cannot be smoothed. By taking general hypersurface sections of such cones and using the above fact, we will construct infinitely many examples of *nonreduced* components of the Hilbert scheme of curves in $\mathbb{P}^{r+1}$, $r \geq 9$ (see Theorem 4.11). Note that so far all the known examples of nonreduced components of the Hilbert scheme are for curves in $\mathbb{P}^3$ ([M], [GP], [K], [E], [Fl]).

Another interesting feature of the coranks of Gaussian maps is that they give the cohomology $h^0(N_C(-k))$, $k \geq 1$, of the normal bundle of C in $\mathbb{P}^r$. This in turn, by a theorem of Zak [Z], governs in many cases the existence of higher dimensional varieties having C as their curve section (see for example [CLM1], [CLM2] for Fano varieties, [W1] for K3 surfaces). In Section 3 we will further exploit the power of the above technique, by applying it to varieties of degree five having elliptic curve sections. In particular we will extend to the singular case the well-known smooth classification ([Sc], [I], [Fu]). We will then consider surfaces in $\mathbb{P}^r$, $r \geq 6$ with curve sections of genus at most three. In these cases a straightforward application of the method of Gaussian maps and Zak's theorem does not work. We will instead recover the known classification by means of a projective technique (the "tetragonal lemma" 3.2).

Acknowledgments. The authors wish to thank the Department of Mathematics of the University of Trento and the CIRM for the warm hospitality and the organization provided in the school-conference "Higher dimensional complex varieties" held in Trento in June 1994. We also thank M. Coppens for some helpful suggestions regarding Lemma 3.2.

2. Global sections of the normal bundle and corank of Gaussian maps

As above we let C be a smooth irreducible curve of genus $g \geq 0$, L be a line bundle on C of degree $d \geq 1$ and $\gamma_{C,L} = \operatorname{corank} \Phi_{\omega_C,L}$. We will first collect some facts about $\gamma_{C,L}$.

Proposition 2.1. *We have the following values for* $\gamma_{C,L}$*:*

(2.2) *If* $g = 0$ *then*

$$\gamma_{C,L} = \begin{cases} 0 & \text{for } d \leq 3 \\ d-3 & \text{for } d \geq 4; \end{cases}$$

(2.3) *If* $g = 1$ *then* $\gamma_{C,L} = d$;

(2.4) *If* $g = 2$ *then* $\gamma_{C,L} = 6$ *for* $d \geq 5$;

(2.5) *If* $g = 3$ *and* C *is not hyperelliptic we have*

$$\gamma_{C,L} = \begin{cases} 0 & \text{for } d \geq 17 \\ 0 & \text{for } L \text{ general and } d \geq 14 \\ \geq 14 - d & \text{for } 6 \leq d \leq 13; \end{cases}$$

(2.6) *If* $g = 4$ *and* C *is not hyperelliptic we have*

$$\gamma_{C,L} = \begin{cases} 0 & \text{for } d \geq 19 \\ 0 & \text{for } L \text{ general and } d \geq 15 \\ \geq 15 - d & \text{for } 8 \leq d \leq 14; \end{cases}$$

(2.7) *If* $g = 5$ *and* C *is not trigonal we have*

$$\gamma_{C,L} = \begin{cases} 0 & \text{for } d \geq 17 \\ 0 & \text{for } L \text{ general and } d \geq 12 \\ \geq 36 - 3d & \text{for } 10 \leq d \leq 11; \end{cases}$$

(2.8) *If* $g \geq 6$ *and both* C *and* L *are general we have*

$$\gamma_{C,L} = 0 \text{ if } d \geq \begin{cases} g + 12 & \text{for } 6 \leq g \leq 8 \\ g + 9 & \text{for } g \geq 9. \end{cases}$$

Proof. If $g = 0, 1$ we have $R(\omega_C, L) = 0$, hence $\gamma_{C,L} = h^0(\omega_C^2 \otimes L)$ and this gives (2.2) and (2.3). If $g = 2$, by the base point free pencil trick, we have $R(\omega_C, L) = H^0(\omega_C^{-1} \otimes L)$ hence $\gamma_{C,L} = h^0(\omega_C^2 \otimes L) - h^0(\omega_C^{-1} \otimes L) + \dim \operatorname{Ker} \Phi_{\omega_C,L} = d+3-(d-3)+\dim \operatorname{Ker} \Phi_{\omega_C,L} = 6+\dim \operatorname{Ker} \Phi_{\omega_C,L}$. A straightforward computation in the diagram

$$\begin{array}{ccc} H^0(\omega_C^{-1} \otimes L) & \xrightarrow{\Phi} & H^0(\omega_C^2 \otimes L) \\ & \searrow\cong \quad \nearrow \Phi_{\omega_C,L} & \\ & R(\omega_C, L) & \end{array}$$

gives that, if $H^0(\omega_C) = \langle s, t \rangle$, we have $\Phi(\sigma) = 2\sigma \Phi_{\omega_C,\omega_C}(s \otimes t)$, and $\Phi_{\omega_C,\omega_C}(s \otimes t) \neq 0$, hence both Φ and $\Phi_{\omega_C,L}$ are injective and (2.4) follows. To see (2.5), (2.6) and (2.7) consider the canonical embedding $C \subset \mathbb{P}^{g-1}$. As is well known (see [W2]), from the exact sequence

$$0 \to N^*_{C/\mathbb{P}^{g-1}} \otimes \omega_C \otimes L \to \Omega^1_{\mathbb{P}^{g-1}|C} \otimes \omega_C \otimes L \to \omega_C^2 \otimes L \to 0$$

we get

$$\text{(2.9)} \quad \begin{aligned} H^0(\Omega^1_{\mathbb{P}^{g-1}|C} \otimes \omega_C \otimes L) &\xrightarrow{\Phi_{\omega_C,L}} H^0(\omega_C^2 \otimes L) \to H^1(N^*_{C/\mathbb{P}^{g-1}} \otimes \omega_C \otimes L) \\ &\to H^1(\Omega^1_{\mathbb{P}^{g-1}|C} \otimes \omega_C \otimes L) \end{aligned}$$

and in the cases at hand we will show that

$$\text{(2.10)} \quad H^1(\Omega^1_{\mathbb{P}^{g-1}|C} \otimes \omega_C \otimes L) = 0$$

and therefore

(2.11) $$\gamma_{C,L} = h^1(N^*_{C/\mathbb{P}^{g-1}} \otimes \omega_C \otimes L).$$

To see (2.10) consider the Euler sequence

$$0 \to \Omega^1_{\mathbb{P}^{g-1}|C} \otimes \omega_C \otimes L \to H^0(\omega_C) \otimes L \to \omega_C \otimes L \to 0$$

and notice that, with the given hypotheses on d, we have $H^1(L) = 0$ and the multiplication map $H^0(\omega_C) \otimes H^0(L) \to H^0(\omega_C \otimes L)$ is surjective (by [C2]), hence (2.10).

When $g = 3$ by (2.11) we see that $\gamma_{C,L} = h^1(L \otimes \omega_C^{-3}) = h^0(L^{-1} \otimes \omega_C^4)$ and this is zero for $d \geq 17$ or is at least $14 - d$ for $6 \leq d \leq 13$. For $14 \leq d \leq 16$ and L general in $\operatorname{Pic}^d C$ we have that $\gamma_{C,L} = 0$ since $\dim \operatorname{Pic}^d C = 3$ while $\dim\{L : h^0(L^{-1} \otimes \omega_C^4) \neq 0\} = 16 - d$. This gives (2.5). When $g = 4$ we get by (2.11) that $\gamma_{C,L} = h^1(L \otimes (\omega_C^{-1} \oplus \omega_C^{-2})) = h^0(L^{-1} \otimes \omega_C^2) + h^0(L^{-1} \otimes \omega_C^3)$ and this is zero for $d \geq 19$ or is at least $15 - d$ for $8 \leq d \leq 14$. For $15 \leq d \leq 18$ and L general in $\operatorname{Pic}^d C$ we have that $\gamma_{C,L} = 0$ since $\dim \operatorname{Pic}^d C = 4$ while $\dim\{L : h^0(L^{-1} \otimes \omega_C^3) \neq 0\} = 18 - d$. This proves (2.6). Similarly when $g = 5$, again by (2.11), we have that $\gamma_{C,L} = h^1(L \otimes (\omega_C^{-1})^{\oplus 3}) = 3h^0(L^{-1} \otimes \omega_C^2)$ and this is zero for $d \geq 17$ or is at least $36 - 3d$ for $10 \leq d \leq 11$. For $12 \leq d \leq 16$ and L general in $\operatorname{Pic}^d C$ we have that $\gamma_{C,L} = 0$ since $\dim \operatorname{Pic}^d C = 5$ while $\dim\{L : h^0(L^{-1} \otimes \omega_C^2) \neq 0\} = 16 - d$. Hence (2.7) follows. Finally (2.8) was proved in [L, Corollary (1.7)]. □

We turn now to the relation between the corank of Gaussian maps and the obstructedness of cones. With notation as above, suppose from now on that L is very ample, $C \subset \mathbb{P}^r = \mathbb{P}H^0(L)^*$ and let $X_{C,L}$ be a cone over C in $\mathbb{P}^{r+1}$ with vertex a point $P \in \mathbb{P}^{r+1} - \mathbb{P}^r$.

We recall for the reader's convenience the connection between cohomology of the normal bundle and corank of Gaussian maps. Let N_C, $N_{X_{C,L}}$ be the normal bundles of $C \subset \mathbb{P}^r$ and of $X_{C,L} \subset \mathbb{P}^{r+1}$, respectively. Then

Proposition 2.12. *Suppose* $g \geq 1$. *Then:*

(2.13) $h^0(N_C(-1)) = r + 1 + \gamma_{C,L}$;

(2.14) $h^0(N_C(-k)) = \dim \operatorname{Coker} \Phi_{\omega_C \otimes L^{k-1}, L}$ *for every* $k \geq 2$;

(2.15) *If* $\gamma_{C,L} = 0$ *then* $h^0(N_C(-k)) = 0$ *for every* $k \geq 2$;

(2.16) $h^0(N_{X_{C,L}}) \leq \sum_{k \geq 0} h^0(N_C(-k))$;

(2.17) *If* $C \subset \mathbb{P}^r = \mathbb{P}H^0(L)^*$ *is projectively normal we have equality in* (2.16).

Proof. (2.13) and (2.14) are in [CM, Proposition 1.2]. Of course (2.15) follows for $k \geq 3$ as soon as we prove it for $k = 2$. To this end from the diagram

$$\begin{array}{ccc} R(\omega_C, L) \otimes H^0(L) & \xrightarrow{\Phi_{\omega_C,L} \otimes \mathrm{id}_{H^0(L)}} & H^0(\omega_C^2 \otimes L) \otimes H^0(L) \\ \downarrow & & \downarrow \mu \\ R(\omega_C \otimes L, L) & \xrightarrow{\Phi_{\omega_C \otimes L, L}} & H^0(\omega_C^2 \otimes L^2) \end{array}$$

we deduce that $\Phi_{\omega_C \otimes L, L}$ is surjective since $\Phi_{\omega_C, L}$ is by hypothesis and so is μ (see for example [G, Theorem (4.e.1)]). Hence $H^0(N_C(-2)) = 0$ by (2.14).
To see (2.16) set $X = X_{C,L}$ and notice that $N_{X|C} \cong N_C$, and therefore the exact sequence

$$0 \to N_X(-h-1) \to N_X(-h) \to N_C(-h) \to 0$$

implies that $h^0(N_X(-h)) \leq h^0(N_X(-h-1)) + h^0(N_C(-h))$, and applying this successively we get (2.16) since $h^0(N_X(-j)) = 0$ for $j \gg 0$. Finally (2.17) follows by standard facts (see for example [CM, (4.2)]). □

Now let W be a component of the Hilbert scheme $H_{d,g,r}$ of curves of degree d and genus g in $\mathbb{P}^r$ such that $W \ni [C]$, the point representing C, and denote by $\mathcal{H}(W)$ the family of cones in $\mathbb{P}^{r+1}$ over curves representing points of W. We have the following general fact

Proposition 2.18. *Suppose that $g \geq 3$ and a general point $[C_\eta]$ of W is a smooth point such that $\gamma_{C_\eta, \mathcal{O}_{C_\eta}(1)} = 0$. Then*

(a) *$\mathcal{H}(W)$ is a generically smooth component of the Hilbert scheme of surfaces of degree d in $\mathbb{P}^{r+1}$ and* $\dim \mathcal{H}(W) = h^0(N_{C_\eta}) + r + 1$. *In particular* $\dim \mathcal{H}(W) = (r+1)(d+1) + (r-3)(1-g)$ *if $H^1(N_{C_\eta}) = 0$;*

(b) *If $\gamma_{C,L} = 0$ and $[C]$ is a smooth point of W, then $X_{C,L}$ is unobstructed;*

(c) *If $C \subset \mathbb{P}^r = \mathbb{P}H^0(L)^*$ is projectively normal and $X_{C,L}$ is unobstructed, then $\gamma_{C,L} = 0$.*

Proof. First suppose that $\gamma_{C,L} = 0$ and that $[C]$ is a smooth point of W. Let V be a component of the Hilbert scheme of surfaces of degree d in $\mathbb{P}^{r+1}$ such that $V \supseteq \mathcal{H}(W)$. We have

$$h^0(N_C)+r+1 = \dim W + r + 1 = \dim \mathcal{H}(W) \leq \dim V \leq h^0(N_{X_{C,L}}) \leq h^0(N_C)+r+1$$

where the last inequality follows by (2.16), (2.15) and (2.13). Hence $V = \mathcal{H}(W)$ and $X_{C,L}$ is unobstructed. This shows (b). Applying this to C_η we get (a) since $h^0(N_{C_\eta}) + r + 1 = (r+1)(d+1) + (r-3)(1-g)$ when $H^1(N_{C_\eta}) = 0$. Now suppose $C \subset \mathbb{P}^r = \mathbb{P}H^0(L)^*$ is projectively normal and $X_{C,L}$ is unobstructed. Since $\mathcal{H}(W)$ is a component of the Hilbert scheme, it must be the only one containing $[X_{C,L}]$. If

$\gamma_{C,L} > 0$ we would have

$$\begin{aligned}\dim \mathcal{H}(W) &= h^0(N_{C_\eta}) + r + 1 \leq h^0(N_C) + r + 1 \\ &< h^0(N_C) + r + 1 + \gamma_{C,L} + \sum_{k\geq 2} h^0(N_C(-k)) \\ &= h^0(N_{X_{C,L}})\end{aligned}$$

by (2.17). But this shows that $X_{C,L}$ is obstructed. Hence (c) is proved. □

Remark 2.19. Even for line bundles of high degree it is possible that $\gamma_{C,L} > 0$ for a given pair (C, L) while $\gamma_{C_\eta, L_\eta} = 0$ for the general pair (C_η, L_η). For example on a smooth nonhyperelliptic curve C of genus $g \geq 9$ take P a general point of C and $L = \omega_C(2P)$. It is shown in [L] that $\gamma_{C,L} > 0$; on the other hand by Proposition 2.1 we have $\gamma_{C_\eta, L_\eta} = 0$ since $\deg L_\eta = 2g \geq g + 9$. Other examples can be deduced from [St], [T].

Pairing Propositions 2.1 and 2.18 we get

Corollary 2.20. *Let C be a smooth irreducible curve of genus $g \geq 3$, L a line bundle on C of degree d such that either*

(a) $d \geq 16g - 2g^2 - 13$ *for* $3 \leq g \leq 5$, *or*

(b) $d \geq 15g - 2g^2 - 13$ *for* $3 \leq g \leq 5$ *and L is general, C is not hyperelliptic and also not trigonal if* $g = 5$, *or*

(c) $d \geq \begin{cases} g + 12 & \text{for } 6 \leq g \leq 8 \\ g + 9 & \text{for } g \geq 9 \end{cases}$ *and both C and L are general.*

Then $X_{C,L}$ is unobstructed and $\mathcal{H}(W)$ is a generically smooth component of the Hilbert scheme of dimension $(r + 1)(d + 1) + (r - 3)(1 - g)$ where $r = d - g$.

Proof. By Proposition 2.1 under hypothesis (a), (b) or (c), we have $\gamma_{C,L} = 0$. Moreover in all cases $H^1(L) = 0$, hence we can apply (a) and (b) of Proposition 2.18. □

3. Remarks on the classification of varieties with elliptic curve sections

Let $X^n \subset \mathbb{P}^N$ be a nondegenerate variety of dimension $n \geq 2$ with $\dim \operatorname{Sing} X^n \leq n - 2$, degree 5, whose curve sections are smooth elliptic normal curves. If H is the hyperplane divisor of X^n we have $N = n + 3$ and $-K_{X^n} = (n - 1)H$ if X^n is smooth.

Well-known examples of such varieties are the linear sections of the Grassmann variety $\mathbb{G}(1, 4) \subset \mathbb{P}^9$ in its Plücker embedding. Just as in [CLM2], we will show that the classification of the varieties X^n as above can be recovered by means of deformations to cones and the theorem of Zak, whose statement we recall here. A smooth nondegenerate variety $Y \subset \mathbb{P}^m$ is said to be *k-extendable* if there exists a variety $Z \subset \mathbb{P}^{m+k}$ that is not a cone and such that $Y = Z \cap \mathbb{P}^m$. Zak's theorem ([Z]) says that if codim $Y \geq 2$ and $h^0(N_Y(-1)) = m + 1$ then Y is not 1-extendable; also if $k \geq 2$, $h^0(N_Y(-1)) \leq m + k$ and $h^0(N_Y(-2)) = 0$, then Y is not k-extendable.

In the Hilbert scheme of nondegenerate varieties $X^n \subset \mathbb{P}^{n+3}$ of dimension $n \geq 2$ with dim Sing $X^n \leq n-2$, degree 5 and such that its general curve sections are smooth elliptic normal curves, we let $\mathcal{X}_n$ be the open subset parametrizing varieties X^n that are not cones over an elliptic curve. Then

Proposition 3.1.

(a) $\mathcal{X}_n \neq \emptyset$ *if and only if* $n \leq 6$;

(b) *For* $n \leq 6$, $\mathcal{X}_n$ *is irreducible and the family of linear sections of* $\mathbb{G}(1,4) \subset \mathbb{P}^9$ *forms a dense open subset of smooth points of* $\mathcal{X}_n$.

Proof. By (2.3) and (2.13) if $E \subset \mathbb{P}^4$ is a smooth curve section of X^n, we have that $h^0(N_{E/\mathbb{P}^4}(-1)) = 10$. Also it is easily seen that $h^0(N_{E/\mathbb{P}^4}(-2)) = 0$, for example using (2.14) and the surjectivity of $\Phi_{L,L}$ (see [BEL]). Hence Zak's theorem (see [Z]) implies that $n \leq 6$. On the other hand dim $\mathbb{G}(1,4) = 6$ and $-K_{\mathbb{G}(1,4)} = 5H$, where H is the Plücker divisor, hence $\mathcal{X}_6 \neq \emptyset$ and so is $\mathcal{X}_n, n \leq 5$, since a hyperplane section of an X^n is a X^{n-1}. This proves (a). To see (b) we will use an argument similar to the one in [CLM2], in the proof of theorems (3.2) and (3.11). Since the varieties $X^n, n \geq 2$, are projectively Cohen–Macaulay (because E is), they flatly degenerate to the n-dimensional cone $\widehat{E}$ over their general curve section E. Of course the locus of such cones is irreducible, hence (b) follows if we show that these cones are smooth points of the closure of $\mathcal{X}_n$. To this end we will prove, as in [CLM2], that $h^0(N_{\widehat{E}})$ is bounded above by the dimension of the family of the known examples, that is linear sections of $\mathbb{G}(1,4) \subset \mathbb{P}^9$. Since $h^0(N_E(-k)) = 0$ for every $k \geq 2$, we have

$$h^0(N_{\widehat{E}}) = h^0(N_E) + (n-1)h^0(N_E(-1)) = 25 + 10(n-1).$$

On the other hand the family of $X^n \subset \mathbb{P}^{n+3}$ obtained by linear sections of $\mathbb{G}(1,4) \subset \mathbb{P}^9$ has dimension

$$\begin{aligned}&\dim \mathbb{G}(n+3,9) + \dim \operatorname{Aut} \mathbb{P}^{n+3} - \dim \operatorname{Aut} \mathbb{G}(1,4)\\ &\qquad = (n+4)(6-n) + (n+4)^2 - 25 = 15 + 10n.\end{aligned}$$ □

Note that the case $n = 2$ of Proposition 3.1 is the one of Del Pezzo surfaces of degree 5 in $\mathbb{P}^5$. Suppose now $S \subset \mathbb{P}^r$, $r \geq 6$ is a (not necessarily smooth) surface whose general hyperplane section is a smooth elliptic normal curve E. We have deg $E = r$, hence S is a surface of degree r in $\mathbb{P}^r$ nonsingular in codimension one. In this case a straightforward application of the methods of [CLM2] does not allow to recover the family of Del Pezzo surfaces. In fact by (2.3) and (2.13) we have $h^0(N_E(-1)) = 2r$, hence Zak's theorem does not rule out the existence of such surfaces $S \subset \mathbb{P}^r$ for $r \geq 10$. Moreover even for $6 \leq r \leq 9$ the upper bound on $h^0(N_{\widehat{E}})$ is larger than the dimension of the family of Del Pezzo surfaces. Also note that this implies that the cones over elliptic normal curves are obstructed for $6 \leq r \leq 9$; in Section 4 we will see that they are also obstructed for $r \geq 10$.

We will then use another strategy to recover the classification. Recall that E is projectively Cohen–Macaulay, its homogeneous ideal is generated by quadrics, the relations among them being minimally generated by linear ones. The same then holds for S. Let $C = S \cap Q$ be a smooth quadric section of S. We have that C is a canonical curve whose homogeneous ideal is generated by quadrics and therefore by Petri's theorem C is not trigonal nor isomorphic to a plane quintic. Hence $\text{Cliff}(C) \geq 2$, where $\text{Cliff}(C)$ is the Clifford index of C. Moreover the Koszul relations among Q and the quadrics generating the ideal of S give rise to nonlinear syzygies among the generators of the ideal of C, relations that do not depend on the linear ones (or in other words $h^0(N_C(-2)) \neq 0$). By a result of Schreyer [S1] and Voisin [V], we must have $\text{Cliff}(C) = 2$, that is C is either tetragonal or isomorphic to a plane sextic.

This fact will allow us to classify S. The key point is given by the following

Lemma 3.2 (tetragonal lemma). *Let $S \subset \mathbb{P}^r, r \geq 5$ be a surface, nonsingular in codimension one, not a cone, whose general hyperplane section has genus g, degree $d \geq 2g+3$ and is linearly normal. If there is a linear system of curves on S, of dimension at least* 1, *whose general element D is smooth, irreducible, special and tetragonal, then S is either a rational normal scroll (i.e. $g = 0$) or it lies on a rational normal threefold scroll of planes, each meeting S in a conic.*

Proof. First notice that S is cut out by quadrics because its general hyperplane section is projectively normal and is cut out by quadrics, since $d \geq 2g+3$ [G]. Take $p_1+\cdots+p_4$ a general divisor of the g^1_4 on D. The span of this divisor in $\mathbb{P}^r$ is at most a 2-plane π. If it is a line, then S is a rational normal scroll. Suppose now π is a 2-plane. Consider the hyperplane sections $\{H_t\}$ of S through π. Suppose that a general H_t is irreducible. By monodromy the behaviour of H_t at the points $p_1, \dots, p_4$ is the same, and we claim that $p_1, \dots, p_4$ are smooth points of H_t. Otherwise H_t would be tangent at $p_1, \dots, p_4$ and therefore π would be tangent to S at these four points. Now let D and p_1 vary. Since p_1 describes the whole surface, we conclude that the general tangent plane to the surface is tangent at four points, a contradiction, since the Gauss map is birational. By the hypothesis $d \geq 2g+3$, if H_t is irreducible, we have that $H^1(\mathcal{O}_{H_t}(1)(-p_1-\cdots-p_4)) = 0$ and hence $\dim|\mathcal{O}_{H_t}(1)(-p_1-\cdots-p_4)| = d-4-g$ while the $H_{t'}$ cut out, away from $p_1, \dots, p_4$, a linear series on H_t of dimension at least $r-3 = d-g-3$. This contradiction shows that a general H_t must be reducible. Therefore, by Bertini's theorem, there is a fixed component F of the linear system $\{H_t\}$, and $F \subset \pi$ unless $\{H_t\}$ is composed with a pencil. In the latter case, since the dimension of $\{H_t\}$ is at least $r-3 \geq 2$, we have that H_t has to contain the tangent plane to S at each point p_i, $i = 1, \dots, 4$, for all t's and this, as we saw, is impossible.

Since S is cut out by quadrics, then the degree of F is at most 2 and F has to contain $p_1, \dots, p_4$. If F is a line we conclude that S is a rational normal scroll. If F is a conic, and is reducible, then S is a scroll. Arguing as we did above, we see that the lines of the scroll are pairwise coplanar, hence S is a cone, which we excluded. If F is an irreducible conic, then S is contained in the rational normal threefold scroll of planes

$$\bigcup_{p_1+\cdots+p_4 \in g^1_4} \langle p_1 + \cdots + p_4 \rangle. \qquad \square$$

We will use the tetragonal lemma to recover the classification of Del Pezzo surfaces.

Proposition 3.3. *Let $S \subset \mathbb{P}^r$, $r \geq 6$ be a surface whose general hyperplane section is a smooth elliptic normal curve E. Then S is either an elliptic normal cone or $r \leq 8$ and S is a divisor of class $2H + (4 - r)R$ on a rational normal threefold scroll of planes $X \subset \mathbb{P}^r$, with hyperplane section H and ruling R, or S is the Veronese embedding $v_3(\mathbb{P}^2) \subset \mathbb{P}^9$ given by the cubics.*

Proof. Notice that if S is a cone then it is an elliptic normal cone. From now on we suppose that S is not a cone. We claim that either S lies on a rational normal threefold scroll of planes, each meeting S in a conic, or $r = 9$ and S is smooth. In fact if S has a singular point P, since the ideal of S is generated by quadrics, the projection $S' \subset \mathbb{P}^{r-1}$ of S from P is a surface of degree at most $r - 2$, hence $\deg S' = r - 2$ and S' contains a one dimensional family of lines $\{L_t\}$. Set $\pi_t = \langle P, L_t \rangle$ and $F_t = \pi_t \cap S$. Then $\deg F_t \leq 2$ since the ideal of S is generated by quadrics and S lies on a rational normal threefold scroll of planes, each meeting S in a conic. If S is nonsingular, by the discussion at the beginning of this section, we know that a smooth quadric section C of S is either tetragonal or isomorphic to a plane sextic. In the former case we can apply the tetragonal lemma since $H^1(\mathcal{O}_C(1)) = H^1(\omega_C) \neq 0$, concluding again that S lies on a rational normal threefold scroll of planes, each meeting S in a conic. When S lies on a scroll as above, pulling back to the normalization, we have $S \sim aH + bR$ and $2 = \deg R \cap S = R \cdot S \cdot H = R \cdot (aH^2 + bH \cdot R) = a$ since $R^2 = 0$, $H^2 \cdot R = 1$. Moreover $r = \deg S = S \cdot H^2 = 2(r-2) + b$, hence $b = 4 - r$. By [S2, (6.3)] we have $2 \deg X + 3b \geq 0$, that is $r \leq 8$. It remains the case where smooth quadric sections C of S are isomorphic to a plane sextic. Then $r = 9$ and by what we just showed, S must be smooth. Moreover $K_S = -H$, $P_2(S) = h^0(\mathcal{O}_S(-2)) = 0$ and $q(S) = h^1(\mathcal{O}_S(-1)) = 0$, therefore S is rational. Also S is minimal, since by Noether's formula we get $b_2(S) = h^{1,1}(S) = 1$, hence $S = v_3(\mathbb{P}^2)$. $\square$

Remarks 3.4. (i) With the tetragonal lemma it is easily recovered also the classification of surfaces with sectional genus 2. Let $S \subset \mathbb{P}^r$, $r \geq 6$ be a surface whose general hyperplane section is a smooth linearly normal curve C_1 of genus 2. Then S is either a cone over C_1, or $r \leq 11$ and S is a divisor of class $2H + (5 - r)R$ on a rational normal threefold scroll of planes $X \subset \mathbb{P}^r$, with hyperplane section H and ruling R.

Proof. Let $C = S \cap Q$ be a smooth quadric section of S. We have $\deg C = 2r + 2$, $g(C) = r+4$ and C is linearly normal and special since $h^1(\mathcal{O}_C(1)) = 2$. Moreover $\deg \omega_C(-1) = 4$ and therefore S has a linear system of dimension at least 2, whose general element is smooth, irreducible, special and tetragonal (of course C is not trigonal, since its ideal is generated by quadrics). Since $\deg S = r+1 \geq 7$, by Lemma 3.2 we have that, if S is not a cone, then $S \sim aH + bR$ on a rational normal threefold scroll of planes $X \subset \mathbb{P}^r$ of degree $r-2$, and each plane meets S in a conic. Therefore, as in the proof of Proposition 3.3, we get $a = 2$ and $b = 5 - r$. By [S2, (6.3)] we have $2 \deg X + 3b \geq 0$, that is $r \leq 11$. □

(ii) Now let $S \subset \mathbb{P}^r$, $r \geq 6$ be a surface whose general hyperplane section is a smooth linearly normal curve C_1 of genus 3. Then S is either a cone over C_1, or $r \leq 14$ and S is the Veronese embedding $v_4(\mathbb{P}^2) \subset \mathbb{P}^{14}$ given by the quartics or a projection of it in $\mathbb{P}^{14-r}$ or S is a quadric section of the cone in $\mathbb{P}^6$ over the Veronese surface $v_2(\mathbb{P}^2) \subset \mathbb{P}^5$. The proof of this is similar to the one above and will be omitted.

(iii) Suppose now $S \subset \mathbb{P}^r$, $r \geq 4$, is a surface of degree d whose general hyperplane section is a smooth linearly normal curve C of genus $g \geq 3$. As already observed by many authors ([BEL], [BC]), by using Zak's theorem ([Z]), one can obtain an upper bound for d. If C is hyperelliptic or trigonal this is well known (see [Se], [Fa]). If $\mathrm{Cliff}(C) = 2$ then $d \leq 4g - 4$, while if $\mathrm{Cliff}(C) \geq 3$ then $d \leq 4g - 3\,\mathrm{Cliff}(C)$. The above bounds follow by [BEL], since, if d is larger, then $\gamma_{C,\mathcal{O}_C(1)} = 0$.

4. Nonreduced components of the Hilbert scheme of curves

Let C be a smooth irreducible curve of genus $g \geq 1$ and L a line bundle on C of degree d such that $d \geq 10$ for $g = 1$, $d \geq 4g + 5$ for $g \geq 2$. As opposite to the non hyperelliptic case $g \geq 3$, we saw in Proposition 2.1 that in case $g = 1, 2$, we have $\gamma_{C,L} > 0$. Even though Proposition 2.18 does not apply, we will see here that in fact the cone $X = X_{C,L}$ over C with vertex a point $P \in \mathbb{P}^{r+1} - \mathbb{P}^r$ is obstructed if $g = 1, 2$.

Note that L is very ample, nonspecial, and, by well-known results, $C \subset \mathbb{P}^r = \mathbb{P}H^0(L)^*$ is projectively normal, its ideal is generated by quadrics and the relations among them are linear ([G]). Now consider a general hypersurface F_n in $\mathbb{P}^{r+1}$ of degree $n \geq 4$ and let $\Gamma^n = X \cap F_n$. The nonreduced components of the Hilbert scheme will be obtained by the family of such Γ^n.

Let us first collect some information on X and Γ^n.

Proposition 4.1.

(4.2) Γ^n *is a projectively normal smooth curve of degree* nd *and genus* $p_n = ng + d\frac{n(n-1)}{2} - n + 1$;

(4.3) *Let $\mathcal{F} = \{\Gamma_t\}_{t\in D}$ be a flat family of projective curves parametrized by a smooth irreducible variety D such that:*

(i) *$\mathcal{F}$ is a projective family of curves in $\mathbb{P}^{r+1}$;*

(ii) *there exists a closed point t_0 in D such that Γ_{t_0} is a smooth curve of degree nd and genus $p_n = ng + d\frac{n(n-1)}{2} - n + 1$, complete intersection of a hypersurface of degree $n \geq 4$ with a cone over a projectively normal, nonspecial curve C of degree d and genus g.*

Then there is a neighborhood U of t_0 in D in the Zariski topology such that, for all closed points $t \in U$, Γ_t is again the complete intersection of a hypersurface of degree n with a cone over a projectively normal, nonspecial curve of degree d and genus g.

Proof. Let Δ be a general hyperplane section of Γ^n; we may think of Δ as a divisor on C belonging to the linear system $|\mathcal{O}_C(n)|$. If we denote by $h_\Delta(t)$ the Hilbert function of Δ, we find

$$h_\Delta(t) = \begin{cases} h^0(C, \mathcal{O}_C(t)) = td - g + 1, & \text{for } t = 1, \dots, n-1 \\ h^0(C, \mathcal{O}_C(n)) - 1 = nd - g & \text{for } t = n \\ nd & \text{for } t \geq n+1. \end{cases}$$

Therefore we have

$$\sum_{t=1}^{n}(nd - h_\Delta(t)) = \sum_{t=1}^{n}[g + (n-t)d - 1] + 1 = ng + d\frac{n(n-1)}{2} - n + 1 = p_n$$

which, in view of [C1], remark (1.8), (ii), implies that Γ^n is projectively normal. To see (4.3) we can assume D to be affine, thus $D = \mathrm{Spec}(A)$. We have then a family $Y \subset \mathbb{P}^{r+1}_A = \mathrm{Proj}\, A[x_0, \dots, x_{r+1}]$ over $\mathrm{Spec}(A)$. If $\mathcal{I}_Y$ is the ideal sheaf of Y in $\mathbb{P}^{r+1}$, then $\mathcal{I}(Y) = \bigoplus_{n\in\mathbb{N}} H^0(\mathcal{I}_Y(n))$ is the homogeneous ideal of Y in $S = A[x_0, \dots, x_{r+1}]$. Consider the homogeneous ideal $\mathcal{I} \subset \mathcal{I}(Y)$ generated by $\mathcal{I}(Y)_2 = H^0(\mathcal{I}_Y(2))$, and look at the scheme $W = \mathrm{Proj}(S/\mathcal{I})$. There is a commutative diagram

$$\begin{array}{ccc} Y \hookrightarrow W \subset & \mathbb{P}^{r+1}_A = \mathrm{Proj}\, A[x_0, \dots, x_{r+1}] \\ \searrow \ \downarrow & \\ \mathrm{Spec}(A) & \end{array}$$

and restricting to the local ring of D at t_0 we find that the scheme W is flat over the generic point ξ of D (the proof of this is similar to the one of proposition (1.6) of [C1] and will be omitted). Therefore W is flat over a neighborhood U of t_0 in D, thus giving a flat family of surfaces X_t, with central fiber the cone X. Notice that since C is smooth and projectively normal, then X is normal at the vertex P. For a general point ξ of D, X_ξ has to be singular (see [P, theorem (7.5)]), and has normal singularities (see [EGA, theorem 12.2.4]). On the other hand, by Theorem 2, Chapter 1 of Horowitz's thesis [Ho], X_ξ (which is irreducible and nondegenerate, as well as X) is a scroll. Therefore X_ξ has to be a cone, because these are the only scrolls which may have normal singularities as we will see in the Claim 4.4 below. In any case, since the family $\{\Gamma_t\}$ is very flat, lifting the equation of the hypersurface F_n, we see that also Γ_t is the complete intersection of a hypersurface of degree n with a cone over a projectively normal, nonspecial curve of degree d and genus g.

Claim 4.4. *Let T be an irreducible, nondegenerate normal surface in $\mathbb{P}^r$, $r \geq 3$ and assume that T is a scroll (that is T is ruled by lines). If T is singular then T is a cone over a smooth curve.*

Proof of 4.4. Let Q be a singular point of T and let H be the hyperplane section of T with a general hyperplane passing through Q. We shall separately discuss two cases.

Case 1: H is reducible.
Let $\pi : \widetilde{T} \to T$ be the minimal desingularization of T and let E be the divisor contracted by π in Q. Let moreover Σ be the linear system of divisors on $\widetilde{T}$ which are the pull-backs on $\widetilde{T}$ of the hyperplanes of $\mathbb{P}^r$ passing through Q. Clearly the base locus of Σ coincides with E and our hypothesis on H yields that the movable part of Σ is composed with a pencil $\mathcal{P}$. Now let R be a general point of T and $\widetilde{R}$ the corresponding point of $\widetilde{T}$. All curves of Σ passing through $\widetilde{R}$ must contain the curve of $\mathcal{P}$ passing through $\widetilde{R}$. This curve, in turn, has to be the pull-back on $\widetilde{T}$ of some curve on T contained in all the hyperplanes passing through Q and R. Whence clearly the general curve in $\mathcal{P}$ is the pull-back on $\widetilde{T}$ of a line contained in T and passing through Q. This implies that T is a cone with Q as vertex. The directrix of the cone has to be smooth by the normality of T.

Case 2: H is irreducible.
It is easy to construct in this case a flat family $\{H_t\}_{t \in B}$ such that:

(i) B is a smooth curve;

(ii) for every $t \in B$, H_t is a hyperplane section of T;

(iii) there exists a closed point $t_1 \in B$ such that for all $t \in B - \{t_1\}$, H_t is a smooth irreducible hyperplane section of T;

(iv) in correspondence with the point $t_1 \in B$ one has that H_{t_1} is a general hyperplane section of T through Q;

(v) H_{t_1} is reduced (because of the S_2 property of T at Q) irreducible and singular only at Q.

In view of our hypothesis on H, the lines of the scroll do not all pass through Q (otherwise T would be a cone and H reducible). But this clearly implies that H_t, for $t \neq t_1$, is birational to H_{t_1}, a contradiction since the geometric genus of H_{t_1} has to be strictly less than the one of H_t (see [Hi]).

This proves Claim 4.4 and hence also concludes the proof of Proposition 4.1. □

Now let N_X and N_{Γ^n} be the normal bundles of X and Γ^n, respectively in $\mathbb{P}^{r+1}$; $N_{\Gamma^n/X}$ the normal bundle of Γ^n in X. We have

Proposition 4.5. *Let $C \subset \mathbb{P}^r = \mathbb{P}H^0(L)^*$ be a smooth irreducible curve of genus $g \geq 1$, L a line bundle on C of degree $d \geq \begin{cases} 10 & \text{for } g = 1 \\ 4g+5 & \text{for } g \geq 2 \end{cases}$ and let X be the cone over C with vertex a point $P \in \mathbb{P}^{r+1} - \mathbb{P}^r$. Then*

$$H^0(X, N_X(-k)) = 0 \text{ for all } k \geq 2; \tag{4.6}$$

$$H^1(\Gamma^n, N_{\Gamma^n/X}) = 0; \tag{4.7}$$

$$h^0(\Gamma^n, N_{\Gamma^n}) = h^0(X, N_X) + h^0(\Gamma^n, N_{\Gamma^n/X}). \tag{4.8}$$

Proof. To see (4.6) first notice that by standard facts (for example applying (2.14) and [BEL]) it follows that

$$H^0(C, N_C(-k)) = 0 \quad \text{for all } k \geq 2. \tag{4.9}$$

Of course the assertion (4.6) is true for $k \gg 0$. So we may argue by descending induction. From the exact sequence

$$0 \to N_X(-k-1) \to N_X(-k) \to N_C(-k) \to 0$$

we see that (4.6) follows by (4.9). Since $N_{\Gamma^n/X} \cong \mathcal{O}_{\Gamma^n}(n)$, we have

$$\deg N_{\Gamma^n/X} = n^2 d > n^2 d - n(d - 2g + 2) = 2gn + dn(n-1) - 2n = 2p_n - 2$$

hence (4.7). To show (4.8) observe that by (4.6) we have $H^0(X, N_X(-n)) = 0$, hence there is an injection $H^0(X, N_X) \xrightarrow{h} H^0(\Gamma^n, N_{X|\Gamma^n})$. Now consider the exact sequence

$$0 \to H^0(\Gamma^n, N_{\Gamma^n/X}) \to H^0(\Gamma^n, N_{\Gamma^n}) \xrightarrow{f} H^0(\Gamma^n, N_{X|\Gamma^n}) \to 0 \tag{4.10}$$

the map f being surjective by (4.7). By the proof of (4.3) we can deduce the existence of a map ϕ which makes the following diagram commutative

$$\begin{array}{ccc} H^0(\Gamma^n, N_{\Gamma^n}) & \xrightarrow{f} & H^0(\Gamma^n, N_{X|\Gamma^n}) \\ & \searrow \phi \quad \nearrow h & \\ & H^0(X, N_X) & \end{array}$$

as one can see by a local computation. This implies that h is also surjective. The assertion then follows by (4.10). □

We shall finally prove the announced result on nonreduced components of the Hilbert scheme.

Theorem 4.11. *Let Γ^n be a smooth complete intersection with a hypersurface of degree $n \geq 4$ of a cone $X = X_{C,L}$ over a smooth irreducible curve C of genus $g = 1, 2$, embedded in $\mathbb{P}^r = \mathbb{P}H^0(L)^*$ by a line bundle L of degree $d \geq 3g+7$. Then Γ^n belongs to a unique irreducible component $\mathcal{V}$ of the Hilbert scheme $H_{nd,p_n,r+1}$, $r = d - g$,* *such that $\mathcal{V}$ is nonreduced and*

$$\dim \mathcal{V} = (r+1)(d+1) + (r-3+n)(1-g) + d\frac{n(n+1)}{2}.$$

Proof. Let $\mathcal{V}$ be a component of the Hilbert scheme containing the point $[\Gamma^n]$ and W the component of the Hilbert scheme containing $[C]$. The proof of (4.3) yields the existence of a morphism $\psi : \mathcal{V}_{\rm red} \to \mathcal{H}(W)_{\rm red}$, which is clearly surjective, its (set-theoretical) general fiber being nothing else than the projective space corresponding to the linear system of curves cut out on a cone by the hypersurfaces of degree n in $\mathbb{P}^{r+1}$. Since for any such smooth curve Γ' on a cone X' we have, as we saw in (4.7), that $H^1(\Gamma', N_{\Gamma'/X'}) = 0$, then $h^0(\Gamma', N_{\Gamma'/X'}) = n^2 d - (ng + d\frac{n(n-1)}{2} - n + 1) + 1 = d\frac{n(n+1)}{2} - (g-1)n$. Therefore we find

$$\begin{aligned}\dim \mathcal{V} &= \dim \mathcal{H}(W) + d\frac{n(n+1)}{2} - (g-1)n \\ &= \dim W + r + 1 + d\frac{n(n+1)}{2} - (g-1)n \\ &= (r+1)(d+1) + (r-3+n)(1-g) + d\frac{n(n+1)}{2}.\end{aligned}$$

Thus the part of the statement concerning the uniqueness of $\mathcal{V}$ and its dimension is proved. Now we notice that, by Proposition 2.1 and by (2.13) and (4.8), we have

$$\begin{aligned}&\dim \mathcal{V} = h^0(\Gamma^n, N_{\Gamma^n/X}) + \dim \mathcal{H}(W) = h^0(\Gamma^n, N_{\Gamma^n/X}) + h^0(C, N_C) + r + 1 \\ &< h^0(\Gamma^n, N_{\Gamma^n/X}) + h^0(C, N_C) + r + 1 + \gamma_{C,L} \le h^0(\Gamma^n, N_{\Gamma^n/X}) + h^0(X, N_X) \\ &= h^0(\Gamma^n, N_{\Gamma^n}) = \dim T_{[\Gamma^n]}\mathcal{V}\end{aligned}$$

where $T_{[\Gamma^n]}\mathcal{V}$ is the Zariski tangent space to $\mathcal{V}$ at $[\Gamma^n]$. □

Remark 4.12. We limited ourselves to constructing nonreduced components of the Hilbert scheme of curves that are complete intersection of cones with hypersurfaces. Similar results can be proved also for all curves of sufficiently high degree on such cones and for curves which are algebraically equivalent to a hypersurface intersection plus a line.

References

[BC] Ballico, E., Ciliberto, C., On gaussian maps for projective varieties, in: Geometry of complex projective varieties, Cetraro (Italy), June 1990. Seminars and Conferences 9, Mediterranean Press 1993, 35–54.

[BEL] Bertram, A., Ein, L., Lazarsfeld, R., Surjectivity of Gaussian maps for line bundles of large degree on curves, in: Algebraic Geometry, Proceedings Chicago 1989, Lecture Notes in Math. 1479, Springer, Berlin–New York 1991, 15–25.

[C1] Ciliberto, C., On the Hilbert scheme of curves of maximal genus in a projective space, Math. Z. 194 (1987), 351–363.

[C2] Ciliberto, C., Sul grado dei generatori dell'anello canonico di una superficie di tipo generale, Rend. Sem. Mat. Univ. Politecn. Torino 41 (3) (1983), 83–111.

[CLM1] Ciliberto, C., Lopez, A. F., Miranda, R., Projective degenerations of K3 surfaces, Gaussian maps and Fano threefolds, Invent. Math. 114 (1993), 641–667.

[CLM2] Ciliberto, C., Lopez, A. F., Miranda, R., Classification of varieties with canonical curve section via Gaussian maps on canonical curves, Preprint 1994.

[CM] Ciliberto, C., Miranda, R., On the Gaussian map for canonical curves of low genus, Duke Math. J. 61 (1990), 417–443.

[E] Ellia,P., D'autres composantes non rèduites de Hilb $\mathbb{P}^3$, Math. Ann. 277 (1987), 433–446.

[EGA] Grothendieck, A., Dieudonné, J., EGA IV, Étude locale des schémas et des morphismes de schémas, Publ. Math. IHES 28 (1966).

[Fa] Fania, M. L., Trigonal hyperplane sections of projective surfaces, Manuscripta Math. 68 (1990), 17–34.

[Fl] Fløystad, G., Determining obstructions for space curves, with applications to non-reduced components of the Hilbert scheme, J. Reine Angew. Math. 439 (1993), 11–44.

[Fu] Fujita, T., On the structure of polarized manifolds with total deficiency one II, J. Math. Soc. Japan 33-3 (1981), 415–434.

[G] Green, M., Koszul cohomology and the geometry of projective varieties, J. Differential Geom. 19 (1984), 125–171.

[GP] Gruson, L., Peskine, C., Genre des courbes dans l'espace projectif (II), Ann. Sci. École Norm. Sup. 15 (1982), 401–418.

[Ho] Horowitz, T., Varieties of low degree, Brown University Ph. D. Thesis (1982), and Varieties of low Δ-genus, Duke Math. J. 50 (1983), 667–683.

[Hi] Hironaka, H., On the arithmetic genera and the effective genera of algebraic curves, Mem. of the Coll. of Sci., Univ. of Kyoto, Ser. A 30 (1957), 177–195.

[I] Ionescu, P., Variétés projectives lisses de degrés 5 et 6, C. R. Acad. Sci. Paris 293 (1981), 685–687.

[K] Kleppe, J. O., Non-reduced components of the Hilbert scheme of smooth space curves, in: Proceedings Rocca di Papa 1982, Lecture Notes in Math. 1266, Springer, Berlin–New York 1987, 180–207.

[L] Lopez, A. F., Surjectivity of Gaussian maps on curves in $\mathbb{P}^r$ with general moduli, preprint.

[M] Mumford, D., Further pathologies in algebraic geometry, Amer. J. Math. 84 (1962), 642–648.

[P] Pinkham, H., Deformations of algebraic varieties with G_m action, Harvard University Ph. D. Thesis (1974), and Astérisque 20 (1974).

[S1] Schreyer, F. O., A standard basis approach to syzygies of canonical curves, J. Reine Angew. Math. 421 (1991), 83–123.

[S2] Schreyer, F. O., Syzygies of canonical curves and special linear series, Math. Ann. 275 (1986), 105–137.

[Sc] Scorza, G., Le varietà a curve sezioni ellittiche, Ann. di Mat. (3) 15 (1908), 217–273.

[Se] Serrano, F., The adjunction mapping and hyperelliptic divisors on a surface, J. Reine Angew. Math. 381 (1987), 90–109.

[St] Stevens, J., Deformations of cones over hyperelliptic curves, preprint.

[T] Tendian, S., Deformations of cones over curves of high degree, Ph. D. Thesis, Univ. of North Carolina 1990.

[V] Voisin, C., Courbes tetragonales et cohomologie de Koszul, J. Reine Angew. Math. 387 (1988), 111–121.

[W1] Wahl, J., The Jacobian algebra of a graded Gorenstein singularity, Duke Math. J. 55 (1987), 843–871.

[W2] Wahl, J., Introduction to Gaussian maps on an algebraic curve, in: Complex Projective Geometry, Trieste–Bergen 1989, London Math. Soc. Lecture Notes Ser. 179, Cambridge Univ. Press 1992, 304–323.

[Z] Zak, F. L., Some properties of dual varieties and their application in projective geometry, in: Algebraic Geometry, Proceedings Chicago 1989, Lecture Notes in Math. 1479, Springer, Berlin–New York 1991, 273–280.

Zero-estimates, intersection theory, and a theorem of Demailly

Lawrence Ein, Robert Lazarsfeld**, and Michael Nakamaye****

§1. Introduction

Let X be a smooth complex projective variety of dimension n, and let A be an ample line bundle on X. Fujita has conjectured that the adjoint bundle $\mathcal{O}_X(K_X + mA)$ is basepoint-free if $m \geq n + 1$ and very ample if $m \geq n + 2$. These conjectures and related questions have attracted considerable attention in recent years because they would extend to varieties of arbitrary dimension the most elementary facts about linear series on curves.

In the seminal paper [De1], Demailly drew on deep analytic tools to make the first serious attack on Fujita's conjectures. More generally, starting with a nef line bundle L, Demailly gives numerical and geometric positivity conditions on L to guarantee that the adjoint linear series $|K_X + L|$ be free or very ample or separate any given number of points and jets. He deduces for example that if A is ample, then $\mathcal{O}_X(2K_X + 12n^n A)$ is very ample. While the numbers are rather far from Fujita's predictions, this was the first effective criterion for very ampleness, and it represented a real breakthrough.

Starting with [EL1], it soon became clear that one could also obtain effective (and eventually stronger) results using algebraic methods. Specifically, in [EL1], [Kol1], [Siu1], [Siu2] [De2] and most recently in [AS] (as explained and extended by Kollár [Kol2]), the cohomological ideas pioneered by Kawamata, Reid, Shokurov and others in connection with the minimal model program are applied to obtain effective results on adjoint and pluricanonical linear series. It seems fair to say that by now it is relatively quick to obtain many of the statements occuring in [De1].

---

* Partially supported by NSF Grant DMS 93-02512

** Partially supported by NSF Grant DMS 94-00815

*** Partially supported by an NSF Graduate Fellowship

But if recent work has largely superceeded many of the specific results of [De1], the geometric ideas underlying that paper remain exceedingly interesting, and they are likely to have other applications in the future. However the apparently analytic nature of Demailly's techniques have limited their defusion in the algebro-geometric community. Our purpose here is to explain in algebro-geometric language the geometric underpinnings of [De1], and to develop algebro-geometric analogues of some of the analytic parts of Demailly's argument.

One of the fascinating aspects of this story is that Demailly's work is very close to some ideas that come up in diophantine approximation and transcendence theory. Specifically, one can see the crucial self-intersection formula of [De1] as proving a very refined and novel sort of *zero estimate*. This viewpoint has been very helpful in our thinking about these matters, and we attempt to emphasize the connection here. We also present a conjecture that would put the formula in question into the context of contemporary intersection theory. Some related ideas are applied in a deformation theoretic setting in the paper [EKL].

There is very little here for which we claim any real novelty. We offer these notes in the spirit of *pro bono* work, in thehope that an explanation of the geometric ideas of [De1] may be of some interest to algebraic geometers who — like us — are less than fluent in the analytic language in which that paper is written. We hope that they may contribute to the fruitful interchange that has developed between the algebraic and analytic viewpoints on these matters. The application to adjoint series that we give is slightly weaker than Demailly's because we were unable to prove in general a certain inequality on the self-intersection of the moving part of a linear series. We circumvent this problem with a hyperplane section argument, but this leads to a slight degrading of the numbers. Since the bulk of these notes were written, we have learned that Fujita has proven the general case of the inequality in question in his very interesting note [Fuj] (apparently also motivated by understanding [De1] algebro-geometrically). This means one could give an algebraic proof of the the full statements of [De1]. But since it is in any event the methods rather than the specific results that are of interest here, we have decided not to incorporate Fujita's theorem in our presentation. Finally, we refer the reader to [Laz], §7, for an account of Demailly's approach in the particularly elementary and transparent two dimensional case, and to Demailly's notes [De4] for an overview of the analytic techniques.

We work throughout over the complex numbers $\mathbb{C}$. If X is a projective variety, and L is a line bundle on X, we write indifferently $\int_X c_1(L)^n$, $c_1(L)^n$ or simply L^n for the top Chern number of L. However if $V \subset X$ is a subvariety of dimension k, we generally denote the degree $\int_V c_1(L|V)^k$ by the short-hand $L^k \cdot V$. We trust that this variable notation will not cause undue confusion.

We wish to thank F. Angelini and G. Fernandez del Busto for valuable discussions. We are also grateful to the referee for suggesting several improvements and catching a number of inaccuracies in an earlier version of these notes.

§1. Informal overview of strategy

This section is devoted to an informal overview, in algebro-geometric language, of the rough strategy of [De1]. Our hope is that an impression of the overall picture and main issues will render subsequent sections easier to absorb. We ignore here entirely the problem of trying to weld together the various constants that appear in different parts of the discussion. However we do state and prove some Lemmas in the generality required for later use.

To set the stage, we start by describing the cohomological result on which the whole approach depends. Let X be a smooth projective variety of dimension n, let $S \subset X$ be a finite set, and let L be an ample line bundle on X. Fix an integer $s \geq 0$.

Theorem 1.1. *Suppose there exists a divisor $E \in |kL|$ satisfying the following properties:*

(1.1.1) *For every $x \in S$, $\operatorname{mult}_x(E) > k(n+s)$;*

(1.1.2) *There is an open neighborhood $U \subset X$ of S such that*

$$\operatorname{mult}_y(E) \leq k \quad \textit{for all } y \in U - S.$$

Then

$$H^1(X, \mathcal{O}_X(K_X + L) \otimes \mathcal{I}_S^{s+1}) = 0.$$

In particular, $|K_X + L|$ generates s-jets at every point of S, i.e. the evaluation map

$$H^0(X, \mathcal{O}_X(K_X + L)) \longrightarrow \bigoplus_{x \in S} H^0(X, \mathcal{O}_X(K_X + L) \otimes \mathcal{O}_X / \mathcal{I}_x^{s+1})$$

is surjective.

The theorem is proved by applying vanishing for $\mathbb{Q}$-divisors on a blow-up of X, or analytically via multiplier ideals and Nadel's Vanishing theorem. We refer to [EV2], (7.5),(7.7) or [De2], (5.6) for details, or to [Laz], (6.4), for the particularly simple and transparent case of surfaces. We will say that a divisor $E \in |kL|$ has an *almost isolated singularity of index* $> (n+s)$ along S if it satisfies the hypotheses (1.1.1) and (1.1.2) of the Theorem.

Imagine now that we want to apply Theorem 1.1 to show that $|K_X + L|$ generates s-jets at a given point x. Taking $S = \{x\}$, condition (1.1.1) is very easy to arrange. In fact, by Riemann–Roch:

$$h^0(X, \mathcal{O}_X(kL)) = L^n \cdot \frac{k^n}{n!} + o(k^n).$$

On the other hand, it is

$$\binom{k(n+s)+n}{n} = (n+s)^n \cdot \frac{k^n}{n!} + o(k^n)$$

conditions to impose multiplicity $> k(n+s)$ at a given point. Therefore if

$$\int_X c_1(L)^n > (n+s)^n, \tag{1.2}$$

then for $k \gg 0$ there exists a divisor $E \in |kL|$ for which (1.1.1) holds.[1] Note that (1.2) will certainly be satisfied if for instance $L = (n+s+1)A$ for some ample bundle A.

The essential difficulty arises in trying to guarantee (1.1.2): how can one control the singularities of a divisor E obtained by imposing high multiplicity at a given point? A very similar problem comes up in many situations in diophantine approximation and transcendence theory. Typically one constructs a polynomial (or section of a line bundle) σ vanishing to high order at some point, and the hard part is to show that σ does not vanish identically, or to too high order at some other point.[2] One line of attack on these questions goes under the general heading of *zero-estimates*. We will try to give a rough idea of the strategy in the present situation.

As a matter of notation, given a divisor $E \in |kL|$, define the *index* of E at a point $x \in X$ to be the "normalized multiplicity":

$$\operatorname{ind}_x(E) = \frac{\operatorname{mult}_x(E)}{k}.$$

As we will work asymptotically with the linear series $|kL|$ for $k \gg 0$, this is the convenient measure of singularity. For $\sigma \in \mathbb{Q}^+$, put

$$Z_\sigma(E) = \{x \in E \mid \operatorname{ind}_x(E) \geq \sigma\}.$$

Thus $Z_\sigma(E)$ is a Zariski-closed subset of X. It carries a natural scheme structure, locally defined by the vanishing of all partial derivatives of orders $< k\sigma$ of a local equation for E.

To explain the idea of the zero estimates, suppose that $E \in |kL|$ is a divisor, and $x \in X$ is a point such that

$$\operatorname{ind}_x(E) > \alpha$$

for some rational $\alpha > 1$, and assume that x is not an almost isolated singularity of E. Then for elementary dimensional reasons, there must exist an irreducible subvariety $V \subset X$ of positive dimension passing through x such that the index of E "jumps" by at least $\frac{1}{n}$ along V. More precisely, V is an irreducible component of both $Z_\sigma(E)$ and $Z_{\sigma+(1/n)}(E)$ for suitable σ, so that E has index $\geq \sigma + \frac{1}{n}$ along V, but there exists an

1 We remark that it is already in the construction of E — or more precisely, a current T which plays an analogous role — that Demailly's approach begins to become analytic. His idea is to invoke Yau's solution of the Calabi conjecture to produce a sequence of metrics on L with a given amount of mass concentrated closer and closer to the finite set S. Then he passes to a limit to produce T.

2 Schmidt's notes [Sch] contain a very nice overview of how these different steps come into the proof of Roth's theorem.

open set $U \subset X$ meeting V such that E has index $< \sigma$ on $U - V$. This is a typical "gap" argument which we shall formalize and prove in Lemma 1.6 below.

At this point there are two possibilities. If E has an almost isolated singularity of index $> \alpha$ at x, then we are in a position to apply Theorem 1.1. Alternatively, we can find a positive dimensional "multiplicity jumping" subvariety $V \ni x$ as above. In classical situations — e.g. when $X = \mathbb{P}^n$ — one obtains in the latter case an absolute upper bound on the degree of V:

Proposition 1.3. *Let $X = \mathbb{P}^n$ and $L = \mathcal{O}_{\mathbb{P}^n}(1)$, and fix $\epsilon > 0$. Then there is a constant $C = C(n, \epsilon)$ depending only on n and ϵ with the following property. Suppose that*

$$E \in |kL| \ (k \gg 0)$$

is a hypersurface of degree k, and $V \subset E$ is an irreducible component of both $Z_\sigma(E)$ and $Z_{\sigma+\epsilon}(E)$ for some $\sigma > 0$. Then

$$\deg(V) \leq C.$$

One may view Proposition 1.3 as the "one-factor" case of the Faltings Product Theorem [Falt], Theorem 3.1, and we will give the elementary proof shortly.

Now return to an arbitrary smooth projective variety X, and imagine — which unfortunately is not true — that one had an analogous statement for the jumping locus V associated to a divisor $E \in |kL|$ on X. Consider as before a divisor E with index $> \alpha$ a given point $x \in X$. Then if E does not have an almost isolated singularity at x, the gap argument and the imaginary extension of Proposition 1.3 would yield a positive dimensional subvariety $V \subset X$ of bounded degree with respect to L. On the other hand, since we are interested in very positive line bundles L — e.g. $L = mA$ for A ample and large m — we are free to assume that there are no subvarieties of small degree with respect to L. So we would have a mechanism to guarantee the presence of an almost isolated singularity, and we would be done!

In a word, the basic strategy will be to develop an appropriate extension of Proposition 1.3. In order to understand the issues involved, let us recall the

Proof of Proposition 1.3. Let $c = \text{codim}(V, \mathbb{P}^n)$, and denote by

$$F \in H^0(\mathbb{P}^n, \mathcal{O}_{\mathbb{P}^n}(k))$$

the equation defining E. Then $Z_\sigma(E)$ is cut out by homogeneous polynomials of degrees $\leq k$, to wit all the partial derivatives $\{D(F)\}$ of F of order less than σk in the homogeneous coordinates on $\mathbb{P}^n$. Therefore V is generically cut out by the $\{D(F)\}$. On the other hand, being also a component of $Z_{\sigma+\epsilon}(E)$, V is contained in the common zeroes of all the partial derivatives $\{D'(F)\}$ for D' of order $< (\sigma + \epsilon)k$. In particular, each of the $[\sigma k]^{\text{th}}$ partials $D(F)$ must vanish to order $\geq [\epsilon k]$ along V. In other words, V is an irreducible component of the common zeroes of a collection of hypersurfaces of degrees $\leq k$, each of which vanishes to order $\geq k\alpha$ along V, where $\alpha = \frac{[k\epsilon]}{k} \sim \epsilon$.

Lemma 1.4 below then shows that

$$(k\alpha)^c \cdot \deg(V) \leq k^c.$$

The proposition follows. □

Lemma 1.4 (Compare [Falt], Thm.3.1 and [Fult],12.4). *Let X be a smooth projective variety of dimension n, and let M be an ample line bundle on X. Assume given a collection*

$$s_1, \ldots, s_N \in \Gamma(X, M)$$

of sections of M whose common zeroes contain a codimension c subvariety $V \subset X$ as an irreducible component. Suppose moreover that each s_i vanishes to order $\geq a$ along V. Then for any ample line bundle L,

$$\deg_L(V) \leq \frac{(M^c \cdot L^{n-c})}{a^c}.$$

Proof. After possibly replacing each s_i by a linear combination of the sections in question, we can assume that V is an irreducible component of the zero locus $Z = \{s_1 = s_2 = \ldots = s_c = 0\}$. Let $E_i \subset X$ be the divisor of s_i. According to the general theory of [Fult], the intersection product $E_1 \cdot E_2 \ldots E_c = (M^c) \in A^c(X)$ can be decomposed into a sum of classes $\gamma_j \in A^c(X)$ coming from various "distinguished subvarieties" of $E_1 \cap \cdots \cap E_c$ determined by intersecting the regular embedding

$$E_1 \times \cdots \times E_c \subset X \times \cdots \times X$$

with the small diagonal $X \subset X \times \cdots \times X$. Since the E_i intersect properly along V, V itself appears as one of these components, and since $\mathrm{mult}_V(E_i) \geq a$, the cycle $[V]$ occurs with coefficient $\geq a^c$ in the decomposition (cf. [Fult], (12.4.8)). Hence

$$(L^{n-c} \cdot M^c) = a^c \cdot \deg_L(V) + \sum{}' \deg_L(\gamma_j),$$

where the sum on the right is taken over all the distinguished components other than V. Therefore it is enough to show that $\deg_L(\gamma_j) \geq 0$. But this follows from [FL] (cf. [Fult], (12.2)) since the restriction to X of the normal bundle $N_{E_1 \times \cdots \times E_c / X \times \cdots \times X}$ is ample, and hence nef. □

Remark. The Lemma and proof remain valid assuming only that M is nef. Faltings [Falt] (Proposition 2.3) gives an elmentary proof in the case $X = \mathbb{P}^n$ which avoids most of the intersection theory quoted here.

Looking over the proof of Proposition 1.3, one sees that the first essential point about working on $\mathbb{P}^n$ is that one is able to differentiate homogeneous polynomials. In fact, a similar argument works on an arbitrary smooth projective variety X provided that one controls the positivity of the tangent bundle TX of X. The next section is devoted to this rather direct extension of Proposition 1.3 to general X. However the resulting statement (Theorem 2.1) contains a factor of $\int_X c_1(L)^n$ on the right hand side, so one does not directly get an absolute upper bound on the degree of the multiplicity-jumping locus V. The most interesting part of the approach — and this is where Demailly's argument becomes essentially analytic — starts with the observation that the number of degrees of freedom in choosing a divisor E satisfying (1.1.1) grows with $\int_X c_1(L)^n$. So for the application to adjoint linear series, one will want to apply (1.3) or (2.1) to the general divisor E in a rather large linear series. Motivated by some hints in [De1], (11.1), we develop in §3 some machinery for obtaining in effect a bound on the degree of the jumping locus V that takes into account the dimension of the linear series in which E moves. We give the application of this "zero-estimate with moving parts" (Theorem 3.9) in the spirit of [De1] to adjoint series in §4.

We conclude this section with a formulation and proof of the "gap" lemma that was used above. We begin with

Lemma 1.5. *Let Y be an irreducible variety of dimension n. Suppose that*

$$V_{m+1} \subseteq V_m \subseteq \cdots \subseteq V_1 \subseteq V_0 = Y. \tag{1.5.1}$$

is a chain of irreducible subvarities of Y, with $m \leq n$. Assume that the codimension of V_{m+1} is $< m+1$. Then there is an index c $(0 \leq c \leq m)$ such that V_c and V_{c+1} share a common irreducible component of codimension c.

Proof. Induction on m, the case $m = 0$ being clear. So we assume $m > 0$. If the codimension of V_m is $< m$, then we are done by induction. If the codimension of V_m is $\geq m$, then $\dim V_m = \dim V_{m+1}$. □

By applying above Lemma to jumping locus, we obtain the following Lemma.

Lemma 1.6. *Let X be a smooth irreducible vairiety of dimension n and let $E \in |kL|$ be a divisor on X with $\operatorname{ind}_x(E) > \alpha > n + s$ for some $x \in X$. Choose rational numbers*

$$0 = \beta_0 < \beta_1 \leq \cdots \leq \beta_n = 1.$$

Set

$$Z_j = Z_{\beta_j}(E) = \{y \in X \mid \operatorname{ind}_y(E) \geq \beta_j\} \ (0 \leq j \leq n).$$

If the dimension of Z_n is greater than zero at x, then there exists an irreducible codimension c subvariety $V \subset X$ passing through x $(1 \leq c \leq n-1)$, which is a common irreducible component of both Z_c and Z_{c+1}.

In other words the index of E jumps by at least $\beta_c - \beta_{c+1}$ along V. For example, assuming $\beta_j = \frac{j-1}{n-1}$ for $2 \leq j \leq n$ and $\beta_1 = t$ where $\frac{1}{n-1} \gg t > 0$. Then Z_1 is not equal to X. We deduce the fact stated above that if x is not an almost isolated singularity of index $> \alpha$, then for some σ, Z_σ and $Z_{\sigma+\frac{1}{n-1}}$ or $Z_{\sigma+\frac{1}{n-1}-t}$ share an irreducible component which is a proper subvariety of positive dimension.

Proof. The sets Z_j lie in a chain:

$$Z_n \subseteq Z_{n-1} \subseteq \cdots \subseteq Z_1 \subseteq Z_0 = X.$$

Starting with Z_n and working up in dimension, we can choose irreducible components V_j of Z_j passing through x such that $V_{j+1} \subseteq V_j$. Under the stated hypothesis, we can assume furthermore that codim $V_n < n$. So we arrive at a chain of irreducible varieties:

$$V_n \subseteq V_{n-1} \subseteq \cdots \subseteq V_1 \subseteq V_0 = X,$$

with codim $V_n < n$. The statement then follows from (1.5). □

§2. Derivatives and the self-intersection inequality

By way of warm-up for §3, we prove in this section a simple special case of Demailly's self-intersection inequality [De1], (10.7), (10.8), [De3]. The result in question gives a first generalization of Proposition 1.3 to arbitrary smooth projective varieties, and — while elementary — it is already interesting from the standpoint of enumerative geometry.[3] Our impression is that even this clasically oriented part of Demailly's theorem isn't well known in the algebro-geometric community, so we felt it worthwhile to write out a proof here (which differs only in language from [De1]). The argument follows closely the model of Proposition 1.3, the main new point being that now one has to keep track of the positivity of the tangent bundle in order to make sense of differentiating sections of a line bundle. It is natural to expect that one should also be able to give a proof of Theorem 2.1 based on the general machinery of contemporary intersection theory. This leads us to formulate a conjecture of a local nature concerning normal cones associated to hypersurfaces. We present this conjecture at the end of the section, and we hope it may be of some interest to intersection theorists.

We start with some formalism concerning vector bundles and $\mathbb{Q}$-divisors. Let E be a vector bundle on a smooth projective variety X, and let A be a $\mathbb{Q}$-divisor on X, or more generally an element of $\mathrm{Pic}(X) \otimes \mathbb{Q}$. As in [Myk], it will be convenient to deal with formal expressions of the form $E(A)$, which we think of as the "$\mathbb{Q}$-vector bundle" resulting from twisting E by A. We will want to discuss in particular the positivity of such objects, and this reduces in the usual way to statements about positivity of

3 We stress however that the statement here ignores the most interesting and subtle part of Demailly's inequality, to wit the presence of a contribution from the absolutely continuous part T_{abc} of the current with which he works. The purpose of §3 will be to develop an algebro-geometric analogue of this more delicate result.

$\mathbb{Q}$-divisors (whose meaning is clear). Specifically, we say that $E(A)$ is *ample* if the $\mathbb{Q}$-divisor class

$$c_1(\mathcal{O}_{\mathbb{P}(E)})(1) + \pi^*A \in NS(\mathbb{P}(E)) \otimes \mathbb{Q}$$

is ample on the projectivization $\pi : \mathbb{P}(E) \longrightarrow X$. Most of the elementary formal properties of ample vector bundles hold in this context. For example, if $E(A)$ is ample, then so too is $\mathrm{Sym}^{\ell}(E)(\ell A)$. Or again, given an exact sequence $0 \longrightarrow E' \longrightarrow E \longrightarrow E'' \longrightarrow 0$ of vector bundles, if $E'(A)$ and $E''(A)$ are ample, then $E(A)$ is ample. These (and similar asertions) are most easily proved by taking a finite flat covering $f : Y \longrightarrow X$ such that f^*A is (numerically equivalent to) a genuine line bundle, and reducing to the analogous statements for ample vector bundles on Y. Nefness of $E(A)$ is defined similarly, and in fact for nefness there is no difficulty in allowing $A \in \mathrm{Pic}(X) \otimes \mathbb{R}$.

Our goal now is to prove the following version of Demailly's self-intersection inequality. We keep the notation of §1 concerning the index of a divisor and the corresponding multiplicity loci.

Theorem 2.1 ([De1], Cor. 10.8). *Let X be a smooth projective variety of dimension n, let L be an ample line bundle on X, and assume that $\delta \geq 0$ is a real number such that the twisted tangent bundle $TX(\delta L)$ is nef. Given rational numbers $\epsilon, \sigma > 0$, suppose that*

$$E \in |kL|$$

is a divisor containing a subvariety V, of codimension c in X, such that V is an irreducible component of both $Z_{\sigma}(E)$ and $Z_{\sigma+\epsilon}(E)$. Then

$$\deg_L(V) \leq \frac{(1+\sigma\delta)^c}{\epsilon^c} \int_X c_1(L)^n.$$

Observe that when $X = \mathbb{P}^n$ one can take $\delta = 0$, and the statement reduces to Proposition 1.3. As in the proof of that Proposition, the essential point is to study the equations defining various multiplicity loci of E. So we start with some remarks about differentiating sections of a line bundle.

Let B be any line bundle on the smooth projective variety X. Recall that there are so-called bundles of principal parts P^{ℓ}_B associated to B: the fibre of P^{ℓ}_B at a given point $x \in X$ is the space $B \otimes \mathcal{O}_X/\mathcal{I}_x^{\ell+1}$ of ℓ-jets at x of sections of B. These jet bundles are connected by exact sequences:

$$0 \longrightarrow \mathrm{Sym}^{\ell}(\Omega^1_X) \otimes B \longrightarrow P^{\ell}_B \longrightarrow P^{\ell-1}_B \longrightarrow 0 \tag{2.2}$$

and a basic fact is that any section $\phi \in \Gamma(X, B)$ lifts to a section $j_{\ell}(\phi) \in \Gamma(X, P^{\ell}_B)$. The vector bundle

$$\mathcal{D}^{\ell}_B = \mathrm{Hom}(P^{\ell}_B, B) = \check{P}^{\ell}_B \otimes B$$

is then the sheaf of differential operators of order $\leq \ell$ on B. They sit in exact sequences

$$0 \longrightarrow \mathcal{D}_B^{\ell-1} \longrightarrow \mathcal{D}_B^{\ell} \longrightarrow \mathrm{Sym}^{\ell}(TX) \longrightarrow 0 \tag{2.3}$$

dual to (2.2). A section $\phi \in \Gamma(X, B)$ determines a homomorphism of vector bundles

$$d_\ell(\phi) : \mathcal{D}_B^{\ell} \longrightarrow B$$

defined for example via the transpose of $j_\ell(\phi)$. Locally, representing ϕ by a function f, $d_\ell(\phi)$ is just the map which takes a differential operator D to the function $D(f)$. It follows that $d_\ell(\phi)$ is zero precisely at the locus where ϕ vanishes to order $> \ell$. More precisely, let E be the divisor of ϕ, and consider the multiplicity locus

$$\Sigma_\ell = \Sigma_\ell(E) = \{x \in X \mid \mathrm{mult}_x(E) > \ell\},$$

with its natural scheme structure. Then the image of $d_\ell(\phi)$ is just the ideal sheaf of this scheme, i.e. one has a surjective sheaf homomorphism:

$$\mathcal{D}_B^{\ell} \longrightarrow B \otimes \mathcal{I}_{\Sigma_\ell} \tag{2.4.}$$

Now suppose that $A \in \mathrm{Pic}(X) \otimes \mathbb{Q}$ is an ample $\mathbb{Q}$-line bundle such that the $\mathbb{Q}$-vector bundle $TX(A)$ is ample.

Lemma 2.5. *If ℓ is any sufficiently large and divisible integer such that ℓA is a genuine line bundle (i.e. integral), then*

$$\mathcal{D}_B^{\ell} \otimes \mathcal{O}_X(\ell A)$$

is globally generated for all line bundles B.

Proof. Let H be a fixed very ample divisor on X. We shall show that if ℓ is any sufficiently large and divisble integer $\mathcal{D}_B^{\ell} \otimes \mathcal{O}_X(\ell A)$ is a 0-regular sheaf with respect to $\mathcal{O}_X(H)$. It is well known that such a 0-regular sheaf is generated by its global sections. First we will show that there is an integer ℓ_0 such that for all integers $\ell \geq \ell_0$, the sheaves $\mathrm{Sym}^{\ell}\, TX \otimes \mathcal{O}_X(kA)$ are 0-regular with respect to H, for all k such that $k \geq \ell$ and kA is a genuine line bundle. This means we will need to show that

$$H^i(\mathrm{Sym}^{\ell}\, TX \otimes \mathcal{O}_X(kA) \otimes \mathcal{O}_X(-iH)) = 0, \quad \text{for all } i > 0.$$

Let $\pi : \mathbb{P}(TX) \longrightarrow X$ be the projectivization of TX. Then we observe that

$$\begin{aligned} &H^i(\mathrm{Sym}^{\ell}\, TX \otimes \mathcal{O}_X(kA - iH)) \\ &\quad = H^i(K_{\mathbb{P}(TX)} \otimes \mathcal{O}_{\mathbb{P}(TX)}(\ell) \otimes \pi^*\mathcal{O}_X(kA - iH) \otimes K_{\mathbb{P}(TX)}^{-1}). \end{aligned} \tag{2.6}$$

Since $\mathcal{O}_{\mathbb{P}(TX)}(1) \otimes \pi^* A$ is ample, we see that for sufficiently large ℓ_0 the cohomology groups in (2.6) vanish if $\ell \geq \ell_0$, $k \geq \ell$ and kA is a geniune line bundle by Kodaira's theorem. By Serre's vanishing theorem, we can find an integer ℓ_1 such that for all $k \geq \ell_1$ the sheaves $\mathrm{Sym}^j\, TX \otimes \mathcal{O}_X(kA)$ are 0-regular when kA is a genuine line bundle, and $j = 0, 1, \ldots, \ell_0$. By (2.3), we see that $\mathcal{D}_B^{\ell} \otimes \mathcal{O}_X(kA)$ is 0-regular if $\ell \geq Max(\ell_0, \ell_1)$ and $k \geq \ell$. □

Suppose now given as above a divisor $E \in |B|$. Then for sufficiently large and divisible ℓ one has a surjective map of sheaves

$$\mathcal{D}_B^\ell \otimes \mathcal{O}_X(\ell A) \longrightarrow \mathcal{O}_X(B + \ell A) \otimes \mathcal{I}_{\Sigma_\ell}.$$

Therefore $\mathcal{I}_{\Sigma_\ell}(B + \ell A)$ is generated by its global sections. We will refer to sections in the image of

$$H^0(X, \mathcal{D}_B^\ell \otimes \mathcal{O}_X(\ell A)) \longrightarrow H^0(X, \mathcal{O}_X(B + \ell A))$$

as *differential sections* of order ℓ since they arise by a process of differentiation. For instance suppose $X = \mathbb{P}^n$, $L = \mathcal{O}_{\mathbb{P}^n}(1)$ and $B = \mathcal{O}_{\mathbb{P}^n}(k)$. Then $\mathcal{D}_B^\ell = \oplus \mathcal{O}_{\mathbb{P}^n}(\ell)$. We can take $\delta = -1$ and then the differential sections are exactly the partial derivatives of order ℓ of the homogeneous polynomial defining E.

Now we give the

Proof of Theorem 2.1. We can increase δ slightly and assume that $\delta \in \mathbb{Q}$, and that $TX(\delta L)$ is ample, for then the general statement of the Theorem follows by passing to the limit in δ. Similarly, we may decrease ϵ slightly, so we may assume that $\mathrm{mult}_V(E) > k(\sigma + \epsilon)$. Note also that it suffices to prove the theorem for $aE \in |akL|$ for any positive $a \in \mathbb{Z}$. Set

$$\ell =_{\text{def}} \sigma k, \quad m =_{\text{def}} \epsilon k.$$

Replacing k by ak where a is suffiently divisble, we may assume that ℓ, m and $\ell\delta$ are all positive integers. Futhermore, by (2.5) we may assume that $\mathcal{D}_{kL}^\ell \otimes \mathcal{O}_X(\ell\delta L)$ is generated by its sections. The hypothesis then is that V is a common irreducible component of the two multiplicity loci $\Sigma_{\ell-1} = \Sigma_{\ell-1}(E)$ and $\Sigma_{m+\ell} = \Sigma_{m+\ell}(E)$.

We claim first that

$$\mathcal{I}_{\Sigma_l} \subseteq \mathcal{I}_V^{(m)}, \tag{*}$$

where $\mathcal{I}_V^{(m)}$ denotes the symbolic power sheaf of all functions vanishing to order $\geq m$ on V. This follows from the hypothesis on V just as in (1.3). In fact, let f be a local equation for E on some open set U. Then $\mathcal{I}_{\Sigma_l}$ is locally generated by all the functions $\{D(f)\}$ for $D \in \mathcal{D}^\ell(U)$ a differential operator of order $\leq \ell$. On the other hand, since V is an irreducible component of $\Sigma_{\ell+m}$, $D'(F) \in \mathcal{I}_V(U)$ for every $D' \in \mathcal{D}^{\ell+m}(U)$. Hence all the $D(f)$ $(D \in \mathcal{D}^\ell)$ vanish to order $\geq m$ on V, as asserted.

We now apply the discussion following the statement of Theorem 2.1 with $B = kL$ and $A = \delta L$. The sheaf I_{Σ_ℓ} is generated by finitely many differential sections of order ℓ:

$$s_1, \ldots, s_N \in \Gamma(X, \mathcal{O}_X(kL + \ell\delta L)).$$

In particular, the s_i cut out Σ_ℓ as a set, and hence contain V as an irreducible component of their common zeroes. On the other hand, by (*), the s_i vanish to order $\geq m$ on V.

Then Lemma 1.4 implies that

$$\begin{aligned}\deg_L(V) &\le \frac{((k+\ell\delta)L)^c}{(m)^c}\cdot L^{n-c}\\ &= \frac{(k+\ell\delta)^c}{m^c}\int_X c_1(L)^n\\ &= \frac{(1+\sigma\delta)^c}{\epsilon^c}\int_X c_1(L)^n.\end{aligned}$$

The theorem follows. □

Remark 2.7. F. Angelini has verified that a similar argument proves the following more precise statement of [De1], Cor. 10.8. Given a divisor $E \in |kL|$ and a subvariety $V \subset X$, define the *jumping values* $b_1, \dots, b_{n+1}$ of E with respect to V to be the integers

$$b_p = b_p(E) = \min\left\{\mu > 0 \mid \operatorname{codim}_x(\Sigma_\mu(E), X) \ge p \ \forall x \in V\right\}.$$

Thus

$$0 = b_1 \le b_2 \le \cdots \le b_{n+1} = \max_{x\in V}\ \operatorname{mult}_x(E).$$

For each $1 \le p \le n$ let $\{T_{p,\alpha}\}$ be the irreducible components of $\Sigma_{b_p}(E)$ meeting V, and denote by $\nu_{p,\alpha}$ the generic multiplicity of E along $T_{p,\alpha}$. Then

$$\sum_{\alpha=1}^{\infty}(\nu_{p,\alpha}-b_1)\dots(\nu_{p,\alpha}-b_p)\cdot\deg_L(T_{p,\alpha}) \le (k+b_1\delta)\dots(k+b_p\delta)\cdot\int_X c_1(L)^n. \tag{2.7.1}$$

The idea, roughly speaking, is to argue as above but using differential sections of orders $b_1, \dots, b_p$.

As a particularly interesting example, due also to Angelini, let $(X, \mathcal{O}_X(\Theta))$ be a principally polarized abelian variety of dimension n, with $\Theta \subset X$ the theta divisor on X. Kollár has shown that codim $\Sigma_p(\Theta) \ge p+1$, and so in particular Θ has no $(n+1)$-fold points. Suppose that Θ has an n-fold point at x, and consider the jumping numbers with respect to $\{x\}$. By Kollár's theorem,

$$b_i \le i-1 \ \ (1 \le i \le n).$$

Applying (2.7.1) with $p = n$ we find:

$$(n-b_1)\dots(n-b_n) \le \int_X c_1(L)^n = n!.$$

Therefore $b_i = i-1$ for all i. Let $\{T_\alpha\}$ be the irreducible components of Σ_{n-2}. Applying (2.7.1) with $p = n-1$ then gives $\sum \deg T_\alpha \le n$. If $\cup T_\alpha$ generates X, then it follows from the Matsusaka–Ran criterion that X is a Jacobian, in which case it is easy to see that in fact $(X, \mathcal{O}_X(\Theta))$ must split as a product of elliptic curves. This proves for example that if X is simple, then Θ can have no n-fold points. By a similar sort of argument, Smith and Varley [SV] have recently established that the presence of an n-fold point implies in general that $(X, \mathcal{O}_X(\Theta))$ is a product of elliptic curves, the additional point being to show that Σ_{n-2} generates X.

We conclude this section with a conjecture suggesting how Theorem 2.1 should fit into the framework of contemporary intersection theory, as in [Fult]. Let X be any smooth irreducible variety of dimension n, and let $E \subset X$ be any effective divisor on X. Suppose that $V \subset X$ is an irreducible subvariety of codimension c in X which is simultaneously an irreducible component of the two multiplicity loci $\Sigma_\ell(E)$ and $\Sigma_{\ell+m}(E)$. Our conjecture asserts roughly speaking that $[V]$ always occurs with multiplicity at least m^c as a distinguished component of the Fulton–MacPherson c-fold self-intersection class of E in X.

More precisely, consider the c-fold products

$$E^c = E \times \cdots \times E \subset X \times \cdots \times X = X^c.$$

Let $C = NC(E, E^c)$ denote the (affine) normal cone to the diagonal embedding $E \subset E^c$, so that C has pure dimension $c(n-1)$. Similarly, let $T = N(X, X^c) \cong \oplus^{c-1} TX$ be the normal bundle to the diagonal $X \subset X^c$, with projection map $p : T \longrightarrow X$. Then $C \subset T$, and the Fulton–MacPherson class $E \cdot \ldots \cdot E$ (c times) is obtained roughly speaking by intersecting $[C]$ with the zero-section in T.

Conjecture 2.8. The inverse image $p^{-1}(V)$ is an irreducible component of C, and moreover $[p^{-1}(V)]$ appears with coefficient $\geq m^c$ in the fundamental class $[C]$.

One may view this as a local analogue of Theorem 2.1. The global statement should then follow by using the results of [FL] to control the other contributions to $E \ldots E$ in terms of the nefness of the vector bundle T. One expects a similar statement in the setting of (2.7), with the multiplicity of $[p^{-1}V]$ being estimated in the spirit of (2.7.1).

§3. Graded linear series

This section contains the main technical result of the paper, the self-intersection theorem with moving parts (Theorem 3.9). This is an algebro-geometric analogue of the corresponding inequality of Demailly. Although the techniques employed in [De1] are largely analytic, the basic idea here is suggested by Demailly in the remark following [De1] Corollary 10.8. The reader is also encouraged to compare this section with the proof of Dyson's Lemma by Esnault and Viehweg [EV1]. Although Esnault and Viehweg ignore the moving part in the proof of their main theorem, they indicate how to improve Dyson's Lemma by accounting for the moving part in [EV1] §10. The theorem of Esnault and Viehweg is revisited from this point of view in [Nak].

As suggested by the overview in §1, the main theorem will involve an "asymptotic" analysis of a linear system defined by imposing an index condition on a complete linear series $|kL|$ for $k \gg 0$. We start by introducing some formalism designed to facilitate the discussion of such linear series.

Let X be a smooth projective variety and let L be an ample line bundle on X. Consider the graded algebra

$$R = \bigoplus_{k=0}^{\infty} H^0(\mathcal{O}_X(kL)).$$

Suppose that

$$A \subseteq R$$

is a graded subalgebra; we will call A a *graded linear series*. For technical reasons, we will often assume that A_k is nonzero for all sufficiently large k.

Example 3.1. (a) Let B be a big divisor such that $L(-B)$ is effective. The choice of an effective divisor in $|L(-B)|$ determines for all k an embedding $H^0(\mathcal{O}_X(kB)) \subset H^0(kL)$, and $\bigoplus_{k=0}^{\infty} H^0(\mathcal{O}_X(kB))$ is a graded linear series.

(b) Let V be a subvariety of X and α a positive real number. Let

$$A_k = \{s \in H^0(\mathcal{O}_X(kL)) \mid \operatorname{mult}_x(s) \geq k\alpha \text{ for all } x \in V\}$$

and define

$$A^{\alpha}(V, X) = \bigoplus_{k=0}^{\infty} A_k.$$

One checks that $A^{\alpha}(V, X)$ is a graded linear series and it is the primary example which will appear in the sequel.

(c) Suppose $Y \subset X$ is a smooth subvariety and let

$$R_Y = \bigoplus_{k=0}^{\infty} H^0(\mathcal{O}_Y(kL)).$$

Given a graded linear series $A \subset R$ on X, let A_Y denote the restriction of A to Y. Then it is easy to verify that $A_Y \subset R_Y$ is a graded linear series.

Definition 3.2. Let A be a graded linear series and let

$$a(k) = \dim(A_k) \quad \text{for } k \geq 0.$$

Define the degree of A by

$$\rho(A) = \limsup_{k \to \infty} \frac{a(k)n!}{k^n}.$$

The invariant $\rho(A)$ measures the moving part of the graded linear series A or equivalently the degree of freedom in choosing a divisor $E_k \in |A_k|$ for $k \gg 0$.

Example 3.3. (a) $\rho(R) = c_1(L)^n$.

(b) Let x be a point of X. Then

$$\rho(A^\alpha(x, X)) \geq c_1(L)^n - \alpha^n,$$

as one verifies by counting the number of linear conditions imposed by requiring multplicity $\geq k\alpha$ at x.

The next two lemmas demonstrate that the invariant ρ behaves well under such operations as taking a Veronese subalgebra or restricting the graded linear series to a hyperplane.

Lemma 3.4. *Let $A \subset R$ be a graded linear series such that $A_k \neq 0$ for all $k \gg 0$. Fix a positive integer d and consider the Veronese subalgebra: $V_d = \bigoplus_{k=0}^{\infty} A_{kd} \subset \bigoplus_{k=0}^{\infty} H^0(\mathcal{O}_X(kdL))$. Then*

$$\rho(V_d) = d^n \rho(A). \tag{3.4.1}$$

Proof. It is clear from Definition 3.2 that $\rho(V_d) \leq d^n \rho(A)$. Let $V_d(k)$ denote the kth graded piece of V_d and let

$$v_d(k) = \dim\ V_d(k).$$

Write $\rho = \rho(A)$. We will prove Lemma 3.4 by showing that for any $\epsilon > 0$ there exists a sequence of integers k_i such that $k_i \to \infty$ and

$$v_d(k_i) \geq \frac{(k_i d)^n}{n!}(\rho(1-\epsilon)) \quad \text{for all } i.$$

Fix a large positive integer c such that $a(m) > 0$ for all $m \geq c$. It follows from the definition of ρ that there exists an infinite increasing sequence of integers $\{m_i\}$ such that

$$\frac{a(m_i)n!}{m_i^n} \geq \rho\left(1 - \frac{\epsilon}{2}\right) \quad \text{for all } i.$$

Replacing $\{m_i\}$ by a subsequence if necesary, we may also assume that all the m_i's are sufficiently large so that

$$\left(\frac{m_i}{m_i + c + d}\right)^n \geq 1 - \frac{\epsilon}{2}.$$

Next observe that there is a unique integer b_i such that $m_i + c < b_i \leq m_i + c + d$ and b_i is divisible by d. We will write b_i as $k_i d$. Since $k_i d - m_i > c$ it follows that there is a non-zero section $\sigma \in A_{k_i d - m_i}$. Multiplication by σ gives an injection

$A_{m_i} \hookrightarrow A_{k_i d} = V_d(k_i)$ and consequently $v_d(k_i) \geq a(m_i)$. Thus

$$\begin{aligned}\frac{v_d(k_i)n!}{(k_i d)^n} &\geq \frac{a(m_i)n!}{m_i^n}\frac{m_i^n}{(k_i d)^n} \\ &\geq \rho\left(1-\frac{\epsilon}{2}\right)\left(1-\frac{\epsilon}{2}\right) \\ &\geq \rho(1-\epsilon).\end{aligned}$$

This completes the proof of Lemma 3.4. □

Next we investigate the behaviour of ρ under restriction to hyperplanes. For this we focus only on graded linear series of the form $A^\alpha(V, X)$.

Lemma 3.5. *Let $A = A^\alpha(V, X)$ be the graded linear series defined in* 3.1(b). *Suppose that $|m_0 L|$ is free and choose a general divisor $H \in |m_0 L|$. Set $W = V \cap H$. Then*

$$\rho(A^\alpha(W, H)) \geq m_0 \rho(A).$$

Proof. As above, put

$$R_H = \bigoplus_k H^0(\mathcal{O}_H(kL)).$$

Let $B = B_H \subset R_H$ be the restriction of A to H (cf. Example 3.1 (c)). There is a natural surjective map $A \longrightarrow B$ and we denote the kernel by C. Thus for all $k \geq 0$ there is an exact sequence

$$0 \longrightarrow C_k \longrightarrow A_k \longrightarrow B_k \longrightarrow 0.$$

Let $a(k) = \dim A_k$, $b(k) = \dim B_k$, and $c(k) = \dim C_k$. Observe that $B \subset A^\alpha(W, H)$. In order to prove Lemma 3.5, it will be sufficent to show that

$$\rho(B) = \limsup_{k\to\infty} \frac{b(k)(n-1)!}{k^{n-1}} \geq m_0\rho(A). \tag{3.5.1}$$

Let s_0 be the section of $m_0 L$ defining H. It follows from the definition of C_k that every $t \in C_k$ can be expressed as $t = s_0 s'$ with $s' \in H^0(X, \mathcal{O}_X((k-m_0)L))$. Furthermore, since H is general, we can assume that it does not contain any irreducible component of V and hence

$$\operatorname{mult}_x(s') \geq k\alpha \quad \text{for all } x \in V.$$

Let

$$C'_k = \{s \in H^0(X, \mathcal{O}_X((k-m_0)L)) \mid \operatorname{mult}_x(s) \geq k\alpha \text{ for all } x \in V\}.$$

The above remarks show that there is an injection $C_k \hookrightarrow C'_k$. Since $C'_k \subset A_{k-m_0}$ it follows that $c(k) \leq a(k-m_0)$ and hence $b(k) \geq a(k) - a(k-m_0)$ for all $k \geq 0$. In particular, when k is divisible by m_0, this yields

$$b(tm_0) \geq a(tm_0) - a((t-1)m_0).$$

Summing the above inequalities, we obtain

$$\sum_{t=0}^{N} b(tm_0) \geq a(Nm_0).$$

By Lemma 3.4

$$\limsup_{N\to\infty} \left(\frac{a(Nm_0)n!}{(Nm_0)^n}\right) = \rho(A).$$

Hence Lemma 3.5 follows from the following simple lemma on partial sums:

Lemma 3.6. *Let $g(k)$ be a sequence of real numbers. Let $f(N) = \sum_{j=0}^{N} g(j)$ and suppose that* $\limsup_k (f(k)/k^n) = \rho$. *Then* $\limsup_k \left(g(k)/k^{n-1}\right) \geq n\rho$.

Proof. Suppose Lemma 3.6 is false. Then there exists a real number $\epsilon > 0$ and a positive integer $N_0 > 0$ such that $g(j)/j^{n-1} \leq n\rho - \epsilon$ for all $j \geq N_0$. Then for $k \gg 0$

$$f(k) = \sum_{j=0}^{k} g(j) \leq (n\rho - \epsilon)\frac{k^n}{n} + O(k^{n-1})$$

which contradicts our assumption on $f(k)$. □

Applying Lemma 3.5 inductively gives

Corollary 3.7. *Let $V \subset X$ and α be as in Lemma 3.5. Suppose that $|m_0 L|$ is free and that Λ is a complete intersection of $n - c$ general divisors in $|m_0 L|$. Let $S = \Lambda \cap V$. Then*

$$\rho(A^\alpha(S, \Lambda)) \geq m_0^{n-c} \rho(A^\alpha(V, X)).$$ □

Let

$$A = \bigoplus_{k=0}^{\infty} A_k \subseteq \bigoplus_{k=0}^{\infty} H^0(X, kL)$$

be a graded linear series and choose for each k a general divisor $E_k \in |A_k|$. For any rational number $\sigma > 0$ define

$$Z_\sigma(k) = \{x \in X \mid \text{mult}_x(E) \geq k\sigma, \text{ for all } E \in |A_k|\}.$$

The following lemma shows that these loci stabilize for $k \gg 0$.

Lemma 3.8. *There is a positive integer k_0 such that*

$$Z_\sigma(k_1) = Z_\sigma(k_2) \quad \textit{for all } k_1, k_2 \geq k_0.$$

Proof of Lemma 3.8. As σ is fixed in Lemma 3.8, we will write $Z(k)$ for $Z_\sigma(k)$ in order to alleviate notation. Suppose $a \geq 2$ is a positive integer. We claim that there exists a positive integer $k(a)$ such that

$$Z(c) \subseteq Z(a) \quad \text{whenever } c \geq k(a). \tag{3.8.1}$$

To prove the claim, suppose that $x \notin Z(a)$ so that there exists $\eta > 0$ satisfying

$$\text{mult}_x(E_a) \leq a\sigma - \eta.$$

Note that since the index is a discrete invariant, η is bounded below independently of x; in fact if $m\sigma \in \mathbb{Z}$ then $\eta \geq 1/m$. Suppose b is a positive integer relatively prime to a. Then any integer $c \geq ab$ can be expressed as

$$c = \alpha a + \beta b, \quad \alpha, \beta \in \mathbb{Z} \text{ and } 0 \leq \beta \leq a.$$

Consider the divisor

$$F_c = \alpha E_a + \beta E_b \in |A_c|.$$

Then

$$\begin{aligned}\text{mult}_x(F_c) &= \alpha \cdot \text{mult}_x(E_a) + \beta \cdot \text{mult}_x(E_b)\\ &\leq \alpha a\sigma - \alpha\eta + \beta \cdot \text{mult}_x(E_b)\\ &\leq c\left(\sigma - \frac{\eta\left(1 - \frac{\beta b}{c}\right)}{a} + \frac{\beta \cdot \text{mult}_x(E_b)}{c}\right).\end{aligned}$$

Since η, β, and b are bounded independent of c, it follows that $\text{mult}_x(F_c) < c\sigma$ for $c \gg 0$. Hence $x \notin Z(c)$ for all sufficiently large c as claimed.

If $Z(c) = Z(a)$ for all $c \gg 0$ then we are finished. If not, then by (3.8.1) there exists $a' > 0$ such that $Z(a') \subsetneq Z(a)$. The argument of (3.8.1) can then be repeated with a' in place of a. This process cannot go on indefinitely and this establishes Lemma 3.8. □

Using Lemma 3.8 we can define a subvariety $Z_\sigma(A) \subset X$ by

$$Z_\sigma(A) = Z_\sigma(k) \text{ for } k \gg 0.$$

The main result of this section is the "one factor" product theorem with moving parts:

Theorem 3.9. *Let X be a smooth projective variety of dimension n. Let L be an ample line bundle on X and assume that δ is a non-negative real number such that $TX(\delta L)$ is nef. Let*

$$A \subset \bigoplus H^0(\mathcal{O}_X(kL))$$

be a graded linear series. Suppose that there are rational numbers $\sigma, \epsilon > 0$ and an irreducible subvariety $V \subset X$ of codimension $c < n$ which is an irreducible component of both $Z_\sigma(A)$ and $Z_{\sigma+\epsilon}(A)$. Then

$$\deg_L(V) \leq \left(\frac{1+\sigma\delta}{\epsilon}\right)^c (c_1(L)^n - \rho(A)).$$

Proof of Theorem 3.9. As in the proof of Theorem 2.1, we may assume that $TX(\delta L)$ is ample and δ is rational by replacing δ with a slightly larger rational number. Fix a sufficiently large and divisible integer k_0 such that $k_0\sigma$, $k_0\epsilon$, $k_0\delta\sigma$, and $k_0\epsilon/(1+\sigma\delta)$ are all integers. By Lemma 3.8, we may assume that for all $k \geq k_0$,

$$Z_\sigma(A) = Z_\sigma(A_k),$$
$$Z_{\sigma+\epsilon}(A) = Z_{\sigma+\epsilon}(A_k).$$

By Lemma 2.2, we may also assume that the sheaves $\mathcal{D}_{kL}^{k\sigma} \otimes \mathcal{O}_X(k\delta\sigma L)$ are generated by global sections for all k divisble by k_0. Let $s_k \in A_k$ be a general section and E_k its divisor of zeroes. If $x \notin Z_\sigma(A)$ then we may assume that $\mathrm{mult}_x(E_k) < k\sigma$. Thus there exists a differential operator

$$D \in H^0\left(\mathcal{D}_{kL}^{k\sigma} \otimes \mathcal{O}_X(k\delta\sigma L)\right)$$

such that the differential section $D(s_k) \in H^0(\mathcal{O}_X(k(1+\sigma\delta)L))$ does not vanish at x. Furthermore, since the order of D is $\leq k\sigma$,

$$\mathrm{mult}_V(D(s_k)) \geq k(\sigma+\epsilon) - k\sigma = k\epsilon.$$

Since V is an irreducible component of $Z_\sigma(A)$, V is also an irreducible component of the base locus of the linear system

$$\left|\mathcal{I}_V^{(k\epsilon)} \otimes \mathcal{O}_X(k(1+\delta\sigma)L)\right| \quad \text{for all } k \text{ divisble by } k_0.$$

Letting $\alpha = \epsilon/(1+\delta\sigma)$, this shows that V is an irreducible component of $|A^\alpha(V,X)_m|$ for all m divisible by $k_0(1+\sigma\delta)$. Lemma 3.10 below then implies that

$$\deg_L(V) \leq \frac{c_1(L)^n - \rho}{\alpha^c} = \left(\frac{1+\sigma\delta}{\epsilon}\right)^c \left(c_1(L)^n - \rho(A^\alpha(V,X))\right). \tag{3.9.1}$$

But since $\alpha \leq \sigma + \epsilon$ and since $V \subset Z_{\sigma+\epsilon}(A)$ it follows that

$$A_k \subset A^\alpha(V,X)_k \text{ for all } k \text{ divisible by } k_0(1+\sigma\delta).$$

Hence, using Lemma 3.4, $\deg(A^\alpha(V,X)) \geq \rho(A) = \rho$; combining this with 3.9.1 completes the proof of Theorem 3.9. □

Lemma 3.10. *Let $V \subset X$ be an irreducible subvariety of codimension $c < n$ and let α be a positive real number. Let $\rho = \rho(A^\alpha(V,X))$. Suppose that there is a positive integer k_0 such that V is an irreducible component of the base locus of the linear* *system* $|A^\alpha(V,X)_k|$ *for all* $k \gg 0$ *divisible by* k_0. *Then*

$$\deg_L(V) \le \left(\frac{1}{\alpha}\right)^c \left(c_1(L)^n - \rho\right)$$

Proof. Fix a positive integer m_0 such that $|m_0L|$ is free of base points. Choose general divisors $D_1, \ldots, D_{n-c} \in |m_0L|$ and let

$$\Lambda = \bigcap_{i=1}^{n-c} D_i.$$

Also define

$$S = \Lambda \cap V.$$

Since the divisors D_i are general, S is a finite set of $\deg_{m_0L}(V)$ distinct points. The hypotheses of Lemma 3.10 imply that for all k sufficiently large and divisible there is a divisor $E_k \in H^0(\Lambda, \mathcal{O}(k\Lambda))$ with an almost isolated singularity at every point of S. The key observation in the proof of Lemma 3.10 is that Theorem 1.1 gives surjectivity statements onto skyscraper sheaves and this provides a means of bounding the size of S, or equivalently, of bounding $\deg_L(V)$; the moving part ρ enters by considering the kernel of these surjective maps.

We will show that

$$\operatorname{card}(S) \le \frac{m_0^{n-c}}{\alpha^c}(c_1(L)^n - \rho). \tag{3.10.1}$$

It is clear, in light of fact that $\operatorname{card}(S) = \deg_{m_0L}(V)$, that (3.10.1) implies Lemma 3.10. Choose two positive rational numbers α_1, α_2 slightly less than α, such that $\alpha_1 < \alpha_2 < \alpha$. Replacing k_0 by a suitably high multiple, we may assume that V is an irreducible component of the base locus of $|A^\alpha(V,X)_d|$ for all d divisible by k_0. Next fix a sufficiently divisible positive integer k such that $k\alpha_1$, $k\alpha_2$, $k\alpha$, $k\alpha_2/k_0\alpha$ and k/k_0 are all integers. In order to apply Theorem 1.1, we also require k to be sufficiently large so that

$$k\left(1 - \frac{\alpha_2}{\alpha}\right)L - K_\Lambda \quad \text{is ample on } \Lambda \tag{3.10.2}$$

and

$$k(\alpha_2 - \alpha_1) > c. \tag{3.10.3}$$

Choose a general divisor $D \in |A^\alpha(V,X)_{k\alpha_2/\alpha}|$ and let

$$D_\Lambda = D \cap \Lambda.$$

Since V is an irreducible component of the base locus of $|A^\alpha(V,X)_{k\alpha_2/\alpha}|$ and since D is general, it follows that D_Λ has isolated singularities along S. Moreover,

$$\operatorname{mult}_S(D_\Lambda) \ge k\alpha\frac{\alpha_2}{\alpha} = k\alpha_2 > k\alpha_1 + c,$$

where the last inequality follows by (3.10.3). Applying Theorem 1.1 to D_Λ gives

$$H^1\left(\mathcal{I}_S^{k\alpha_1}\otimes\mathcal{O}\left(K_\Lambda+\frac{k\alpha_2}{\alpha}L\right)\right)=0. \tag{3.10.4}$$

By (3.10.2), $kL-(K_\Lambda+k\alpha_2/\alpha L)$ is ample. This, together with (3.10.4) implies that

$$H^1\left(\mathcal{I}_S^{k\alpha_1}\otimes\mathcal{O}_\Lambda(kL)\right)=0.$$

Hence the following sequence is exact:

$$0\longrightarrow H^0\left(\mathcal{I}_S^{k\alpha_1}\otimes\mathcal{O}_\Lambda(kL)\right)\longrightarrow H^0(\mathcal{O}_\Lambda(kL))\longrightarrow H^0\left(\mathcal{O}_\Lambda(kL)/\mathcal{I}_S^{k\alpha_1}\right)\longrightarrow 0. \tag{3.10.5}$$

To prove (3.10.1) it remains to compute the dimension of the three terms in (3.10.5). Since $\alpha_1\le\alpha$, $\dim(A^{\alpha_1}(S,\Lambda)_k)\ge\dim(A^{\alpha}(S,\Lambda)_k))$. Lemma 3.4 and Lemma 3.7 imply that

$$\limsup_{k_0|k}\ \dim(A^{\alpha_1}(S,\Lambda)_k)\frac{c!}{k^c}\ge m_0^{n-c}\rho.$$

By the Riemann–Roch theorem,

$$h^0(\mathcal{O}_\Lambda(kL))=m_0^{n-c}(c_1(L)^n)\frac{k^c}{c!}+O(k^{c-1}).$$

Also

$$h^0\left(\mathcal{O}_\Lambda(kL)/\mathcal{I}_S^{k\alpha_1}\right)=\operatorname{card}(S)\frac{(k\alpha_1)^c}{c!}+O(k^{c-1}).$$

We deduce from (3.10.5), that

$$\operatorname{card}(S)\le\frac{(m_0)^{n-c}}{(\alpha_1)^c}(c_1(L)^n-\rho).$$

As α_1 approaches α we obtain the inequlaity (3.10.1) and this concludes the proof of Lemma 3.10. □

§4. Application to adjoint series

In this section, following [De1], we address the problem of producing sections of $|K_X+L|$ for a suitably positive line bundle L. The logic of the argument is simple: as outlined in §1, we choose a divisor $E_k\in|kL|$ with very high multiplicity at a point $x\in X$. If E_k has an almost isolated singularity of index $n+s$ at x then Theorem 1.1 guarantees that $|K_X+L|$ separates s-jets at x. If the singularity of E_k at x is not isolated, then E_k has large index along some positive dimensional subvariety V containing x; the techniques developed in §3 will allow us to choose V in such a fashion that there is an absolute upper bound for $\deg_L(V)$. Thus if L is sufficiently positive to rule out the latter case, $|K_X+L|$ must separate s-jets at x.

Theorem 4.1. *Let X be a smooth projective variety of dimension n and fix a point $x \in X$. Let L be an ample line bundle on X satisfying $c_1(L)^n > (n+s)^n$. Suppose that δ is a non-negative real number such that $TX(\delta L)$ is nef. Then either:*

(a) *the linear system $|K_X + L|$ separates s-jets at x, or*

(b) *there is a subvariety V of codimension c containing x such that*

$$\deg_L V \leq (n - 1 + (c-1)\delta)^c (n+s)^n.$$

Proof. Fix a rational number α satisfying

$$(n+s)^n < \alpha^n < c_1(L)^n.$$

Let $A = A^\alpha(x, X)$. Note that $|A^\alpha(x, X)_k| \neq 0$ for all $k \gg 0$ and by Example 3.3 (b), $\rho(A) \geq c_1(L)^n - \alpha^n$. For each $k > 0$ choose a general divisor $E_k \in |A^\alpha(x, X)|$. Let

$$\beta_j = \frac{j-1}{n-1} \quad \text{for } 2 \leq j \leq n,$$

$$\beta_1 = t, \quad \text{where } \frac{1}{n-1} \gg t > 0.$$

and $\beta_0 = 0$. Set $Z_j = Z_{\beta_j}(A)$. If the the dimension of Z_n is equal to zero at x, then for all $k \gg 0$, E_k has an isolated singularity of index $\geq n+s$ at x and consequently Theorem 1.1 implies that $|K_X + L|$ separates s-jets at x. Hence we may assume that the codimension of Z_n is less than n. By Lemma 1.6, there is an irreducible subvaritey V of of codimension c which is a common irreducible component of Z_c and Z_{c+1} for some c where $1 \leq c \leq n-1$. Theorem 3.9 applies in this situation with $\sigma = \frac{c-1}{n-1}$ (or t), and $\epsilon = \frac{1}{n-1}$ (or $\epsilon = \frac{1}{n-1} - t$) and $\rho = c_1(L)^n - \alpha^n$. As α approaches $(n+s)$ and t approaches 0, we get

$$\deg_L V \leq (n - 1 + (c-1)\delta)^c (n+s)^n,$$

as asserted. □

Theorem 4.1 is similar to the main theorem of [De1] although the upper bound for $\deg_L V$ is worse and there are no β_i's in the statement because (to simplify the exposition) we chose explicit values in the proof. The weaker upper bound on $\deg_L V$ stems from the fact that we passed to hyperplane sections in the proof of Theorem 3.9. This could be avoided by using an interesting theorem of Fujita [Fuj], which would allow one to recover Demailly's statements *in toto*. However the basic idea is not much different.

A negative feature of Theorem 4.1 is the dependence on the positivity of TX; Demailly removes this dependence through an iteration argument which we present here. The argument allows one to prove a very ampleness criterion for line bundles of the form $2K_X + B$ where B is a suitably positive ample line bundle.

Proposition 4.2. *Let X be a smooth projective variety dimension n and let B be an ample line bundle on X. Assume that $A = K_X \otimes B$ is also ample and that for all subvarieties V of codimension c:*

$$\deg_A(V) > (n-1+(c-1)(2n+1))^c(n+1)^n, \quad 1 \leq c \leq n-1. \tag{4.2.1}$$

Then $K_X + A = 2K_X + B$ is very ample.

Proof. We will use Theorem 4.1 to show the following inductive statement:

$$\text{If } 2K_X+2^k B \text{ is very ample, then } 2K_X+2^{k-1}B \text{ is also very ample for } k \geq 1. \tag{4.2.2}$$

Since $2K_X + 2^m B$ is very ample for sufficiently large m, (4.2.2) and descending induction on m show that $2K_X + B$ is very ample establishing Proposition 4.2. It remains to show (4.2.2). It is well known that for any very ample line bundle L, $TX(K_X + nL)$ is generated by its global sections. In particular, since $2K_X + 2^k B$ is very ample by assumption, this implies that $TX\left(K_X + n(2K_X + 2^k B)\right)$ is *nef*. Hence we can apply Theorem 4.1 with $L = K_X + 2^{k-1}B$ and $\delta = (2n+1)$. Since $k-1 \geq 0$, (4.2.1) rules out case (b) of Theorem 4.1 and hence $|2K_X + 2^{k-1}B|$ separates 1-jets at every point $x \in X$. Similarly, one can show that $|2K_X + 2^{k-1}B|$ separates any two distinct points of X. Thus $2K_X + 2^{k-1}B$ is very ample and (4.4.2) is established. □

Recall that $K_X + (n+1)L$ is *nef* for any ample line bundle L on X by a theorem of Mori. Now the hypothesis (4.2.1) of Proposition 4.2 can of course be guaranteed by choosing $B = mL$ for an ample line bundle L and a suitably positive integer m:

Corollary 4.3. *Let X be a smooth projective variety of dimension n and let L be an ample line bundle on X. Then $2K_X + mL$ is very ample for*

$$m > n+1+(2n^2-2n-3)^{n-1}(n+1)^n. \quad \square$$

Finally we state without proof a variant of Theorem 4.1 which depends on the positivity of K_X rather than the positivity of TX.

Theorem 4.4. *Let X be a smooth projective variety of dimension n and let L be an ample line bundle on X satisfying $c_1(L)^n > (n+s)^n$. Let b be a non-negative number such that $-K_X + bL$ is nef. Suppose that m_0 is a positive integer such that $|m_0 L|$ is free. Then for any point $x \in X$ either*

(a) *$|K_X + L|$ separates s-jets at x, or*

(b) *there exists a codimension c subvariety V containing x and satisfying*

$$\deg_L V \leq \left(b + m_0(n-c) + \frac{n!}{c!}\right)^c (n+s)^n. \quad \square$$

Remark 4.5. (a) We can make theorem 4.4 effective by using Kollár's effective base point free theorem. According to this result, $|m_0 L|$ is base point free for

$$m_0 = 2(n+2)!(n + \ulcorner b \urcorner).$$

(In some preliminary notes on (4.4) we overlooked the necessity to invoke Kollár's theorem at this point, and accordingly claimed a somewhat better result. We apologize for any confusion that may have resulted from this error.)

(b) As Demailly pointed out, one can also deduce Theorem 4.4 from Theorem 4.1 and Corollary 4.3. Under the same assumption as in Theorem 4.4, we can apply 4.3 to deduce that $TX(\delta L)$ is *nef*, if $\delta > (2n+1)b + n(2n^2 - 2n - 3)^{n-1}(n+1)^{n+1}$. Now we can apply Theorem 4.1 to deduce a result similar to Theorem 4.4.

References

[AS] U. Angehrn and Y.-T. Siu, Effective freeness and point separation for adjoint bundles, to appear.

[De1] J.-P. Demailly, A numerical criterion for very ample line bundles, J. Differential Geom. 37 (1993), 323–374.

[De2] J.-P. Demailly, Effective bounds for very ample line bundles, to appear.

[De3] J.-P. Demailly, Regularization of closed positive currents and intersection theory, J. Alg. Geom. 1 (1992), 361–409.

[De4] J.-P. Demailly, L^2 vanishing theorems for positive line bundles, and adjunction theory, to appear.

[EKL] L. Ein, O. Küchle, R. Lazarsfeld, Local positivity of ample line bundles, J. Differential. Geom. 42 (1995), 193–219.

[EL1] L. Ein and R. Lazarsfeld, Global Generation of Pluricanonical and Adjoint Linear Series on Smooth Projective Threefolds, J.Amer. Math. Soc. 6 (1993), 875–903.

[EV1] H. Esnault and E. Viehweg, Dyson's Lemma for polynomials in several variables (and the Theorem of Roth), Invent. Math. 78 (1984), 445–490.

[EV2] H. Esnault and E. Viehweg, Lectures on Vanishing Theorems, DMV Sem. 20, Birkhäuser 1992.

[Falt] G. Faltings, Diophantine Approximation on Abelian Varieties, Ann. of Math. 133 (1991), 549–576.

[Fuj] T. Fujita, Approximating Zariski decomposition of big line bundles, Kodai Math. J. 17 (1994), 1–3.

[Fult] W. Fulton, Intersection Theory, Springer 1984.

[FL] W. Fulton and R. Lazarsfeld, Positivity and excess intersection, in: Enumerative Geometry and Classical Algebraic Geometry, Progr. in Math. Vol. 24, Birkhäuser 1982.

[Koll] J. Kollár, Effective basepoint freeness, Math. Ann. 296 (1993), 595–605.

[Kol2] J. Kollár, Unpublished notes on Siu's theorem.

[Laz] R. Lazarsfeld, Lectures on linear series, to appear.

[Nak] M. Nakamaye, in preparation.

[Myk] Y. Miyaoka, The Chern classes and Kodaira dimension of a minimal variety, Adv. Stud. Pure Math. 10, T. Oda (ed.), 1987, 449–476.

[Sch] W. Schmidt, Diophantine Approximations and Diophantine Equations, Lecture Notes in Math. 1467, Springer 1991.

[Siu1] Y.-T. Siu, Effective very ampleness, to appear.

[Siu2] Y.-T. Siu, Very ampleness criterion of double adjoint of ample line bundles, to appear.

[SV] R. Smith and R. Varley, Multiplicity *g* points on theta divisors, to appear.

Limits of joins and intersections

Hubert Flenner and Wolfgang Vogel

1. Introduction

The aim of this note is to characterize distinguished subvarieties of the intersection of closed subschemes $X, Y \subseteq \mathbb{P}^n = \mathbb{P}^n_K$ in terms of limit of join varieties. We recall that a distinguished variety of the intersection of X and Y (in the sense of [8]) is by definition a distinguished variety of $X \cap Y \subseteq X \times Y$, see 2.5.

Let $C_{X \cap Y}(X \times Y)$ denote the normal cone of $X \times Y$ along $X \cap Y$ where $X \cap Y$ is diagonally embedded into $X \times Y$. With other words,

$$C_{X\cap Y}(X \times Y) = \operatorname{Spec}(\operatorname{gr}_{\mathcal{I}} \mathcal{O}_{X\times Y})$$

where $\mathcal{I} \subseteq \mathcal{O}_{X\times Y}$ is the ideal sheaf of $X \cap Y$. The *limit of join* variety $\operatorname{LJoin}_C(X, Y)$ of X, Y at C will be introduced as the fibre of $C_{X\cap Y}(X \times Y) \to X \cap Y$ over the generic point of C. In the special case that the base field is $\mathbb{C}$ and $C = \{p\}$ is a closed point it may be visualized as the union of all limits $\ell = \lim(\overline{x_\nu y_\nu})$ where $x_\nu \in X$, $y_\nu \in Y$ are sequences tending to 0, see 3.3. If $X = Y$ then $\operatorname{LJoin}_C(X, X)$ will also be denoted by $\operatorname{LSec}_C X$ and called the *limit of secant variety* of X at C. For closed points $C = \{c\}$ the limit of secant variety is a classical object, which is one of the cones studied by H. Whitney [16]. The limit of join variety appears in [2] and is studied there in the case that X and Y meet in one point.

If $X, Y \subseteq \mathbb{P}^n$ are projective schemes this yields the classical join variety XY. Recall that the join variety is by definition the closure of the union of $X \cap Y$ and all lines $\overline{xy}$ with $x \in X$, $y \in Y$, $x \neq y$. Then set theoretically the affine cone $(XY)\hat{} \subseteq \mathbb{A}^{n+1}$ of XY is just the limit of join variety $\operatorname{LJoin}_0(\hat{X}, \hat{Y})$ of the affine cones $\hat{X}, \hat{Y}$ over X, Y, see 3.9. In particular, this construction provides a natural scheme structure on XY. The join variety and its connection with intersection theory were studied in several papers, see e.g. [3], [4], [6], [10].

One of the main results of L.van Gastel [10] is concerned with the characterization of the K-rational components of the intersection cycle $v = v(X, Y)$ in the sense of [14], [15] of projective varieties $X, Y \subseteq \mathbb{P}^n$: A subvariety C of $X \cap Y$ appears in v iff C is a distinguished subvariety of $X \cap Y$. We will state this result in terms of limit of join varieties, see 3.12. Moreover we will characterize the coefficient of C in the v-cycle as the degree of a refined limit of join variety.

This note is organized as follows. In section 2 we will review the intersection algorithm first introduced in [14], [15] and in a more general context in [9], [10]. Moreover, we will describe some of the main results of [10] and give an interpretation of the intersection multiplicity of distinguished varieties in terms of a refined normal cone. In section 3 we will introduce limits of joins and study basic properties. In particular we will give bounds for their dimensions. The main result is stated in 3.12 which shows that a subvariety C of $X \cap Y$ is distinguished iff $\mathrm{LJoin}_C(X, Y)$ has maximal dimension.

Throughout the paper K is an arbitrary field if not otherwise stated. If X is a scheme and C is a subvariety then we denote by η_C its generic point and by $k(\eta_C)$ its residue field. A scheme X will be called *strictly equidimensional* if X is equidimensional without embedded components. If C is an irreducible component of X then C inherits a scheme structure from X given by the primary ideal belonging to C. Let X_0 be an open subset of X and Z_0 a closed subscheme of X_0. Then the scheme theoretic closure, say Z, of Z_0 in X is given by the largest quasicoherent ideal sheaf in $\mathcal{O}_X$ which restricts to the ideal sheaf of Z_0 in $\mathcal{O}_{X_0}$. Otherwise we use the standard notations of [11].

2. Normal cones and intersections

2.1. Let V be an equidimensional algebraic K-scheme. For a line bundle $\mathcal{L}$ over V and $d \geq 0$ consider sections

$$\sigma_0, \ldots, \sigma_d \in H^0(V, \mathcal{L}).$$

Let $W := \{\sigma_0 = \cdots = \sigma_d = 0\}$ be the set of zeros of $\sigma_0, \ldots, \sigma_d$. We will construct from these data cycles $v_i = v_i(\underline{\sigma}, V)$ and $\beta_i = \beta_i(\underline{\sigma}, V)$, $i \geq 0$, of dimension i on W, V resp. as follows.

We introduce indeterminates U_{ij} and set

$$L := K(U_{ij})_{0 \leq i, j \leq d},$$

$$D_i := \{\sum_{j=0}^{d} U_{ij}\sigma_j = 0\}.$$

Then D_i is a subscheme of $V_L := V \bigotimes_K L$ of codimension ≤ 1. We start by decomposing $[V]$ as a direct sum

$$[V] = \beta_{\dim V} + v_{\dim V},$$

where $\mathrm{supp}(v_{\dim V}) \subseteq W$ and no (reduced) component of $\beta_{\dim V}$ is contained in W. If β_k has already been constructed then we write again the intersection $\beta_k \cap D_{\dim V-k}$

as a sum

$$\beta_k \cap D_{\dim V-k} = \beta_{k-1} + v_{k-1}$$

where as above $\mathrm{supp}(v_{k-1}) \subseteq W_L$ and no component of β_{k-1} is contained in W_L. Observe that every component of β_k meets $D_{\dim V-k}$ properly by construction, see [7].

In the following it will be important to compare this algorithm on V with that on the associated normal cone

$$C := C_W V = \mathrm{Spec}\left(\bigoplus_{n=0}^{\infty} \mathcal{J}^n/\mathcal{J}^{n+1}\right),$$

where $\mathcal{J} \subseteq \mathcal{O}_V$ is the ideal sheaf of $W \subseteq V$.

2.2. Let $\sigma_0^*, \dots, \sigma_d^* \in H^0(W, \mathcal{J}/\mathcal{J}^2 \otimes \mathcal{L})$ be the initial forms of $\sigma_0, \dots, \sigma_d$. We will consider $\sigma_0^*, \dots, \sigma_d^*$ as sections in $H^0(C, \mathcal{L}^*)$, where $\mathcal{L}^*$ is the line bundle on C associated to $\bigoplus \mathcal{J}^n/\mathcal{J}^{n+1} \otimes \mathcal{L}$. Applying the algorithm in (1.1) to C and $\sigma_0^*, \dots, \sigma_d^*$ we get cycles

$$v_0^*, \dots, v_{\dim W}^* \quad \text{and} \quad \beta_0^*, \dots, \beta_{\dim V}^*$$

on W_L, C_L respectively.

A basic fact of this construction is the following result, which is due to L. van Gastel, see [10], (3.2), see also [7] for another proof.

Theorem 2.3. $v_i = v_i^*$ *for all* $i \geq 0$.

Remark 2.4. (1) Let $Z \subseteq W$ be an irreducible component. Then it is known (see [15]) that Z_L appears in the v-cycle with coefficient $e(I, \mathcal{O}_{V,Z})$ where $I \subseteq \mathcal{O}_{V,Z}$ is the ideal of W at Z and e denotes the multiplicity.

(2) Assume that $[V] = \sum m_i [V_i]$ as a cycle. Then it follows from the construction that $v(\underline{\sigma}^*, V) = \sum m_i v(\underline{\sigma}^*, V_i)$, see e.g. [7].

As an application of 2.3 we will deduce in 2.9 a characterization of distinguished components in the sense of Fulton [8], (6.1.2). Slightly more generally we will introduce distinguished varieties as follows.

Definition 2.5. Let $\pi : C := C_W V \to W$ be the normal cone of W in V. A subvariety $Z \subseteq W$ is called a *distinguished component* for $W \subseteq V$ if there is an irreducible component C' of C with $\pi(C') = Z$.

Assume that V is a subscheme of an algebraic K-scheme X. Let $\tilde{\mathcal{L}} \in \mathrm{Pic}(X)$ be a locally free sheaf restricting to $\mathcal{L}$ on V, and $\tilde{\sigma}_i \in H^0(X, \tilde{\mathcal{L}})$ sections with $\tilde{\sigma}_i | V = \sigma_i$. If $Y \subseteq X$ is the complete intersection of the divisors $\{\tilde{\sigma}_i = 0\}$, $0 \leq i \leq d$, then the distinguished components of W in V as above introduced are just the distinguished components of the intersection $X \cap V$ in the sense of Fulton [8].

In 2.9 below we will give a description of the coefficient of distinguished subvarieties in the v-cycle. For this we need the fibre and refined fibre of the normal cone at Z:

Definition 2.6. (1) For a subvariety Z of W with generic point η_Z we denote by $(C_W V)_Z$ the fibre $\pi^{-1}(\eta_Z) \subseteq C_W V$, where $\pi : C_W V \to W$ is the projection.

(2) Let $C^* \subseteq C_W V$ the scheme theoretic union of all irreducible components C_i with $\pi(C_i) = Z$, where we equip C_i with the structure inherited from $C_W V$. Then $(C_W V)^*_Z := C^* \times_V \operatorname{Spec} \mathcal{O}_{V,Z}$ is called the *refined fibre* of the normal cone at Z.

Clearly $(C_W V)^*_Z = \emptyset$ unless Z is distinguished. In algebraic terms, let I be the ideal of W in $A := \mathcal{O}_{V,Z}$ and $\operatorname{gr}_I A := \bigoplus_{n\geq 0} I^n/I^{n+1}$ the associated graded ring. Then $(C_W V)_Z$ is the spectrum of the ring $A/\mathfrak{m}_A \otimes_A \operatorname{gr}_I A$. Similarly, $(C_W V)^*_Z$ is the scheme theoretic union of all irreducible components of $\operatorname{Spec}(\operatorname{gr}_I A)$ which map to the closed point in $\operatorname{Spec} A/I$. With other words, the ideal of $(C_W V)^*_Z$ in $\operatorname{gr}_I A$ is the intersection of all primary ideals, say $\mathfrak{q}$, of 0 in $\operatorname{gr}_I A$ such that the intersection of $\mathfrak{q}$ with A/I is $\mathfrak{m}_{A/I}$-primary.

The following lemma gives a simple estimate for the dimension of $(C_W V)_Z$.

Lemma 2.7. *Let $I \subseteq A := \mathcal{O}_{V,Z}$ be the ideal of W at Z and $\mu(I)$ its minimal number of generators. Then the following holds.*

(1) $\dim \mathcal{O}_{V,Z} - \dim \mathcal{O}_{W,Z} \leq \dim(C_W V)_Z \leq \min\{\mu(I), \dim \mathcal{O}_{V,Z}\}$.

(2) *If* $\dim(C_W V)_Z = \dim \mathcal{O}_{V,Z}$ *then Z is distinguished.*

(3) *If* $\dim(C_W V)_Z = \mu := \mu(I)$ *then* $(C_W V)_Z \cong \mathbb{A}^{\mu}_{k(\eta_Z)}$.

Proof. By definition $(C_W V)_Z$ is the spectrum of the ring $B := A/\mathfrak{m}_A \otimes_A \operatorname{gr}_I A$. As a graded $A/\mathfrak{m}_A$-algebra B is generated by $\mu(I)$ elements and so $\dim B \leq \mu(I)$ with equality iff B is a polynomial ring proving (3). Moreover

$$\dim B \leq \dim \operatorname{gr}_I A = \dim A,$$

which gives the second inequality in (1). For the first one we apply the dimension formula to the morphism of rings $A/I \to \operatorname{gr}_I A$ and obtain

$$\dim \operatorname{gr}_I A \leq \dim A/I + \dim B,$$

see [12], 15.1 (i).

Assume that $\dim B = \dim A$. Then every irreducible component of $\operatorname{Spec} B$ of maximal dimension is an irreducible component of $\operatorname{Spec}(\operatorname{gr}_I A)$ and so corresponds to an irreducible component, say C_i, of $C_W V$ with $\pi(C_i) = Z$ proving (2). □

Remark 2.8. (1) With the notations as above $(C_W V)_Z$ is a cone over $\eta_Z = \operatorname{Spec} A/\mathfrak{m}_A$, and

$$\mathbb{P}(C_W V)_Z = \operatorname{Proj}(A/\mathfrak{m}_A \otimes_A \operatorname{gr}_I A)$$

is a projective $k(\eta_Z)$-scheme. Similarly, $(C_W V)^*_Z$ is the spectrum of a graded ring, say G, such that G_0 is a Artinian algebra with residue field $k(\eta_Z)$. In particular, its associated projective scheme has a well defined degree, which is in algebraic terms the multiplicity of G with respect to the ideal G_+ generated by the elements of positive degree.

(2) The dimension of $A/\mathfrak{m}_A \otimes_A \mathrm{gr}_I A$ or, equivalently, of $(C_W V)_Z$ is often called the *analytic spread* of the ideal I in A.

One of the main problems concerning the intersection cycle is to characterize subvarieties belonging to $v = v(\underline{\sigma}, V)$. The following geometric criterion is often very useful and describes the K-rational components of v.

Theorem 2.9. *Assume that V is strictly equidimensional and $Z \subseteq W_L$ is a subvariety. The following conditions are equivalent.*

(1) *Z is a K-rational component of v.*

(2) *Z is, up to the field extension $K \to L$, a distinguished subvariety.*

(3) $\dim(C_{W_L} V_L)_Z = \dim \mathcal{O}_{V_L, Z}$.

Moreover, if $Z = Z' \otimes_K L$ is K-rational then the coefficient of Z in the v-cycle is given by the degree of the projective scheme $\mathbb{P}(C_W V)^*_{Z'}$.

Proof. The equivalence of (1), (2) and (3) is due to [10], see also [1] for the equivalence of (1) and (3). We will give an alternative proof: Write $[C_W V] = \sum m_i [C_i]$, where $C_i \subseteq C := C_W V$ are the irreducible components. It is clear that every distinguished subvariety is K-rational and that it appears in the v-cycle because of 2.3, 2.4 (1) and the fact that

$$v(\underline{\sigma}^*, C) = \sum m_i v(\underline{\sigma}^*, C_i), \qquad (*)$$

see 2.4 (2). Now assume that $Z \subseteq W$ is a K-rational component of the v-cycle. Let I be the ideal of W_L in $A := \mathcal{O}_{V_L, Z}$ and set $k := \dim A$. The elements $s_0, \ldots, s_d$ induce elements of A also denoted by $s_0, \ldots, s_d$. Let G be the ring $A/\mathfrak{m}_A \otimes_A \mathrm{gr}_I A$ so that $(C_{W_L} V_L)_Z$ is the spectrum of G. By construction, the initial forms of $s_0, \ldots, s_{\dim G - 1}$ form a system of parameters in G. As Z is in the v-cycle, the set $V(s_0, \ldots, s_{k-2})$ is not contained in $V(I)$. Thus by the subsequent lemma $\dim G \geq k$ proving (3). That (3) implies (2) follows from 2.7.

Finally assume that $Z = Z' \otimes_K L$ is K-rational. We need to show that the coefficient of Z in v is given by $\deg \mathbb{P}(C_W V)^*_{Z'}$. First we remark that this degree is just $\sum m_i \deg \mathbb{P}(C_i)^*_{Z'}$, where the sum is taken over all i such that $\dim(C_i)_{Z'} = \dim(C_W V)_{Z'} = \dim \mathcal{O}_{V, Z'}$ or, equivalently, $\pi(C_i) = Z'$. Using 2.4 and comparing with $(*)$ it remains to show:

(a) For every i with $\pi(C_i) = Z'$ we have

$$\deg \mathbb{P}(C_i)^*_{Z'} = e(I_{C_i, Z'}, \mathcal{O}_{C_i, Z'}),$$

where $I_{C_i,Z'} \subseteq \mathcal{O}_{C_i,Z'}$ is the ideal of $Z' \hookrightarrow C_i$.

(b) For every i with $\pi(C_i) \neq Z'$ the coefficient of Z' in $v(\underline{\sigma}^*, C_i)$ is zero.

(a) follows from the fact that the multiplicity $e(B_+, B)$ of a graded ring, say B, is just the degree of $\operatorname{Proj}(B)$. In order to prove (b) it suffices to remark that Z' is not a distinguished component for C_i and so by the equivalence of (1) and (2) does not appear in the cycle $v(\underline{\sigma}^*, C_i)$. □

In the above proof we have used the following lemma.

Lemma 2.10. *Let* $(A, \mathfrak{m})$ *be a local Noetherian ring and* $I \subseteq A$ *an ideal generated by* $x_1, \ldots, x_n$. *Set* $G := A/\mathfrak{m} \otimes \operatorname{gr}_I A$ *and assume that the initial forms of* $x_1, \ldots, x_{\dim G}$ *form a system of parameters in* G. *Suppose moreover that for some* k *with* $0 \le k \le n+1$ *the set* $V(x_1, \ldots, x_{k-1})$ *is not contained in* $V(I)$. *Then* $\dim G \ge k$.

Proof. Assume that $\dim G < k$. Let Y be the blowing up of Spec A along I, i.e. Y is the scheme theoretic closure of $D(I)$ under the rational map

$$(x_1 : \cdots : x_n) : \operatorname{Spec} A \dashrightarrow \mathbb{P}_A^{n-1}.$$

The exceptional set , say E, is just Proj G. The intersection of E with the proper transform of $V(x_1, \ldots, x_{k-1})$ is contained in the set of zeros of the initial forms of $x_1, \ldots, x_{k-1}$ in G and so is empty by assumption. Hence $V(x_1, \ldots, x_{k-1}) \subseteq V(I)$, which is a contradiction. □

Remark 2.11. Assume that $Z = Z' \otimes_K L$ is K-rational and that as a cycle

$$[C_W V] = \sum m_i [C_i],$$

where C_i are the irreducible components of $C_W V$. Then the proof of 2.9 shows that the coefficient of Z in v may also be written as

$$\sum m_i e(I_{C_i,Z'}, \mathcal{O}_{C_i,Z'}),$$

where the sum is taken over all i such that $\pi(C_i) = Z'$ and where $I_{C_i,Z'} \subseteq \mathcal{O}_{C_i,Z'}$ denotes the ideal of $Z' \hookrightarrow C_i$.

3. Limit of joins and intersections

We begin with the following definition.

Definition 3.1. Let Z be an algebraic K-scheme and $X, Y \subseteq Z$ closed subschemes. Let

$$W := X \cap Y \hookrightarrow V := X \times Y$$

be the diagonal embedding and $C \subseteq W$ a subvariety. Then the scheme

$$\operatorname{LJoin}_C(X, Y) := (C_W V)_C$$

is called the *limit of join variety* of X, Y at C, see 2.6. Similarly by

$$\operatorname{LJoin}^*_C(X, Y) := (C_W V)^*_C$$

we denote the refined fibre of the normal cone of $W \subseteq V$ at C which we also call the *refined* limit of join variety of X, Y at C.

In the special case $X = Y$ we denote $\operatorname{LJoin}_C(X, Y)$ resp. $\operatorname{LJoin}^*_C(X, Y)$ by $\operatorname{LSec}_C X$ resp. $\operatorname{LSec}^*_C X$ and call this the (refined) *limit of secant variety* of X at C.

We note that the limit of join variety is neither irreducible nor reduced, in general, even in the case that X and Y are varieties. The motivation for introducing refined limit of join varieties is that they allow a simple description of the coefficients of distinguished subvarieties in terms of the degrees of this projective scheme, see 3.12 below.

Remark 3.2. (1) $\operatorname{LJoin}_C(X, Y)$ as well as $\operatorname{LJoin}^*_C(X, Y)$ only depend on the scheme-theoretic union of X and Y given by the intersection of the ideal sheaves of X and Y in $\mathcal{O}_Z$, but not on the embedding in Z. In particular $\operatorname{LSec}_C X$ and $\operatorname{LSec}^*_C X$ do not depend on embeddings.

(2) By definition, $\operatorname{LJoin}_C(X, Y)$ is a cone over $k(\eta_C)$, where η_C is the generic point of C. With other words, we may write

$$\operatorname{LJoin}_C(X, Y) = \operatorname{Spec} A,$$

where A is a finitely generated graded $k(\eta_C)$-algebra with $A_0 = k(\eta_C)$. In particular $\mathbb{P}\operatorname{LJoin}_C(X, Y) = \operatorname{Proj} A$ is a projective $k(\eta_C)$-scheme. Similarly, $\mathbb{P}\operatorname{LJoin}^*_C(X, Y) = \operatorname{Proj} B$ for some finitely generated graded B_0-algebra B, where B_0 is an Artinian local ring with residue field $k(\eta_C)$, see 2.8 (1).

(3) Let X', Y' be subvarieties of X, Y respectively. Then it follows immediately from the definition that $\operatorname{LJoin}(X', Y')$ is a closed subscheme of $\operatorname{LJoin}(X, Y)$.

The geometric interpretation of the limit of join variety is described in the following result.

Proposition 3.3. *Let $X, Y \subseteq \mathbb{C}^n$ be locally closed complex subspaces with $0 \in X \cap Y$. Then set-theoretically $\operatorname{LJoin}_0(X, Y)$ is the union of 0 and of all limits of lines*

$$\ell = \lim_{\nu \to \infty} \overline{x_\nu y_\nu}, \qquad x_\nu \neq y_\nu$$

with sequences $x_\nu \in X$, $y_\nu \in Y$ converging to zero.

Proof. By definition $\operatorname{LJoin}_0(X, Y) = (C_W V)_0$, where $W := X \cap Y \hookrightarrow V := X \times Y$ is diagonally embedded. Let $(x_1, \ldots, x_n, y_1, \ldots, y_n)$ be the coordinate functions of

$\mathbb{C}^n \times \mathbb{C}^n$ so that $W \subseteq V$ is given by the equations $x_i - y_i = 0$. By 3.4 below $\mathrm{LJoin}_0(X, Y)$ is given by the union of 0 and all limits

$$\ell = \lim_{\nu \to \infty} \overline{0(x_\nu - y_\nu)}, \qquad (x_\nu, y_\nu) \in X \times Y \setminus W.$$

As such a line is just the limit $\lim \overline{x_\nu y_\nu}$ this proves the desired result. □

It remains to show the following well known proposition.

Proposition 3.4. *Assume that $V \subseteq \mathbb{C}^N$ is a locally closed complex subspace and $W \subseteq V$ is a closed subspace with $0 \in W$, given by the equations $f_1, \dots, f_m \in \Gamma(V, \mathcal{O}_V)$. Consider $f = (f_1, \dots, f_m)$ as a morphism $V \to \mathbb{C}^m$. Then set theoretically $(C_W V)_0 \subseteq \mathbb{C}^m$ is the union of 0 and all limits*

$$\ell = \lim_{\nu \to \infty} \overline{0 f(v_\nu)},$$

with $v_\nu \in V \setminus W$ converging to zero.

Proof. In lack of a reference we give a simple proof. Let $\pi : \hat{V} \to V$ be the blowing up of V along W, i.e. $\hat{V}$ is the closure of the image of the map

$$(1, f) : V \setminus W \to V \times \mathbb{P}^m.$$

The exceptional set $\pi^{-1}(W)$ is given by $\mathbb{P}C_W V$. Hence $\pi^{-1}(0) = \mathbb{P}(C_W V)_0$ is just the set of all lines $\ell = \lim \overline{0 f(v_\nu)}$ where v_ν is a sequence in $V \setminus W$ converging to 0.□

In our next result we will give important estimates for the dimension of these varieties. For subschemes X, Y of a scheme Z we will denote by $X \cup Y$ the scheme theoretic union of X and Y, i.e. the ideal sheaf of $X \cup Y$ in $\mathcal{O}_Z$ is the intersection of the ideal sheaves of X and Y.

Proposition 3.5. *Let Z be an algebraic K-scheme and $X, Y \subseteq Z$ closed subschemes. Then for a subvariety $C \subseteq X \cap Y$*

$$\dim \mathrm{LJoin}_C(X, Y) \leq \min\{\mu(\Omega^1_{X \cup Y/K, C}), \dim \mathcal{O}_{X,C} + \dim \mathcal{O}_{Y,C} + \dim C\}$$

and

$$\dim \mathrm{LJoin}_C(X, Y) \geq \dim \mathcal{O}_{X,C} + \dim \mathcal{O}_{Y,C} - \dim \mathcal{O}_{X \cap Y, C} + \dim C.$$

Moreover, if $\dim \mathrm{LJoin}_C(X, Y) = \mu := (\Omega^1_{X \cup Y, C})$ *then* $\mathrm{LJoin}_C(X, Y) \cong \mathbb{A}^\mu_{k(\eta_C)}$ *where η_C denotes the generic point of C.*

Proof. This follows from 2.7 using the following two lemmata. □

Lemma 3.6. *Let $X, Y \subseteq Z$ and C be as in* 3.5. *Then*

$$\dim \mathcal{O}_{X \times Y, C} = \dim \mathcal{O}_{X,C} + \dim \mathcal{O}_{Y,C} + \dim C.$$

Proof. Shrinking X and Y as neighborhoods of the generic point η_C of C we may assume that all irreducible components of X and Y pass through η_C. But then

$$\begin{aligned}\dim \mathcal{O}_{X,C} &= \dim X - \dim C\\ \dim \mathcal{O}_{Y,C} &= \dim Y - \dim C\\ \dim \mathcal{O}_{X\times Y,C} &= \dim X \times Y - \dim C\end{aligned}$$

which gives the desired result. □

The next lemma, which was also used in the proof of 3.5, is stated in terms of commutative algebra.

Lemma 3.7. *Let R be a K-algebra, $\mathfrak{a}, \mathfrak{b} \subseteq R$ ideals and denote by $I \subseteq R/\mathfrak{a} \otimes_K R/\mathfrak{b}$ the kernel of the multiplication map $R/\mathfrak{a} \otimes_K R/\mathfrak{b} \to R/\mathfrak{a}+\mathfrak{b}$. Then*

$$I/I^2 \cong \Omega^1_{R/\mathfrak{a}\cap\mathfrak{b}}/(\mathfrak{a}+\mathfrak{b})\Omega^1_{R/\mathfrak{a}\cap\mathfrak{b}}.$$

Proof. Clearly we may replace R by $R/\mathfrak{a}\cap\mathfrak{b}$ and thus are reduced to the case $\mathfrak{a}\cap\mathfrak{b} = 0$. Tensoring the exact sequence

$$0 \to \mathfrak{a} \to R \to R/\mathfrak{a} \to 0$$

with $R/\mathfrak{b}$ gives

$$0 \to \operatorname{Tor}_1^R(R/\mathfrak{a}, R/\mathfrak{b}) \to \mathfrak{a}/\mathfrak{a}\mathfrak{b} \to R/\mathfrak{b} \to R/\mathfrak{a}+\mathfrak{b} \to 0,$$

and so

$$(*) \quad \operatorname{Tor}_1^R(R/\mathfrak{a}, R/\mathfrak{b}) = \mathfrak{a}\cap\mathfrak{b}/\mathfrak{a}\mathfrak{b} = 0.$$

Set $J := \operatorname{Ker}(R \otimes_K R/\mathfrak{a} \to R/\mathfrak{a})$. Then $J/J^2 \cong \Omega^1_R/\mathfrak{a}\Omega^1_R$. Considering $R \otimes_K R/\mathfrak{a}$ as an R-module with respect to the first component and tensoring

$$0 \to J \to R \otimes_K R/\mathfrak{a} \to R/\mathfrak{a} \to 0$$

with $R/\mathfrak{b}$ gives because of $(*)$

$$0 \to R/\mathfrak{b} \otimes_R J \to R/\mathfrak{b} \otimes_K R/\mathfrak{a} \to R/\mathfrak{a}+\mathfrak{b} \to 0,$$

i.e. $I = R/\mathfrak{b} \otimes_R J$. Thus

$$I/I^2 \cong J/J^2 \otimes_R R/\mathfrak{b} \cong \Omega^1_R/(\mathfrak{a}+\mathfrak{b})\Omega^1_R.$$

□

Remark 3.8. If K is a perfect field then

$$\mu(\Omega^1_{X\cup Y/K,C}) = \operatorname{emdim} \mathcal{O}_{X\cup Y,C} + \dim C,$$

see [13], and so the upper bound for $\dim \mathrm{LJoin}_C(X, Y)$ in 3.5 becomes

$$\dim \mathrm{LJoin}_C(X, Y) \leq \min\left\{\dim \mathcal{O}_{X,C} + \dim \mathcal{O}_{Y,C} + \dim C, \operatorname{emdim} \mathcal{O}_{X\cup Y,C} + \dim C\right\}.$$

There is a close connection of the join variety of projective schemes and the limit of join variety of their affine cones which we describe in the following result.

Proposition 3.9. *Let $X, Y \subseteq \mathbb{P}^n$ be closed subschemes and $\hat{X}, \hat{Y} \subseteq \mathbb{A}^{n+1}$ the affine cones. Then set theoretically the affine cone $(XY)\hat{}$ over the join variety XY is just $\mathrm{LJoin}_0(\hat{X}, \hat{Y})$.*

Proof. We may assume that K is algebraically closed. Let $x \in X$, $y \in Y$ be closed points and $\hat{x} = \ell_x$, $\hat{y} = \ell_y$ the corresponding lines in $\mathbb{A}^{n+1}$. It is easy to check that $\mathrm{LJoin}_0(\ell_x, \ell_y)$ is the plane $\ell_x + \ell_y \subseteq \mathbb{A}^{n+1}$. Thus, if $\ell \subseteq \mathbb{P}^n$ is the line joining x and y then

$$\hat{\ell} = \ell_x + \ell_y \subseteq \mathrm{LJoin}_0(X, Y).$$

It follows that $(XY)\hat{} \subseteq \mathrm{LJoin}_0(X, Y)$. For the converse inclusion let $\hat{J} = \hat{X} \times \hat{Y}$ be the cone over the ruled join $J := J(X, Y)$, see [4]. If $\pi : \hat{J}_1 \to \hat{J}$ is the blowing up along $\hat{X} \cap \hat{Y}$ then $\pi^{-1}(0) \cong \mathbb{P}\,\mathrm{LJoin}_0(\hat{X}, \hat{Y})$. Let $(x_0, \ldots, x_n, y_0, \ldots, y_n)$ be coordinates on $\mathbb{A}^{n+1} \times \mathbb{A}^{n+1}$. The rational map

$$\sigma := (x_0 - y_0 : \cdots : x_n - y_n) : \hat{J} \longrightarrow \mathbb{P}^n$$

lifts to a regular map $\sigma_1 : \hat{J}_1 \to \mathbb{P}^n$, and the restriction of σ_1 to $\mathbb{P}\,\mathrm{LJoin}_0(\hat{X}, \hat{Y})$ is just the canonical inclusion. Thus

$$\sigma_1(\mathbb{P}\,\mathrm{LJoin}_0(\hat{X}, \hat{Y})) \subseteq \sigma_1(\hat{J}_1) = XY$$

or, equivalently, $\mathbb{P}\,\mathrm{LJoin}_0(\hat{X}, \hat{Y}) \subseteq (XY)\hat{}$ set theoretically. □

Remark 3.10. The above proposition allows to define a canonical scheme structure on the join variety XY of projective schemes.

Before proving the main result of this section we need a description of the (refined) limit of join variety for projective schemes in terms of the normal cone of the ruled join.

Lemma 3.11. *Let $X, Y \subseteq \mathbb{P}^n$ be closed subschemes and $J := J(X, Y) \subseteq \mathbb{P}^{2n+1}$ the ruled join. Let $X \cap Y \hookrightarrow J$ be the diagonal embedding. Then for every subvariety $C \subseteq X \cap Y$*

$$(C_{X\cap Y} J)_C \cong \mathrm{LJoin}_C(X, Y) \times \mathbb{A}^1_{k(\eta_C)}$$

$$(C_{X\cap Y} J)^*_C \cong \mathrm{LJoin}^*_C(X, Y) \times \mathbb{A}^1_{k(\eta_C)},$$

where η_C is the generic point of C.

Proof. Let $U \subseteq J$ be the subset of all $(x_0 : \cdots : x_n : y_0 : \cdots : y_n)$ such that $(x_0 : \cdots : x_n) \neq 0$ and $(y_0 : \cdots : y_n) \neq 0$. Clearly $X \cap Y \subseteq U$, and the morphism

$$U \longrightarrow X \times Y$$

$$(x : y) \longmapsto (x, y)$$

defines a regular map which is a locally trivial $\mathbb{A}^1 \setminus \{0\}$-bundle, and restricts to the diagonal embedding on $X \cap Y$. It follows that the induced map

$$C_{X \cap Y} J \longrightarrow C_{X \cap Y}(X \times Y)$$

is a locally trivial $\mathbb{A}^1$-bundle proving the lemma. □

Theorem 3.12. *Let $X, Y \subseteq \mathbb{P}^n$ be strictly equidimensional closed subschemes and $C \subseteq X \cap Y$ a subvariety. Then C appears in the v-cycle $v = v(X, Y)$ (up to a pure transcendental field extension) iff*

$$\dim \mathrm{LJoin}_C(X, Y) = \dim \mathcal{O}_{X,C} + \dim \mathcal{O}_{Y,C} + \dim C.$$

i.e. C is a distinguished component of the intersection $X \cap Y$. Moreover, then the coefficient of C in v is given by $\deg(\mathbb{P}\,\mathrm{LJoin}_C^*(X, Y))$.

Proof. Let $J := J(X, Y)$ be the ruled join and $X \cap Y \hookrightarrow J$the diagonal embedding. By the preceding lemma

$$\dim \mathrm{LJoin}_C(X, Y) = \dim (C_{X \cap Y} J)_C - 1,$$
$$\deg \mathbb{P}\,\mathrm{LJoin}_C^*(X, Y) = \deg \mathbb{P}(C_{X \cap Y} J)_C^*$$

Moreover $\dim \mathcal{O}_{X \times Y, C} = \dim \mathcal{O}_{J,C} - 1$, and so by 3.6

$$\dim \mathcal{O}_{X,C} + \dim \mathcal{O}_{Y,C} + \dim C = \dim \mathcal{O}_{J,C} - 1.$$

Now the theorem follows from 2.9. □

Remark 3.13. (1) In particular it follows that the coefficient of a distinguished subvariety C is bounded below by the degree of $\mathbb{P}\,\mathrm{LJoin}_C(X, Y)$. It is not known to us whether there is a relationship when C is not distinguished.

(2) It would be interesting to know whether there is a relation between the k_C-multiplicity as introduced in [5] and the degree of the variety $\mathbb{P}\,\mathrm{LJoin}_C(X, Y)$, for an irreducible component C of $X \cap Y$. Another interesting question is whether (4.1) of [5] remains true if one replaces the k-multiplicity by the degree of $\mathbb{P}\,\mathrm{LJoin}_C(X, Y)$.

References

[1] Achilles, R., Manaresi, M., An algebraic characterization of distinguished varieties of intersection, Rev. Roumaine Math. Pures Appl. 38 (1993), 569–578.

[2] Achilles, R., Tworzewski, P., Winiarski, T., Improper isolated intersection in complex analytic geometry, Ann. Polon. Math. 51 (1990), 21–36.

[3] Ådlandsvik, B., Joins and higher secant varieties, Math. Scand. 61 (1987), 213–222.

[4] Flenner, H., Gastel, L. J. van, Vogel, W., Joins and intersections, Math. Ann. 291 (1991), 691–704.

[5] Flenner, H., Vogel, W., Improper intersections and a converse to Bezout's theorem, J. Algebra 159 (1993), 460–476.

[6] Flenner, H., Vogel, W., Joins, tangencies and intersections, Math. Ann. 302 (1995), 489–505.

[7] Flenner, H., O'Carroll, L., Vogel, W., Algebraic and geometric refined intersections, Cambridge University Press, in preparation.

[8] Fulton, W., Intersection theory, Ergeb. Math. Grenzgeb. (3) 2, Springer, Berlin–Heidelberg–New York 1984.

[9] Fulton, W., Algebraic refined intersections, Notes Math. 271, Providence, R.I., Brown University 1986.

[10] Gastel, L. J. van, Excess intersections and a correspondence principle, Invent. Math. 103 (1991), 197–221.

[11] Hartshorne, R., Algebraic geometry, Graduate Texts in Math. 52, Springer, Berlin–Heidelberg–New York 1977.

[12] Matsumura, H., Commutative ring theory, Cambridge University Press 1989.

[13] Scheja, G., Storch, U., Differentielle Eigenschaften der Lokalisierungen analytischer Algebren, Math. Ann. 197 (1972), 137–170.

[14] Stückrad, J., Vogel, W., An algebraic approach to the intersection theory, in: The curves seminar at Queen's Vol. II, 1–32, Queen's papers in Pure and Appl. Math. 61, Kingston, Ontario 1982.

[15] Vogel, W., Lectures on results on Bezout's theorem, (Tata Inst. Lectures on Math. and Phys. 74, Bombay), Springer, Berlin–Heidelberg–New York 1984.

[16] Whitney, H., Complex analytic varieties, Addison–Wesley, Reading, MA 1972.

The McKay correspondence for finite subgroups of SL(3, $\mathbb{C}$)

Yukari Ito and Miles Reid

Abstract. Let $G \subset \mathrm{SL}(n, \mathbb{C})$ be a finite subgroup and $\varphi: Y \to X = \mathbb{C}^n/G$ any resolution of singularities of the quotient space. We prove that crepant exceptional prime divisors of Y correspond one-to-one with "junior" conjugacy classes of G. When $n = 2$ this is a version of the McKay correspondence (with irreducible representations of G replaced by conjugacy classes). In the case $n = 3$, a resolution with $K_Y = 0$ is known to exist by work of Roan and others; we prove the existence of a basis of $H^*(Y, \mathbb{Q})$ by algebraic cycles in one-to-one correspondence with conjugacy classes of G. Our treatment leaves lots of open problems.

Contents

1. Statement of the results

Let $G \subset \mathrm{SL}(n, \mathbb{C})$ be a finite subgroup and $X = \mathbb{C}^n/G$ the quotient space, an affine variety with $K_X = 0$. A *crepant resolution* $f: Y \to X$ is a resolution of singularities such that $K_Y = f^*K_X = 0$. A crepant resolution does not necessarily exist in dimension ≥ 4 (see 4.5); it is known to exist in dimension 2 (classical, Du Val), and in dimension 3 by work of a number of people (see for example Markushevich [M], Ito [I1]–[I3] and Roan [R]), but the proofs are computational rather than conceptual.

This paper contributes some very easy remarks to the following question raised by Dixon, Harvey, Vafa and Witten [DHVW], worked out by Hirzebruch and Höfer [HH], and now famous among algebraic geometers as the "Physicists' Euler number conjecture". See for example [R] for the background.

Conjecture 1.1. $G \subset \mathrm{SL}(n, \mathbb{C})$ is a finite subgroup, $X = \mathbb{C}^n/G$ the quotient space and $f: Y \to X$ a crepant resolution. Then there exists a basis of $H^*(Y, \mathbb{Q})$ consisting of algebraic cycles in one-to-one correspondence with conjugacy classes of G.

It is an elementary fact that Y has no odd-dimensional cohomology, and that $H^{2i}(Y, \mathbb{Q})$ is spanned by algebraic cycles (see 4.1).

Definition 1.2. Let G be a finite group. Write $\Gamma = \mathrm{Hom}(\mu_R, G)$, where R is any common multiple of the orders of all $g \in G$ and μ_R is the group of complex Rth roots of 1; thus Γ is a set which becomes a group isomorphic to G after making a choice of roots of 1. In colourful language, you can think of Γ as the "Tate twist" $\Gamma = G(-1) = \mathrm{Hom}(\mu, G)$, where $\mu = \widehat{\mathbb{Z}}(1) = \varprojlim \mu_r$.

Theorem 1.3 (age grading). *Let $G \subset \mathrm{SL}(n, \mathbb{C})$ be a finite subgroup and Γ as above. Then Γ has a canonical grading*

$$\Gamma = \sqcup_{i=1}^{n-1} \Gamma_i, \qquad \textit{with} \quad \mathrm{age}\, \Gamma_i = i.$$

(Here $\Gamma_0 = \{e\}$, the identity element of Γ.)

Putting a grading on G is a well-known idea in toric geometry; see for example Reid [R1], Theorem 3.1, and 2.2 below, and compare Batyrev and Dais [BD]. The point of the definition of Γ is to make the grading independent of the choice of roots of 1. If you believe Conjecture 1.1, you expect conjugacy classes in Γ_i to give the basis of H^{2i}, so that the age grading of Γ is an analog of the usual weight filtration of $H^*(Y)$, with weight $= 2 \times$ age.

The *junior elements* of Γ (or of G) are the elements $g \in \Gamma_1$ of minimal age 1; the set Γ_1 is called the *junior simplex* of Γ (or of G). The grading of Theorem 1.3 is invariant under conjugacy by G (or by Γ, it's the same thing). The notation Γ_i/G always means conjugacy classes of elements of age i.

Theorem 1.4. *There is a canonical one-to-one correspondence between junior conjugacy classes in Γ and crepant discrete valuations of X.*

The statement is explained in 2.3 below. Roughly, a discrete valuation of X is a prime divisor $E \subset V$ on any normal model V of X up to birational equivalence. Discrete valuations enable us to state Theorem 1.4 in a birationally invariant language, without any assumptions on crepant resolutions or minimal models of X.

A *minimal model* of X (in the sense of Mori theory) is a projective morphism $f: Y \to X$, where f is crepant (that is, $K_Y = f^* K_X$) and Y has $\mathbb{Q}$-factorial terminal singularities.

Corollary 1.5. *Suppose that there exists a minimal model of* $X = \mathbb{C}^n/G$. *Then there are canonical one-to-one correspondences*

$$\Gamma_1/G \longleftrightarrow \text{exceptional prime divisors of } f \longleftrightarrow \text{basis of } H^2(Y, \mathbb{Q}).$$

Now suppose that $n = 3$. Then it is known by work of Mori and others (these are general arguments, applicable to any 3-fold) that a minimal model $f: Y \to X = \mathbb{C}^3/G$ exists. Moreover, it has factorial cDV singularities (see Kollár [K], 2.1.7.6), and is therefore a $\mathbb{Q}$-cohomology manifold, so that Poincaré duality applies between $H^4(Y, \mathbb{Q})$ and cohomology with compact support $H^2_c(Y, \mathbb{Q})$.

Theorem 1.6. *Let* $f: Y \to X$ *be as just described. Then* $H^2_c(Y, \mathbb{Q})$ *is based by the compactly supported exceptional prime divisors of* Y, *that is, the divisors* $E \subset f^{-1}(0)$ *lying over* 0. *Write* $\Gamma_1^{(0)}$ *for the junior elements of* Γ *corresponding to these (see also* 2.10 *below).*

Then there are canonical one-to-one correspondences

$$\Gamma_2/G \longleftrightarrow \Gamma_1^{(0)}/G \longleftrightarrow \text{basis of } H^2_c(Y, \mathbb{Q}) \longleftrightarrow \text{dual basis of } H^4(Y, \mathbb{Q}).$$

These results are proved in §2. Some features and implications are discussed in §4.

Acknowledgements. We are grateful to the organisers of the Trento conference (Jun 1994), where a seminar was held which discussed lots of this stuff (see for example Batyrev [B]). A. Corti pointed out that 3-folds with $\mathbb{Q}$-factorial terminal singularities are $\mathbb{Q}$-homology manifolds, so that Poincaré duality holds. Questions by John Moody have helped us to improve the argument in a number of places. The second author wishes to thank S. Mori and Kyoto Univ. Koenkai for generous invitation and financial support. We are also indebted to David Morrison for pointing out that physicists are human beings and have individual names.

2. Proofs

2.1. Proof of Theorem 1.3. Let $G \subset \mathrm{SL}(n, \mathbb{C})$ be a finite group. Any element $g \in G$ has n eigenvalues $\lambda_1, \dots, \lambda_n$, which are rth roots of 1 if $g^r = 1$. In the normal run of things, in order to write $\lambda_i = \varepsilon^{a_i}$, we have to choose ε a primitive rth root of 1. However, the present set-up side-steps this problem by considering elements $g \in \Gamma = \mathrm{Hom}(\mu_R, G)$; then $\varepsilon \to g(\varepsilon) \to \lambda_i(g(\varepsilon))$ is an unordered set of n characters of μ_R, and thus $\lambda_i(g(\varepsilon)) = \varepsilon^{a_i}$. If g has order r then $a_1, \dots, a_n$ correspond to the characters $\varepsilon \mapsto \varepsilon^{a_i}$ of order r, and we write them as the fractional expression $\frac{1}{r}(a_1, \dots, a_n)$ with integers $a_i \in [0, r)$.

Now, since $G \subset \mathrm{SL}(n)$ it follows that $\sum a_i \equiv 0 \mod r$, and hence $\frac{1}{r}\sum a_i = \alpha$ for some integer $\alpha = \operatorname{age} g \in [0, n)$. This defines the age grading of Γ, and so completes the proof. Q.E.D.

Remarks. 1. In fixed coordinates $x_1, \dots, x_n$ on $\mathbb{C}^n$, the expression $g = \frac{1}{r}(a_1, \dots, a_n)$ is a well-defined homomorphism $\mu_r \to \mathrm{GL}(n, \mathbb{C})$, namely

$$\varepsilon \mapsto \operatorname{diag}(\varepsilon^{a_1}, \dots, \varepsilon^{a_n}).$$

2. Note that we didn't need to make an explicit choice of eigencoordinates, since the characters ε^{a_i} are obtained as the set of eigenvalues of matrixes, and so are invariant under conjugacy by $\mathrm{GL}(n, \mathbb{C})$. We chose an order $\frac{1}{r}(a_1, \dots, a_n)$ of the characters for convenience, but the order disappears at the end, since the age is the elementary symmetric function $\frac{1}{r}\sum a_i$.

3. The group G itself does not have a well-defined grading: for example, if ε is a primitive 7th root of 1 and $G = \mathbb{Z}/7$ is generated by $\operatorname{diag}(\varepsilon, \varepsilon^2, \varepsilon^4) \in \mathrm{SL}(3, \mathbb{C})$, then choosing the same ε gives

$$\operatorname{age}(\overline{1}, \overline{2}, \overline{4}) = 1 \quad \text{and} \quad \operatorname{age}(\overline{3}, \overline{5}, \overline{6}) = 2,$$

whereas the choice of $\eta = \varepsilon^{-1}$ gives the opposite grading.

2.2. Notation: the unit box $\square$ and junior simplex Δ_1. We now go through the same proof in more conventional terms, and, at the same time, introduce notation that we need later.

Let $G \subset \mathrm{SL}(n, \mathbb{C})$ be as usual and $A \subset G$ an Abelian group; write r for some exponent of A. Choose coordinates $x_1, \dots, x_n$ on $\mathbb{C}^n$ to diagonalise the action, so that A acts on the $x_1, \dots, x_n$ by diagonal matrixes:

$$g \mapsto \operatorname{diag}(\lambda_1(g), \dots, \lambda_n(g)),$$

where $\lambda_i \colon A \to \mu_r$ are characters of A. Then the Pontryagin dual of $A \hookrightarrow \mu_r^n \subset (\mathbb{C}^\times)^n$ is a surjective homomorphism

$$h \colon \mathbb{Z}^n \to A^\vee \qquad \text{defined by } \prod x_i^{m_i} \mapsto \sum m_i \lambda_i.$$

Here $A^\vee = \operatorname{Hom}(A, \mathbb{C}^\times)$ is the character group of A, and we identify $\mathbb{Z}^n$ with the set of Laurent monomials $x^m = \prod x_i^{m_i}$, as usual in toric geometry. The kernel of h is the set of invariant monomials $M \subset \mathbb{Z}^n$, a sublattice of finite index. The dual lattice of M is an overlattice $L \supset \mathbb{Z}^n$. For example, if A is cyclic with generator $\operatorname{diag}(\varepsilon^{a_1}, \dots, \varepsilon^{a_n})$ then $L = \mathbb{Z}^n + \mathbb{Z} \cdot \frac{1}{r}(a_1, \dots, a_n)$. To distinguish the two, M is sometimes called the *lattice of monomials*, and L the *lattice of weights*.

Now in the copy of $\mathbb{R}^n$ corresponding to $\mathbb{Z}^n \subset L$, write

$$\square = \prod [0, 1) = \left\{ (\alpha_1, \dots, \alpha_n) \in \mathbb{R}^n \;\middle|\; 0 \le \alpha_i < 1 \right\}$$

for the unit cube, and

$$\Delta_1 = \left\{(\alpha_1, \dots, \alpha_n) \in \square \;\middle|\; \sum \alpha_i = 1\right\}$$

for the unit simplex (see Figure 1). The point is that, provided roots of 1 are chosen, the quotient $L/\mathbb{Z}^n$ is naturally isomorphic to A, and since $\square$ is a fundamental domain for the action of $\mathbb{Z}^n$ on $\mathbb{R}^n$, every element of $L/\mathbb{Z}^n \cong A$ has a unique representative in $\square$, and so $A = L \cap \square$. Thus the age grading of A is just the natural slicing of $\square$ by the planes $\sum \alpha_i = j$ for $j = 1, \dots, n-1$. In particular the set of junior elements of A equals $L \cap \Delta_1$.

Lemma. *There is a canonical isomorphism* $L/\mathbb{Z}^n = A(1) = \mathrm{Hom}(\mu_r, A)$.

Proof. Given the coordinate system $x_1, \dots, x_n$, the exact sequence

$$0 \to M \to \mathbb{Z}^n \to A^\vee \to 0$$

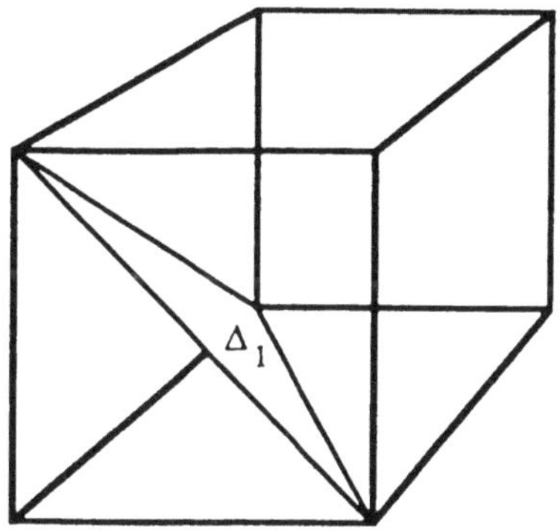

Figure 1. The unit box $\square$ and junior simplex Δ_1.

is intrinsic, and, at first sight, it seems that the isomorphism $L/\mathbb{Z}^n \to A$ should also be: in fact, lattice theory gives a canonical duality pairing $\mathbb{Z}^n/M \times L/\mathbb{Z}^n \to \mathbb{Q}/\mathbb{Z}$. The subtle point is that this is duality with coefficients in $\mathbb{Q}/\mathbb{Z}$, giving

$$L/\mathbb{Z}^n = \mathrm{Hom}(A^\vee, \mathbb{Q}/Z),$$

whereas the character group $A^\vee = \mathrm{Hom}(A, \mathbb{C}^\times)$ is the dual with coefficients in $\mathbb{C}^\times$. Now obviously

$$A(1) = \mathrm{Hom}(\mu_r, A) = \mathrm{Hom}(A^\vee, \mathbb{Q}/\mathbb{Z}).$$

An alternative, slightly bizarre, proof is to apply $\mathrm{Hom}_{\mathbb{Z}}(\text{blank}, \mathbb{Z})$ to the exact sequence $0 \to M \to \mathbb{Z}^n \to A^\vee \to 0$, to get

$$0 = \mathrm{Hom}(A^\vee, \mathbb{Z}) \to \mathbb{Z}^n \to L \to \mathrm{Ext}^1_{\mathbb{Z}}(A^\vee, \mathbb{Z}) \to 0, \text{ that is, } L/\mathbb{Z}^n = \mathrm{Ext}^1_{\mathbb{Z}}(A^\vee, \mathbb{Z}),$$

and finally to argue that

$$\mathrm{Ext}^1(A^\vee, \mathbb{Z}) = A \otimes \mathrm{Ext}^1(\widehat{\mathbb{Z}}, \mathbb{Z}) = A \otimes \widehat{\mathbb{Z}}(1). \qquad \text{Q.E.D.}$$

2.3. Overview of proof of Theorem 1.4. We start by explaining the statement and the strategy. Let $f\colon V \to X$ be any resolution of singularities, with exceptional divisors $\{E\}$. Write $K_V = f^*K_X + \sum a_E E$. The discrepancy a_E is independent of V. It is known that $a_E \geq 0$ for all E (see 2.4 below, and compare [R1], Theorem 3.1). The *crepant* divisors E are those with $a_E = 0$. Every crepant E must appear on any resolution V. The theorem states that the crepant divisors E for any V correspond one-to-one with Γ_1/G.

Let V be a normal variety (in applications, a blowup of $\mathbb{C}^n$ or a partial resolution of X). Counting the zeros and poles of rational functions $h \in k(V)$ along a prime divisor $E \subset V$ defines a discrete valuation of the function field $k(V)$, which is a homomorphism $v_E\colon k(V)^\times \to \mathbb{Z}$ satisfying a suitable compatibility with the additive structure of $k(V)$. The theory of discrete valuations is a device for discussing properties of $E \subset V$ in a birational way. In what follows, although not absolutely essential, it is convenient to use the definition of discrete valuation v of $k(V)$, and the fact that every geometric discrete valuation of a function field $k(X)$ is of the form v_E for some prime divisor E on a partial resolution $V \to X$. (*Geometric* means that the residue field k_v has transcendence degree $n-1$; this is a very easy case of Zariski's "local uniformisation", see Zariski and Samuel [ZS], Vol. II, Chap. VI, §14, Theorem 31.)

It turns out that the only exceptional divisors of partial resolutions $V \to X$ we need arise from weighted blowups of $\mathbb{C}^n$: for each $g \in G$, take eigencoordinates $x_1, \dots, x_n$ for g, and a weighting $\beta = (b_1, \dots, b_n)$ on $x_1, \dots, x_n$ (usually closely related to the eigenvalues of g; see 2.7 for more precise conditions). The exceptional divisor of the weighted blowup $W_\beta \to \mathbb{C}^n$ is a weighted projective space $F_\beta = \mathbb{P}(b_1, \dots, b_n)$, and the valuation $v_\beta = v_{F_\beta}$ of $k(\mathbb{C}^n)$ along $F_\beta \subset W_\beta$ takes x_i to b_i. The corresponding exceptional divisor E of $X = \mathbb{C}^n/G$ is somewhat messy to describe geometrically (see 2.6 below), but the valuation v_E is just the restriction of v_β to the subfield $k(X) = k(\mathbb{C}^n)^G \subset k(\mathbb{C}^n)$, divided by the ramification degree. A valuation of this form is called a *monomial valuation* of $k(X)$.

Theorem 1.4 is a consequence of two easy tricks: (I) if $X = \mathbb{C}^n/G$ is a quotient singularity and $E \subset V \to X$ a prime divisor on a partial resolution, *ramification theory* reduces the calculation of the discrepancy of E to the case of a cyclic subgroup $\mathrm{Ram}\, F = \mathbb{Z}/r \subset G$ (see 2.5–6). (II) A cyclic quotient singularity $\mathbb{C}^n/(\mathbb{Z}/r)$ is a toric variety, so has a toric resolution (see 2.4); every valuation appearing on it is monomial in the sense just discussed. Therefore every nonmonomial valuation of X is discrepant.

The rest of the proof just boils down to calculating monomial valuations and the effect of the group action on them.

2.4. The cyclic case. Consider first the case of a cyclic group $\langle g\rangle$ with $g \in \Gamma$ an element of order r. Choose eigencoordinates to make $g = \frac{1}{r}(a_1, \dots, a_n)$ as described in 2.2. The following toric construction is standard. (For details, compare, for example, the proof of [R1], Theorem 3.1 or [R2], §4.)

The barycentric subdivision of the first octant of $\mathbb{R}^n$ at $\frac{1}{r}(a_1, \dots, a_n)$ (with respect to the two different lattices $\mathbb{Z}^n$ and L) defines two blowups

$$\begin{array}{ccccc} F & \subset & W & \to & \mathbb{C}^n \\ & & \downarrow & & \downarrow \\ E & \subset & V & \to & \mathbb{C}^n/\langle g\rangle. \end{array}$$

Here W is the weighted blowup of $\mathbb{C}^n$ with weights $(a_1, \dots, a_n)$. This has a single exceptional component F, which is a weighted projective space $F = \mathbb{P}(a_1, \dots, a_n)$, and $V = W/\langle g\rangle$. It is easy to see that F has discrepancy $a_F = \sum a_i - 1 = \alpha r - 1$ with $\alpha = \operatorname{age} g$ (see [R2], Prop. 4.8), and that $\langle g\rangle$ acts biregularly on W, fixing F pointwise, and multiplying a normal coordinate to F by a primitive rth root of 1. Thus the cyclic cover $W \to V$ has ramification of order r along E, and therefore E has discrepancy $\frac{1}{r}(a_F - (r-1)) = \alpha - 1$. Therefore $a_E \geq 0$, and E is crepant if and only if $\alpha = \operatorname{age} g = 1$, that is, g is junior.

In this case, the quotient X is a toric variety, so has a toric resolution $V_\Sigma \to X$, obtained by subdividing the first octant of $\mathbb{R}^n$ into a fan Σ of basic simplexes for the lattice $L = \mathbb{Z}^n + \mathbb{Z} \cdot \frac{1}{r}(a_1, \dots, a_n)$. The exceptional divisors of V_Σ correspond to the 1-skeleton of Σ; if $e = \frac{1}{r'}(b_1, \dots, b_n) \in L$ (for some $r' \mid r$) is in the positive octant then its discrepancy is $\frac{1}{r'} \sum b_i - 1$. Thus (again because $g \in \mathrm{SL}(n, \mathbb{C})$, so that $\sum b_i \equiv 0 \mod r'$), every exceptional divisor has discrepancy ≥ 0, and the crepant divisors correspond to $\Delta_1 \cap L$, that is, to the junior elements Γ_1.

2.5. Ramification groups. The ramification theory of discrete valuations in a Galois field extension $K \subset L$ is very similar to the factorisation of prime ideals in a Galois extension of number fields. (See 2.6 below for the birational geometric picture.) Let $K \subset L$ with $G = \mathrm{Gal}(L/K)$ be a finite Galois extension of function fields. A discrete valuation $v\colon L^\times \to \mathbb{Z}$ of L restricts to a map $v_{|K}\colon K^\times \to Z$ which obviously satisfies all the conditions for a discrete valuation, except that it only maps onto a subgroup $r\mathbb{Z} \subset \mathbb{Z}$ of finite index. We call r the *ramification degree* of v and write $w = \frac{1}{r}(v_{|K})$. Conversely, a discrete valuation w of K extends in finitely many ways to discrete valuations v of L. These extensions are permuted transitively by the Galois group, which acts on the valuations of L by

$$g\colon v \mapsto v \circ g \qquad \text{for } g \in \mathrm{Gal}(L/K).$$

Corresponding to a valuation v of L, there is a tower of subgroups

$$\operatorname{Ram} v \subset \operatorname{Stab} v \subset \operatorname{Gal}(L/K).$$

The *stabiliser group* Stab v and the *ramification group* Ram v are defined by

$$\operatorname{Stab} v = \{g \in G \mid v \circ g = v\} \quad \text{and} \quad \operatorname{Ram} v = \{g \in \operatorname{Stab} v \mid g^* = \text{id on } k(v)\},$$

where $k(v) = A_v/m_v$ is the residue field of v. (Compare [ZS], Vol. II, Chap. VI, §12; these are also traditionally called the *decomposition subgroup* G_Z (Zerlegungsgruppe) and the *inertia subgroup* G_T (Trägheitsgruppe) respectively.)

A crucial point for us is that Ram v is a cyclic group. The complex analysis argument is transparent and is given at the end of 2.6. For convenience, we spell out the algebraic proof of Zariski and Samuel: in their notation, by [ZS], Vol. II, Chap. VI, §12, Theorem 24, Corollary, we have $G_V = 1$; moreover (same book, pp. 75–76), G_T/G_V is dual to the extension of the value groups Γ^*/Γ, which for discrete valuations is a finite cyclic group of order r, where r is the ramification degree. In either case, characteristic zero is used in an essential way.

2.6. The geometric picture. For a finite subgroup $G \subset \mathrm{GL}(n, \mathbb{C})$, the set-up is as follows. Write $L = k(\mathbb{C}^n)$ and $K = k(X) = L^G$. A discrete valuation v of L is of the form $v = v_F$, where $F \subset W$ is a prime divisor on a model $W \to \mathbb{C}^n$. There is a partial resolution $V \to X$ of X such that F maps generically onto a prime divisor $E \subset V$. You can prove this by applying the local uniformisation result quoted in 2.3. The alternative is to construct V as a quotient of some blowup of W on which G acts biregularly. For this, let $\varphi_g\colon W^g \to \mathbb{C}^n$ be the morphism with $W^g = W$ and φ_g the composite $W \to \mathbb{C}^n \xrightarrow{g} \mathbb{C}^n$, and define W' to be the birational fibre product of all the W^g. In other words, take the closure in the fibre product of the graph of the locus where all the g are regular, normalised if necessary.

Thus we can assume that $E \subset V \to X$ is a prime divisor on a partial resolution of X, and that W is a blowup of $\mathbb{C}^n$ that dominates V:

$$\begin{array}{ccc} W & \longrightarrow & \mathbb{C}^n \\ \downarrow & & \downarrow \\ V & \to & X = \mathbb{C}^n/G \end{array}$$

(see Figure 2). Write $\mathcal{F} = \{F\}$ for the set of all prime divisors $F \subset W$ which dominate $E \subset V$. Under these assumptions, the valuations extending $w = v_E$ are of the form v_F where $F \in \mathcal{F}$. Then G acts transitively on $\mathcal{F}$, and if $v = v_F$ then

$$\operatorname{Stab} v = \operatorname{Stab} F = \{g \in G \mid g(F) = F\}$$

and

$$\operatorname{Ram} v = \operatorname{Ram} F = \{g \in G \mid g_{|F} = \mathrm{id}_F\}.$$

Note: the action of G on W is only birational; however, it is regular at the generic point of every component F, and $g(F) = F$ and $g_{|F} = \mathrm{id}_F$ are to be interpreted in this sense.

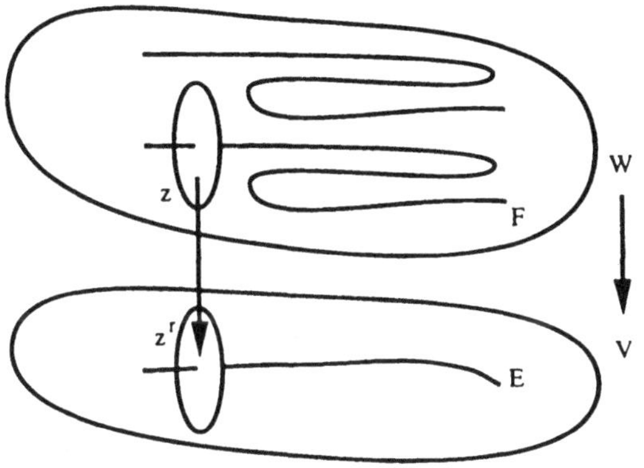

Figure 2. The cover $W \to V$ over the generic point of E.

Now since G acts transitively, the orbit space $G/\operatorname{Stab} F$ equals $\mathcal{F}$. Moreover, $\operatorname{Stab} F$ acts birationally on F with birational quotient E, so that $\operatorname{Ram} F \lhd \operatorname{Stab} F$ is a normal subgroup, and

$$\operatorname{Stab} F/\operatorname{Ram} F = \operatorname{Gal}(F/E) = \operatorname{Gal}(k(F)/k(E)).$$

Finally, $\operatorname{Ram} F$ is the cyclic group describing the behaviour of the cover $W \to V$ along the general point of F. Since we are dealing there with a codimension 1 submanifold of a complex manifold, the covering is locally of the form $z \mapsto w = z^r$, so that $\operatorname{Ram} F = \mathbb{Z}/r$ is cyclic.

Ramification and discrepancy. If s is a canonical form (that is, n-form) on V with $v_E(s) = m$ then

$$s = (\text{unit}) \cdot w^m \mathrm{d}x_1 \wedge \cdots \wedge \mathrm{d}x_{n-1} \wedge \mathrm{d}w$$

with $x_1, \ldots, x_{n-1}$ local coordinates along E, and its pullback to W is of the form

$$\psi^* s = (\text{unit}) \cdot z^{mr+r-1} \mathrm{d}x_1 \wedge \cdots \wedge \mathrm{d}x_{n-1} \wedge \mathrm{d}z,$$

so that $v_F(\psi^* s) = mr + r - 1 = r v_E(s) + r - 1$. Therefore $a_E = \frac{1}{r}(a_F - (r-1))$.

As discussed in 2.3, a crucial point for us is that the discrepancy of E is already present in the cyclic quotient $\mathbb{C}^n/\operatorname{Ram} F$. In our picture, the quotient $W/\operatorname{Ram} F$ contains a divisor $\overline{F}$, and the subsequent map $W/\operatorname{Ram} F \to V$ is etale at a general point of $\overline{F}$.

2.7. Ramification for monomial valuations. The definitions of Stab v and Ram v for a discrete valuation v of $k(\mathbb{C}^n)$ extend to the whole of $\mathrm{GL}(n, \mathbb{C})$ (without reference to a finite subgroup) as follows:

$$\mathrm{Stab}'\, v = \mathrm{Stab}'\, F = \{g \in \mathrm{GL}(n, \mathbb{C}) \mid v \circ g = v\} = \{g \in \mathrm{GL}(n, \mathbb{C}) \mid g(F) = F\}$$

and

$$\mathrm{Ram}'\, v = \{g \in \mathrm{Stab}'\, v \mid g^* = \mathrm{id} \text{ on } k(v)\} = \{g \in \mathrm{Stab}'\, F \mid g_{|F} = \mathrm{id}_F\},$$

where $F \subset W$ is a prime divisor of a blowup $W \to \mathbb{C}^n$ such that $v = v_F$. Here $\mathrm{Stab}'\, v = \mathrm{Stab}_{\mathrm{GL}}\, v$ denotes (temporarily) the stabiliser of v in GL, and similarly for Ram$'$.

We now determine these subgroups for monomial valuations. We first fix some ideas and notation. Consider $\mathbb{C}^n$ with a fixed coordinate system $x_1, \dots, x_n$. A primitive nonzero vector $\beta = (b_1, \dots, b_n) \in \mathbb{Z}^n$ with all $b_i \geq 0$ corresponds to the $\mathbb{C}^\times$ action on $\mathbb{C}^n$ given by

$$\lambda\colon (x_1, \dots, x_n) \mapsto (\lambda^{b_1} x_1, \dots, \lambda^{b_n} x_n).$$

The action defines a decomposition of $\mathbb{C}^n$ as a direct sum of eigenspaces; we simply group together the x_i with the same b_i:

$$\mathbb{C}^n = \bigoplus_{b \in \mathbb{Z}} S_b, \quad \text{where} \quad S_b = \operatorname{Spec} k\big[x_i \mid b_i = b\big]. \tag{2.7.1}$$

There is a corresponding filtration by increasing b_i:

$$T_0 \subset \cdots \subset T_c \subset \cdots \subset \mathbb{C}^n, \quad \text{where} \quad T_c = \operatorname{Spec} k\big[x_i \mid b_i \leq c\big] = \bigoplus_{b \leq c} S_b, \tag{2.7.2}$$

and, of course, $\mathrm{Gr}_b\, T = T_b / T_{b-1} = S_b$.

The weighting $\beta = (b_1, \dots, b_n)$ defines a grading on $k[x_1, \dots, x_n]$, and a weighted blowup $\sigma_\beta\colon W_\beta \to \mathbb{C}^n$, whose exceptional locus $F_\beta \subset W_\beta$ is $\operatorname{Proj} k[x_1, \dots, x_n] = \mathbb{P}(b_1, \dots, b_n)$. Note that some of the b_i may be zero, and then Proj must be taken as a product of a conventional weighted projective space with the affine space S_0, the centre of the blowup in $\mathbb{C}^n$. The weighting $(b_1, \dots, b_n)$ also defines a monomial discrete valuation $v_{F_\beta} = v_\beta$, taking $x_i \mapsto b_i$.

Lemma. *As above, $\beta = (b_1, \dots, b_n)$ is a primitive nonzero vector in the positive octant; suppose that $g \in \mathrm{GL}(n, \mathbb{C})$.*

(1) *$g \in \mathrm{Stab}'\, v_\beta$ if and only if it preserves the filtration* (2.7.2), *that is, g^* takes each x_i to a linear combination of x_j with $b_j \leq b_i$:*

$$g^*(x_i) = \sum m_{ij}(g) x_j \quad \textit{with } m_{ij} = 0 \textit{ whenever } b_i > b_j. \tag{2.7.3}$$

(2) *Suppose that* $g \in \operatorname{Stab}' v_\beta$; *then the action of* g *on* $F_\beta = \mathbb{P}(b_1, \ldots, b_n)$ *is given by the induced map* $\overline{g}$ *on the graded vector space* $\bigoplus T_b/T_{b-1}$, *that is,*

$$\overline{g}^*(x_i) = \sum \overline{m}_{ij}(g)x_j \quad \textit{with } \overline{m}_{ij} = \begin{cases} m_{ij} & \textit{if } b_j = b_i, \\ 0 & \textit{if } b_j < b_i. \end{cases} \tag{2.7.4}$$

In particular, $g \in \operatorname{Ram}' v_\beta$ *if and only if* $\overline{g} = \operatorname{diag}(\lambda^{b_1}, \ldots, \lambda^{b_n})$ *for some* $\lambda \in \mathbb{C}^\times$.

Proof. This is really obvious, because the valuation v_β determines the filtration (2.7.2). Indeed, the valuation of a linear form $\sum \alpha_i x_i$ for v_β is just the smallest b_i for which $\alpha_i \neq 0$. For (1), $g \in GL(n, \mathbb{C})$ preserves the space of linear forms, and if it also preserves v_β then it preserves the filtration (2.7.2).

If g preserves the filtration (2.7.2), it also preserves the filtration of $k[x_1, \ldots, x_n]$ by the ideals of polynomials vanishing k times along F_β:

$$I_k = \{f \mid v_\beta(f) \geq k\}.$$

Therefore its action on $\mathbb{C}^n$ extends to a biregular action on the weighted blowup $W_\beta = \operatorname{Proj} \bigoplus_{k \geq 0} I_k \to \mathbb{C}^n$. The restriction to $F_\beta = \mathbb{P}(b_1, \ldots, b_n)$ is in terms of the associated $\operatorname{Gr} = \bigoplus_{k \geq 0} I_k/I_{k+1}$, so the first part of (2) is clear. For the final part, the condition $g \in \operatorname{Ram}' v_\beta$ is that g fixes F_β pointwise (on a dense open set). Thus this is the usual homogeneous coordinate business: $(x_1 : \cdots : x_n) = (y_1 : \cdots : y_n)$ in $\mathbb{P}(b_1, \ldots, b_n)$ if and only if $x_i = \lambda^{b_i} y_i$ for some $\lambda \in \mathbb{C}^\times$. Q.E.D.

Corollary. *Let* $G \subset GL(n, \mathbb{C})$ *be a finite group, and* $g \in G$ *an element of order* r, *written as usual* $g = \frac{1}{r}(b_1, \ldots, b_n)$ *in eigencoordinates* $x_1, \ldots, x_n$. *Assume that* $\beta = (b_1, \ldots, b_n)$ *is a primitive vector of* $\mathbb{Z}^n$, *and let* $v = v_\beta$ *be the corresponding monomial valuation.*

Then an element $h \in G$ *is in* $\operatorname{Stab} v_\beta$ *if and only if it preserves the direct sum decomposition* (2.7.1). *Moreover,* $h \in \operatorname{Ram} v_\beta$ *if and only if* $h = \operatorname{diag}(\varepsilon^{b_1}, \ldots, \varepsilon^{b_n})$, *where* ε *is a root of* 1.

Proof. Choose a G-invariant Hermitian metric on $\mathbb{C}^n$, so that all the elements of G are unitary. Then the direct sum (2.7.1) is the eigenspace decomposition of the unitary operator g, and is thus orthogonal. Moreover, h is also a unitary operator, so it preserves the filtration (2.7.2) if and only if it preserves the direct sum decomposition (2.7.1). The final part follows directly from (2) of the lemma. Q.E.D.

2.8. Conclusion of proof of Theorem 1.4. We divide up the proof into 5 easy steps.

Step 1. *Construction of* $g \mapsto v_g$. As we hinted above, the correspondence

$$\Gamma_1 \to \{\text{valuations of } k(X)\},$$

sends g to the monomial valuation v_g of $k(X)$ corresponding to g. That is, as described in 2.2, write $g = \frac{1}{r}(a_1, \ldots, a_n) \in \square$, to mean that g has order exactly r, and is

$\varepsilon \mapsto \operatorname{diag}(\varepsilon^{a_1}, \dots, \varepsilon^{a_n})$ in eigencoordinates $x_1, \dots, x_n$. Now set $\widetilde{v}_g = v_\beta$, where $\beta = (a_1, \dots, a_n)$, and

$$v_g = \frac{1}{r}(\widetilde{v}_{g|k(X)}) \colon k(X) \to \mathbb{Z} \cup \{\infty\}.$$

It's clear from the definition of discrete valuation that v_g is (possibly a multiple of) a discrete valuation of $k(X)$.

Step 2. *Ramification and discrepancy.* We claim that for $g \in \Gamma_1$, the ramification group of $\widetilde{v}_g = v_\beta$ in G is exactly $\langle g \rangle$, and the ramification degree is r. It follows from this that v_g is in fact a discrete valuation of $k(X)$, and, by the calculation in 2.4, that its discrepancy is age $g - 1 = 0$.

Since $g = \frac{1}{r}(a_1, \dots, a_n)$, Corollary 2.7 says that the ramification group of $\widetilde{v}_g$ is the subgroup of G of elements of the form $h = \operatorname{diag}(\varepsilon^{a_1}, \dots, \varepsilon^{a_n})$, where ε is a root of 1. If the order of ε divides r then $h \in \langle g \rangle$, which is what we want. If not, replacing h by a suitable combination of g, h, we can assume that ε is a root of unity of order larger than r; but then, because $\sum a_i = r$, such an h is not in $\mathrm{SL}(n, \mathbb{C})$.

Step 3. *v_g is well defined on conjugacy classes.* If $g_1 = hgh^{-1}$, and $x_1, \dots, x_n$ are eigencoordinates for g then h^*x_i are eigencoordinates for g_1 with the same eigenvalues. Since $\widetilde{v}_g$ and $\widetilde{v}_{g_1}$ are constructed as monomial valuations with respect to these eigencoordinates, it follows at once that $\widetilde{v}_g \circ h = \widetilde{v}_{g_1}$, and therefore they restrict to the same valuations of $k(X) = k(\mathbb{C}^n)^G$.

In geometric terms, the weighted blowup of $\mathbb{C}^n$ corresponding to g and g_1 fit together into a diagram of morphisms:

$$\begin{array}{ccc} W_g & \xrightarrow{h} & W_{g_1} \\ \downarrow & & \downarrow \\ \mathbb{C}^n & \xrightarrow[h]{} & \mathbb{C}^n \end{array}$$

That is, h takes the exceptional prime divisor F_g into F_{g_1}, so that they both map birationally to the same exceptional divisor of X.

Step 4. *Injectivity.* This is the converse of Step 3: if $g, g' \in \Gamma_1$ and $v_g = v_{g'}$ then g and g' are conjugate in G. The assumption is that $\widetilde{v}_g$ and $\widetilde{v}_{g'}$ restrict to the same valuation of $k(X)$. Recall that the Galois group $G = \operatorname{Gal}(k(\mathbb{C}^n)/X)$ acts transitively on the set of extensions of a valuation of $k(X)$. Therefore $\widetilde{v}_{g'} = \widetilde{v}_g \circ h$ for some $h \in G$. We've seen in Step 3 that $\widetilde{v}_g \circ h = \widetilde{v}_{g_1}$ where $g_1 = hgh^{-1}$. Now $\widetilde{v}_{g_1} = \widetilde{v}_{g'}$ clearly implies that $g' = g_1$. (Because they both induce the same filtration of the vector space of linear forms, and give x_i the same weights b_i.)

Step 5. *Surjectivity: every crepant divisor of X comes from this construction.* If v_E is a crepant valuation of $k(X)$ and $\widetilde{v} = v_F$ a valuation of $k(\mathbb{C}^n)$ extending v then $\operatorname{Ram} \widetilde{v}$ is a cyclic group of order r, the ramification degree. The discrepancy of E depends only on r and that of F, and is given by

$$a_E = \frac{1}{r}(a_F - (r-1)).$$

As explained in 2.3, only monomial valuations of $\mathbb{C}^n$ with nontrivial ramification groups can give crepant valuations of X. These are all of the form v_g with $g \in \Gamma$, and only those with $g \in \Gamma_1$ are crepant.

This completes the proof of Theorem 1.4. Q.E.D.

Remark. One can also consider monomial valuations v_g for conjugacy classes $g \in \Gamma_i/G$ with $i \geq 2$. To get the right discrepancy $a_{E_g} = \text{age} - 1$, and to keep the injectivity proved in Step 3, we restrict to *primitive* elements g (corresponding to a primitive vector $g = \frac{1}{r}(a_1, \ldots, a_n) \in \square$). It's an exercise to generalise Steps 2–4 to these valuations. For example, the cyclic group $\frac{1}{4}(1,1,1,1)$ (the Gorenstein cone on the 4-fold Veronese embedding of $\mathbb{P}^3$) has only one crepant divisor.

2.9. Proof of Corollary 1.5. This is clear. If a minimal model Y exists then the exceptional divisors occurring on it are the crepant divisors and no others. Next, $H^2(Y, \mathbb{Q})$ is spanned by algebraic cycles, the first Chern classes of divisors (just use the exponential sequence and $R^1 f_* \mathcal{O}_Y = 0$). But every divisor D on Y pushes down to divisor on X some multiple of which is linearly equivalent to zero, because a quotient singularity is analytically $\mathbb{Q}$-factorial. Therefore some multiple of D is linearly equivalent to an exceptional divisor.

Q.E.D.

2.10. Proof of Theorem 1.6. For $g = \frac{1}{r}(a_1, \ldots, a_n) \in \square$, the fixed locus $\operatorname{Fix} g \subset \mathbb{C}^n$ is the linear subspace S_0 corresponding to the coordinates x_i with $a_i = 0$. In particular, 0 is an isolated point of $\operatorname{Fix} g$ if and only if all $a_i > 0$, or equivalently, g is in the interior of $\square$. Thus we set

$$\Gamma_1^{(0)} = \{g \mid \operatorname{Fix} g = \{0\}\}.$$

Note in passing that if $n = 3$ and $g = \frac{1}{r}(a_1, a_2, a_3)$ then $a_1 = 0$ implies $a_2 + a_3 = r$, so that

$$\operatorname{Fix} g \neq \{0\} \implies g \text{ is junior.}$$

We now restrict attention to the case $n = 3$. Then $g \in \Gamma_1^{(0)}$ is equivalent to $g^{-1} \in \Gamma_2$: in eigencoordinates, if $g = \frac{1}{r}(a_1, a_2, a_3)$ with all $a_i > 0$ then

$$g^{-1} = (1,1,1) - \frac{1}{r}(a_1, a_2, a_3).$$

Thus $g \mapsto g^{-1}$ gives the bijection $\Gamma_1^{(0)} \to \Gamma_2$, which is clearly compatible with conjugacy. This proves the first bijection in Theorem 1.6.

The second bijection is clear, since a divisor E has compactly supported class in H^2 if and only if E contracts to $0 \in X$. The third bijection is just Poincaré duality.

Q.E.D.

3. Examples

We start with the examples of the Klein quotient singularities or Du Val surface singularities, and explain why our result Theorem 1.4 gives the exceptional components *together with their configuration*. It is perhaps remarkable that our methods lead naturally to a "McKay quiver" structure on the conjugacy classes of G (rather than irreducible representations).

The Du Val singularity D_{n+2} is the quotient $\mathbb{C}^2/\,\mathrm{BD}_{4n}$, where BD_{4n} is the binary dihedral group of order $4n$ generated by

$$A = \begin{pmatrix} \varepsilon & 0 \\ 0 & \varepsilon^{-1} \end{pmatrix} \quad \text{and} \quad B = \begin{pmatrix} 0 & 1 \\ -1 & 0 \end{pmatrix},$$

and ε is a primitive $2n$th root of 1. Thus $A^n = B^2 = -1$, and the elements of BD_{4n} are the $2n$ "rotations" A^i, and the $2n$ "reflections" $A^iB = BA^{2n-i}$.

Conjugacy in G gives $A^i \sim A^{2n-i}$, so there are n conjugacy classes of nonidentity rotations; and $ABA^{-1} = A^2B$, so that $B \sim A^{2i}B$ and $AB \sim A^{2i+1}B$. The effect of conjugacy on the cyclic subgroups $\langle A^iB\rangle \cong \mathbb{Z}/4$ depends on n even or odd: if n is even then $\langle B\rangle$ and $\langle AB\rangle$ are not conjugate, but $B \sim A^nB = -B = B^3$ and $AB \sim (AB)^3 = BA$. If n is odd then all the subgroups $\langle A^iB\rangle$ are conjugate, but the elements of order 4 within any subgroup are not conjugate.

Example 1 (Binary dihedral group BD_8, singularity D_4). In this case the two matrixes are

$$A = \begin{pmatrix} i & 0 \\ 0 & -i \end{pmatrix} \quad \text{and} \quad B = \begin{pmatrix} 0 & 1 \\ -1 & 0 \end{pmatrix} \in \mathrm{SL}(2, \mathbb{C}),$$

where $i = \sqrt{-1}$. Thus G has three cyclic subgroups of order 4, $\langle A\rangle$, $\langle B\rangle$ and $\langle AB\rangle$, with $A^2 = B^2 = (AB)^2 = -1$, which we draw as follows:

$$\begin{array}{rccccc}
 & A & & A^2 & & A^3 \\
\langle A\rangle : & \bigcirc & \text{---} & \bigcirc & \text{---} & \bigcirc \\[1em]
 & B & & B^2 & & B^3 \\
\langle B\rangle : & \bigcirc & \text{---} & \bigcirc & \text{---} & \bigcirc \\[1em]
 & AB & & (AB)^2 & & (AB)^3 \\
\langle AB\rangle : & \bigcirc & \text{---} & \bigcirc & \text{---} & \bigcirc
\end{array} \tag{1}$$

The pictures are the exceptional divisors in the minimal resolution of $\mathbb{C}^2/\langle g\rangle$ as given by toric geometry. The nodes represent elements of $\langle g\rangle$, and they are joined by an edge if they are neighbouring points in the junior simplex (interval) Δ_1, that is, if the corresponding curves intersect on the minimal resolution.

Conjugacy in G gives

$$A \sim A^3, \quad B \sim B^3, \quad AB \sim (AB)^3.$$

Thus identifying equal and conjugate group elements gives the familiar D_4 diagram:

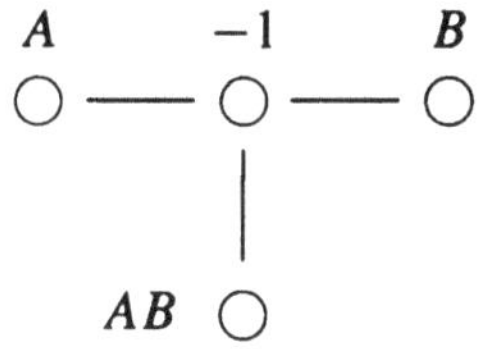

(2)

Example 2 (Binary dihedral group BD_{8l+4}, singularity D_{2l+3}). In this case A has order $4l+2$. We should draw all the maximal cyclic subgroups, then consider the effect of conjugacy. As noted above, the cyclic subgroups $\langle A^i B\rangle$ are all conjugate, so we only draw $\langle A\rangle$ and $\langle B\rangle$:

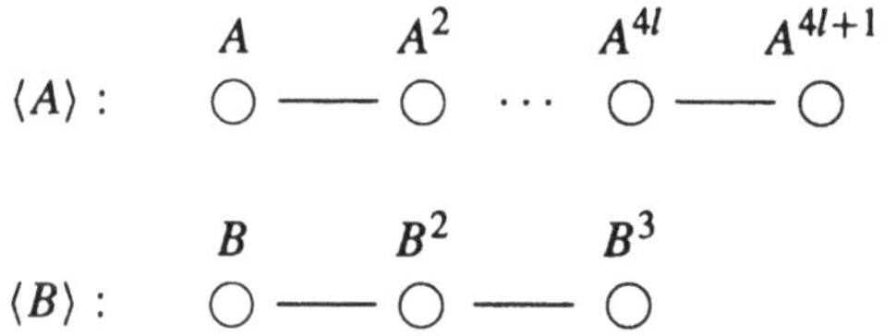

(3)

Now $A^{2l+1} = B^2 = -1$, and conjugacy gives $A^i \sim A^{4l+2-i}$, so that identifying equal and conjugate elements gives

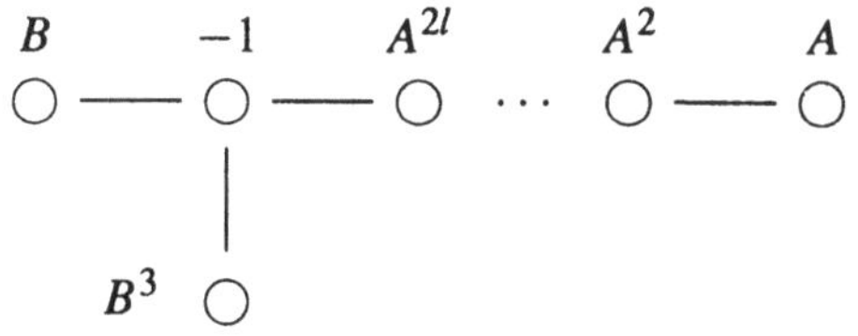

(4)

The case of the binary dihedral group BD_{8l}, giving the singularity D_{2l+2} is similar. The only difference is that to see the action of conjugacy, we have to draw both $\langle B\rangle$ and $\langle AB\rangle$, and then fold each of them in two.

Example 3 (Binary tetrahedral group BT_{48}, singularity E_6). The singularity E_6 is the quotient of $\mathbb{C}^2$ by the group BT_{48} generated by the three matrixes

$$A = \begin{pmatrix} i & 0 \\ 0 & -i \end{pmatrix}, \quad B = \begin{pmatrix} 0 & 1 \\ -1 & 0 \end{pmatrix}, \quad \text{and} \quad C = \frac{1}{\sqrt{2}}\begin{pmatrix} e^7 & e^7 \\ e^5 & e \end{pmatrix} \in \mathrm{SL}(2, \mathbb{C}),$$

where e is a primitive 8th root of 1 (for example, $e = \frac{1+i}{\sqrt{2}}$). Thus

$$\mathrm{BT}_{48} = \langle A, B, C \mid A^2 = B^2 = C^3 = -1,\ BA^iB^{-1} = A^{-i},\ CAC^{-1} = B^3 \rangle.$$

We draw the cyclic subgroups $\langle A \rangle$ and $\langle C \rangle$ of BT_{48} following the previous recipe:

$$\begin{array}{lccccccccc} & A & & A^2 & & A^3 & & & & \\ \langle A \rangle : & \bigcirc & — & \bigcirc & — & \bigcirc & & & & \\ & & & & & & & & & \\ & C & & C^2 & & C^3 & & C^4 & & C^5 \\ \langle C \rangle : & \bigcirc & — & \bigcirc & — & \bigcirc & — & \bigcirc & — & \bigcirc \end{array} \qquad (5)$$

It is easy to see the effect of conjugacy in BT_{48}:

$$A \sim A^3 \sim B \sim B^3 \sim AB \sim BA, \quad \text{and} \quad A^kC^j \sim C^j \text{ for all } j, k,$$

and the elements of $\langle C \rangle$ are not conjugate to one another. Thus identifying equal and conjugate elements of these two groups gives the resolution graph of E_6:

$$\begin{array}{ccccccccc} C & & C^2 & & -1 & & C^4 & & C^5 \\ \bigcirc & — & \bigcirc & — & \bigcirc & — & \bigcirc & — & \bigcirc \\ & & & & | & & & & \\ & & & A & \bigcirc & & & & \end{array} \qquad (6)$$

Now we see some 3-dimensional examples. (More examples are furnished by the crepant resolutions of $\mathbb{C}^3/G$ of [R], [I1]–[I3] and Markushevich [M].)

Example 4 (Trihedral group). A *trihedral* group is a subgroup G generated by a diagonal subgroup $H \subset \mathrm{SL}(3, \mathbb{C})$, together with the cyclic permutation matrix

$$T = \begin{pmatrix} 0 & 1 & 0 \\ 0 & 0 & 1 \\ 1 & 0 & 0 \end{pmatrix}.$$

The Abelian subgroup H is generated by one or more diagonal matrixes

$$A_i = \frac{1}{r_i}(a_i, b_i, c_i), \quad \text{with} \quad a_i + b_i + c_i = r_i,$$

meaning as usual that $A_i = \mathrm{diag}(\varepsilon^{a_i}, \varepsilon^{b_i}, \varepsilon^{c_i})$ where ε is a primitive r_ith root of 1. We can of course assume that H has cyclic symmetry, and then it is a normal subgroup $H \triangleleft G$. The quotient of $\mathbb{C}^3$ by G is a *trihedral quotient singularity*. A crepant resolution of singularities $Y \to X$ is described in [I1]–[I2].

We work out the special case $H = \left\langle \frac{1}{3}(0, 1, 2), \frac{1}{3}(1, 2, 0), \frac{1}{3}(2, 0, 1) \right\rangle$. Obviously, $G = \left\langle \frac{1}{3}(0, 1, 2), T \right\rangle$, and is a group of order 27.

It turns out that G has only one senior element $\frac{1}{3}(2,2,2)$, and 25 juniors. Note that, as we saw in 2.10, for subgroups of SL(3, $\mathbb{C}$), all elements with nonisolated fixed points are junior; this applies to every element of G except $\frac{1}{3}(1,1,1)$ and $\frac{1}{3}(2,2,2)$ because, for example, Fix $\frac{1}{3}(0,1,2) = x$-axis, and every element hT is obviously conjugate in GL(3, $\mathbb{C}$) to T, which fixes the line $\{(x,x,x)\} \subset \mathbb{C}^3$. Thus

$$\Gamma_1 = \left\{ \begin{matrix} \frac{1}{3}(1,1,1), \ \frac{1}{3}(0,1,2), \ \frac{1}{3}(1,2,0), \ \frac{1}{3}(2,0,1), \\ \frac{1}{3}(0,2,1), \ \frac{1}{3}(2,1,0), \ \frac{1}{3}(1,0,2) \end{matrix} \right\} \cup HT \cup HT^2$$

Under conjugacy in G,

$$\frac{1}{3}(a,b,c) \sim \frac{1}{3}(b,c,a), \quad \text{for all } a,b,c \in \{0,1,2\},$$

and

$$T \sim \frac{1}{3}(a,b,c)T, \quad \text{and} \quad T^2 \sim \frac{1}{3}(a,b,c)T^2 \quad \text{for all } a \neq b.$$

Thus the exceptional divisors in the minimal resolution $Y \to X = \mathbb{C}^3/G$ correspond one-to-one with the following junior conjugacy classes:

$$\begin{gathered} \tfrac{1}{3}(0,1,2), \ \tfrac{1}{3}(0,2,1), \ \tfrac{1}{3}(1,1,1), \\ T, \ \tfrac{1}{3}(1,1,1)T, \ \tfrac{1}{3}(2,2,2)T, \\ T^2, \ \tfrac{1}{3}(1,1,1)T^2, \ \tfrac{1}{3}(2,2,2)T^2. \end{gathered}$$

These elements correspond to a basis of $H^2(Y,\mathbb{Q})$, so that $h^2(Y,\mathbb{Q}) = 9$. On the other hand, by Corollary 1.6, $H^4(Y,\mathbb{Q})$ has a basis corresponding to the single element $\Gamma_2 = \{\frac{1}{3}(2,2,2)\}$, and its Poincaré dual $H_c^2(Y,\mathbb{Q})$ has a basis corresponding to $\Gamma_1^{(0)} = \{\frac{1}{3}(1,1,1)\}$. In particular the Euler number $e(Y)$ is given by

$$\begin{aligned} e(Y) &= h^0(Y,\mathbb{Q}) + h^2(Y,\mathbb{Q}) + h^4(Y,\mathbb{Q}) = 1+9+1 = 11 \\ &= \sharp\{\text{conjugacy class in } G\}. \end{aligned}$$

Example 5 (Icosahedral group). $G_{60} \subset$ SO(3) is the group of rotations of the icosahedron, acting on $\mathbb{C}^3$ by the inclusion SO(3) $\subset$ SL(3). Generating matrixes and defining relations are given for example in Miller, Blichtfeld and Dickson [MBD], p. 250. The reader may eke out our poor words of explanation by imagining the picture of the icosahedron.

It is easy to see that G_{60} has just 5 conjugacy classes:

(1) the identity;

(2) a rotation through π about an axis through the midpoint of an edge;

(3) a rotation through $2\pi/3$ about an axis through the midpoint of a face;

(4) a rotation through $2\pi/5$ about an axis through a vertex;

(5) a rotation through $4\pi/5$ about the same axis.

Each rotation fixes an axis, and is therefore junior by the remark in 2.10. The quotient $\mathbb{C}^3/G_{60}$ has 3 curves of transversal Du Val singularities: a curve of A_1 under the axis of (2); a curve of A_2 under the axis of (3), and a curve C of transversal A_4 singularities under the common axis of (4) and (5). Resolving these curves gives 4 crepant divisors. For example, blowing up the curve C of A_4 singularities gives two surfaces which (outside the origin) are $\mathbb{P}^1$ bundles over a curve that is a double cover of C.

4. Discussion

4.1. Theorem 1.6 implies directly Conjecture 1.1 for $n = 3$, with the proviso that you have to fix roots of 1 to get a canonical correspondence. To see that Y cannot have any odd-dimensional homology, it is enough to note that there exists a resolution of X dominated by a variety W obtained by a sequence of blow-ups of $\mathbb{C}^n$ in linear subspaces. Then $H^*(W)$ has no odd-dimensional cohomology, and $H^*(Y) \subset H^*(W)$.

4.2. The minimal model is in fact nonsingular: indeed, by standard results of Mori theory, although the minimal model is usually not unique, all minimal models have the same local singularities. Moreover, Roan and others have proved that a smooth crepant resolution exists in all cases. Unfortunately, the present abstract point of view does not (as yet) offer any clues why this should be so.

4.3. There should be a theoretical proof that the minimal model $Y \to X$ is nonsingular, even though we are not clever enough to find it at present. For example, if Y has factorial cDV points, then these have local deformations. It should follow from Friedman's local-to-global spectral sequence that Y has deformations inducing all possible deformations of the local singularities, and in particular, should have a smoothing Y_t; then Y_t also contracts to a variety X_t. Alternatively the Milnor fibre shows that Y_t must have a middle cohomology of Hodge type $H^{1,2} \oplus H^{2,1}$—but so what? Or maybe the singularities would have homotopy properties that contradict that the link of X is 2-connected.

4.4. The shape of Theorems 1.4 and 1.6 is very suggestive: the canonical basis for $H^2(Y, \mathbb{Z})$ involves junior conjugacy classes of $\Gamma = G(-1)$, which is probably the only thing that makes sense in terms of motivic weights. The basis of $H^4(Y, \mathbb{Q})$ is dual to a basis of $H^2_c(Y, \mathbb{Q})$, so to take account of the roots of 1 correctly, you presumably have to take Hom into $\mathbb{Z}(-3)$, and end up with another Tate twist of the set of conjugacy classes Γ_2/G before getting a canonical identification with the basis of $H^4(Y, \mathbb{Q})$ (?).

4.5. Crepant resolutions in higher dimension. In dimension ≥ 4, when does $X = \mathbb{C}^n/G$ have a crepant resolution $Y \to X$? The known counterexamples are the terminal quotient singularities, for example, $\frac{1}{r}(i, r-i, j, r-j)$ (see also Mori, Morrison and Morrison [MMM]). For these, there are not enough junior elements: the unit cube $\square$ in $\mathbb{R}^4$ contains $r-1$ elements on the Γ_2 hyperplane $\sum \alpha_i = 2$, and no junior elements at all. Thus there are no crepant divisors to pull out.

It seems plausible that the following 2 conditions to the effect that G has lots of junior elements are necessary conditions: for each *maximal* cyclic subgroup $\mathbb{Z}/r \subset G$, choose eigencoordinates and write $L = \mathbb{Z}^n + \mathbb{Z} \cdot \frac{1}{r}(a_1, \ldots, a_n)$ as usual. Then

(i) every point of $L \cap \square$ is a positive integral combination of points of $L \cap \Delta_1$.

(ii) the positive octant of L has a subdivision into basic cones with 1-skeleton contained in Δ_1.

Obviously (ii) implies (i), and, for Abelian groups, by toric geometry, (ii) is a necessary and sufficient condition.

4.6. Multiplicative structure. Assume that a crepant resolution $Y \to X$ exists, and that $n = 4$ or 5, to fix ideas. There is one respect in which Conjecture 1.1 looks bad: to get a canonical correspondence, as we have seen, you have to introduce an age grading, and this will only be canonical if you twist G to $G(-1)$. No other twist has a canonical age grading. But then the theory of weights in the philosophy of motives make it ungrammatical to put elements of $G(-1)$ in correspondence with elements of H^4.

One way of trying to rescue the conjecture might be to put elements of H^4 in correspondence with quadratic monomials in Γ_1, maybe with relations between them deduced from a multiplicative structure; this seems to be good for Abelian quotients (compare [BD]). Some natural questions arise in this connection. Presumably it is true that the cohomology ring $H^*(Y, \mathbb{Q})$ is generated by H^2? Is there a reasonable multiplicative structure involving the conjugacy classes of G?

4.7. Remark. The easy arguments in this paper have left to one side the most important aspect of the McKay correspondence for SL(2, $\mathbb{C}$), namely, the relation with irreducible representations. Understanding this point might begin to explain a lot of things. Some indications:

(1) In 2.5 we needed to find an extraction $V \to X$ pulling out exactly the divisor E corresponding to a conjugacy class, by analogy with the weighted blowup of 2.3 in the cyclic case. This could be provided by quiver-theoretic methods: view $X = \mathbb{C}^n/G$ as a moduli space of representations of a quiver (e.g., the McKay quiver of G). Then X is a categorical quotient of a big vector space by a big algebraic group, and birational modifications of X are naturally obtained by "varying the linearisation".

(2) The age of $g \in \Gamma$ is defined as the elementary symmetric function age $g = \frac{1}{r}\sum a_i$. Do the other elementary symmetric functions (e.g. $\frac{1}{r^2}\sum a_i a_j$) also have some significance for the McKay correspondence?

References

[B] V. Batyrev, Strong McKay correspondence, lecture at Trento conference, June 1994.

[BD] Victor V. Batyrev and Dimitrios I. Dais, Strong McKay correspondence, String-theoretic Hodge Numbers and Mirror Symmetry, Duke server alg-geom/9410001, to appear in Topology 35 (1996).

[D] V. Danilov, The geometry of toric varieties, Uspekhi Mat. Nauk 33:2 (1978) 85–134 = Russian Math. Surveys 33:2 (1978) 97–154.

[D] V. Danilov, The birational geometry of toric 3-folds, Izv. Akad. Nauk SSSR Ser. Mat. 46 (1982), 972-981 = Math. USSR - Izv. 21 (1983).

[DHVW] L. Dixon, J. Harvey, C. Vafa and E. Witten, Strings on orbifolds, Nucl. Phys. B 261 (1985) 678–686. II, Nucl. Phys. B 274 (1986) 285–314.

[GSV] G. González-Sprinberg and J.-L. Verdier, Points doubles rationnels et représentations de groupes, C. R. Acad. Sci. Paris, Sér. I, Math. 293 (1981), 111–113.

[GSV1] G. González-Sprinberg and J.-L. Verdier, Construction géométrique de la correspondance de McKay, Ann. Sci. Ecole Norm. Sup. (4) 16 (1983), 409–449.

[HH] F. Hirzebruch and T. Höfer, On the Euler number of an orbifold, Math. Ann. 286 (1990), 255–260.

[I1] Y. Ito, Crepant resolutions of trihedral singularities, Proc. Japan Acad. 70 (1994), 131–136.

[I2] Y. Ito, Crepant resolution of trihedral singularities and the orbifold Euler characteristic, Internat. J. Math. 6:1 (1995), 33–43.

[I3] Y. Ito, Gorenstein quotient singularities of monomial type in dimension three, J. Math. Sci. Univ. Tokyo 2 (1995), 419–440, Duke server alg-geom/9406001.

[K] J. Kollár, Flips, flops etc., in: Surveys in differential geometry 1 (1991) 113–199.

[M] D. Markushevich, Resolution of $\mathbb{C}^3/H_{168}$, Technion preprint, 1993.

[MBD] G. A. Miller, H. F. Blichtfeldt and L. E. Dickson, Theory and applications of finite groups, Wiley, London 1917, reprinted Dover 1961.

[Mc] J. McKay, Graphs, singularities, and finite group, in: Santa Cruz conference on finite groups (Santa Cruz, 1979), Proc. Sympos. Pure Math. 37 (1980), 183–186.

[MMM] S. Mori, D. R. Morrison and I. Morrison, On four-dimensional terminal singularities, Math. Comp. 51 (1988), 769–786.

[R] S.-S. Roan, Minimal resolutions of Gorenstein orbifolds in dimension three, Preprint R940606–1, Acad. Sinica, Taibei, June 1994, to appear in Topology 35 (1996).

[R1] M. Reid, Canonical 3-folds, in: Géométrie algébrique (Angers, 1979), (A. Beauville, ed.), Sijthoff and Noordhoof, Alphen aan den Rijn, 1980, 273–310.

[R2] M. Reid, Young person's guide to canonical singularities, in: Algebraic geometry (Bowdoin, 1985), Proc. Symp. Pure Math. 46 (1987), 345–416.

[ZS] O. Zariski and P. Samuel, Commutative Algebra, Vols. I and II, Van Nostrand, 1960.

Divisorial contractions to 3-dimensional terminal quotient singularities

Yujiro Kawamata

Abstract. We prove that a divisorial contraction to any given 3-dimensional terminal quotient singularitiy is unique.

Let Y be a projective algebraic variety with only $\mathbb{Q}$-factorial terminal singularities defined over a complex number field $\mathbb{C}$. If the canonical divisor K_X is not nef, then there exists an elementary contraction $\mu : Y \to X$ (cf. [KMM]). μ is a projective surjective morphism onto a normal variety with connected fibers such that (1) the relative Picard number $\rho(Y/X) = 1$, and (2) $-K_Y$ is μ-ample. μ is called a Mori fiber space (resp. a divisorial contraction, a small contraction), if $\dim X < \dim Y$ (resp. μ is birational and the exceptional locus $E = \mathrm{Exc}(\mu)$ is a prime divisor, μ is birational and E has codimension at least 2).

We are concerned with divisorial contractions in dimension 3 in this paper. There is a classification of terminal singularities in dimension 3 ([MS], [M]), and it might be possible to classify all the divisorial contractions if $\dim Y = 3$. It is desirable not only in its own right but also by the following reasons: (1) a flip seems to be decomposed into divisorial contractions and simpler flips ([K2]), (2) divisorial contractions are the leading ingredients of the elementary links between Mori fiber spaces ([Co]).

A divisorial contraction will be called *type* O (resp. I) in this paper, if the image of the exceptional divisor is a point (resp. a curve).

We have already following results for the elementary contractions in dimension 3. (1) Cutkosky [Cu] classified them under the additional assumption that K_Y is a Cartier divisor. (2) Kollár and Mori [KM] investigated those μ which have 1-dimensional fibers including divisorial contractions of type I. They proved that a general member of the anti-canonical linear system $|-K_Y|$ has only rational double points as singularities when X is replaced by a germ of an analytic space.

Let $\mu : Y \to X$ be a divisorial contraction in dimension 3. We shall fix a point $P \in \mu(E)$, restrict X to its analytic neighborhood of P, and replace Y by the inverse image of X by μ.

Since we start with X instead of Y, one can call μ a *divisorial extraction.*

Then the problem is formulated as follows: given a terminal singularity (X, P), we should classify the projective birational morphism $\mu : Y \to X$ from a variety Y

with only terminal singularities whose exceptional locus E is a $\mathbb{Q}$-Cartier divisor such that $-E$ is μ-ample and that (1) E is irreducible, or (2) all the irreducible components of E are mapped to curves by μ, which coincides with the blowup of $\mu(E)$ along its generic points.

We note that the condition $\rho(Y/X) = 1$ and the $\mathbb{Q}$-factoriality of Y are lost under this replacement. $-K_Y$ is automatically μ-ample, because X is terminal.

In this paper we shall prove that a divisorial contraction to a given terminal quotient singularity of dimension 3 is unique.

We begin with fixing the notation. Let $n, r, s_1, \dots, s_n$ be integers such that $n, r > 0$. A cyclic quotient singularity $\mathbb{C}^n/\mathbb{Z}_r(s_1, \dots, s_n)$ is defined as the singular point at the image of the origin of the quotient space of $\mathbb{C}^n$ by a group $\mathbb{Z}_r = \mathbb{Z}/r\mathbb{Z} = \langle g \rangle$ whose action on the coordinates $(x_1, \dots, x_n)$ is given by $g(x_i) = \exp(2\pi\sqrt{-1}s_i/r)x_i$.

It is known that $\mathbb{C}^3/\mathbb{Z}_r(s_1, s_2, s_3)$ is a terminal singularity if and only if $s_1 + s_2 \equiv 0 \mod r$ and $(s_i, r) = 1$ $(\forall i)$ up to the permutation of the coordinates ([MS]). Note that one may assume $s_3 = 1$ by a choice of the generator g.

Let $(X, P) = \mathbb{C}^n/\mathbb{Z}_r(s_1, \dots, s_n)$, and let $a, t_1, \dots, t_n$ be nonnegative integers such that $t_i \equiv as_i \mod r$. Then we can define a *weighted blowup* $\mu : Y \to X$ with the system of weights $(t_1/r, \dots, t_n/r)$ by giving weights t_i/r to the x_i.

For example, let $(X, P) = \mathbb{C}^3/\mathbb{Z}_r(s, -s, 1)$ with $0 < s < r$ and $(r, s) = 1$, and $(t_1, t_2, t_3) = (s, r-s, 1)$. Then Y has 2 quotient singularities of types $\mathbb{C}^3/\mathbb{Z}_s(r, -r, 1)$ and $\mathbb{C}^3/\mathbb{Z}_{r-s}(r, -r, 1)$, the exceptional locus of μ is a prime divisor E, and $K_Y = \mu^*K_X + \frac{1}{r}E$ ([K1]). In particular, such a singularity X can be resolved by a succession of weighted blowups.

Example 1. Let $X = \mathbb{C}^3$. Then a weighted blowup $\mu : Y \to X$ is a divisorial contraction if and only if the weights are given by $(1, a, b)$ up to the permutation for nonnegative integers a, b such that $(a, b) = 1$. In this case, we have $K_Y = \mu^*K_X + (a+b)E$. We note that there are infinite choices of coordinates.

Example 2. Let (X, P) be an isolated hypersurface singularity in $\mathbb{C}^4/\mathbb{Z}_r(s, -s, 1, 0)$ given by an equation $xy + f(z^r, w) = 0$, where $(r, s) = 1$ ([M]).

Let $\mu : Y \to X$ be the weighted blowup with weights $(t_1/r, t_2/r, d/r, 1)$, where t_1, t_2, d are positive integers which satisfy the following conditions:

(1) $t_1 + t_2 = dnr$ and $t_1 \equiv sd \mod r$,

(2) if the weights are given by $\mathrm{wt}(x, y, z, w) = (t_1/r, t_2/r, d/r, 1)$, then $\mathrm{wt}(f) = dn$ for a positive integer n,

(3) $(d, r) = (d, t_1) = 1$.

Then μ is a divisorial contraction for a generic f. We have $K_Y = \mu^*K_X + \frac{d}{r}E$. For example, we can take $f = z^{nr} + w^m$ with $m \geq dn$. Therefore, we have divisorial contractions with arbitrarily large discrepancy of canonical divisors.

Example 3. Let (X, P) be an isolated hypersurface singularity in $\mathbb{C}^4/\mathbb{Z}_2(1,1,0,1)$ given by an equation $w^2 + f(x,y,z) = 0$, where $f \in (x,y,z)^3$ and f contains x^2z or xyz ([M]).

Let $\mu : Y \to X$ be the weighted blowup with weights $(1/2, 1/2, 1, 3/2)$. Then μ is a divisorial contraction whose exceptional divisor is *not normal* for a generic f.

For example, we can take $f = x^2z - y^4 + z^3$. Then Y has an affine open subset given by an equation $zw^2 + x^2 - y^4 + z = 0$ in $\mathbb{C}^4/\mathbb{Z}_2(1,1,0,1)$ on which the exceptional divisor is defined by $z = 0$. The 2 components $x - \pm y^2 = z = 0$ are interchanged by the group action of $\mathbb{Z}_2$, which is free outside the locus $x = y = w = 0$, and the exceptional divisor is singular along the locus $x = y = z = 0$.

Example 4. Let $X = \mathbb{C}^3$, and C a curve on X which is a complete intersection of a smooth surface with another surface. Then the blowup $\mu : Y \to X$ along C is a divisorial contraction. In fact, this type of blowups exhaust all the divisorial contractions of type I such that K_Y are Cartier divisors ([Cu]). Note that μ is not necessarily a weighted blowup.

Now we state the main theorem of this paper.

Theorem 5. *Let (X, P) be a quotient singularity $\mathbb{C}^3/\mathbb{Z}_r(s, -s, 1)$ with $0 < s < r$ and $(r, s) = 1$, and $\mu : Y \to X$ a projective birational morphism from a variety Y with only terminal singularities. Assume that the exceptional locus E of μ is a $\mathbb{Q}$-Cartier divisor, $-E$ is μ-ample, $P \in \mu(E)$, and that* (1) *E is irreducible, or* (2) *all the irreducible components of E are mapped to curves by μ, which coincides with the blowup of $\mu(E)$ along its generic points. Then μ coincides with the weighted blowup with the system of weights $(s/r, 1 - s/r, 1/r)$. In particular, $\mu(E) = \{P\}$, i.e, μ is of type O.*

We note that the weighted blowup given in the above theorem is independent of the choice of the coordinates and unique for the given singularity (X, P). This is proved in the course of the proof.

Let M be an effective Weil divisor on $(X, P) = \mathbb{C}^3/\mathbb{Z}_r(s, -s, 1)$. The *weighted multiplicity* of M at P is defined by $\mathrm{w} - \mathrm{mult}_P(M) = mr$ if $\mu^*M = \mu^{-1}M + mE$ for the weighted blowup μ as in the theorem. The weighted multiplicity at a nonsingular point is the same as the usual multiplicity.

Lemma 6. *Let $\mu : Y \to X$ be a weighted blowup of a terminal quotient singularity as above, and M an effective Weil divisor on X. Then*

$$\mathrm{w} - \mathrm{mult}_Q(\mu^{-1}M) \leq \mathrm{w} - \mathrm{mult}_P(M) \quad \textit{for any } Q \in E.$$

Proof. Let $F = \sum_{i,j,k} c_{ijk}x^iy^jz^k$ be an equation of M. Then $\mathrm{w} - \mathrm{mult}_P(M) = mr = \min\{is + j(r-s) + k;\, c_{ijk} \neq 0\}$.

Y is covered by 3 open subsets U_x, U_y, U_z which are the complements of the strict transforms of the divisors defined by $x = 0, y = 0, z = 0$, respectively. We consider the case in which $Q \in U_x$. Other cases are treated similarly. U_x has a quotient singularity of type $\mathbb{C}^3/\mathbb{Z}_s(-r, r, 1)$ with semi-invariant coordinates $(\bar{x}, \bar{y}, \bar{z})$ given by $\mu^*x = \bar{x}^{s/r}, \mu^*y = \bar{x}^{1-s/r}\bar{y}, \mu^*z = \bar{x}^{1/r}\bar{z}$. Then the equation of $\mu^{-1}M$ on U_x becomes

$$\bar{F} = \sum_{i,j,k} c_{ijk}\bar{x}^{(is+j(r-s)+k-mr)/r}\bar{y}^j\bar{z}^k$$

If $s \geq 2$ and $r = st + u$ with $0 < u < s$, then at the point $P_x \in U_x$ defined by $\bar{x} = \bar{y} = \bar{z} = 0$, we have

$$\begin{aligned} &\mathrm{w-mult}_{P_x}(\mu^{-1}M) \\ &= \min\{(is/r + j(r-s)/r + k/r - m)(s-u) + ju + k; c_{ijk} \neq 0\} \\ &\leq \min\{ju + k; c_{ijk} \neq 0, is + j(r-s) + k = mr\} \\ &\leq \mathrm{w-mult}_P(M) \end{aligned}$$

If Q is a nonsingular point of U_x, then

$$\begin{aligned} &\mathrm{mult}_Q(\mu^{-1}M) \leq \mathrm{mult}_Q(\mu^{-1}M|_E) \leq \deg(\bar{F}|_E) \\ &= \max\{j + k; c_{ijk} \neq 0, is + j(r-s) + k = mr\} \\ &\leq \mathrm{w-mult}_P(M) \end{aligned}$$

□

Lemma 7. *Let $\sigma : Z \to X$ be a projective birational morphism from a nonsingular variety Z to a quotient singularity $(X, P) = \mathbb{C}^3/\mathbb{Z}_r(s, -s, 1)$, and M an effective Weil divisor on X. Let $K_Z = \sigma^*K_X + \sum_j a_jE_j$ and $\sigma^*M = \sigma^{-1}M + \sum_j m_jE_j$ for the exceptional divisors E_j of σ. Then*

$$m_j/a_j \leq \mathrm{w-mult}_P(M) \quad \textit{for all } j.$$

Proof. We may assume that σ is a composition of morphisms $Z = Y_n \xrightarrow{\mu_n} Y_{n-1} \to \cdots \xrightarrow{\mu_1} Y_0 = X$, where μ_j is a weighted blowup for $1 \leq j \leq n_0$ with Y_{n_0} nonsingular and μ_j is a permissible blowup in the sense of Hironaka for $n_0 < j \leq n$ ([H]). Let $\sigma_j : Z \to Y_j$ and $\tau_j : Y_j \to X$ be the induced morphisms. We may assume that $E_j = \sigma_j^{-1}\bar{E}_j$ for the exceptional divisor $\bar{E}_j$ of μ_j.

We write $K_{Y_j} = \mu_j^*K_{Y_{j-1}} + \bar{a}_j\bar{E}_j$ and $\mu_j^*\tau_{j-1}^{-1}M = \tau_j^{-1}M + \bar{m}_j\bar{E}_j$. We have $\bar{a}_j = 1/r_j$ (resp. 2, 1) if μ_j is a weighted blowup of a quotient terminal singulatity of index r_j (resp. a blowup of a nonsingular point, a blowup of a nonsingular curve). In the latter cases, we set $r_j = 1$.

Let $\sigma_j^*\bar{E}_j = E_j + \sum_{j<k}\alpha_{jk}E_k$. Then $a_j = \bar{a}_j + \sum_{i<j}\bar{a}_i\alpha_{ij}$ and $m_j = \bar{m}_j + \sum_{i<j}\bar{m}_i\alpha_{ij}$. By Lemma 6, $\bar{m}_j \leq mr/r_j \leq mr\bar{a}_j$, hence $m_j/a_j \leq mr$. □

Proof of Theorem 5. Let $\mu_1 : Y_1 \to X$ be the weighted blowup, and E_1 its exceptional divisor. Supposing that $\mu \neq \mu_1$, we shall derive a contradiction. By resolution of singularities, we obtain a nonsingular variety Z with projective birational morphisms $\nu : Z \to Y$ and $\nu_1 : Z \to Y_1$. We define $\sigma = \mu \circ \nu = \mu_1 \circ \nu_1$. Since $\rho(Y_1/X) = 1$ and Y_1 is analytically $\mathbb{Q}$-factorial, we have $\nu^{-1}E \neq \nu_1^{-1}E_1$.

Let L be a general member of a very ample linear system $|-mE|$ on Y for a large and divisible integer m, and $M = \mu(L) \ni P$. We can write

$$K_Z = \sigma^* K_X + a\nu^{-1}E + \frac{1}{r}\nu_1^{-1}E_1 + \cdots$$

$$\sigma^* M = \sigma^{-1}M + m\nu^{-1}E + m_1\nu_1^{-1}E_1 + \cdots$$

$$\nu^* E = \nu^{-1}E + b\nu_1^{-1}E_1 + \cdots$$

with $m_1, a, b > 0$, where $\cdots$ stands for the sum of irreducible components different from $\nu^{-1}E$ and $\nu_1^{-1}E_1$. Since $\nu^* L = \nu^{-1}L$, we have

$$\sigma^* M = \nu^*(L + mE) = \nu^{-1}L + m\nu^{-1}E + bm\nu_1^{-1}E_1 + \cdots$$

Since Y has only terminal singularities,

$$\begin{aligned} K_Z &= \nu^*(\mu^* K_X + aE) + c\nu_1^{-1}E_1 + \cdots \\ &= \sigma^* K_X + a\nu^{-1}E + (ab + c)\nu_1^{-1}E_1 + \cdots \end{aligned}$$

for a positive rational number c. Therefore, $m_1 = bm$ and $1/r = ab + c$, hence $m/a = m_1/(1/r - c) > m_1 r$, a contradiction to Lemma 7. □

In the above proof, $\max_j\{m_j/a_j\}$ is equal to the inverse of the *canonical threshold* for the pair (X, M), the maximum number c such that $K_X + cM$ is canonical. It should be compared with the *log canonical threshold* which was used in the proof of the base point free theorem.

References

[Co] A. Corti, Factoring birational maps of threefolds after Sarkisov, J. Alg. Geom. 4 (1995), 223–254.

[Cu] S. Cutkosky, Elementary contractions of Gorenstein threefolds, Math. Ann. 280 (1988), 521–525.

[H] H. Hironaka, Resolution of singularities of an algebraic variety over a field of characteristic zero, Ann. of Math. 79 (1964), 109–326.

[K1] Y. Kawamata, On the plurigenera of minimal algebraic 3-folds with $K_X \equiv 0$, Math. Ann. 275 (1986), 539–546.

[K2] —, Semistable minimal models of threefolds in positive or mixed characteristic, J. Alg. Geom. 3 (1994), 463–491.

[KMM] Y. Kawamata, K. Matsuda and K. Matsuki, Introduction to the minimal model problem, Adv. Stud. Pure Math. 10 (1987), 283–360.

[KM] J. Kollár and S. Mori, Classification of three dimensional flips, J. Amer. Math. Soc. 5 (1992), 533–706.

[M] S. Mori, On 3-dimensional terminal singularities, Nagoya Math. J. 98 (1985), 43–66.

[MS] D. Morrison and G. Stevens, Terminal quotient singularities in dimension three and four, Proc. Amer. Math. Soc. 90 (1984), 15–20.

Ample vector bundle characterizations of projective bundles and quadric fibrations over curves

Antonio Lanteri and Hidetoshi Maeda

Abstract. Let $\mathcal{E}$ be an ample vector bundle of rank $r \geq 2$ on a smooth complex projective n-fold X having a section whose zero locus Z is a smooth submanifold of the expected dimension $n - r \geq 3$. Pairs $(X, \mathcal{E})$ as above are classified in the following cases: a) Z is a $\mathbb{P}^{n-r-1}$-bundle over a smooth curve of positive genus, b) (Z, H_Z) is a scroll over a smooth curve B, and c) (Z, H_Z) is a quadric fibration over B, for some ample line bundle H on X.

1991 Mathematics Subject Classification: Primary 14F05; Secondary 14J40

0. Introduction and statement of the results

In the eighties the study of some special varieties as ample divisors attracted the attention of several authors. In [LM1] we started to reconsider results on ample divisors in the setting of ample vector bundles by investigating the following situation:

0.1. X is a complex projective manifold of dimension n and $\mathcal{E}$ is an ample vector bundle of rank $r \geq 2$ on X such that there exists a section $s \in \Gamma(\mathcal{E})$ whose zero locus $Z = (s)_0$ is a smooth submanifold of the expected dimension $n - r$.

In [LM1] we studied the case when Z is a projective space or a quadric. Here we deal with other special varieties. First of all, we are interested in the following case, which represents the ample vector bundle analog of a study of $\mathbb{P}^k$-bundles over smooth curves made by Bădescu in [B1], [B2]:

(a) Z is a $\mathbb{P}^k$-bundle over a smooth curve B.

In [LM2] this was settled in the case when $k = 1$. Here we deal with case $k \geq 2$ when the genus $g(B)$ of B is positive. Our first result is the following

Theorem A. *Let $(X, \mathcal{E})$ be as in (0.1) and assume that Z is a $\mathbb{P}^k$-bundle ($k \geq 2$) over a smooth curve B of positive genus. Then X is a $\mathbb{P}^{n-1}$-bundle over B and $\mathcal{E}_F \cong \mathcal{O}_{\mathbb{P}^{n-1}}(1)^{\oplus r}$ for every fibre F of the bundle projection $\pi : X \to B$.*

The idea of the proof is very simple. First we show that the Albanese map of X fibres X over B, which can be identified with the image of the Albanese map of Z, and then we apply [LM1, Theorem A] to our ample vector bundle $\mathcal{E}$ restricted to the fibres. The fact that X is a fibration over B comes from the Lefschetz–Sommese theorem which we state in a complete form in Section 1.

If Z is as in (a), then there exists a vector bundle $\mathcal{F}$ of rank $k+1$ on B such that $Z = \mathbb{P}_B(\mathcal{F})$ and, up to twisting $\mathcal{F}$ by a line bundle of sufficiently high degree, we can assume that $\mathcal{F}$ is ample. So, if $h \in \mathrm{Pic}(Z)$ denotes the tautological bundle of $\mathcal{F}$, (Z, h) is a scroll over B. Thus the assumption in the Theorem A is equivalent to saying that (Z, h) is a scroll over a smooth curve of positive genus for some ample line bundle $h \in \mathrm{Pic}(Z)$. Now assume that h is induced by a global polarization on X, i.e.,

(b) (Z, H_Z) is a scroll over B for some ample $H \in \mathrm{Pic}(X)$.

Note that the result of the Theorem A itself makes us think that (b) is realized at least when $n - r \geq 3$ and $g(B) > 0$. However for $n - r = 2$, even when the restriction homomorphism $\mathrm{Pic}(X) \to \mathrm{Pic}(Z)$ is an isomorphism, we cannot exclude the existence of a rational scroll (Z, h) whose polarization is induced by no ample line bundle on X (e.g. see [LM2, Proposition]).

As to case (b) we prove the following result.

Theorem B. *Let $(X, \mathcal{E})$ be as in (0.1) with $n - r \geq 3$ and assume that (Z, H_Z) is a scroll over a smooth curve B for some ample line bundle $H \in \mathrm{Pic}(X)$. Then (X, H) is a scroll over B and $\mathcal{E}_F \cong \mathcal{O}_{\mathbb{P}^{n-1}}(1)^{\oplus r}$ for every fibre F of the scroll projection $\pi : X \to B$.*

Theorem B can be regarded as a polarized version of the Theorem A. Note that the assumption on the existence of a global polarization is compensated by the fact that no condition is required on the base curve B.

The proof of the Theorem B does not rely on the Lefschetz-Sommese theorem, but is based on a completely different argument. In fact, we consider the ample vector bundle on X of rank $\dim X - 1$ given by $\mathcal{E}' = \mathcal{E} \oplus H^{\oplus(\dim Z - 1)}$ and we prove that its adjoint bundle $K_X + \det \mathcal{E}'$ is not numerically effective. Then we apply a result of Ye-Zhang [YZ] to $(X, \mathcal{E}')$, to get our conclusion. The same approach outlined above also allows us to give a new proof of the easy part of our results in [LM1] concerning the case when $Z = \mathbb{P}^{n-r}$ or $Z = \mathbb{Q}^{n-r}$, a smooth quadric hypersurface in $\mathbb{P}^{n-r+1}$. This is done at the end of Section 3.

Moreover, the same method, combined with a result of Zhang improved by the second author [M] (for a related result see also [AM]), allows us to study also the case when

(c) (Z, H_Z) is a quadric fibration over B for some ample $H \in \mathrm{Pic}(X)$.

(for our definition of quadric fibrations, see Section 1.) In this case our result is the following

Theorem C. *Let* $(X, \mathcal{E})$ *be as in* (0.1) *with* $n - r \geq 3$ *and assume that* (Z, H_Z) *is a quadric fibration over a smooth curve* B *for some ample line bundle* $H \in \mathrm{Pic}(X)$. *Then* $(X, H, \mathcal{E})$ *is one of the following:*

(1) (X, H) *is a scroll over* B *and* $\mathcal{E}_F \cong \mathcal{O}_{\mathbb{P}^{n-1}}(2) \oplus \mathcal{O}_{\mathbb{P}^{n-1}}(1)^{\oplus(r-1)}$ *for every fibre* F *of the scroll projection* $\pi : X \to B$,

(2) *there is a surjective morphism* $\varphi : X \to B$ *such that any general fibre* F *of* φ *is a smooth hyperquadric* $\mathbb{Q}^{n-1}$ *in* $\mathbb{P}^n$ *with* $H_F \cong \mathcal{O}_{\mathbb{Q}^{n-1}}(1)$ *and* $\mathcal{E}_F \cong \mathcal{O}_{\mathbb{Q}^{n-1}}(1)^{\oplus r}$, *or, possibly*

(3) $n - r = 3$, (X, H) *is a scroll over a geometrically ruled surface* S *over* B *and* $\mathcal{E}_F \cong \mathcal{O}_{\mathbb{P}^{n-2}}(1)^{\oplus(n-3)}$ *for any fibre* F *of the scroll projection* $\pi : X \to S$.

To prove the Theorem C, we have to study five polarized pairs and case (3) is the most complicated. In case (3), the quadric fibration morphism $p : Z \to B$ factors through $\pi_{|Z}$ and the ruling projection of S except the following case.

$$Z = \mathbb{P}^1 \times \mathbb{P}^1 \times \mathbb{P}^1, \ B = \mathbb{P}^1 \text{ and } S = \mathbb{P}^1 \times \mathbb{P}^1.$$

In this case, any fibre of p dominates S. The effectiveness of this last case looks uncertain. As a matter of fact, it is well known that $\mathbb{P}^1 \times \mathbb{P}^1 \times \mathbb{P}^1$ cannot be contained as an ample divisor in any projective manifold.

Our Theorems B and C can be also regarded as generalizations of some classical results on very ample line bundles (e.g. [I, Proposition 1.8 and Remark 1.12.c]).

The present paper was written when the second author was visiting the University of Milan in the winter of 1994. He would like to thank the Italian C.N.R. and the J.S.P.S. for financial support and for making this collaboration possible. The first author was partially supported by the M.U.R.S.T. of the Italian Government in the framework of the 40% research project "Algebraic Geometry". Both authors would like to thank the C.I.R.M. and the University of Trento for their hospitality at different stages of this research. They are also grateful to A. J. Sommese for his useful correspondence on the Theorem A.

1. Preliminaries

In this paper we will work over the complex number field $\mathbb{C}$. We use the standard notation from algebraic geometry. The words "vector bundle" and "locally free sheaf" are used interchangeably. The tensor products of line bundles are denoted additively, while we use multiplicative notation for intersection products in Chow rings. The pull-back $\iota^*\mathcal{E}$ of a vector bundle $\mathcal{E}$ on X by an embedding $\iota : Y \hookrightarrow X$ is denoted by $\mathcal{E}_Y$. The canonical bundle of a smooth variety X is denoted by K_X.

A *polarized variety* is a pair (X, L) consisting of a projective variety X and an ample line bundle L on X. The Δ*-genus* $\Delta(X, L)$ of a polarized variety (X, L) is defined by the formula $\Delta(X, L) = n + L^n - h^0(X, L)$, where $n = \dim X$. A polarized manifold (X, L) is said to be a *scroll* over a smooth variety W if $(X, L) \cong (\mathbb{P}_W(\mathcal{F}), H(\mathcal{F}))$ for some ample vector bundle $\mathcal{F}$ on W, where $H(\mathcal{F})$ is the tautological line bundle on the projective space bundle $\mathbb{P}_W(\mathcal{F})$ associated to $\mathcal{F}$. A polarized manifold (X, L) is called a *Del Pezzo manifold* if $K_X = (1-n)L$, where $n = \dim X$. A polarized manifold (X, L) is said to be a *quadric fibration* over a smooth curve C if there is a surjective morphism $\varphi : X \to C$ such that any fibre of φ is irreducible and reduced and that any smooth fibre F of φ is a hyperquadric $\mathbb{Q}^{n-1}$ in $\mathbb{P}^n$ with $L_F \cong \mathcal{O}_{\mathbb{Q}^{n-1}}(1)$, where $n = \dim X$. Now we recall one formula. Let F be a general fibre of φ. Then F is a smooth hyperquadric $\mathbb{Q}^{n-1}$ in $\mathbb{P}^n$ with $L_F \cong \mathcal{O}_{\mathbb{Q}^{n-1}}(1)$. Let G be any fibre of φ. Then G is irreducible and reduced, and $L_G^{n-1} = L_F^{n-1} = 2$. Moreover, $\Delta(G, L_G) \le \Delta(F, L_F) = 0$ by the upper-semicontinuity theorem. Thus it follows from [F1, Theorem 4.8] that G is a normal hyperquadric and that L_G is very ample. Of course $h^0(G, L_G) = n + 1$, so that $\mathcal{F} := \varphi_* L$ is a vector bundle of rank $n+1$ on C. Since the natural homomorphism $\varphi^*\mathcal{F} = \varphi^*\varphi_* L \to L$ is surjective, there exists an embedding

$$i : X \to P := \mathbb{P}_C(\mathcal{F}) \text{ with } i^*H = L \text{ and } \varphi = \pi \circ i,$$

where H is the tautological line bundle on P and π is the projection $P \to C$. We note that $X \in | 2H + \pi^* B |$ for some $B \in \operatorname{Pic}(C)$. Now, since $K_P = -(n+1)H + \pi^*(K_C + \det \mathcal{F})$,

$$K_P + \mathcal{O}_P(X) = -(n-1)H + \pi^*\mathcal{L}$$

for some $\mathcal{L} \in \operatorname{Pic}(C)$. By the adjunction formula,

$$K_X + (n-1)L = \varphi^*\mathcal{L}. \tag{1.0.1}$$

Note that our definition of quadric fibrations is different from the adjunction theoretic one [BS, p. 81], since $\mathcal{L}$ is not necessarily ample in (1.0.1), e.g., the case when $\mathcal{L} = \mathcal{O}_C$ may also occur. For instance, according to our definition, the Del Pezzo 3-fold $(\mathbb{P}^1 \times \mathbb{P}^1 \times \mathbb{P}^1, \mathcal{O}(1,1,1))$ is a quadric fibration over $\mathbb{P}^1$ via the projection onto any factor.

The main result of this section is a Lefschetz-type theorem; the topological part of it, essentially due to Sommese [S], was proved in [LM1, Section 1] under slightly weaker assumptions than those of [S, Proposition 1.16, case $k = 0$].

Theorem 1.1 (Lefschetz–Sommese). *Let X, $\mathcal{E}$ and Z be as in* (0.1). *Let $\iota : Z \hookrightarrow X$ be the embedding. Then*

(1.1.1) $H^i(\iota) : H^i(X, \mathbb{Z}) \to H^i(Z, \mathbb{Z})$ *is an isomorphism for* $i \le \dim Z - 1$.

(1.1.2) $H^i(\iota)$ *is injective and its cokernel is torsion free for* $i = \dim Z$.

(1.1.3) $H^{p,q}(\iota) : H^q(X, \Omega_X^p) \to H^q(Z, \Omega_Z^p)$ *is an isomorphism for* $p+q \le \dim Z - 1$.

(1.1.4) $H^{p,q}(\iota)$ *is injective for* $p + q = \dim Z$.

(1.1.5) $\mathrm{Pic}(\iota) : \mathrm{Pic}(X) \to \mathrm{Pic}(Z)$ *is an isomorphism for* $\dim Z \geq 3$.

(1.1.6) $\mathrm{Pic}(\iota)$ *is injective and its cokernel is torsion free for* $\dim Z = 2$.

(1.1.7) $\mathrm{Pic}^0(\iota) : \mathrm{Pic}^0(X) \to \mathrm{Pic}^0(Z)$ *is an isomorphism for* $\dim Z \geq 2$.

(1.1.8) $\mathrm{Alb}(\iota) : \mathrm{Alb}(Z) \to \mathrm{Alb}(X)$ *is an isomorphism for* $\dim Z \geq 2$.

Proof. (1.1.1) and (1.1.2) follow from [LM1, Theorem 1.3]. The Hodge decomposition on cohomology gives (1.1.3) and (1.1.4). (1.1.5) (resp. (1.1.6)) follows from (1.1.1) and (1.1.3) (resp. (1.1.1), (1.1.2), (1.1.3) and (1.1.4)) by the exact cohomology sequences arising from the exponential sheaf sequences on X and Z. (1.1.7) follows from (1.1.1) and (1.1.3). Since the Picard variety and the Albanese variety are, in a natural way, dual complex tori, (1.1.8) follows from (1.1.7). □

2. Proof of the Theorem A

Let $\iota : Z \hookrightarrow X$ be the inclusion and let $\alpha : X \to \mathrm{Alb}(X)$ and $\alpha' : Z \to \mathrm{Alb}(Z)$ be the Albanese morphisms of X and Z, respectively. We have $\mathrm{Alb}(Z) \cong \mathrm{Alb}(X)$ by (1.1.8) because $\dim Z \geq 3$. In view of the commutativity of the diagram

$$\begin{array}{ccc} Z & \xrightarrow{\iota} & X \\ \downarrow \alpha' & & \downarrow \alpha \\ \mathrm{Alb}(Z) & \xrightarrow[\mathrm{Alb}(\iota)]{} & \mathrm{Alb}(X), \end{array}$$

we have $\alpha_{|Z} = \alpha'$. So $\alpha(X) \supseteq \alpha(Z) = \alpha'(Z)$, which is a smooth curve isomorphic to B, since by assumption Z is a $\mathbb{P}^k$-bundle over an irrational curve B. We claim that

$$\alpha(X) = \alpha(Z). \tag{2.1.1}$$

To prove this, recall that, since $\dim Z \geq 3$, by (1.1.3) the restriction homomorphism $\Gamma(\Omega^2_X) \to \Gamma(\Omega^2_Z)$ is an isomorphism. On the other hand, we have $h^0(\Omega^2_Z) = 0$, since Z is a $\mathbb{P}^k$-bundle over B. Therefore $\Gamma(\Omega^2_X) = 0$, so that $\omega_1 \wedge \omega_2 = 0$ for any two 1-forms $\omega_1, \omega_2 \in \Gamma(\Omega^1_X)$. This shows that $\alpha(X)$ is 1-dimensional, hence (2.1.1) is proved. Thus $\alpha : X \to \alpha(Z) \cong B$ fibres X over B. Let F be any smooth fibre of α and consider $\mathcal{E}_F$, which is an ample vector bundle of rank r on F. Its section $s_F \in \Gamma(\mathcal{E}_F)$ vanishes on $Z \cap F = \mathbb{P}^k$ whose dimension is the expected one, since

$$k = \dim Z - 1 = n - r - 1 = \dim F - r.$$

So [LM1, Theorem A] applies and we conclude that

$$(F, \mathcal{E}_F) \cong (\mathbb{P}^{n-1}, \mathcal{O}_{\mathbb{P}^{n-1}}(1)^{\oplus r}). \tag{2.1.2}$$

Since $H^2(Z, \mathbb{Z}) \cong \mathbb{Z} \times \mathbb{Z}$ and the restriction homomorphism $H^2(X, \mathbb{Z}) \to H^2(Z, \mathbb{Z})$ is an isomorphism by (1.1.1), we have $H^2(X, \mathbb{Z}) \cong \mathbb{Z} \times \mathbb{Z}$. Thus $h^2(\mathcal{O}_X) = 0$, and so

$\mathrm{Pic}(X) \to H^2(X, \mathbb{Z})$ is surjective. Let h be the tautological line bundle on Z. Since the restriction homomorphism $\mathrm{Pic}(X) \to \mathrm{Pic}(Z)$ is an isomorphism by (1.1.5), there exists a line bundle H on X such that $H_Z = h$. Let F be any smooth fibre of α. Then $f = Z \cap F (\cong \mathbb{P}^k)$ is a fibre of α'. Now

$$(H_F)_f = (H_Z)_f = h_f = \mathcal{O}_{\mathbb{P}^k}(1),$$

and so

$$(F, H_F) \cong (\mathbb{P}^{n-1}, \mathcal{O}_{\mathbb{P}^{n-1}}(1)). \tag{2.1.3}$$

Thus the integral cohomology classes $[H]$ and $[F]$ of H and F generate $H^2(X, \mathbb{Z})$. Note that

$$\text{every fibre of } \alpha \text{ is irreducible and reduced.} \tag{2.1.4}$$

Actually, assume that there is a fibre F_0 of α with irreducible components A_i, $i = 1, \dots, s$ with multiplicities m_i's. Then $F_0 = \sum_{i=1}^{s} m_i A_i$. For any $i = 1, \dots, s$ we can write $[A_i] = a_i[H] + b_i[F]$ for some integers a_i, b_i. Now

$$0 = [A_i]_F = (a_i[H] + b_i[F])_F = a_i[H]_F,$$

and so $a_i = 0$ for all i's. Thus $[A_i] = b_i[F]$ for some positive integer b_i. But then

$$[F] = [F_0] = \sum_{i=1}^{s} m_i[A_i] = (\sum_{i=1}^{s} m_i b_i)[F],$$

which clearly gives a contradiction unless $s = 1$ and $m_1 = 1$. This proves (2.1.4). Now by (2.1.2) and (2.1.3) we have $\det \mathcal{E} \equiv rH$ modulo fibres of α, so that H is α-ample. Let G be any fibre of α. Then $H_G^{n-1} = H_F^{n-1} = 1$. Moreover, $\Delta(G, H_G) \leq \Delta(F, H_F) = 0$ by (2.1.3) and the upper-semicontinuity theorem. Thus $(G, H_G) \cong (\mathbb{P}^{n-1}, \mathcal{O}_{\mathbb{P}^{n-1}}(1))$ by [F1, Theorem 2.1], and so there are no singular fibres. Consequently X is a $\mathbb{P}^{n-1}$-bundle over B and $\mathcal{E}_F \cong \mathcal{O}_{\mathbb{P}^{n-1}}(1)^{\oplus r}$ for any fibre F of α. □

3. Proof of the Theorem B and related results

3.1. *Proof of the Theorem* B. First of all, note that, since Z is a smooth variety of the expected dimension $n - r$, $\mathcal{E}_Z$ is the normal bundle of Z in X [Fu, Example 6.3.4]. Hence, by adjunction, we get

$$K_Z = (K_X + \det \mathcal{E})_Z. \tag{3.1.1}$$

Now let (Z, H_Z) be a scroll over B with the projection $p : Z \to B$. Then

$$K_Z = -(\dim Z) H_Z + p^* L$$

for some $L \in \mathrm{Pic}(B)$. Therefore

$$K_Z + (\dim Z - 1)H_Z = -H_Z + p^*L$$

is not nef. By using (3.1.1) we see that

$$K_Z + (\dim Z - 1)H_Z = (K_X + \det \mathcal{E} + (\dim Z - 1)H)_Z,$$

and so we conclude that

$$K_X + \det \mathcal{E} + (\dim Z - 1)H$$

itself is not nef. Set

$$\mathcal{E}' := \mathcal{E} \oplus H^{\oplus(\dim Z - 1)}. \tag{3.1.2}$$

Then $\mathcal{E}'$ is an ample vector bundle on X of rank $r + \dim Z - 1 = n - 1$ and $K_X + \det \mathcal{E}' = K_X + \det \mathcal{E} + (\dim Z - 1)H$ is not nef. It thus follows from [YZ, Theorem 3] that $(X, \mathcal{E}')$ is one of the following pairs.

(1) $(\mathbb{P}^n, \mathcal{O}_{\mathbb{P}^n}(1)^{\oplus(n-1)})$.

(2) $(\mathbb{P}^n, \mathcal{O}_{\mathbb{P}^n}(2) \oplus \mathcal{O}_{\mathbb{P}^n}(1)^{\oplus(n-2)})$.

(3) $(\mathbb{Q}^n, \mathcal{O}_{\mathbb{Q}^n}(1)^{\oplus(n-1)})$.

(4) $X \cong \mathbb{P}_C(\mathcal{F})$ for some vector bundle $\mathcal{F}$ of rank n on a smooth curve C and $\mathcal{E}'_F \cong \mathcal{O}_{\mathbb{P}^{n-1}}(1)^{\oplus(n-1)}$ for any fibre F of the bundle projection $\pi : X \to C$.

By (3.1.2) and noting that $\dim Z - 1 = n - r - 1 \geq 2$, this implies that $(X, \mathcal{E})$ is one of the following.

(1) $(\mathbb{P}^n, \mathcal{O}_{\mathbb{P}^n}(1)^{\oplus r})$.

(2) $(\mathbb{P}^n, \mathcal{O}_{\mathbb{P}^n}(2) \oplus \mathcal{O}_{\mathbb{P}^n}(1)^{\oplus(r-1)})$.

(3) $(\mathbb{Q}^n, \mathcal{O}_{\mathbb{Q}^n}(1)^{\oplus r})$.

(4) $X \cong \mathbb{P}_C(\mathcal{F})$ and $\mathcal{E}_F \cong \mathcal{O}_{\mathbb{P}^{n-1}}(1)^{\oplus r}$ for any fibre F of $\pi : X \to C$.

In case (1) any section $s \in \Gamma(\mathcal{E})$ vanishes on a linear space, which contradicts our assumption on Z. In case (2) or (3), since Z is a smooth variety of dimension $n - r$, Z must be a quadric; this is compatible with our assumption on Z only if $n - r = 2$, a contradiction. In case (4), since $H_F \cong \mathcal{O}_{\mathbb{P}^{n-1}}(1)$ for any fibre F of π, (X, H) is a scroll over C. Let $s \in \Gamma(\mathcal{E})$ be a section such that $(s)_0 = Z$ and let s_F denote its restriction to any fibre F. Then $s_F \in \Gamma(\mathcal{O}_{\mathbb{P}^{n-1}}(1)^{\oplus r})$, and so $(s_F)_0 = Z \cap F$ is a linear space of dimension $\geq \dim Z - 1$, hence $= \dim Z - 1$ because of the irreducibility of Z. Note that, since $\dim Z \geq 3$, this linear space can only be a fibre of $p : Z \to B$. Thus for any fibre F of π we have $Z \cap F = f$, a fibre of p. This means that $\pi_{|Z} = p$ and that $C = B$. This proves the assertion. □

Remark 3.2. The same proof as above works also when $\dim Z = 2$, but in the statement of the Theorem B we have to insert the following cases for $(X, H, \mathcal{E})$:

B1) $(\mathbb{P}^n, \mathcal{O}_{\mathbb{P}^n}(1), \mathcal{O}_{\mathbb{P}^n}(2) \oplus \mathcal{O}_{\mathbb{P}^n}(1)^{\oplus(n-3)})$.

B2) $(\mathbb{Q}^n, \mathcal{O}_{\mathbb{Q}^n}(1), \mathcal{O}_{\mathbb{Q}^n}(1)^{\oplus(n-2)})$.

Of course B1) and B2) come from cases (2) and (3), respectively. As to case (4), when $\dim Z = 2$, it may happen that Z has two $\mathbb{P}$-bundle structures over curves, but then $Z = \mathbb{P}^1 \times \mathbb{P}^1$ and so the assertion of the Theorem B is valid because $B = C = \mathbb{P}^1$.

A method similar to the one used to prove Theorem B allows us to recover the easy part of our results in [LM1] in a very short way.

Proposition 3.3. *Let* $(X, \mathcal{E})$ *be as in* (0.1).

(i) *Assume that* $Z = \mathbb{P}^{n-r}$ *with* $n - r \geq 2$. *Then* $(X, \mathcal{E}) \cong (\mathbb{P}^n, \mathcal{O}_{\mathbb{P}^n}(1)^{\oplus r})$.

(ii) *Assume that* $Z = \mathbb{Q}^{n-r}$ *with* $n - r \geq 3$. *Then* $(X, \mathcal{E})$ *is either* $(\mathbb{P}^n, \mathcal{O}_{\mathbb{P}^n}(2) \oplus \mathcal{O}_{\mathbb{P}^n}(1)^{\oplus(r-1)})$ *or* $(\mathbb{Q}^n, \mathcal{O}_{\mathbb{Q}^n}(1)^{\oplus r})$.

Proof. (i) Since $\mathrm{Pic}(Z) \cong \mathbb{Z}$, by (1.1.5) and (1.1.6) the restriction homomorphism $\mathrm{Pic}(X) \to \mathrm{Pic}(Z)$ is an isomorphism, so there exists an ample line bundle $H \in \mathrm{Pic}(X)$ such that $(Z, H_Z) \cong (\mathbb{P}^{n-r}, \mathcal{O}_{\mathbb{P}^{n-r}}(1))$. Since Z is a smooth variety of the expected dimension, by adjunction, we have

$$\mathcal{O}_Z = K_Z + (\dim Z + 1)H_Z = (K_X + \det \mathcal{E} + (\dim Z + 1)H)_Z. \tag{3.3.1}$$

Consider the ample vector bundle

$$\mathcal{G} := \mathcal{E} \oplus H^{\oplus(\dim Z+1)} \tag{3.3.2}$$

of rank $r + \dim Z + 1 = n + 1$ on X. By (3.3.1) its adjoint bundle

$$K_X + \det \mathcal{G} = K_X + \det \mathcal{E} + (\dim Z + 1)H$$

is not ample. Thus by [F2, Main Theorem] we see that $(X, \mathcal{G}) \cong (\mathbb{P}^n, \mathcal{O}_{\mathbb{P}^n}(1)^{\oplus(n+1)})$. Then the assertion follows from (3.3.2).

(ii) Since $\mathrm{Pic}(Z) \cong \mathbb{Z}$ and the restriction homomorphism $\mathrm{Pic}(X) \to \mathrm{Pic}(Z)$ is an isomorphism by (1.1.5), there exists an ample line bundle $H \in \mathrm{Pic}(X)$ such that $(Z, H_Z) \cong (\mathbb{Q}^{n-r}, \mathcal{O}_{\mathbb{Q}^{n-r}}(1))$. If we set $\mathcal{G} := \mathcal{E} \oplus H^{\oplus \dim Z}$, then $\mathcal{G}$ is an ample vector bundle of rank n and we have

$$\mathcal{O}_Z = K_Z + (\dim Z)H_Z = (K_X + \det \mathcal{E} + (\dim Z)H)_Z = (K_X + \det \mathcal{G})_Z. \tag{3.3.3}$$

Thus $K_X + \det \mathcal{G}$ is not ample. Since $\mathrm{Pic}(X) \cong \mathbb{Z}$ and $\mathcal{G}$ is decomposable, we use [F2, Main Theorem] to see that $(X, \mathcal{G})$ is one of the following.

(1) $(\mathbb{P}^n, \mathcal{O}_{\mathbb{P}^n}(1)^{\oplus n})$.

(2) $(\mathbb{P}^n, \mathcal{O}_{\mathbb{P}^n}(2) \oplus \mathcal{O}_{\mathbb{P}^n}(1)^{\oplus(n-1)})$.

(3) $(\mathbb{Q}^n, \mathcal{O}_{\mathbb{Q}^n}(1)^{\oplus n})$.

But in case (1) we have $K_X + \det \mathcal{G} = \mathcal{O}_{\mathbb{P}^n}(-1)$, which contradicts (3.3.3). The result follows from (2) and (3). □

4. Proof of the Theorem C

4.1. By our formula (1.0.1),

$$K_Z + (\dim Z - 1)H_Z = p^*L$$

for some $L \in \mathrm{Pic}(B)$, where p is the projection $Z \to B$. Thus $K_Z + (\dim Z - 2)H_Z = -H_Z + p^*L$ is not nef. Now let

$$\mathcal{E}' = \mathcal{E} \oplus H^{\oplus(\dim Z - 2)}. \tag{4.1.0}$$

Then $\mathcal{E}'$ is an ample vector bundle on X of rank $r + \dim Z - 2 = n - 2$. Since $(K_X + \det \mathcal{E})_Z = K_Z$,

$$\begin{aligned}(K_X + \det \mathcal{E}')_Z &= (K_X + \det \mathcal{E} + (\dim Z - 2)H)_Z \\ &= K_Z + (\dim Z - 2)H_Z;\end{aligned}$$

thus $K_X + \det \mathcal{E}'$ is not nef and [M, Theorem] applies. By (1.1.1) we have $H^2(X, \mathbb{Z}) \cong H^2(Z, \mathbb{Z})$ because $\dim Z = n - r \geq 3$. Thus $b_2(X) \geq 2$ and consequently $(X, \mathcal{E}')$ is one of the following.

(4.1.1) There exists an effective divisor E on X such that $E \cong \mathbb{P}^{n-1}$, $\mathcal{O}_E(E) \cong \mathcal{O}_{\mathbb{P}^{n-1}}(-1)$ and $\mathcal{E}'_E \cong \mathcal{O}_{\mathbb{P}^{n-1}}(1)^{\oplus(n-2)}$.

(4.1.2) There is a vector bundle $\mathcal{F}$ on a smooth curve C such that $X \cong \mathbb{P}_C(\mathcal{F})$ and $\mathcal{E}'_F \cong \mathcal{O}_{\mathbb{P}^{n-1}}(1)^{\oplus(n-2)}$ for any fibre $F(\cong \mathbb{P}^{n-1})$ of $\pi : X \to C$.

(4.1.3) There is a vector bundle $\mathcal{F}$ on a smooth curve C such that $X \cong \mathbb{P}_C(\mathcal{F})$ and $\mathcal{E}'_F \cong \mathcal{O}_{\mathbb{P}^{n-1}}(2) \oplus \mathcal{O}_{\mathbb{P}^{n-1}}(1)^{\oplus(n-3)}$ for any fibre $F(\cong \mathbb{P}^{n-1})$ of $\pi : X \to C$.

(4.1.4) There is a surjective morphism $\varphi : X \to C$ onto a smooth curve C such that any general fibre F of φ is a smooth hyperquadric $\mathbb{Q}^{n-1}$ in $\mathbb{P}^n$ with $\mathcal{E}'_F \cong \mathcal{O}_{\mathbb{Q}^{n-1}}(1)^{\oplus(n-2)}$.

(4.1.5) There is a vector bundle $\mathcal{F}$ on a smooth surface S such that $X \cong \mathbb{P}_S(\mathcal{F})$ and $\mathcal{E}'_F \cong \mathcal{O}_{\mathbb{P}^{n-2}}(1)^{\oplus(n-2)}$ for any fibre $F(\cong \mathbb{P}^{n-2})$ of $\pi : X \to S$.

We proceed now by cases.

4.2. Case (4.1.1).

We have $H_E \cong \mathcal{O}_{\mathbb{P}^{n-1}}(1)$ and $\mathcal{E}_E \cong \mathcal{O}_{\mathbb{P}^{n-1}}(1)^{\oplus r}$ by (4.1.0). Since the restriction s_E of s to E is an element of $\Gamma(\mathcal{O}_{\mathbb{P}^{n-1}}(1)^{\oplus r})$, $D := (s_E)_0 = Z \cap E$ is a linear subspace of E and $\dim D \geq n - r - 1 \geq 2$. If $Z \subset E$, then $D = Z \cap E = Z$, and so Z is linear in E, a contradiction. Assume that $Z \not\subset E$. Then $\dim D = n - r - 1$, i.e., $D \cong \mathbb{P}^{n-r-1}$. Since $n - r - 1 \geq 2$, $p(D)$ must be a point of B. This implies that D is a fibre of the quadric fibration $p : Z \to B$, a contradiction.

4.3. Case (4.1.2).

By (4.1.0) we have $H_F \cong \mathcal{O}_{\mathbb{P}^{n-1}}(1)$ and $\mathcal{E}_F \cong \mathcal{O}_{\mathbb{P}^{n-1}}(1)^{\oplus r}$ for any fibre F of $\pi : X \to C$. Consider $s_F \in \Gamma(\mathcal{O}_{\mathbb{P}^{n-1}}(1)^{\oplus r})$. Then $D := (s_F)_0 = Z \cap F$ is a linear subspace of F and $\dim D \geq n - r - 1 \geq 2$. Since Z is irreducible, we have $\dim D = n - r - 1$, and hence D is a fibre of p. This is impossible.

4.4. Case (4.1.3).

By (4.1.0) H_F is either $\mathcal{O}_{\mathbb{P}^{n-1}}(2)$ or $\mathcal{O}_{\mathbb{P}^{n-1}}(1)$ for any fibre F of π. Assume that $H_F \cong \mathcal{O}_{\mathbb{P}^{n-1}}(2)$ for some fibre F of π. Then $\mathcal{E}_F \cong \mathcal{O}_{\mathbb{P}^{n-1}}(1)^{\oplus r}$. Let $D = (s_F)_0$. Then $D = Z \cap F$ is a linear subspace of F and $\dim D \geq n - r - 1 \geq 2$ because $s_F \in \Gamma(\mathcal{O}_{\mathbb{P}^{n-1}}(1)^{\oplus r})$. If $Z \subset F$, then $D = Z \cap F = Z$, and so Z is linear in F, a contradiction. Therefore $Z \not\subset F$. But then the same argument as in (4.3) gives a contradiction. Thus $H_F \cong \mathcal{O}_{\mathbb{P}^{n-1}}(1)$ and $\mathcal{E}_F \cong \mathcal{O}_{\mathbb{P}^{n-1}}(2) \oplus \mathcal{O}_{\mathbb{P}^{n-1}}(1)^{\oplus(r-1)}$ for any fibre F of π. In particular, (X, H) is a scroll over C. Since $s_F \in \Gamma(\mathcal{O}_{\mathbb{P}^{n-1}}(2) \oplus \mathcal{O}_{\mathbb{P}^{n-1}}(1)^{\oplus(r-1)})$ and Z is irreducible, $D := (s_F)_0 = Z \cap F$ is a smooth hyperquadric of dimension $n-r-1 \geq 2$ for any general fibre F of π. This implies that $\pi_{|Z} : Z \to C$ is surjective.

There are two possibilities: either $n - r - 1 \geq 3$ or $n - r - 1 = 2$. We consider these separately:

Case 1: $n - r - 1 \geq 3$

In this case $p(D)$ is a point of B. Thus D is a fibre of the quadric fibration $p : Z \to B$, and so $\pi_{|Z} = p$. Consequently $C = B$.

Case 2: $n - r - 1 = 2$

If $p(D)$ is a point of B, then the same argument as above shows that $C = B$, hence we may assume that $p(D) = B$ for some fibre F. Then $B \cong \mathbb{P}^1$ because $D \cong \mathbb{P}^1 \times \mathbb{P}^1$. Now take an arbitrary smooth fibre G of p. Then $G \cap D \neq \emptyset$ and of course $G \neq D$, so that $\pi_{|Z}(G) = C$. Since $G \cong \mathbb{P}^1 \times \mathbb{P}^1$, we have $C \cong \mathbb{P}^1$.

So $C = B$ in either case. Thus we are in the case (1) of the Theorem C.

4.5. Case (4.1.4).

We have $H_F \cong \mathcal{O}_{\mathbb{Q}^{n-1}}(1)$ and $\mathcal{E}_F \cong \mathcal{O}_{\mathbb{Q}^{n-1}}(1)^{\oplus r}$ for any general fibre F of $\varphi : X \to C$. The same argument as in (4.4) gives $C = B$, and so we are in the case (2) of the Theorem C.

4.6. Case (4.1.5).

In this case, $H_F \cong \mathcal{O}_{\mathbb{P}^{n-2}}(1)$ and $\mathcal{E}_F \cong \mathcal{O}_{\mathbb{P}^{n-2}}(1)^{\oplus r}$ for any fibre F of $\pi : X \to S$. In particular, (X, H) is a scroll over S. Consider $s_F \in \Gamma(\mathcal{O}_{\mathbb{P}^{n-2}}(1)^{\oplus r})$. Then $D := (s_F)_0 = Z \cap F$ is a linear subspace of F and $\dim D \geq n - r - 2 \geq 1$. Thus $\pi_{|Z} : Z \to S$ is surjective. On the other hand, since Z is irreducible, we have $n - r - 1 \geq \dim D$. Assume that $\dim D = n - r - 1$ for some fibre F of π. Then $p(D)$ is a point of B and $D \cong \mathbb{P}^{n-r-1}$ is a fibre of the quadric fibration $p : Z \to B$, a contradiction. Thus $D \cong \mathbb{P}^{n-r-2}$ for any fibre F of π.

Suppose that $n - r - 2 \geq 2$. Then, since every D is sent to a point of B via p, D lies in a smooth fibre $\mathbb{Q}^{n-r-1}$ of p for some fibre F of π. This is impossible. Thus $\dim Z = n - r = 3$ and $\mathcal{E}_F \cong \mathcal{O}_{\mathbb{P}^{n-2}}(1)^{\oplus(n-3)}$ for any fibre F of π.

Let f be a fibre of $\pi_Z := \pi_{|Z}$. Then, since $f = D$, we have $f \cong \mathbb{P}^1$. Moreover, $(H_Z)_f = (H_F)_f \cong \mathcal{O}_{\mathbb{P}^1}(1)$ because $H_F \cong \mathcal{O}_{\mathbb{P}^{n-2}}(1)$ and f is a line of F. Thus (Z, H_Z) is a scroll over S and we can write $(Z, H_Z) = (\mathbb{P}_S(\mathcal{G}), H(\mathcal{G}))$ for some ample vector bundle $\mathcal{G}$ of rank 2 on S. Then we know

$$K_Z + 2H_Z = \pi_Z^*(K_S + \det \mathcal{G}). \tag{4.6.1}$$

On the other hand, since (Z, H_Z) is also a quadric fibration over B, it follows from (1.0.1) that

$$K_Z + 2H_Z = p^*L \tag{4.6.2}$$

for some $L \in \mathrm{Pic}(B)$.

We set $\mathbb{Q}_b = p^{-1}(b)(b \in B)$. Then $\dim \pi_Z(\mathbb{Q}_b) = 1$ or 2 for all $b \in B$. The argument now splits up into two cases.

4.7. Assume that $\dim \pi_Z(\mathbb{Q}_b) = 1$ for any smooth fibre $\mathbb{Q}_b$. Let $C_b = \pi_Z(\mathbb{Q}_b)$. Then $\mathbb{Q}_b = \mathbb{P}_{C_b}(\mathcal{G}_{C_b})$. This implies that C_b is a smooth curve, and so $C_b \cong \mathbb{P}^1$, since $\mathbb{Q}_b \cong \mathbb{P}^1 \times \mathbb{P}^1$. Let $\mathbb{Q}_{b'}$ be a singular fibre. Assume that $\dim \pi_Z(\mathbb{Q}_{b'}) = 2$. Then $\pi_Z^{-1}(c) \cap \mathbb{Q}_{b'} \neq \emptyset$ for any $c \in C_b$, which implies that $\mathbb{Q}_b \cap \mathbb{Q}_{b'} \neq \emptyset$. This is impossible. Thus $\dim \pi_Z(\mathbb{Q}_{b'}) = 1$. Let $C_{b'} = \pi_Z(\mathbb{Q}_{b'})$. Then $\mathbb{Q}_{b'} = \mathbb{P}_{C_{b'}}(\mathcal{G}_{C_{b'}})$. If $C_{b'}$ has a singular point, then $\dim(\mathrm{Sing}(\mathbb{Q}_{b'})) = 1$, where $\mathrm{Sing}(\mathbb{Q}_{b'})$ is the singular locus of $\mathbb{Q}_{b'}$. This is impossible because $\mathbb{Q}_{b'}$ is normal. Thus $C_{b'}$ and $\mathbb{Q}_{b'}$ are smooth. Consequently $C_{b'} \cong \mathbb{P}^1$ and we conclude that S is a $\mathbb{P}^1$-bundle over the smooth curve B. In particular, p factors through π_Z and the ruling of S. We are in the case (3) of the Theorem C.

4.8. Assume that $\dim \pi_Z(\mathbb{Q}_b) = 2$ for some smooth fibre $\mathbb{Q}_b \cong \mathbb{P}^1 \times \mathbb{P}^1$. Then $\pi_{\mathbb{Q}_b} : \mathbb{Q}_b \to S$ is finite and $S \cong \mathbb{P}^1 \times \mathbb{P}^1$ or $\mathbb{P}^2$. Take an arbitrary fibre f of π_Z. Assume that $f \subset \mathbb{Q}_{b'}$ for some $b' \in B$. Then $\mathbb{Q}_b \cap \mathbb{Q}_{b'} \neq \emptyset$ because $f \cap \mathbb{Q}_b \neq \emptyset$, a contradiction. Thus $p(f) = B$, so $B \cong \mathbb{P}^1$. By virtue of (4.6.2) we may write $K_Z + 2H_Z = p^*\mathcal{O}_{\mathbb{P}^1}(t)$ for some integer t. If $t \neq 0$, then

$$\begin{aligned}\deg \pi_{\mathbb{Q}_b} &= \mathbb{Q}_b f = (p^*\mathcal{O}_{\mathbb{P}^1}(1))f \\ &= (\frac{1}{t}(K_Z + 2H_Z))f = 0\end{aligned}$$

by (4.6.1), a contradiction. Thus $t = 0$ and $K_Z + 2H_Z = 0$. By (4.6.1) we have $K_S + \det \mathcal{G} = 0$. If $S = \mathbb{P}^1 \times \mathbb{P}^1$, then $\det \mathcal{G} = \mathcal{O}_S(2, 2)$. Since $\mathcal{G} \otimes \mathcal{O}_S(-1, -1)$ is trivial on any line on $\mathbb{P}^1 \times \mathbb{P}^1$, $\mathcal{G} \otimes \mathcal{O}_S(-1, -1)$ is globally trivial by virtue of [W, Lemma 3.6.1]. Thus $\mathcal{G} = \mathcal{O}_S(1, 1)^{\oplus 2}$ and $Z = \mathbb{P}_S(\mathcal{G}) = \mathbb{P}^1 \times \mathbb{P}^1 \times \mathbb{P}^1$. If $S = \mathbb{P}^2$, then $\det \mathcal{G} = \mathcal{O}_{\mathbb{P}^2}(3)$ because $K_S + \det \mathcal{G} = 0$, so by the theory of uniform vector bundles $\mathcal{G}$ can only be either $T_{\mathbb{P}^2}$ or $\mathcal{O}_{\mathbb{P}^2}(2) \oplus \mathcal{O}_{\mathbb{P}^2}(1)$. These two cases could be ruled out by an *ad hoc* argument, but we take this opportunity for proving the following more general result regardless of the special form of $\mathcal{G}$.

Lemma 4.9. *If* (Z, H_Z) *satisfies the assumption in* (4.8), *then the case* $S = \mathbb{P}^2$ *cannot occur.*

Proof. Suppose to the contrary that $S = \mathbb{P}^2$. Since (Z, H_Z) is a scroll over $\mathbb{P}^2$, we have $\mathrm{Pic}(Z) \cong \mathbb{Z} \times \mathbb{Z}$, a basis being given by H_Z and $\pi_Z^*\mathcal{O}_{\mathbb{P}^2}(1)$. Now we can find a smooth fibre $\mathbb{Q} \cong \mathbb{P}^1 \times \mathbb{P}^1$ of p such that $\pi_{\mathbb{Q}} : \mathbb{Q} \to \mathbb{P}^2$ is finite, as has been seen in (4.8). Since (Z, H_Z) is also a quadric fibration over $\mathbb{P}^1$, H_Z and $\mathcal{O}_Z(\mathbb{Q})$ are another basis of $\mathrm{Pic}(Z)$. Write $\mathcal{O}_Z(\mathbb{Q}) = \alpha H_Z + \beta\pi_Z^*\mathcal{O}_{\mathbb{P}^2}(1)$ for some integers α, β. Then the matrix expressing H_Z and $\mathcal{O}_Z(\mathbb{Q})$ in terms of H_Z and $\pi_Z^*\mathcal{O}_{\mathbb{P}^2}(1)$ is

$$\begin{pmatrix} 1 & \alpha \\ 0 & \beta \end{pmatrix}.$$

Thus

$$\beta = \pm 1. \tag{4.9.1}$$

Moreover, for any fibre f of π_Z,

$$\deg \pi_{\mathbb{Q}} = \mathbb{Q}f = (\alpha H_Z + \beta\pi_Z^*\mathcal{O}_{\mathbb{P}^2}(1))f = \alpha. \tag{4.9.2}$$

This implies that

$$\alpha \geq 2 \tag{4.9.3}$$

because $\pi_{\mathbb{Q}}$ is not birational. Let

$$\mathcal{O}_{\mathbb{Q}}(u, v) = \pi_{\mathbb{Q}}^*\mathcal{O}_{\mathbb{P}^2}(1). \tag{4.9.4}$$

Then, since $\pi_{\mathbb{Q}}$ is finite, $\mathcal{O}_{\mathbb{Q}}(u, v)$ is ample. Thus

$$u > 0 \text{ and } v > 0. \tag{4.9.5}$$

On the other hand, since $(H_Z)_{\mathbb{Q}} \cong \mathcal{O}_{\mathbb{Q}}(1, 1)$, we have

$$\begin{aligned} \mathcal{O}_{\mathbb{Q}} = (\mathcal{O}_Z(\mathbb{Q}))_{\mathbb{Q}} &= (\alpha H_Z + \beta\pi_Z^*\mathcal{O}_{\mathbb{P}^2}(1))_{\mathbb{Q}} \\ &= \alpha\mathcal{O}_{\mathbb{Q}}(1, 1) + \beta\mathcal{O}_{\mathbb{Q}}(u, v) \\ &= \mathcal{O}_{\mathbb{Q}}(\alpha + \beta u, \alpha + \beta v), \end{aligned}$$

so

$$\alpha + \beta u = \alpha + \beta v = 0. \tag{4.9.6}$$

Putting together (4.9.1), (4.9.3), (4.9.5) and (4.9.6), we have

$$\beta = -1$$

and

$$u = v = \alpha.$$

Thus by (4.9.4) $\pi_{\mathbb{Q}}^*\mathcal{O}_{\mathbb{P}^2}(1) = \mathcal{O}_{\mathbb{Q}}(\alpha, \alpha)$, and so

$$\deg \pi_{\mathbb{Q}} = (\pi_{\mathbb{Q}}^*\mathcal{O}_{\mathbb{P}^2}(1))^2 = \mathcal{O}_{\mathbb{Q}}(\alpha, \alpha)^2 = 2\alpha^2.$$

Combining this with (4.9.2), we have $2\alpha^2 = \alpha$, a contradiction. □

Thus in case (4.8) we have $S = \mathbb{P}^1 \times \mathbb{P}^1$. Of course, this conforms to case (3) of the Theorem C because $B = \mathbb{P}^1$. This completes the proof of the Theorem C.

References

[AM] M. Andreatta and M. Mella, Contractions on a manifold polarized by an ample vector bundle, preprint, 1994.

[B1] L. Bădescu, On ample divisors, II, in: Week of Algebraic Geometry, Proc. Bucharest 1980, Texte Math. 40, Teubner, Leipzig 1981, 12–32.

[B2] —–, The projective plane blown up at a point as an ample divisor, Atti Accad. Ligure Sci. Lett. 38 (1982), 88–92.

[BS] M. C. Beltrametti and A. J. Sommese, The Adjunction Theory of Complex Projective Varieties, de Gruyter Exp. Math. 16, Walter de Gruyter, Berlin–New York 1995.

[F1] T. Fujita, On the structure of polarized varieties with Δ-genera zero, J. Fac. Sci. Univ. Tokyo Sect. IA Math. 22 (1975), 103–115.

[F2] —–, On adjoint bundles of ample vector bundles, in: Complex Algebraic Varieties, Proc. Bayreuth 1990 (K. Hulek, T. Peternell, M. Schneider and F.-O. Schreyer, eds.), Lecture Notes in Math. 1507, Springer-Verlag, Berlin–Heidelberg 1992, 105–112.

[Fu] W. Fulton, Intersection Theory, Ergeb. Math. Grenzgeb. (3) 2, Springer-Verlag, Berlin–Heidelberg–New York–Tokyo 1984.

[I] P. Ionescu, Embedded projective varieties of small invariants, in: Algebraic Geometry, Proc. Bucharest 1982 (L. Bădescu and D. Popescu, eds.), Lecture Notes in Math. 1056, Springer-Verlag, Berlin–Heidelberg 1984, 142–186.

[LM1] A. Lanteri and H. Maeda, Ample vector bundles with sections vanishing on projective spaces or quadrics, Internat. J. Math. 6 (1995), 587–600.

[LM2] —–, Geometrically ruled surfaces as zero loci of ample vector bundles, Forum Math., to appear.

[M] H. Maeda, Nefness of adjoint bundles for ample vector bundles, Matematiche (Catania), to appear.

[S] A. J. Sommese, Submanifolds of Abelian varieties, Math. Ann. 233 (1978), 229–256.

[W] J. A. Wiśniewski, Length of extremal rays and generalized adjunction, Math. Z. 200 (1989), 409–427.

[YZ] Y.-G. Ye and Q. Zhang, On ample vector bundles whose adjunction bundles are not numerically effective, Duke Math. J. 60 (1990), 671–687.

The Diagram Method for 3-folds and its application to the Kähler cone and Picard number of Calabi–Yau 3-folds. I

With an Appendix by Vyacheslav V. Shokurov: Anticanonical boundedness for curves

*Viacheslav V. Nikulin**

Abstract. We prove a general "Diagram Method" theorem valid for a quite large class of 3-folds with $\mathbb{Q}$-factorial singularities (see Basic Theorems 1.3.2 and 3.2 and also Theorem 2.2.6). This generalizes our results on Fano 3-folds with $\mathbb{Q}$-factorial terminal singularities [N8].

As an application, we get the following result about Calabi–Yau 3-folds X: Assume that the Picard number $\rho(X) > 40$. Then one of two cases (i) or (ii) holds: (i) There exists a small extremal ray on X. (ii) There exists a nef element h such that $h^3 = 0$ (thus, the nef cone NEF(X) and the cubic intersection hypersurface $\mathcal{W}_X$ have a common point; here, we don't claim that h is rational!).

As a corollary, we get: Let X be a Calabi–Yau 3-fold. Assume that the nef cone NEF(X) is finite polyhedral and X does not have a small extremal ray. Then there exists a rational nef element h with $h^3 = 0$ if $\rho(X) > 40$.

To prove these results on Calabi–Yau manifolds, we also use a result by V. V. Shokurov on the length of divisorial extremal rays (see the Appendix by V. V. Shokurov). Thus the above results on Calabi–Yau 3-folds are joint work with V. V. Shokurov.

We also discuss the generalization of the above results to so called $\mathbb{Q}$-factorial models of Calabi–Yau 3-folds. This sometimes enables us to play the same game even when the Mori cone is not polyhedral, or when there are small extremal rays.

0. Introduction

We consider projective algebraic varieties over the field $\mathbb{C}$ of complex numbers.

In our paper [N8], we developed for Fano 3-folds X with terminal $\mathbb{Q}$-factorial singularities the so called Diagram Method (for divisorial case). As an application,

---

* Partially supported by Grant of Russian Fund of Fundamental Research; Grant of AMS; Grant of ISF MI6000; Grant of ISF and Russian Government MI6300.

we proved that if the Picard number $\rho(X) > 7$, then X has either a small extremal ray or a nef rational element h with $h^3 = 0$.

Here, we generalize this method for arbitrary 3-folds (also for divisorial case). It was very surprising for us that this is possible. See Basic Theorems 1.3.2 and 3.2.

As an application, we use this method for Calabi–Yau manifolds. We refer to the very important papers of P. M. H. Wilson [W1], [W2] for terminology and basic results on Calabi–Yau 3-folds we need (we cite these results in Section 4.1). Using a result of V. V. Shokurov about the length of divisorial extremal rays (see Appendix by V. V. Shokurov), we get the following result:

Theorem 0.1 (joint work with V. V. Shokurov). *Let X be a Calabi–Yau 3-fold and suppose that the Picard number $\rho(X) > 40$.*

Then one of two cases (i) *or* (ii) *below holds:*

(i) *There exists a small extremal ray on X.*

(ii) *There exists a nef element h such that $h^3 = 0$ (thus, the nef cone* NEF(X) *and the cubic intersection hypersurface $\mathcal{W}_X$ have a common point; here, we don't claim that h is rational!).*

As a corollary, we get

Theorem 0.1′ (joint work with V. V. Shokurov). *Let X be a Calabi–Yau 3-fold. Assume that the nef cone* NEF(X) *(equivalently, the Mori cone* $\overline{\text{NE}}(X)$*) is finite polyhedral and X does not have a small extremal ray.*

Then there exists a rational nef element h with $h^3 = 0$ if $\rho(X) > 40$.

Proof. See Theorems 4.3.1 and 4.3.1′ for more exact statements and proof. □

It seems that the existence of a nef rational element h with $h^3 = 0$ is very important for Calabi–Yau 3-folds. See [D-Gro], [Gro], [Gra], [Hu], [O], [W1], [W2], [W3]. By I. I. Piatetsky-Shapiro and I. R. Shafarevich [PŠ-Š], an algebraic K3 surface has a nef rational element E with $E^2 = 0$ if its Picard number ≥ 5. See Section 6 where we discuss these results and their connection with our results.

In Section 5 we consider one possibility to extend our results for cases when either the Mori cone $\overline{\text{NE}}(X)$ is not finite polyhedral or X has small extremal rays. It is connected with considering so called *$\mathbb{Q}$-factorial models Y* of a Calabi–Yau manifold X, which one gets as a sequence of contractions of divisorial extremal rays and flops in small extremal rays. These models have $\mathbb{Q}$-factorial canonical singularities, and we can apply the Diagram Method to these models obtaining results similar to Theorems 0.1 and 0.1′ (replacing the constant 40 by another constant). For some very special class of $\mathbb{Q}$-factorial models (which we call *very good*) we prove results similar to Theorems 0.1 and 0.1′ (with the constant 163 instead of 40). See Theorem 5.5 and Corollaries 5.6 and 5.9. We conjecture that the analogs of Theorems 0.1 and 0.1′

are valid for arbitrary $\mathbb{Q}$-factorial models (one should replace 40 by another absolute constant). See Conjecture 5.3.

In particular, this preprint contains results announced in [N10].

I am grateful to I. Dolgachev, A. Grassi, M. Gross, D. Morrison, M. Reid and V. V. Shokurov for useful discussions, and to the referee for useful comments. I am grateful to Professor I. R. Shafarevich for supporting my studies.

1. An effective variant of the Diagram Method for the divisorial case

1.1. Reminder. First, we recall a combinatorial result from [N8].

Let X be a projective algebraic variety with $\mathbb{Q}$-factorial singularities over an algebraically closed field. Let $N_1(X)$ be the $\mathbb{R}$-vector space generated by all algebraic curves on X modulo numerical equivalence, and let $N^1(X)$ be the $\mathbb{R}$-vector space generated by all Cartier (or Weil) divisors on X modulo numerical equivalence. The vector spaces $N_1(X)$ and $N^1(X)$ are dual to one another by the intersection pairing. Let $\mathrm{NE}(X)$ be the convex cone in $N_1(X)$ generated by all effective curves on X. Let $\overline{\mathrm{NE}}(X)$ be the closure of the cone $\mathrm{NE}(X)$ in $N_1(X)$. It is called the *Mori cone (*or *polyhedron)* of X. A non-zero element $x \in N^1(X)$ is called *nef* if $x \cdot \overline{\mathrm{NE}}(X) \geq 0$. Let $\mathrm{NEF}(X)$ be the set of all nef elements of X and the zero. It is the convex cone in $N^1(X)$ dual to the Mori cone $\overline{\mathrm{NE}}(X)$. A ray $R \subset \overline{\mathrm{NE}}(X)$ with origin 0 is called *extremal* if from $C_1 \in \overline{\mathrm{NE}}(X)$, $C_2 \in \overline{\mathrm{NE}}(X)$ and $C_1 + C_2 \in R$ it follows that $C_1 \in R$ and $C_2 \in R$.

For a set $\mathcal{R}$ of extremal rays of X, consider the following condition:

(i) If $R \in \mathcal{R}$, then all curves $C \in R$ fill out an irreducible divisor $D(R)$ on X. We call this extremal ray *divisorial.*

Under this assumption, we use the following construction to assign an oriented graph $G(\mathcal{R})$ to $\mathcal{R}$ (or to subsets of $\mathcal{R}$): Two different rays R_1 and R_2 are joined by an arrow R_1R_2 with the beginning in R_1 and the end in R_2 if $R_1 \cdot D(R_2) > 0$. Here and in what follows, for an extremal ray R and a divisor D we write $R \cdot D > 0$ if $r \cdot D > 0$ for $r \in R$ and $r \neq 0$. (The same for the symbols $\leq$, $\geq$ and $<$.)

A set $\mathcal{E}$ of extremal rays is called *extremal* if it is contained in a face of $\overline{\mathrm{NE}}(X)$. Equivalently, there exists a nef element $H \in N^1(X)$ such that $\mathcal{E} \cdot H = 0$. Clearly, a subset of an extremal set is extremal too.

We consider the following condition (ii) for extremal sets $\mathcal{E}$ of extremal rays.

(ii) An extremal set $\mathcal{E} = \{R_1, \dots, R_n\}$ satisfies the condition (i), and for any real numbers $m_1 \geq 0, \dots, m_n \geq 0$, not all 0, there exists a ray $R_j \in \mathcal{E}$ such that $R_j \cdot (m_1D(R_1) + m_2D(R_2) + \cdots + m_nD(R_n)) < 0$. In particular, the effective divisor $m_1D(R_1) + m_2D(R_2) + \cdots + m_nD(R_n)$ is not nef.

A set $\mathcal{L}$ of extremal rays is called an *E-set* (extremal in a different sense) if $\mathcal{L}$ is not extremal but every proper subset of $\mathcal{L}$ is extremal. Thus $\mathcal{L}$ is a minimal non-extremal set of extremal rays. Clearly, an E-set $\mathcal{L}$ contains at least two elements.

We consider the following condition (iii) for E-sets $\mathcal{L}$.

(iii) Any proper subset of an E-set $\mathcal{L} = \{Q_1, \dots, Q_m\}$ satisfies the condition (ii), and there exists a non-zero effective nef divisor $D(\mathcal{L}) = a_1 D(Q_1) + a_2 D(Q_2) + \dots + a_m D(Q_m)$.

We have the following statement:

Lemma 1.1.1. *An E-set $\mathcal{L}$ satisfying the condition* (iii) *is connected in the following sense: For any decomposition $\mathcal{L} = \mathcal{L}_1 \coprod \mathcal{L}_2$, where $\mathcal{L}_1$ and $\mathcal{L}_2$ are non-empty, there exists an arrow Q_1Q_2 such that $Q_1 \in \mathcal{L}_1$ and $Q_2 \in \mathcal{L}_2$.*

If $\mathcal{L}$ and $\mathcal{M}$ are two different E-sets satisfying the condition (iii), *then there exists an arrow LM where $L \in \mathcal{L}$ and $M \in \mathcal{M}$.*

Proof. See [N8, Lemma 1.1]. □

Let $\mathrm{NEF}(X) = \overline{\mathrm{NE}}(X)^* \subset N^1(X)$ be the cone of nef elements of X and $\mathcal{M}(X) = \mathrm{NEF}(X)/\mathbb{R}^+$ its projectivization. We use usual relations of orthogonality between subsets of $\mathcal{M}(X)$ and $\overline{\mathrm{NE}}(X)$. So for $U \subset \mathcal{M}(X)$ and $V \subset \overline{\mathrm{NE}}(X)$ we write $U \perp V$ if $x \cdot y = 0$ for any $\mathbb{R}^+x \in U$ and any $y \in V$. Thus for $U \subset \mathcal{M}(X)$, $V \subset \overline{\mathrm{NE}}(X)$ we denote

$$U^\perp = \{y \in \overline{\mathrm{NE}}(X) \mid U \perp y\}, \quad V^\perp = \{x \in \mathcal{M}(X) \mid x \perp V\}.$$

A subset $\gamma \subset \mathcal{M}(X)$ is called a *face* of $\mathcal{M}(X)$ if there exists a non-zero element $r \in \overline{\mathrm{NE}}(X)$ such that $\gamma = r^\perp$. Similarly, a subset $\alpha \subset \overline{\mathrm{NE}}(X)$ is called a *face* of $\overline{\mathrm{NE}}(X)$ if there exists a non-zero element $h \in \mathcal{M}(X)$ such that $\alpha = h^\perp$.

A convex set is called *a closed polyhedron* (equivalently, *finite polyhedral*) if it is the convex hull of a finite set of points. A convex closed polyhedron is called *simplicial* if all its proper faces are simplexes. A convex closed polyhedron is called *simple* (equivalently, it has *simplicial angles*) if it is dual to a simplicial one. In other words, any face of codimension k is contained in exactly k faces of the highest dimension. Obviously, $\mathcal{M}(X)$ is simple if the Mori cone $\overline{\mathrm{NE}}(X)$ is finite polyhedral (has a finite set of extremal rays) and is simplicial (all proper faces of $\overline{\mathrm{NE}}(X)$ are cones over simplexes). Obviously, the last property is equivalent to the fact that any extremal set of extremal rays on X is linearly independent.

Let A, B be two vertices of an oriented graph G. The *distance* $\rho(A, B)$ in G is the length (the number of links) of a shortest oriented path of the graph G with the beginning in A and the end in B. The distance is $+\infty$ if this path does not exist. The *diameter* diam G of an oriented graph G is the maximum distance between ordered pairs of its vertices. By Lemma 1.1.1, the diameter of an E-set is a finite number if this set satisfies the condition (iii).

We have the following result.

Theorem 1.1.2. *Let X be a projective algebraic variety with $\mathbb{Q}$-factorial singularities and* dim $X \geq 2$. *Suppose that $\mathcal{M}(X)$ is closed and simple (equivalently, the Mori cone $\overline{\mathrm{NE}}(X)$ is a finite polyhedral simplicial cone).*

Assume that all extremal rays on X are divisorial (satisfy the condition (i) *above), each extremal subset of extremal rays satisfies the condition* (ii), *and each E-set of extremal rays satisfies the condition* (iii). *Assume that there are some constants d, C_1, C_2 such that the conditions* (a) *and* (b) *below hold:*

(a) *any E-set $\mathcal{L}$ of extremal rays on X satisfies*

$$\operatorname{diam} G(\mathcal{L}) \leq d,$$

(b) *any extremal set $\mathcal{E}$ of extremal rays on X satisfies*

$$\sharp\{(R_1, R_2) \in \mathcal{E} \times \mathcal{E} \mid 1 \leq \rho(R_1, R_2) \leq d\} \leq C_1 \sharp \mathcal{E},$$

and

$$\sharp\{(R_1, R_2) \in \mathcal{E} \times \mathcal{E} \mid d+1 \leq \rho(R_1, R_2) \leq 2d+1\} \leq C_2 \sharp \mathcal{E}.$$

Then $\dim N_1(X) = \dim \overline{\mathrm{NE}}(X) \leq (16/3)C_1 + 4C_2 + 6.$

Proof. This is a particular case of [N8, Theorem 1.2]. □

1.2. General results on divisorial extremal rays for 3-folds. Most of the results here were contained in [N8]. We restrict to normal projective 3-folds X with $\mathbb{Q}$-factorial singularities.

Let R be an extremal ray of the Mori polyhedron $\overline{\mathrm{NE}}(X)$ of X. A morphism $f: X \to Y$ on a normal projective variety Y is called the *contraction* of the ray R if for an irreducible curve C of X the image $f(C)$ is a point iff $C \in R$. The contraction f is defined by a linear system H on X (H gives the nef element of $N^1(X)$, which we denote by H also). It follows that an irreducible curve C is contracted iff $C \cdot H = 0$. We assume that the contraction f has properties: $f_*\mathcal{O}_X = \mathcal{O}_Y$ and the sequence

$$0 \to \mathbb{R}R \to N_1(X) \xrightarrow{f_*} N_1(Y) \to 0 \tag{1-2-1}$$

is exact. An extremal ray R is called *contractible* if it has a contraction f with these properties.

The number $\kappa(R) = \dim Y$ is called the *Kodaira dimension* of the contractible extremal ray R.

We recall (see above) that a subset γ of $\overline{\mathrm{NE}}(X)$ is called a *face* if there exists a non-zero nef element $r \in \overline{\mathrm{NE}}(X)$ such that $\gamma = r^{\perp}$. A face γ of $\overline{\mathrm{NE}}(X)$ is called *contractible* if there exists a morphism $f: X \to Y$ on a normal projective variety Y such that $f_*\gamma = 0$, $f_*\mathcal{O}_X = \mathcal{O}_Y$ and f contracts curves lying in γ only. The $\kappa(\gamma) = \dim Y$ is called the *Kodaira dimension* of γ.

Let H be a general nef element orthogonal to a face γ of the Mori polyhedron. The *numerical Kodaira dimension of* γ is defined by the formula

$$\kappa_{\text{num}}(\gamma) = \begin{cases} 3, & \text{if } H^3 > 0; \\ 2, & \text{if } H^3 = 0 \text{ and } H^2 \not\equiv 0; \\ 1, & \text{if } H^2 \equiv 0. \end{cases}$$

It is obvious that for a contractible face γ we have $\kappa_{\text{num}}(\gamma) \geq \kappa(\gamma)$. In particular, $\kappa_{\text{num}}(\gamma) = \kappa(\gamma)$ for a contractible face γ of Kodaira dimension $\kappa(\gamma) = 3$.

We will use the following statement which (in different variants) is standard:

Proposition 1.2.1. *Let X be a projective 3-fold with $\mathbb{Q}$-factorial singularities, $D_1, \ldots, D_m$ irreducible Weil divisors on X and $f: X \to Y$ a surjective morphism such that* $\dim Y = \dim X$ *and* $\dim f(D_i) < \dim D_i$. *Let $y \in f(D_1) \cap \cdots \cap f(D_m)$.*

Then there are real numbers $a_1 > 0, \ldots, a_m > 0$ and an open set U with $y \in U \subset f(D_1) \cup \cdots \cup f(D_m)$, such that

$$C \cdot (a_1 D_1 + \cdots + a_m D_m) < 0$$

if a curve $C \subset D_1 \cup \cdots \cup D_m$ belongs to a non-trivial algebraic family of curves on $D_1 \cup \cdots \cup D_m$ and $f(C) =$ point $\in U$.

Proof. See Proposition 2.2.6 in [N8] □

By this Proposition, we have

Lemma 1.2.2. *Let R be a contractible extremal ray of Kodaira dimension 3 and $f: X \to Y$ its contraction. Then there are three possibilities:*

(I) *All curves $C \in R$ fill an irreducible Weil divisor $D(R)$, the contraction f contracts $D(R)$ to a point and $R \cdot D(R) < 0$.*

(II) *All curves $C \in R$ fill an irreducible Weil divisor $D(R)$, the contraction f contracts $D(R)$ to an irreducible curve and $R \cdot D(R) < 0$.*

(III) *(small extremal ray) All curves $C \in R$ give a finite set of irreducible curves and the contraction f contracts these curves to points.*

Proof. Assume that some curves of R fill an irreducible divisor D. Then $R \cdot D < 0$, by Proposition 1.2.1. Suppose that $C \in R$ and D does not contain C. It follows that $R \cdot D \geq 0$. We get a contradiction. The lemma follows. □

According to the Lemma 1.2.2, we say that an extremal ray R has *type* (I), (II) *or* (III) *(small)* if it is contractible of Kodaira dimension 3 and the statements (I), (II) or (III) respectively hold. Extremal rays of type (I) and (II) are also called *divisorial.*

For a divisor D on X, let

$$\overline{\text{NE}}(X, D) = (\text{image } \overline{\text{NE}}(D)) \subset \overline{\text{NE}}(X).$$

Lemma 1.2.3. *Let R be a divisorial extremal ray of type* (I). *Then* $\overline{\mathrm{NE}}(X, D(R))$ *is a ray* $\overline{\mathrm{NE}}(X, D(R)) = R$.

Let R be a divisorial extremal ray of type (II), *and let f be its contraction. Then* $\overline{\mathrm{NE}}(X, D(R))$ *is an angle* $\overline{\mathrm{NE}}(X, D(R)) = R + \mathbb{R}^+ S$ *with edges R and S, where* $f_* \mathbb{R}^+ S = \mathbb{R}^+(f(D))$.

Proof. This follows at once from the exact sequence (1-2-1). □

Using Lemma 1.2.3, we get Lemmas 1.2.4 and 1.2.5 below.

Lemma 1.2.4. *Let* R_1 *and* R_2 *be two different extremal rays of type* (II) *having the same divisor* $D(R_1) = D(R_2)$.

Then for $D = D(R_1) = D(R_2)$ *we have:*

$$\overline{\mathrm{NE}}(X, D) = R_1 + R_2.$$

In particular, there do not exist three different extremal rays of type (II) *having the same divisor.*

If for two different extremal rays R_1, R_2 of type (II), $D(R_1) = D(R_2)$ (thus, we have the case of the Lemma 1.2.4 above), we say that the set $\{R_1, R_2\}$ of extremal rays has *type* $\mathfrak{B}_2$.

Lemma 1.2.5. *The divisors* $D(R_1)$ *and* $D(R_2)$ *of two different extremal rays of type* (I) *do not intersect one another.*

The divisors of an extremal ray of type (I) *and a pair of type* $\mathfrak{B}_2$ *do not intersect one another.*

The divisors of two different pairs of type $\mathfrak{B}_2$ *are disjoint.*

The next lemma was proved in [N8, Theorem 2.3.3], but we give the proof since this statement is very important.

Lemma 1.2.6. *Suppose that a pair* $\{R_1, R_2\}$ *has type* $\mathfrak{B}_2$. *Then*

$$\overline{\mathrm{NE}}(X, D(R_1)) = \overline{\mathrm{NE}}(X, D(R_2)) = R_1 + R_2$$

is a 2-dimensional face of the Mori polyhedron of numerical Kodaira dimension 3 *and such that* $(R_1 + R_2)^\perp$ *is a face of* NEF(X) *of codimension* 2.

Proof. Since the rays R_1, R_2 are extremal of Kodaira dimension 3, there are nef elements H_1, H_2 such that $H_1 \cdot R_1 = H_2 \cdot R_2 = 0$, ${H_1}^3 > 0$, ${H_2}^3 > 0$. Let $0 \neq C_1 \in R_1$ and $0 \neq C_2 \in R_2$. Let D be the divisor of the rays R_1 and R_2. Let us consider the map

$$(H_1, H_2) \to H \tag{1-2-2}$$
$$= (-D \cdot C_2)(H_2 \cdot C_1)H_1 + (-D \cdot C_1)(H_1 \cdot C_2)H_2 + (H_2 \cdot C_1)(H_1 \cdot C_2)D.$$

For fixed H_1, we get a linear map $H_2 \to H$ of the set of nef elements H_2 orthogonal to R_2 into the set of nef elements H orthogonal to R_1 and R_2. This map has a one dimensional kernel, generated by $(-D \cdot C_2)H_1 + (H_1 \cdot C_2)D$. It follows that $R_1 + R_2$ is a 2-dimensional face of $\overline{\mathrm{NE}}(X)$.

For a general nef element $H = a_1 H_1 + a_2 H_2 + bD$ orthogonal to this face, where $a_1, a_2, b > 0$, we have

$$\begin{aligned} H^3 &= (a_1 H_1 + a_2 H_2 + bD)^3 \\ &\geq (a_1 H_1 + a_2 H_2 + bD)^2 \cdot (a_1 H_1 + a_2 H_2) \\ &= (a_1 H_1 + a_2 H_2 + bD) \cdot (a_1 H_1 + a_2 H_2 + bD) \cdot (a_1 H_1 + a_2 H_2) \\ &\geq (a_1 H_1 + a_2 H_2)^2 \cdot (a_1 H_1 + a_2 H_2 + bD) \\ &\geq (a_1 H_1 + a_2 H_2)^3 > 0, \end{aligned}$$

since $a_1 H_1 + a_2 H_2 + bD$ and $a_1 H_1 + a_2 H_2$ are nef. It follows that the face $R_1 + R_2$ has numerical Kodaira dimension 3.

The last statement follows by construction. □

By Proposition 1.2.1, we have

Lemma 1.2.7. *Let* $\mathcal{E} = \{R_1, \ldots, R_n\}$ *be a set of divisorial extremal rays of type* (I) *or* (II) *on* X *and suppose that* $\mathcal{E}$ *is contained in a face of* $\overline{\mathrm{NE}}(X)$ *of Kodaira dimension* 3.

Then there are real numbers $a_1 > 0, \ldots, a_n > 0$ *such that*

$$R_i \cdot (a_1 D(R_1) + \cdots + a_n D(R_n)) < 0 \tag{1-2-3}$$

for all $R_i \in \mathcal{E}$.

It follows that $\mathcal{E}$ *is linearly independent if all the divisors* $D(R_1), \ldots, D(R_n)$ *are different.*

Proof. We only need to prove the last statement. Let us assume that we have a linear dependence condition $c_1 R_1 + \cdots + c_s R_s + c_{s+1} R_{s+1} + \cdots + c_n R_n = 0$. We can suppose that $c_1, \ldots, c_s$ are positive and $c_{s+1}, \ldots, c_n$ are non-positive. By (1-2-3) and our condition,

$$(c_1 R_1 + \cdots + c_s R_s) \cdot (a_1 D(R_1) + \cdots + a_s D(R_s)) < 0$$

and

$$(c_{s+1} R_{s+1} + \cdots + c_n R_n) \cdot (a_1 D(R_1) + \cdots + a_s D(R_s)) \leq 0.$$

We get a contradiction. □

We have the following converse statement based on standard arguments connected with the Perron–Frobenius Theorem.

Lemma 1.2.8. *Let $\mathcal{E} = \{R_1, \ldots, R_n\}$ be a set of divisorial extremal rays of type* (I) *or* (II) *on X, and suppose that all the divisors $D(R_1), \ldots, D(R_n)$ are different. Assume that there are positive $a_1, \ldots, a_n$ such that*

$$R_i \cdot (a_1 D(R_1) + \cdots + a_n D(R_n)) < 0$$

for all $R_i \in \mathcal{E}$.

Then for any $b_1 \geq 0, \ldots, b_n \geq 0$, not all 0, there exists $1 \leq j \leq n$ such that

$$R_j \cdot (b_1 D(R_1) + \cdots + b_n D(R_n)) < 0.$$

For each $1 \leq k \leq n$, there are non-negative $u_{1k}, \ldots, u_{nk}$ such that $R_k \cdot (u_{1k} D(R_1) + \cdots + u_{nk} D(R_n)) < 0$ and $R_j \cdot (u_{1k} D(R_1) + \cdots + u_{nk} D(R_n)) = 0$ if $j \neq k$ and $1 \leq j \leq n$. In particular, $\mathcal{E}$ is linearly independent.

Proof. We can find $\lambda > 0$ such that $\lambda a_i \geq b_i$ for all $1 \leq i \leq n$ and one of these inequalities is equality. Suppose that $\lambda a_j = b_j$ for $1 \leq j \leq n$. Then

$$\begin{aligned} R_j \cdot (b_1 D(R_1) + \cdots + b_n D(R_n)) &= R_j \cdot \lambda (a_1 D(R_1) + \cdots + a_n D(R_n)) \\ &\quad + R_j \cdot ((b_1 - \lambda a_1) D(R_1) + \cdots + (b_n - \lambda a_n) D(R_n)) < 0. \end{aligned}$$

Obviously, $R_i \cdot (a_1 D(R_1) + \cdots + a_{n-1} D(R_{n-1})) < 0$ for $1 \leq i \leq n-1$. Let us choose generators $r_i \in R_i$. Since $r_i \cdot D(R_n) \geq 0$ for $1 \leq i \leq n-1$, using induction, we can find $u_1 \geq 0, \ldots, u_{n-1} \geq 0$ such that $-r_i \cdot D(R_n) = r_i \cdot (u_1 D(R_1) + \cdots + u_{n-1} D(R_{n-1}))$ for all $1 \leq i \leq n-1$. Thus $u_1 D(R_1) + \cdots + u_{n-1} D(R_{n-1}) + D(R_n)$ is orthogonal to $R_1, \ldots, R_{n-1}$. By the statement proved above, then $R_n \cdot (u_1 D(R_1) + \cdots + u_{n-1} D(R_{n-1}) + D(R_n)) < 0$.

This finishes the proof. □

Using Lemma 1.2.8, we get

Lemma 1.2.9. *Let $\mathcal{E} = \{R_1, \ldots, R_n\}$ be a set of divisorial extremal rays of type* (I) *or* (II) *on X and suppose that all the divisors $D(R_1), \ldots, D(R_n)$ are different. Assume that there are positive $a_1, \ldots, a_n$ such that*

$$R_i \cdot (a_1 D(R_1) + \cdots + a_n D(R_n)) < 0$$

for all $R_i \in \mathcal{E}$. Let us additionally suppose that $C \cdot D(R_i) \geq 0$ for any curve $C \subset D(R_i)$, $C \notin R_i$, and all $1 \leq i \leq n$.

Then $R_1 + \cdots + R_n$ is a face of dimension n (thus, it is a cone over an $(n-1)$-dimensional simplex) and of numerical Kodaira dimension 3 of $\overline{\mathrm{NE}}(X)$ and such that $(R_1 + \cdots + R_n)^{\perp}$ is a face of the cone $\mathrm{NEF}(X)$ of codimension n.

Proof. Let H be a nef element on X. By Lemma 1.2.8, there are non-negative linear functions $b_1(H), \ldots, b_n(H)$ such that $H' = H + \sum_{i=1}^{i=n} b_i(H) D(R_i)$ is orthogonal to $R_1, \ldots, R_n$. By the additional condition, H' is nef. The map $H \to H'$ gives a linear map from the set of nef elements on X to the set of nef elements orthogonal to

$R_1, \dots, R_n$. This map has n-dimensional kernel generated by $D(R_1), \dots, D(R_n)$. It follows that the extremal rays $R_1, \dots, R_n$ belong to a face of $\overline{\text{NE}}(X)$ of dimension $\leq n$. By Lemma 1.2.8, $R_1, \dots, R_n$ are linearly independent. If follows that $R_1, \dots, R_n$ belong to a face γ of $\overline{\text{NE}}(X)$ of dimension n. By induction, we can suppose that any $(n-1)$-element subset of $\{R_1, \dots, R_n\}$ generates a face of $\overline{\text{NE}}(X)$ which is a cone over an $(n-1)$-dimensional simplex. It follows that $\gamma = R_1 + \cdots + R_n$ is a cone over an n-dimensional simplex.

As above, one can prove that $(H')^3 \geq H^3 > 0$ for an ample H. Thus the face $R_1 + \cdots + R_n$ has numerical Kodaira dimension 3. We get the last property by construction. □

Using the Perron–Frobenius Theorem, we get the following result.

Lemma 1.2.10. *Let $\{R_1, \dots, R_n\}$ be a set of divisorial extremal rays of type* (I) *or* (II) *having distinct divisors $D(R_1), \dots, D(R_n)$. Suppose that any proper subset of $\{R_1, \dots, R_n\}$ satisfies the conditions of Lemma* 1.2.8 *but the set $\{R_1, \dots, R_n\}$ itself does not.*

Then there exist positive numbers $a_1, \dots, a_n$ such that

$$R_i \cdot (a_1 D(R_1) + \cdots + a_n D(R_n)) \geq 0 \tag{1-2-4}$$

for all $1 \leq i \leq n$. Moreover, one of the following holds:

(a) $R_i \cdot (a_1 D(R_1) + \cdots + a_n D(R_n)) = 0$ *for all* $1 \leq i \leq n$. *Then the set of positive numbers $(a_1, \dots, a_n)$ with the property* (1-2-4) *is defined uniquely up to multiplication by $\lambda > 0$.*

(b) *There exists $1 \leq j \leq n$ such that the inequality* (1-2-4) *is strict for this j:*

$$R_j \cdot (a_1 D(R_1) + \cdots + a_n D(R_n)) > 0.$$

Proof. Suppose that after renumbering, for some $1 \leq m < n$, we have $R_i \cdot D(R_j) = 0$ for all $1 \leq i \leq m < j \leq n$. By our condition, we can find $a_1 > 0, \dots, a_m > 0$ and $a_{m+1} > 0, \dots, a_n > 0$ such that

$$R_i \cdot (a_1 D(R_1) + \cdots + a_m D(R_m)) < 0$$

for all $1 \leq i \leq m$, and

$$R_j \cdot (a_{m+1} D(R_{m+1}) + \cdots + a_n D(R_n)) < 0$$

for all $m+1 \leq j \leq n$. Obviously, for small $\varepsilon > 0$, we then have

$$R_k \cdot \varepsilon(a_1 D(R_1) + \cdots + a_m D(R_m)) + (a_{m+1} D(R_{m+1}) + \cdots + a_n D(R_n)) < 0$$

for all $1 \leq k \leq n$. We get a contradiction.

Thus the subdivision above is impossible. Then, by the Perron–Frobenius Theorem, there are positive $a_1, \dots, a_n$ with the property (1-2-4) above.

Assume that we have case (a). Suppose that there are positive $b_1, \dots, b_n$ such that $R_i \cdot (b_1 D(R_1) + \cdots + b_n D(R_n)) \geq 0$ for all $1 \leq i \leq n$. There exists $\lambda > 0$ such that $b_i - \lambda a_i \geq 0$ and one of these inequalities is equality. Then

$$R_i \cdot (b_1 D(R_1) + \cdots + b_n D(R_n)) = \lambda R_i \cdot (a_1 D(R_1) + \cdots + a_n D(R_n)) \\ + R_i \cdot ((b_1 - \lambda a_1) D(R_1) + \cdots + (b_n - \lambda a_n) D(R_n)) \geq 0.$$

for all $1 \leq i \leq n$. If at least one $b_i - \lambda a_i > 0$, we get a contradiction with Lemma 1.2.8. □

From Lemma 1.2.10, we get

Lemma 1.2.11. *Let* $\{R_1, \dots, R_n\}$ *be a set of divisorial extremal rays of type* (I) *or* (II) *having distinct divisors* $D(R_1), \dots, D(R_n)$. *Suppose that any proper subset of* $\{R_1, \dots, R_n\}$ *satisfies the conditions of Lemma* 1.2.8 *but the set* $\{R_1, \dots, R_n\}$ *itself does not. Moreover, suppose that* $C \cdot D(R_i) \geq 0$ *for any curve* $C \subset D(R_i)$, $C \notin R_i$, *and all* $1 \leq i \leq n$.

Then, in the notation of Lemma 1.2.10, *the element* $H = a_1 D(R_1) + \cdots + a_n D(R_n)$ *is nef. The set* $\{R_1, \dots, R_n\}$ *is not contained in a face of* $\overline{\mathrm{NE}}(X)$ *of Kodaira dimension* 3. *For case* (a) *of Lemma* 1.2.10, *the set* $\{R_1, \dots, R_n\}$ *is extremal.*

Proof. By the additional condition, H is nef. The set $\{R_1, \dots, R_n\}$ is not contained in a face of $\overline{\mathrm{NE}}(X)$ of Kodaira dimension 3 by Proposition 1.2.1 and Lemma 1.2.8.

If we have the case (a) of Lemma 1.2.10, then $\{R_1, \dots, R_n\}$ is contained in the face of $\overline{\mathrm{NE}}(X)$ orthogonal to H. This finishes the proof. □

As a particular case, we get

Lemma 1.2.12. *Assume that all finite polyhedral faces* γ *of* $\overline{\mathrm{NE}}(X)$ *with the property* $\operatorname{codim} \gamma^{\perp} = \dim \gamma$ *are contractible and have numerical Kodaira dimension equal to their Kodaira dimension (here* $\gamma^{\perp}$ *is the corresponding face of* NEF(X)). *Let* $\mathcal{L} = \{R_1, \dots, R_n\}$ *be an E-set of divisorial extremal rays of type* (I) *or* (II). *Moreover, suppose that* $C \cdot D(R_i) \geq 0$ *for any curve* $C \subset D(R_i)$, $C \notin R_i$, *and all* $1 \leq i \leq n$.

Then there are non-negative $a_1, \dots, a_n$, *not all* 0, *such that* $a_1 D(R_1) + \cdots + a_n D(R_n)$ *is nef.*

Proof. By Proposition 1.2.1 and Lemma 1.2.9, there exists a minimal subset $\mathcal{L}' \subset \mathcal{L}$ such that $\mathcal{L}'$ satisfies the conditions of Lemma 1.2.11. By Lemma 1.2.11, we get the statement. □

Using Theorem 1.1.2 and the above lemmas, we get the following result.

Theorem 1.2.13. *Let* X *be a projective* 3*-fold with* $\mathbb{Q}$*-factorial singularities. Suppose that the Mori cone* $\overline{\mathrm{NE}}(X)$ *is finite polyhedral and any face of* $\overline{\mathrm{NE}}(X)$ *has Kodaira* *dimension* 3. *Assume that all extremal rays on* X *are divisorial of type* (I) *or* (II) *and* X *does not have a pair of extremal rays of type* $\mathfrak{B}_2$.

Then extremal and E*-sets of extremal rays on* X *satisfy the conditions* (i), (ii) *and* (iii) *of Section* 1.1, *and any proper face of* $\overline{\mathrm{NE}}(X)$ *is a cone over a simplex.*

Assume that there are some constants d, C_1, C_2 *such that conditions* (a) *and* (b) *below hold:*

(a) *any* E*-set* $\mathcal{L}$ *of extremal rays on* X *satisfies*

$$\operatorname{diam} G(\mathcal{L}) \leq d;$$

(b) *any extremal set* $\mathcal{E}$ *of extremal rays on* X *satisfies*

$$\sharp\{(R_1, R_2) \in \mathcal{E} \times \mathcal{E} \mid 1 \leq \rho(R_1, R_2) \leq d\} \leq C_1 \sharp\mathcal{E}$$

and

$$\sharp\{(R_1, R_2) \in \mathcal{E} \times \mathcal{E} \mid d + 1 \leq \rho(R_1, R_2) \leq 2d + 1\} \leq C_2 \sharp\mathcal{E}$$

(here we use distance in the graph $G(\mathcal{E})$*).*

Then we have the inequality:

$$\dim N_1(X) = \dim \overline{\mathrm{NE}}(X) \leq (16/3)C_1 + 4C_2 + 6.$$

1.3. Basic Theorem

Definition 1.3.1. We say that an algebraic 3-fold X belongs to the class $\mathcal{LT}$ if the following conditions hold:

(1) X has $\mathbb{Q}$-factorial singularities;

(2) every face γ of $\overline{\mathrm{NE}}(X)$ generated by a finite set of divisorial extremal rays of type (I) or (II) and with the property $\dim \gamma = \operatorname{codim} \gamma^{\perp}$ for the face $\gamma^{\perp}$ of $\mathrm{NEF}(X)$, and of numerical Kodaira dimension 3 is contractible and has Kodaira dimension 3;

(3) any 3-fold obtained from X by successive contractions of extremal rays of type (I) and (II) satisfies the above two conditions.

We say that a 3-fold X of class $\mathcal{LT}$ has constants (bounded by) $q(X), d(X), C_1(X), C_2(X)$ if the following hold:

(1)
$$\sharp\mathcal{E} \leq q(X)$$

for any extremal set $\mathcal{E}$ of Kodaira dimension 3 (i.e. $\mathcal{E}$ is contained in a face of $\overline{\mathrm{NE}}(X)$ of Kodaira dimension 3) of extremal rays of type (II) such that the graph $G(\mathcal{E})$ is full (i.e. any two rays $R_1, R_2 \in \mathcal{E}$ are joined by both arrows R_1R_2 and R_2R_1);

(2) the diameter

$$\operatorname{diam} G(\mathcal{L}) \leq d(X)$$

for any E-set $\mathcal{L}$ of extremal rays of type (I) or (II) such that any proper subset of $\mathcal{L}$ is extremal of Kodaira dimension 3 and $\mathcal{L}$ has the property (iii) of Section 1.1;

(3) $$\sharp\{(R_1, R_2) \in \mathcal{E} \times \mathcal{E} \mid 1 \le \rho(R_1, R_2) \le d(X)\} \le C_1(X)\sharp\mathcal{E};$$

and

$$\sharp\{(R_1, R_2) \in \mathcal{E} \times \mathcal{E} \mid d(X) + 1 \le \rho(R_1, R_2) \le 2d(X) + 1\} \le C_2(X)\sharp\mathcal{E}$$

for any extremal set $\mathcal{E}$ of Kodaira dimension 3 of extremal rays of type (I) or (II) having distinct divisors.

We want to prove the following basic result:

Basic Theorem 1.3.2. *Let X be a 3-fold of class $\mathcal{LT}$. We assume that*

(a) *the Mori cone $\overline{\mathrm{NE}}(X)$ is finite polyhedral;*

(b) *X does not have a small extremal ray;*

(c) *$\overline{\mathrm{NE}}(X)$ does not have a face of numerical Kodaira dimension 1 or 2 (equivalently, there are no nef elements h with $h^3 = 0$).*

(From these conditions, it follows that all extremal rays on X are divisorial of type (I) or (II).)

Then we have the following statements about X with the constants $q(X)$, $d(X)$, $C_1(X)$, $C_2(X)$ above:

(1) *X does not have a pair of extremal rays of type $\mathfrak{B}_2$ and the Mori cone $\overline{\mathrm{NE}}(X)$ is simplicial.*

(2) *The number of extremal rays of type (I) on X is $\le q(X)$.*

(3) *$\rho(X) = \dim N_1(X) \le (16/3)C_1(X) + 4C_2(X) + 6.$*

Proof. Let us prove (1). We need the following analog of [N8, Lemma 2.5.7].

Lemma 1.3.3. *Let X be a 3-fold of class $\mathcal{LT}$, and suppose that the Mori cone $\overline{\mathrm{NE}}(X)$ is finite polyhedral. Let $\mathcal{E}$ be the set of all extremal rays of a proper face $[\mathcal{E}]$ of $\overline{\mathrm{NE}}(X)$. Let*

$$\{R_{11}, R_{12}\} \cup \cdots \cup \{R_{t1}, R_{t2}\}$$

be the set of different pairs of extremal rays of type $\mathfrak{B}_2$. Assume that $R \cdot D(R_{i1}) = 0$ for all $R \in \mathcal{E}$ and all i, $1 \le i \le t$.

Then there are extremal rays $Q_1, \dots, Q_r$ such that the following statements hold:

(a) *$r \le t$.*

(b) *For any i, $1 \le i \le r$, there exists j, $1 \le j \le t$, such that $Q_i \cdot D(R_{j1}) > 0$ (in particular, Q_i is different from the extremal rays of pairs of extremal rays $\{R_{u1}, R_{u2}\}$ of type $\mathfrak{B}_2$).*

(c) *For any j, $1 \le j \le t$, there exists an extremal ray Q_i, $1 \le i \le r$, such that*

$$Q_i \cdot D(R_{j1}) > 0.$$

(d) *The set $\mathcal{E} \cup \{Q_1, \ldots, Q_r\}$ is extremal, and the extremal rays $\{Q_1, \ldots, Q_r\}$ are linearly independent.*

Proof. If $t = 0$, we can take $r = 0$. Thus we assume that $t \ge 1$.

Since $R_{ij} \cdot D(R_{ij}) < 0$, $1 \le i \le t$, $1 \le j \le 2$, the set $\mathcal{E}$ does not contain the rays R_{ij}. Let H be a general nef element orthogonal to $[\mathcal{E}]$. Since $t \ge 1$, there exists $a > 0$ such that $H' = H + aD(R_{11})$ is nef and H' is orthogonal to $\mathcal{E}$ and one of the rays R_{11}, R_{12}. Let this ray be R_{11}. Then the set $\mathcal{E} \cup \{R_{11}\}$ is extremal and is contained in a (proper) face of $\overline{\text{NE}}(X)$. It follows that $\dim[\mathcal{E}] < \dim[\mathcal{E} \cup \{R_{11}\}] < \dim \overline{\text{NE}}(X)$, and $\dim[\mathcal{E}] < \dim \overline{\text{NE}}(X) - 1$. Consider the vector subspace $V(\mathcal{E}) \subset N_1(X)$ spanned by the set of extremal rays $\mathcal{E}$, that is by the face $[\mathcal{E}]$ of $\overline{\text{NE}}(X)$. Consider the quotient map $\pi: N_1(X) \to N_1(X)/V(\mathcal{E})$. Since the cone $\overline{\text{NE}}(X)$ is polyhedral, the cone $\pi(\overline{\text{NE}}(X))$ is generated by images of extremal rays T such that the set $\mathcal{E} \cup \{T\}$ is contained in a face $[\mathcal{E} \cup \{T\}]$ of $\overline{\text{NE}}(X)$ of dimension $\dim[\mathcal{E}] + 1$. In particular, since $\dim[\mathcal{E}] < \dim N_1(X) - 1$, the face $[\mathcal{E} \cup \{T\}]$ is proper, and the set $\mathcal{E} \cup \{T\}$ is extremal.

There exists a curve C on X such that $C \cdot D(R_{11}) > 0$. This curve C (as any element $x \in \overline{\text{NE}}(X)$) is a linear combination of extremal rays T with non-negative coefficients and extremal rays from $\mathcal{E}$ with real coefficients. We have $R \cdot D(R_{11}) = 0$ for any extremal ray $R \in \mathcal{E}$. Thus there exists an extremal ray T as above such that $T \cdot D(R_{11}) > 0$. It follows that T is different from the extremal rays of pairs of type $\mathfrak{B}_2$. We take $Q_1 = T$. By our construction, the set $\mathcal{E} \cup \{Q_1\}$ is extremal. If $Q_1 \cdot D(R_{j1}) > 0$ for any j such that $1 \le j \le t$, then $r = 1$, and the set $\{Q_1\}$ gives the set we are looking for. Otherwise, there exists a minimal j such that $2 \le j \le t$ and $Q_1 \cdot D(R_{j1}) = 0$. Then we replace $\mathcal{E}$ by the set $\mathcal{E}_1$ of all extremal rays in the face $[\mathcal{E} \cup \{Q_1\}]$ of dimension $\dim[\mathcal{E}_1] = \dim[\mathcal{E}] + 1$, and the set

$$\{R_{11}, R_{12}\} \cup \cdots \cup \{R_{t1}, R_{t2}\}$$

by

$$\{R_{j1}, R_{j2} \mid 1 \le j \le t,\ Q_1 \cdot D(R_{j1}) = 0\},$$

and repeat this procedure. □

We also need the following crucial lemma:

Lemma 1.3.4. *Let X be a 3-fold of class $\mathcal{LT}$ and suppose that the Mori cone $\overline{\text{NE}}(X)$ is finite polyhedral. Assume that all extremal rays on X have type* (I) *or* (II). *Let R_{11}, R_{12} be a pair of extremal rays of type $\mathfrak{B}_2$ on X, and assume that $\rho(X) \ge 3$.*

Then there exists an extremal ray R on X such that R does not belong to a pair of extremal rays of type $\mathfrak{B}_2$ and the sets $\{R_{11}, R\}$ and $\{R_{12}, R\}$ generate 2-dimensional faces of $\overline{\text{NE}}(X)$.

Proof. Let R_{21}, R_{22} be another pair of extremal rays of type $\mathfrak{B}_2$. By Lemmas 1.2.5 and 1.2.6, the extremal rays R_{11}, R_{12} generate a 2-dimensional face of $\overline{\text{NE}}(X)$ and the divisors $D(R_{11})$ and $D(R_{21})$ are disjoint (have empty intersection). Applying Lemma 1.3.3 to $\mathcal{E} = \{R_{11}, R_{12}\}$ and the set $\{\{R_{21}, R_{22}\}\}$ of pairs of type $\mathfrak{B}_2$, we find an extremal ray Q such that $Q \cdot D(R_{21}) > 0$. It follows that $D(Q) \cap D(R_{21})$ is a non-empty curve C. By Lemma 1.2.5, Q has type (II). By Lemma 1.2.6, $\overline{\text{NE}}(X, D(R_{21})) = R_{21} + R_{22}$ is a 2-dimensional face of $\overline{\text{NE}}(X)$. By Lemma 1.2.3, the 2-dimensional angle $\overline{\text{NE}}(X, D(Q)) = Q + \mathbb{R}^+C$. Since $\overline{\text{NE}}(X, D(R_{11})) = R_{11} + R_{12}$ is another 2-dimensional face of $\overline{\text{NE}}(X)$ which does not have a common ray with $R_{21} + R_{22}$, the angle $\overline{\text{NE}}(X, D(Q))$ does not have a common ray with the angle $\overline{\text{NE}}(X, D(R_{11}))$. Thus $D(Q) \cap D(R_{11}) = \emptyset$. By Lemma 1.2.5, Q does not belong to a pair of type $\mathfrak{B}_2$.

Let H be a general nef element orthogonal to the 2-dimensional face $R_{11} + R_{12}$. Then there exists $\alpha > 0$ such that the nef element $H' = H + \alpha D(Q)$ is orthogonal to the set of extremal rays R_{11}, R_{12}, Q. If an extremal ray R is different from these three extremal rays, clearly, $H' \cdot R > 0$ since all extremal rays on X are divisorial and Q does not belong to a pair of type $\mathfrak{B}_2$. It follows that $R_{11} + R_{12} + Q$ is a simplicial 3-dimensional face of $\overline{\text{NE}}(X)$. It follows that Q is the extremal ray we are looking for.

Thus we can suppose that the pair R_{11}, R_{12} is the only pair of extremal rays of type $\mathfrak{B}_2$ on X.

If there exists an extremal ray Q such that $D(Q) \cap D(R_{11}) = \emptyset$, as above, Q is the required extremal ray. Thus we can suppose that any extremal ray Q different from R_{11}, R_{12} has the property $D(Q) \cap D(R_{11}) \neq \emptyset$. In particular, $D(R_{11}) \cap D(Q)$ is a non-empty curve.

Since $\overline{\text{NE}}(X)$ is finite polyhedral, there exists an extremal ray $Q_1 \notin \{R_{11}, R_{12}\}$ such that $R_{11} + Q_1$ is a 2-dimensional face of $\overline{\text{NE}}(X)$. If $Q_1 + R_{12}$ is a 2-dimensional face of $\overline{\text{NE}}(X)$, then Q_1 is the required extremal ray. Thus we can suppose that $Q_1 + R_{12}$ is not a 2-dimensional face of $\overline{\text{NE}}(X)$. Similarly, we can find an extremal ray $Q_2 \notin \{R_{11}, R_{12}\}$ such that $R_{12}+Q_2$ is a 2-dimensional face of $\overline{\text{NE}}(X)$ but $R_{11}+Q_2$ is not. Then $Q_1 \neq Q_2$. Recall that we also suppose that $C_1 = D(R_{11}) \cap D(Q_1)$ and $C_2 = D(R_{11}) \cap D(Q_2)$ are non-empty curves.

We will normalize the generator $C \in T$ of a divisorial extremal ray T by the condition $C \cdot D(T) = -2$.

Let r_{ij} be the generator of R_{ij} and q_i of Q_i. Let $t = q_1 \cdot D(R_{11})$ and $t_1 = r_{11} \cdot D(Q_1)$ and $t_2 = r_{12} \cdot D(Q_1)$.

Since $R_{11} + Q_1$ is a 2-dimensional face of $\overline{\text{NE}}(X)$ and all faces of $\overline{\text{NE}}(X)$ have Kodaira dimension 3, by Proposition 1.2.1, there are positive a_1, a_2 such that $r_{11} \cdot (a_1 D(R_{11}) + a_2 D(Q_1)) = -2a_1 + t_1 a_2 < 0$ and $q_1 \cdot (a_1 D(R_{11}) + a_2 D(Q_1)) = ta_1 - 2a_2 < 0$. Thus $tt_1 < 4$. We claim that $t_1 < t_2$. Let us assume that $t_2 \leq t_1$. By the above inequality, $tt_2 < 4$. Let H be a general nef element orthogonal to R_{12}. Then

$$H' = H + \frac{(H \cdot q_1)}{2 - (tt_2)/2}((t_2/2)D(R_{11}) + D(Q_1))$$

is a nef element which is orthogonal to the extremal rays R_{12}, Q_1 only. We use here the inequality $t_2 \le t_1$ to check that $R_{11} \cdot H' > 0$. Thus $R_{12} + Q_1$ is a 2-dimensional face of $\overline{\mathrm{NE}}(X)$. We get a contradiction. Thus we have proved the claim: $t_1 < t_2$.

Let us consider the curve $C_1 = D(R_{11}) \cap D(Q_1)$. Then $C_1 = u_1 r_{11} + u_2 r_{12}$. Let $r_{11} \cdot r_{12} = m$ (equivalently, m is the degree of the maps $f_{11} \mid r_{12}: r_{12} \to f_{11}(D(R_{11}))$ and $f_{12} \mid r_{11}: r_{11} \to f_{12}(D(R_{12}))$, where f_{11} and f_{12} are the contractions of R_{11} and R_{12} respectively). Then

$$(t_2 : t_1) = (r_{12} \cdot D(Q_1) : r_{11} \cdot D(Q_1)) = (r_{12} \cdot C_1 : r_{11} \cdot C_1) = (u_1 m : u_2 m).$$

Thus

$$C_1 = u_1 r_{11} + u_2 r_{12}, \ (u_1 : u_2) = (t_2 : t_1), \ t_2 > t_1. \tag{1-2-5}$$

Similarly, for Q_2 we introduce $s_1 = r_{11} \cdot D(Q_2)$, $s_2 = r_{12} \cdot D(Q_2)$ where $s_2 < s_1$. And we have for $C_2 = D(R_{11}) \cap D(Q_2)$:

$$C_2 = v_1 r_{11} + v_2 r_{12}, \ (v_1 : v_2) = (s_2 : s_1), \ s_2 < s_1. \tag{1-2-6}$$

By (1-2-5) and (1-2-6), $C_1 \cdot D(Q_2) = u_1 s_1 + u_2 s_2 \ge u_1 s_1 > 0$, Thus the divisors $D(Q_1)$ and $D(Q_2)$ have a common curve. It follows that the 2-dimensional angles $\overline{\mathrm{NE}}(X, D(Q_1)) = \mathbb{R}^+ C_1 + Q_1$ and $\overline{\mathrm{NE}}(X, D(Q_2)) = \mathbb{R}^+ C_2 + Q_2$ should have a common ray. By (1-2-5) and (1-2-6), we then get that the rays R_{11}, R_{12}, Q_1, Q_2 generate a 3-dimensional space V, and are extremal rays of the 3-dimensional polyhedral convex cone $\overline{\mathrm{NE}}(X) \cap V$ with 2-dimensional faces $R_{11} + R_{12}$, $R_{11} + Q_1$, $R_{12} + Q_2$. Moreover, the angles $\mathbb{R}^+ C_1 + Q_1$ and $\mathbb{R}^+ C_2 + Q_2$ have a common ray. One should draw the picture to see a contradiction with (1-2-5) and (1-2-6). This finishes the proof of the lemma. □

Now we can prove the statement (1). Thus suppose that X satisfies the conditions of the theorem and has a pair $\{R_{11}, R_{12}\}$ of type $\mathfrak{B}_2$. By Lemma 1.3.4, we can find an extremal ray Q which does not belong to a pair of extremal rays of type $\mathfrak{B}_2$ and $Q + R_{11}$, $Q + R_{12}$ are 2-dimensional faces of $\overline{\mathrm{NE}}(X)$. Let us contract the extremal ray Q. By our conditions, we get a 3-fold X' of class $\mathcal{LT}$ which satisfies conditions (a), (b) and (c) of the theorem, and the image of the pair R_{11}, R_{12} of type $\mathfrak{B}_2$ will be a pair of type $\mathfrak{B}_2$ again. Thus using Lemma 1.3.4 $\rho(X) - 2$ times, we get a 3-fold Y which satisfies the condition of the theorem, has $\rho(Y) = 2$, and has a pair of extremal rays $\{R_{11}, R_{12}\}$ of type $\mathfrak{B}_2$. Then obviously $\overline{\mathrm{NE}}(Y) = R_{11} + R_{12}$. We have $R_{11} \cdot D(R_{11}) < 0$ and $R_{12} \cdot D(R_{12}) < 0$. Thus any curve of Y has negative intersection with the effective divisor $D(R_{11}) = D(R_{12})$. We get a contradiction.

This proves the statement (1).

Now let us prove (2): X does not have more than $q(X)$ extremal rays of type (I).

By Proposition 1.2.1 and Lemma 1.2.5, divisors of different extremal rays of type (I) do not have a common point and their number is finite. By Lemma 1.2.9, the set of extremal rays of type (I) generates a simplicial face of $\overline{\mathrm{NE}}(X)$ of Kodaira dimension

3. Let

$$\{R_1, \ldots, R_s\}$$

be the whole set of extremal rays of type (I) on X. We should prove that $s \leq q(X)$.

We say that two divisorial extremal rays are joined (more formally: divisorially joined) if their divisors have a common point. This defines the connected components of a set of divisorial extremal rays.

Let $\mathcal{E}$ be a maximal extremal set of extremal rays on X containing $\{R_1, \ldots, R_s\}$ and such that each connected component of $\mathcal{E}$ contains at least one of the extremal rays $R_1, \ldots, R_s$. Let T be a connected component of $\mathcal{E}$. Assume that T has two different extremal rays $R_i, R_j \in \{R_1, \ldots, R_s\}$. After contracting all extremal rays of T different from R_i, R_j, we get a 3-fold Y of class $\mathcal{L}T$, and the images of extremal rays R_i, R_j give different extremal rays of type (I) on Y such that their divisors are not disjoint. We get the contradiction with Lemma 1.2.5.

Thus $\mathcal{E}$ has exactly s connected components $T_1, \ldots, T_s$ such that T_i contains the extremal ray R_i. Obviously, the maximal $\mathcal{E}$ exists.

By Proposition 1.2.1, for $1 \leq i \leq s$, there exists an effective divisor $D(T_i)$ which is a linear combination of divisors of rays from T_i with positive coefficients and $R \cdot D(T_i) < 0$ for any $R \in T_i$. Obviously, the contraction of T_i contracts all T_i to a point. Thus any curve of divisors of T_i belongs to the sum of extremal rays of T_i with positive coefficients. Thus for this curve C we also have $C \cdot D(T_i) < 0$. Using the divisors $D(T_i)$, similarly to Lemma 1.3.4, we can find extremal rays

$$\{Q_1, \ldots, Q_r\}$$

with the properties:

(a) $r \leq s$.

(b) For any i, $1 \leq i \leq r$, there exists j, $1 \leq j \leq t$, such that $Q_i \cdot D(T_j) > 0$ (in particular, Q_i is different from the extremal rays of $\mathcal{E}$ and does not have type (I)).

(c) For any j, $1 \leq j \leq s$, there exists an extremal ray Q_i, $1 \leq i \leq r$, such that

$$Q_i \cdot D(T_j) > 0.$$

(d) The set $\{Q_1, \ldots, Q_r\}$ of extremal rays is extremal.

By our conditions, all extremal rays on X are divisorial. Thus by (b), the extremal rays $Q_1, \ldots, Q_r$ have type (II).

Let us take the ray Q_i, and let $Q_i \cdot D(T_j) > 0$. By Lemma 1.2.9, the set T_j generates a simplicial face (i.e. a cone over a simplex) γ_j of $\overline{\mathrm{NE}}(X)$. We have mentioned above that each curve of divisors of rays from T_j belongs to this face. It follows that $\overline{\mathrm{NE}}(X, D(Q_i))$ is a 2-dimensional angle bounded by the ray Q_i and a ray from the face γ_j since the divisor $D(Q_i)$ obviously has a common curve with one of the divisors $D(R)$, $R \in T_j$. Since any two sets from $T_1, \ldots, T_s$ do not have a common extremal ray, the faces $\gamma_1, \ldots, \gamma_s$ do not have a common ray (not necessarily extremal). It follows that the angle $\overline{\mathrm{NE}}(X, D(Q_i))$ does not have a common ray with the face γ_k for

$k \neq j$. Thus the divisor $D(Q_i)$ does not have a common point with divisors of rays T_k. It follows that $r = s$ and we can renumber $Q_1, \ldots, Q_s$ such that $Q_i \cdot D(T_i) > 0$, but $D(Q_i)$ does not have a common point with the divisors of the extremal rays T_j if $j \neq i$.

Let us fix i, $1 \leq i \leq s$. By our construction, the set $\mathcal{E} \cup \{Q_i\}$ has connected components

$$T_1, \ldots, T_{i-1}, \; T_i \cup \{Q_i\}, \; T_{i+1}, \ldots, T_s.$$

By definition of $\mathcal{E}$, then $\mathcal{E} \cup \{Q_i\}$ is not extremal. Thus it contains an E-set (minimal non-extremal) $\mathcal{L}_i$ which contains Q_i. By Lemmas 1.2.12 and 1.1.1, $\mathcal{L}_i$ is connected. Thus $\{Q_i\} \subset \mathcal{L}_i \subset T_i \cup \{Q_i\}$. Let us consider the sets $\mathcal{L}_1, \ldots, \mathcal{L}_s$. By Lemma 1.1.1, $\mathcal{L}_i$, $\mathcal{L}_j$ are joined by arrows in both directions. By our construction, it follows that Q_i, Q_j are joined by arrows $Q_i Q_j$ and $Q_j Q_i$ for any $1 \leq i < j \leq s$. Thus $s \leq q(X)$ because the set $\{Q_1, \ldots, Q_s\}$ is extremal.

This finishes the proof of the statement (2).

From (1) and Theorem 1.2.13, the statement (3) follows. This finishes the proof of Basic Theorem 1.3.2. □

In fact, Basic Theorem 1.3.2 is a generalization of the analogous theorem on Fano 3-folds with terminal $\mathbb{Q}$-factorial singularities proved in [N8]. We will apply Basic Theorem 1.3.2 to Calabi–Yau 3-folds in Section 3.

To apply Theorem 1.3.2, one should work with extremal and E-sets of extremal rays of type (I) or (II). For Fano 3-folds this is done in [N8]. The following Section 2 is, in fact, devoted to generalization of this part of [N8] for 3-folds of class $\mathcal{LT}$ (see Theorem 2.2.6). Roughly speaking, we should reduce the problem to sets of divisorial extremal rays of type (II) which have non-single arrows only. The last case is very similar to surfaces.

2. Sets of divisorial extremal rays

2.1. Special divisorial extremal rays. Here we continue the geometrical study of configurations of divisorial extremal rays of types (I) or (II) for 3-folds started in Section 1.2.

Let T be a set of divisorial extremal rays of types (I) or (II). Recall (see Section 1.1) that we assign to T a graph $G(T)$ as follows: we draw an arrow $R_1 R_2$ from R_1 to R_2 if $R_1 \cdot D(R_2) > 0$. We draw an extremal ray of type (I) (respectively (II)) as a black (respectively white) vertex. A pair of type $\mathfrak{B}_2$ we draw as a pair of white vertices connected by a dotted line. For this pair $\{R_1, R_2\}$ we have $D(R_1) = D(R_2)$ and $R_1 \cdot D(R_1) < 0$, $R_2 \cdot D(R_1) < 0$. Thus two divisorial extremal rays R_1, R_2 give one of the pictures of Figure 1. One should consider a single arrow $R_1 R_2$ (thus, $R_1 \cdot D(R_2) > 0$, but $R_2 \cdot D(R_1) = 0$) as a "weak orthogonality" of extremal rays.

They are completely orthogonal (are not joined by arrows) if their divisors do not have a common point: this follows from Lemma 1.2.2.

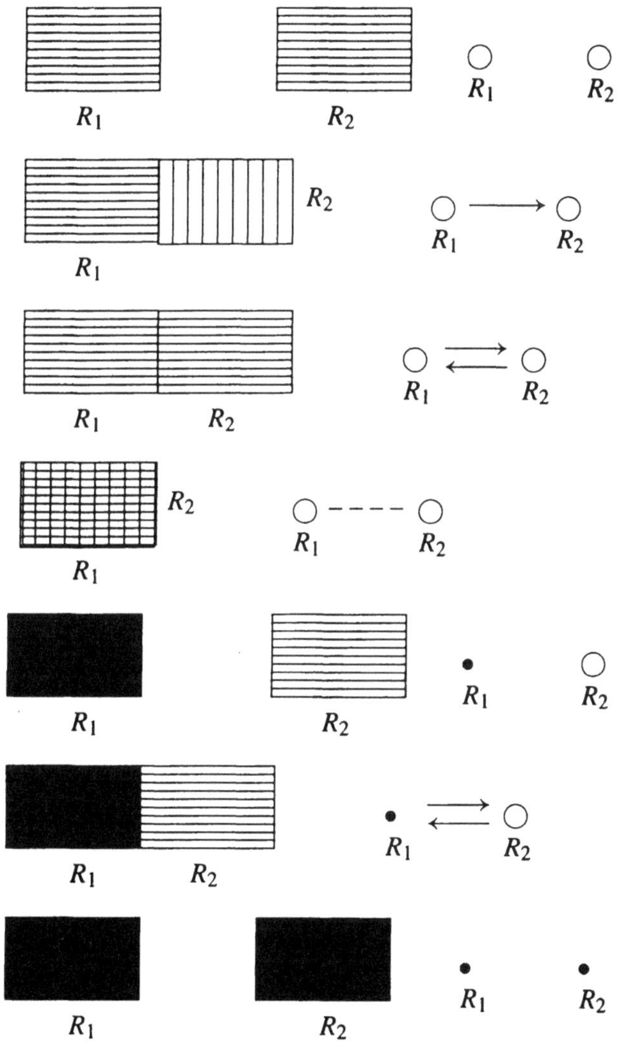

Figure 1

Definition 2.1.1. We say that extremal rays $Q_1, Q_2, \dots, Q_m$, $m \geq 2$, define a configuration of type $\mathfrak{C}_m$ if all these extremal rays have type (II), the extremal rays $Q_2, \dots, Q_m$ are divisorially disjoint (that is, the intersection of any two divisors $D(Q_2)$, $D(Q_3), \dots, D(Q_m)$ is empty) and Q_iQ_1 is a single arrow (thus, $Q_i \cdot D(Q_1) > 0$, $Q_1 \cdot D(Q_i) = 0$) for all $i = 2, \dots, m$.

The following lemma was proved in [N8]. Since it is important, we give the proof here.

Lemma 2.1.2. *Assume that the divisorial extremal rays $Q_1, Q_2, \dots, Q_m$ define a configuration of type $\mathfrak{C}_m$, $m \geq 2$. Then Q_i does not belong to a pair of type $\mathfrak{B}_2$ and $\overline{\mathrm{NE}}(X, D(Q_i)) = Q_1 + Q_i$ is a 2-dimensional face of numerical Kodaira dimension 3 of the Mori polyhedron for any i, $2 \leq i \leq m$.*

Moreover, $Q_1 + Q_2 + \dots + Q_m$ is an m-dimensional face of $\overline{\mathrm{NE}}(X)$ of numerical Kodaira dimension 3 and such that the face $(Q_1 + Q_2 + \dots + Q_m)^{\perp}$ of $\mathrm{NEF}(X)$ has codimension m.

Proof. Let $2 \leq i \leq m$. Clearly, since $Q_1 \cdot D(Q_i) = 0$, the curve $D(Q_i) \cap D(Q_1)$ belongs to Q_1. By Lemma 1.2.3, it follows that $\overline{\mathrm{NE}}(X, D(Q_i)) = Q_1 + Q_i$. If Q_i belongs to a pair of type $\mathfrak{B}_2$, the angle $Q_1 + Q_i$ contains another extremal ray (different from Q_1 and Q_i), which is impossible.

Let H be a nef element orthogonal to the ray Q_1. Let $0 \neq C_i \in Q_i$. Let us consider the map

$$H \to H' = H + \sum_{i=2}^{m} \frac{-(H \cdot C_i)}{(C_i \cdot D(Q_i))} D(Q_i). \tag{2-1-1}$$

It is a linear map of the set of nef elements H orthogonal to Q_1 into the set of nef elements H' orthogonal to the rays $Q_1, Q_2, \dots, Q_m$. Here the element H' is nef because $C \in Q_1 + Q_i$ for any curve $C \subset D(Q_i)$ if $2 \leq i \leq m$. The kernel of the map (2-1-1) has dimension $m - 1$. It follows that the rays $Q_1, Q_2, \dots, Q_m$ belong to a face of $\overline{\mathrm{NE}}(X)$ of dimension $\leq m$. On the other hand, multiplying rays $Q_1, \dots, Q_m$ against the divisors $D(Q_1), \dots, D(Q_m)$, one sees easily that the rays $Q_1, \dots, Q_m$ are linearly independent. Thus they generate an m-dimensional face of $\overline{\mathrm{NE}}(X)$. Let us show that this face is $Q_1 + Q_2 + \dots + Q_m$. To prove this, we show that every $(m-1)$-element subset of $\mathcal{E}$ is contained in a face of $\overline{\mathrm{NE}}(X)$ of dimension $\leq m - 1$.

If this subset contains the ray Q_1, this subset has type $\mathfrak{C}_{m-1}$. By induction, we can suppose that this subset belongs to a face of $\overline{\mathrm{NE}}(X)$ of dimension $m - 1$. Let us consider the subset $\{Q_2, Q_3, \dots, Q_m\}$. Let H be an ample element of X. For the element H, the map (2-1-1) gives an element H' which is orthogonal to the rays $Q_2, \dots, Q_m$, but is not orthogonal to the ray Q_1. It follows that the set $\{Q_2, \dots, Q_m\}$ belongs to a face of the Mori polyhedron of dimension $< m$. As above (see the proof of Lemma 1.2.6), one can see that for a general H orthogonal to Q_1 the element H' has $(H')^3 \geq H^3 > 0$.

The last property follows by construction. □

We use the next statement quite often:

Lemma 2.1.3. *Let X be a 3-fold of class $\mathcal{LT}$ (see Definition 1.3.1). Let R be a divisorial extremal ray of type (I) or (II), and assume that D_1, D_2 are two different irreducible Weil divisors which are different from $D(R)$.*

Then the curve $D_1 \cap D_2$ does not belong to R.

Proof. Let $f: X \to X'$ be the contraction of the ray R. If $D_1 \cap D_2$ belongs to R, then the intersection of Weil divisors $f(D_1)$ and $f(D_2)$ is zero-dimensional. This is impossible if X' is $\mathbb{Q}$-factorial. □

We denote by $\mathbb{P}C$ the projectivization of a cone C with the origin at zero. Let R be an extremal ray of type (II). By Lemma 1.2.3, the projectivization $\mathbb{P}\overline{\mathrm{NE}}(X, D(R))$ is an interval with one of its endpoints $\mathbb{P}R$.

Considering lines generated by these intervals we use the following well-known and elementary result.

Proposition 2.1.4. *Let S be a set of lines any two of which intersect.*

Then there are two cases:

(a) *There exists a 2-dimensional plane Π such that each line of S belongs to Π.*

(b) *There exists a point P which belongs to each line of S.*

The Π and P are unique if S has at least two different lines.

Applying this statement, we get

Lemma 2.1.5. *Let $\mathcal{S}$ be a set of divisorial extremal rays of type (I) or (II) such that the divisors $D(R_i)$, $D(R_j)$ have a common point for any two extremal rays in $\mathcal{S}$ (equivalently, R_i and R_j are connected by non-single, single arrow or dotted line). Suppose that $\mathcal{S}$ contains at least 3 elements. Then $\mathcal{S}$ has one of the following types:*

(a) $\dim[\mathcal{S}] = 3$;

(b) *All the angles $\overline{\mathrm{NE}}(X, D(R_i))$ for $R_i \in \mathcal{S}$ intersect in a common ray Q (not necessarily extremal).*

Let P be a set of divisorial extremal rays. We say that P is *divisorially connected* if there does not exist a decomposition $P = P_1 \cup P_2$ such that both P_1 and P_2 are non-empty and for any $R \in P_1$ and any $Q \in P_2$ the divisors $D(R)$ and $D(Q)$ do not have a common point. It defines *divisorially connected components* of a set of extremal rays. Also, we can say what it means that two sets P_1 and P_2 of divisorial extremal rays are *divisorially joined*: this means that there exist extremal rays $Q_1 \in P_1$ and

$Q_2 \in P_2$ such that divisors $D(Q_1)$ and $D(Q_2)$ have a common point (in particular, these divisors, or even the extremal rays Q_1, Q_2 may coincide).

Definition 2.1.6. A divisorial extremal ray R is called *special* if R satisfies one of conditions:

(1) R has type (I);

(2) R belongs to a pair of extremal rays of type $\mathfrak{B}_2$;

(3) There exists an extremal ray Q such that exactly one pair RQ or QR is an arrow: thus, either $R \cdot D(Q) > 0$ and $Q \cdot D(R) = 0$ or $Q \cdot D(R) > 0$ and $R \cdot D(Q) = 0$ (equivalently, the set $\{R, Q\}$ has type $\mathfrak{C}_2$).

We have the following description of the set of special divisorial extremal rays:

Theorem 2.1.7. *Assume that a 3-fold X belongs to the class $\mathcal{LT}$.*

Then the set of special divisorial extremal rays is finite. Its divisorially connected component is one of the following:

$\mathfrak{A}_1$: *One extremal ray Q of type* (I).

$\mathfrak{B}_2$: *Two different extremal rays Q_{11}, Q_{12} of type* (II) *with the same divisor $D(Q_{11}) = D(Q_{12})$. Then $Q_{11} + Q_{12}$ is a 2-dimensional face of $\overline{\mathrm{NE}}(X)$ of Kodaira dimension* 3.

$\mathfrak{C}_n$, $n \geq 2$: *n extremal rays $Q_1, \dots, Q_n$ of type* (II), *such that divisors $D(Q_2), \dots, D(Q_n)$ do not have a common point, and Q_i and Q_1 are joined by the single arrow Q_iQ_1. For this case, $Q_1 + \cdots + Q_n$ is an n-dimensional face of $\overline{\mathrm{NE}}(X)$ of Kodaira dimension* 3.

$\mathfrak{B}_2\mathfrak{C}_n$, $n \geq 1$: *$n+1$ extremal rays $\{Q_{11}, Q_2, \dots, Q_n, Q_{12}\}$ of type* (II), *such that $Q_1 = Q_{11}, \dots, Q_n$ define the configuration $\mathfrak{C}_n$ (in the above notation), and Q_{11}, Q_{12} is a pair of type $\mathfrak{B}_2$. The extremal rays Q_i, Q_{12} are connected by non-single arrows for $i \geq 2$. For this case, $Q_{11} + Q_2 + \cdots + Q_n$ is a face of $\overline{\mathrm{NE}}(X)$ of dimension n and Kodaira dimension* 3, *and $Q_{11} + Q_{12}$ is a 2-dimensional face of $\overline{\mathrm{NE}}(X)$ of Kodaira dimension* 3.

$\mathfrak{T}_3$ *(triangle): Three extremal rays Q_1, Q_2, Q_3 of type* (II) *connected by single arrows Q_1Q_2, Q_2Q_3 and non-single arrows Q_1Q_3 and Q_3Q_1. For this case, $\overline{\mathrm{NE}}(X, D(Q_1)) = Q_1 + Q_2$ and $\overline{\mathrm{NE}}(X, D(Q_2)) = Q_2 + Q_3$ are 2-dimensional faces of $\overline{\mathrm{NE}}(X)$ of Kodaira dimension* 3. *And $\overline{\mathrm{NE}}(X, D(Q_3)) = T + Q_3$ where $T \subset \overline{\mathrm{NE}}(X, D(Q_1)) = Q_1 + Q_2$.*

$\mathfrak{T}'_3$ *(special triangle): Three extremal rays Q_1, Q_2, Q_3 of type* (II) *connected by single arrows Q_1Q_2, Q_2Q_3 and Q_3Q_1. For this case, $\overline{\mathrm{NE}}(X, D(Q_1)) = Q_1 + Q_2$, $\overline{\mathrm{NE}}(X, D(Q_2)) = Q_2 + Q_3$ and $\overline{\mathrm{NE}}(X, D(Q_3)) = Q_3 + Q_1$ are 2-dimensional faces of $\overline{\mathrm{NE}}(X)$ of Kodaira dimension* 3.

Proof. Assume that extremal rays Q_1, Q_2 of type (II) are joined by the single arrow Q_1Q_2, and extremal rays R_1, R_2 of type (II) are joined by the single arrow R_1R_2. Suppose that the sets $\{Q_1, Q_2\}$ and $\{R_1, R_2\}$ are different and divisorially joined. We consider possible cases below.

Suppose that Q_1, R_1 are divisorially disjoint.

If $Q_2 = R_2$, we get the configuration $\mathfrak{C}_3$. We prove that only this case is possible.

Assume that $Q_2 \neq R_2$. Since the curve $D(R_1) \cap D(R_2)$ belongs to R_2, we have $R_2 \cdot D(Q_1) = 0$ (because $D(R_1) \cap D(Q_1) = \emptyset$). Similarly, $Q_2 \cdot D(R_1) = 0$.

Suppose that $D(Q_2) = D(R_2) = D$. Then any curve of the divisor D belongs to $R_2 + Q_2$ by Lemma 1.2.4. Since $Q_2 \cdot D(R_1) = R_2 \cdot D(R_1) = 0$, the divisors $D(R_2)$ and $D(R_1)$ do not have a common curve. We get a contradiction, because, R_1R_2 is an arrow.

Thus we can suppose that $D(Q_2) \neq D(R_2)$.

Suppose that R_1Q_2 is an arrow. We have proved that Q_2R_1 is not an arrow. It follows that the divisor $D(R_1)$ contains curves of 3 extremal rays: R_1, R_2, Q_2 which is impossible. Thus R_1Q_2 and Q_2R_1 are not arrows. It follows that divisors $D(R_1)$ and $D(Q_2)$ are disjoint. Similarly, the divisors $D(R_2)$ and $D(Q_1)$ are disjoint. Since the curve $D(R_1) \cap D(R_2)$ belongs to R_2, we get that $R_2 \cdot D(Q_2) = 0$. Similarly, $Q_2 \cdot D(R_2) = 0$. It follows that divisors $D(R_2)$ and $D(Q_2)$ are divisorially disjoint. It follows that sets $\{R_1, R_2\}$ and $\{Q_1, Q_2\}$ are divisorially disjoint. We get a contradiction.

Thus we proved that if the rays Q_1 and R_1 are divisorially disjoint, then $Q_2 = R_2$, and we have the configuration $\mathfrak{C}_3$.

Now assume that Q_1 and R_1 are divisorially joined. Then $D(Q_1) \cap D(R_1)$ is non-empty.

If $D(Q_1) = D(R_1)$, obviously, the sets $\{Q_1, Q_2\}$ and $\{R_1, R_2\}$ are equal because the divisor $D(Q_1) = D(R_1)$ cannot contain 3 extremal rays by Lemma 1.2.3.

Thus we assume that $D(Q_1) \cap D(R_1)$ is a non-empty curve. By Lemma 2.1.2, $\overline{\mathrm{NE}}(X, D(Q_1)) = Q_1 + Q_2$ and $\overline{\mathrm{NE}}(X, D(R_1)) = R_1 + R_2$ are 2-dimensional faces of $\overline{\mathrm{NE}}(X)$. It follows that the faces $Q_1 + Q_2$ and $R_1 + R_2$ have a common extremal ray.

Thus the sets $\{Q_1, Q_2\}$ and $\{R_1, R_2\}$ have a common extremal ray.

Assume that $Q_2 = R_1$ (the case $Q_1 = R_2$ is similar). We denote $Q_3 = R_2$. Thus we have 3 extremal rays Q_1, Q_2, Q_3 where Q_1, Q_2 are joined by a single arrow Q_1Q_2 and Q_2, Q_3 are joined by a single arrow Q_2Q_3. The curve $D(Q_1) \cap D(Q_2)$ belongs to the ray Q_2 and $Q_2 \cdot D(Q_3) > 0$. It follows that divisors $D(Q_1)$ and $D(Q_3)$ have a non-empty common curve. The extremal rays Q_1 and Q_3 cannot be joined by a single arrow Q_1Q_3 since then $D(Q_1)$ contains curves of 3 different extremal rays Q_1, Q_2, Q_3. Thus the extremal rays Q_1, Q_3 are joined either by non-single arrows Q_1Q_3 and Q_3Q_1 or by the single arrow Q_3Q_1. Thus we get cases $\mathfrak{T}_3$ and $\mathfrak{T}'_3$.

Now suppose that $Q_2 = R_2$. We denote $R_1 = Q_3$. Thus we have 3 extremal rays Q_1, Q_2, Q_3 such that Q_1, Q_2 are joined by a single arrow Q_1Q_2, and Q_3, Q_2 are joined by a single arrow Q_3Q_2. If Q_1 and Q_3 are divisorially disjoint, we get the case $\mathfrak{C}_3$.

Suppose that $D(Q_1) \cap D(Q_3)$ is a non-empty curve. By Lemma 2.1.2, $\overline{\mathrm{NE}}(X, D(Q_1)) = Q_1 + Q_2$ and $\overline{\mathrm{NE}}(X, D(Q_3)) = Q_3 + Q_2$ are 2-dimensional faces of $\overline{\mathrm{NE}}(X)$. The set of common points of these faces is the extremal ray Q_2. It follows that the curve $D(Q_1) \cap D(Q_3)$ belongs to the ray Q_2. By Lemma 2.1.3, this is impossible.

Thus we have proved that the divisors $D(Q_1)$ and $D(Q_3)$ do not have a common point.

As a result, we have proven that any configuration of two single arrows which are divisorially joined satisfies the statement of the theorem. Thus this is either $\mathfrak{C}_3$ or $\mathfrak{T}_3$ or $\mathfrak{T}'_3$.

Suppose that there exists a single arrow R_1R_2 which is divisorially joined with a configuration $\mathfrak{T}_3$ or $\mathfrak{T}'_3$ of extremal rays and different from arrows of this configurations. By the result above, the set $\{R_1, R_2\}$ has a common extremal ray with both sets $\{Q_1, Q_2\}$ and $\{Q_2, Q_3\}$. Thus either $Q_2 = R_1$ or $Q_2 = R_2$. If $Q_2 = R_1$, the divisor $D(R_1)$ contains 3 extremal rays: R_1, R_2, Q_3, which is impossible. If $Q_2 = R_2$, then R_1, R_2, Q_3 give a configuration $\mathfrak{T}_3$ or $\mathfrak{T}'_3$. Then the angle $\overline{\mathrm{NE}}(X, D(Q_3))$ has a side which simultaneously is a ray of $Q_1 + Q_2$ and $R_1 + Q_2$. Thus this side is Q_2. Then $D(Q_2) = D(Q_3)$, which is impossible.

From the above, it follows that a divisorially connected component S of a finite non-empty configuration of single arrows of extremal rays of type (II) has type either $\mathfrak{C}_n, n \geq 2$ or $\mathfrak{T}_3$ or $\mathfrak{T}'_3$. Using Lemma 2.1.2, one can easily see that if an extremal ray Q of this connected component S belongs to a pair Q, R of type $\mathfrak{B}_2$, then S has type $\mathfrak{C}_n, n \geq 2$ and $Q = Q_1$ in the notation of the theorem. Thus $S \cup \{R\}$ gives rise to a configuration of type $\mathfrak{B}_2\mathfrak{C}_n$.

Now suppose that we have two sets $\{Q_1, Q_2\}$ and $\{R_{11}, R_{12}\}$ of extremal rays such that Q_1Q_2 is a single arrow and $R_{11}R_{12}$ is a dotted line, and these two sets are divisorially joined. If these two sets have a common extremal ray, then they give a configuration $\mathfrak{B}_2\mathfrak{C}_2$ by the above considerations.

Thus assume that these two sets do not have a common extremal ray. By Lemmas 2.1.2 and 1.2.6, $\overline{\mathrm{NE}}(X, D(Q_1)) = Q_1 + Q_2$ and $\overline{\mathrm{NE}}(X, D(R_{11})) = R_{11} + R_{12}$ are 2-dimensional faces of $\overline{\mathrm{NE}}(X)$. It follows that if the divisors $D(Q_1)$, $D(R_{11})$ have a common point, then the sets $\{Q_1, Q_2\}$ and $\{R_{11}, R_{12}\}$ have a common extremal ray. But we assume that this is not the case. Thus we can suppose that divisors $D(Q_1)$ and $D(R_{11})$ do not have a common point. The curve $D(Q_1) \cap D(Q_2)$ belongs to the ray Q_2. Since $D(Q_1) \cap D(R_{11}) = \emptyset$, then $Q_2 \cdot D(R_{11}) = 0$. Thus $R_{11}Q_2$ is a single arrow. This is impossible by Lemma 2.1.2. Thus we get a contradiction.

Thus we have proved that a finite configuration of single arrows and dotted lines has divisorially connected components of type $\mathfrak{B}_2$, $\mathfrak{C}_n$, $\mathfrak{B}_2\mathfrak{C}_n$, $\mathfrak{T}_3$, $\mathfrak{T}'_3$ of the theorem.

Now let R be an extremal ray of type (I). Let Q_1Q_2 be a single arrow. If $D(Q_1) \cap D(R)$ is non-empty, then $D(Q_1)$ contains 3 extremal rays Q_1, Q_2 and R, which is impossible. The curve $D(Q_1) \cap D(Q_2)$ belongs to Q_2. From the above, $Q_2 \cdot D(R) =$

0. Thus RQ_2 is a single arrow, which is impossible for an extremal ray R of type (I). Thus $\{R\}$ and $\{Q_1, Q_2\}$ are divisorially disjoint.

By Lemma 1.2.5 and the above considerations, we have proved that any finite set of special extremal rays has the connected components of the theorem. Suppose that this set S has n extremal rays. We claim that then $n \le 2\rho(X)$, where $\rho(X) = \dim N_1(X)$. From the description of connected components of S, for any connected component S_i of S with m_i elements there exist at least $\max\{1, m_i - 1\} \ge m_i/2$ extremal rays with disjoint divisors. Since any extremal ray $R \in S$ has $R \cdot D(R) < 0$, the above inequality follows .

All other statements of the theorem follow from Lemmas 1.2.6 and 2.1.2.

This finishes the proof of the theorem. □

Now let us consider a divisorial extremal ray S which is not special, but the divisor $D(S)$ has a common point with the divisor of one of the special divisorial extremal rays. Thus S is joined by non-single arrows with a special divisorial extremal ray. These extremal rays also look very special. We describe them below.

Theorem 2.1.8. *Let X be a 3-fold of class $\mathcal{LT}$. Let S be a non-special divisorial extremal ray of X. Let Z_S be the set of special divisorial extremal rays which are divisorially joined with S (by definition, S has type* (II) *and is joined by non-single arrows with all extremal rays from Z_S). Assume that $Z_S \neq \emptyset$.*

Then Z_S is contained in a divisorially connected component $C = C_S$ of the set of special extremal rays, and depending on its type by Theorem 2.1.7, *we have the following description for $Z_S \subset C_S$:*

($\mathfrak{A}_1(S)$): *C_S has type $\mathfrak{A}_1$, $Z_S = Q$;*

($\mathfrak{C}_n^{(1)}(S)$): *C_S has type $\mathfrak{C}_n$, $Z_S = \{Q_i\}$, for one of $i > 1$;*

($\mathfrak{C}_2^{(2)}(S)$): *C_S has type $\mathfrak{C}_2$, $Z_S = \{Q_1, Q_2\}$;*

($\mathfrak{B}_2\mathfrak{C}_n(S)$): *$C_S$ has type $\mathfrak{B}_2\mathfrak{C}_n$, $n > 1$, $Z_S = \{Q_i\}$ for one of $i > 1$;*

($\mathfrak{B}_2(S)$): *C_S has type $\mathfrak{B}_2$, $Z_S = \{Q_{11}, Q_{12}\}$.*

We remark that it follows that C_S cannot have type $\mathfrak{T}_3$ or $\mathfrak{T}'_3$.

Proof. We consider one case only since the others are similar.

Let us assume that S is divisorially joined with a divisorially connected component of type $\mathfrak{C}_n$, $n \ge 2$, of the set of special extremal rays. Thus $D(S)$ has a common curve with one of the divisors $D(Q_i)$, $i = 1, \dots, n$ (we use the notation of Theorem 2.1.7).

Suppose that $D(S)$ has a common curve with the divisor $D(Q_1)$. Then S and Q_1 are joined by non-single arrows. Thus $Q_1 \cdot D(S) > 0$. Since $D(Q_i) \cap D(Q_1)$ belongs to Q_1, we get that $D(S)$ has a common curve with all divisors $D(Q_i)$, $i = 2, \dots, n$, too. By Lemma 2.1.2, $\overline{\mathrm{NE}}(X, D(Q_i)) = Q_1 + Q_i$ is a 2-dimensional face of $\overline{\mathrm{NE}}(X)$. We then get that the 2-dimensional angle $\overline{\mathrm{NE}}(X, D(S))$ has the second side (different from S) which belongs to the face $Q_1 + Q_i$, $i = 2, \dots, n$. If $n > 2$, we then get that the second side of $\overline{\mathrm{NE}}(X, D(S))$ is generated by the curve $D(S) \cap D(Q_i)$ which belongs to Q_1. This is impossible by Lemma 2.1.3. Thus $n = 2$, and we get case $\mathfrak{C}_2^{(2)}(S)$.

Now assume that S is not divisorially joined with Q_1, thus $D(S) \cap D(Q_1) = \emptyset$. Then $D(S) \cap D(Q_i)$ is not empty for some of $i > 1$. Then, as above, $\overline{\mathrm{NE}}(X, D(S))$ is a 2-dimensional angle with second side generated by the curve $D(S) \cap D(Q_i)$ which belongs to the 2-dimensional face $\overline{\mathrm{NE}}(X, D(Q_i)) = Q_i + Q_1$ of $\overline{\mathrm{NE}}(X)$. This curve cannot belong to Q_1 since $D(S) \cap D(Q_1) = \emptyset$. Thus the second side of the angle $\overline{\mathrm{NE}}(X, D(S))$ is the ray of $Q_i + Q_1$ different from Q_1. It follows that the extremal ray Q_i, $i = 2, \dots, n$, such that $D(S) \cap D(Q_i) \neq \emptyset$ is unique. Thus we get the case $\mathfrak{C}_n^{(1)}(S)$ if we additionally prove that S is divisorially disjoint with all special extremal rays different from $Q_1, \dots, Q_n$.

By Theorem 2.1.7, any special extremal ray Q of type (II) together with another special extremal ray Q' of the same component of the set of special extremal rays generate a 2-dimensional face $Q + Q'$ of $\overline{\mathrm{NE}}(X)$, and this 2-dimensional face is either $\overline{\mathrm{NE}}(X, D(Q))$ or $\overline{\mathrm{NE}}(X, D(Q'))$ (for the extremal ray Q of type (I), one should use that $\overline{\mathrm{NE}}(X, D(Q)) = Q$). Using this property, as above, one can see that $D(S) \cap D(Q)$ is empty for any special extremal ray Q different from the extremal rays $Q_1, \dots, Q_n$ above.

One can similarly consider cases when S is divisorially joined with other types of connected components of the set of special extremal rays. □

Now we have the following description for cases when $S \neq S'$ but $Z_S \cap Z_{S'} \neq \emptyset$. These cases are very exceptional.

Theorem 2.1.9. *Let X be a 3-fold of class $\mathcal{LT}$. Let S, S' be non-special divisorial extremal rays such that $Z_S \cap Z_{S'} \neq \emptyset$.*

Then $S = S'$ except for the cases below (in the notation of Theorems 2.1.7 *and* 2.1.8)*:*

(1) $C_S = C_{S'} = \mathfrak{C}_2$, $Z_S = Z_{S'} = \{Q_1, Q_2\}$, S, S' *are joined by non-single arrows,* $\dim [Q_1, Q_2, S, S'] = 3$.

(2) $C_S = C_{S'} = \mathfrak{C}_2$, $Z_S = \{Q_1, Q_2\}$, $Z_{S'} = \{Q_2\}$, S, S' *are joined by non-single arrows,* $\dim [Q_1, Q_2, S, S'] = 3$.

(3) $C_S = C_{S'} = \mathfrak{B}_2$, $Z_S = Z_{S'} = \{Q_{11}, Q_{12}\}$, S, S' *are joined by non-single arrows,* $\dim [Q_{11}, Q_{12}, S, S'] = 3$

(4) $C_S = C_{S'} = \mathfrak{B}_2$, $Z_S = Z_{S'} = \{Q_{11}, Q_{12}\}$, S, S' *are joined by non-single arrows, there exists a ray $T \subset Q_{11} + Q_{12}$ such that $\overline{\mathrm{NE}}(X, D(S)) = T + S$ and $\overline{\mathrm{NE}}(X, D(S')) = T + S'$.*

Proof. Suppose that $S' \neq S$ but $Z_{S'} \cap Z_S \neq \emptyset$. Using Theorem 2.1.8, we consider all possible cases (we use the notation of Theorems 2.1.8 and 2.1.7).

The case $(\mathfrak{A}_1(S))$, $Z_S = Z_{S'} = \{Q\}$. First, remark that $D(S) \cap D(S') \neq \emptyset$ since $D(S)$, $D(S')$ both have non-empty intersection with $D(Q)$ and $Q \cdot D(S) > 0$ and $Q \cdot D(S') > 0$. Moreover, $\overline{\text{NE}}(X, D(S)) = S + Q$ and $\overline{\text{NE}}(X, D(S')) = S' + Q$ are 2-dimensional angles with a common edge Q. If the curve $D(S) \cap D(S')$ has a component which does not belong to Q, then one of these angles is contained in another, which is impossible since S and S' are different extremal rays. Thus the curve $D(S) \cap D(S')$ belongs to the extremal ray Q. This is impossible by Lemma 2.1.3.

The case $(\mathfrak{C}_n^{(1)}(S))$, C_S has type $\mathfrak{C}_n$, $Z_S = Z_{S'} = \{Q_i\}$, for one of $i > 1$.

Let $C = D(S) \cap D(Q_i)$. We have $C \cdot D(Q_1) = 0$ because $D(S) \cap D(Q_1) = \emptyset$. Also, $C \in Q_i + Q_1$ and $Q_i \cdot D(Q_1) > 0$, $Q_1 \cdot D(Q_1) < 0$. Thus the ray $\mathbb{R}^+C \subset Q_i + Q_1$ is defined uniquely by the property $\mathbb{R}^+C \cdot D(Q_1) = 0$.

Now assume that $S \neq S'$ and $Z_{S'} = Z_S = \{Q_i\}$. As above, $D(S') \cap D(Q_i) = C'$ and $\mathbb{R}^+C' = \mathbb{R}^+C$. By Lemma 2.1.2, $\overline{\text{NE}}(X, D(Q_i)) = Q_i + Q_1$ is a 2-dimensional face of $\overline{\text{NE}}(X)$. It follows that $\overline{\text{NE}}(X, D(S)) = \mathbb{R}^+C + S$ and $\overline{\text{NE}}(X, D(S')) = \mathbb{R}^+C + S'$. If the curve $D(S) \cap D(S')$ has a component which does not belong to $D(Q_i)$, we then get that one of angles $\overline{\text{NE}}(X, D(S))$, $\overline{\text{NE}}(X, D(S'))$ contains another, which is impossible since S, S' are different extremal rays of $\overline{\text{NE}}(X)$. Let us consider the contraction $f: X \to X'$ of the extremal ray Q_i. Then the curve $C = f(D(S)) \cap f(D(S'))$ belongs to the extremal ray $f(Q_1)$ of X'. This is impossible by Lemma 2.1.3.

The case $(\mathfrak{C}_2^{(2)}(S))$, C_S has type $\mathfrak{C}_2$, $Z_S = \{Q_1, Q_2\}$. For this case, $D(S) \cap D(Q_1)$ contains a component which does not belong to Q_1 (otherwise, SQ_1 is a single arrow). It follows that the 2-dimensional angles $\overline{\text{NE}}(X, D(S))$ and $\overline{\text{NE}}(X, D(Q_1))$ have a common ray different from the edge Q_1 of $\overline{\text{NE}}(X, D(Q_1))$. Moreover, the curve $D(S) \cap D(Q_2)$ is not empty (since SQ_2 is arrow). Thus the 2-dimensional angle $\overline{\text{NE}}(X, D(S))$ has a common ray with the angle $\overline{\text{NE}}(X, D(Q_2))$. It follows that the angle $\overline{\text{NE}}(X, D(S))$ is contained in the 3-dimensional space V generated by angles $\overline{\text{NE}}(X, D(Q_1))$ and $\overline{\text{NE}}(X, D(Q_2))$ with the common edge Q_1. The same is valid for any S' with $Z_{S'} = Z_S = \{Q_1, Q_2\}$. This gives case (1) of the theorem.

By Theorem 2.3.8, $Z_{S'} = \{Q_2\}$ if $Z_{S'} \cap Z_S \neq \emptyset$ and $Z_{S'} \neq Z_S$. Let $C' = D(S') \cap D(Q_2)$. Since $\overline{\text{NE}}(X, D(Q_2)) = Q_2 + Q_1$, $C' \in Q_2 + Q_1$. Since $Q_2 \cdot D(S) > 0$, $Q_1 \cdot D(S) > 0$, it then follows that $C' \cdot D(S) > 0$ and $D(S') \cap D(S) \neq \emptyset$. Thus the angles $\overline{\text{NE}}(X, D(S))$ and $\overline{\text{NE}}(X, D(S'))$ have a common ray. If $\overline{\text{NE}}(X, D(S)) \cap \overline{\text{NE}}(X, D(S')) = \mathbb{R}^+C'$, as above, considering the contraction of the extremal ray Q_2, we get a contradiction with Lemma 2.1.3. Thus $\overline{\text{NE}}(X, D(S)) \cap \overline{\text{NE}}(X, D(S'))$ contains a ray different from $\mathbb{R}^+C'$. It follows that the angle $\overline{\text{NE}}(X, D(S'))$ is contained in the 3-dimensional space V above which contains extremal rays Q_1, Q_2, S. This gives case (2) of the theorem.

The case $(\mathfrak{B}_2\mathfrak{C}_n(S))$, C_S has type $\mathfrak{B}_2\mathfrak{C}_n$, $n > 1$, $Z_S = Z_{S'} = \{Q_i\}$ for one of $i > 1$. As above for the case $\mathfrak{C}_n$, this case is impossible for $S \neq S'$.

The case $(\mathfrak{B}_2(S))$, C_S has type $\mathfrak{B}_2$, $Z_S = \{Q_{11}, Q_{12}\}$. By Lemma 1.2.4, $\overline{\text{NE}}(X, D(Q_{11})) = Q_{11} + Q_{12}$ is a face of $\overline{\text{NE}}(X)$. It follows that $\overline{\text{NE}}(X, D(S)) = T + S$ where T is a ray of the angle $Q_{11} + Q_{12}$. Let S' be another extremal ray with

$Z_{S'} = \{Q_{11}, Q_{12}\}$. Since $Q_{11} \cdot D(S') > 0$, $Q_{12} \cdot D(S') > 0$, we have $T \cdot D(S') > 0$. It follows that S, S' are connected by non-single arrows. Applying Lemma 2.1.5, we get cases (3) and (4) of the theorem. This finishes the proof. □

2.2. Extremal and E-sets of divisorial extremal rays. We want to apply the results of Section 2.1 to describe extremal and E-sets of extremal rays of type (I) and (II) for 3-folds of class $\mathcal{LT}$. But, for future study, it is important to consider more general elliptic, parabolic and Lannér subsets of divisorial extremal rays, which have in fact appeared in Section 1.2.

Definition 2.2.1. Let $\mathcal{E} = \{R_1, \dots, R_n\}$ be a set of extremal rays of type (I) or (II). Then $\mathcal{E}$ is called *elliptic* if there are $a_1 > 0, \dots, a_n > 0$ such that

$$R_i \cdot (a_1 D(R_1) + \cdots + a_n D(R_n)) < 0$$

for all $1 \le i \le n$.

Definition 2.2.2. Let $\mathcal{P} = \{R_1, \dots, R_n\}$ be a set of extremal rays of type (I) or (II). Then $\mathcal{P}$ is called *connected parabolic* if each proper subset of $\mathcal{P}$ is elliptic and there are $a_1 > 0, \dots, a_n > 0$ such that

$$R_i \cdot (a_1 D(R_1) + \cdots + a_n D(R_n)) = 0$$

for all $1 \le i \le n$.

A set Q of extremal rays of type (I) or (II) is called *parabolic* if each divisorially connected component of Q is connected parabolic.

Definition 2.2.3. Let $\mathcal{L} = \{R_1, \dots, R_n\}$ be a set of extremal rays of type (I) or (II). Then $\mathcal{L}$ is called *Lannér* (respectively, *quasi-Lannér*) if each proper subset of $\mathcal{L}$ is elliptic (respectively, either elliptic or parabolic) and there are $a_1 > 0, \dots, a_n > 0$ such that

$$R_i \cdot (a_1 D(R_1) + \cdots + a_n D(R_n)) \ge 0$$

for all $1 \le i \le n$ and there exists j, $1 \le j \le n$, such that

$$R_j \cdot (a_1 D(R_1) + \cdots + a_n D(R_n)) > 0.$$

The following statement will be useful.

Proposition 2.2.4. *Let $\mathcal{E} = \{R_1, \dots, R_n\}$ be a set of extremal rays of type* (I) *or* (II) *having distinct divisors $D(R_1), \dots, D(R_n)$. Then $\mathcal{E}$ is elliptic if and only if for any non-negative $b_1, \dots, b_n$, not all* 0, *there exists R_i, $1 \le i \le n$, such that*

$$R_i \cdot (a_1 D(R_1) + \cdots + a_n D(R_n)) < 0.$$

Proof. If $\mathcal{E}$ is elliptic, then $\mathcal{E}$ satisfies the above condition by Lemma 1.2.10.

Now assume that $\mathcal{E}$ satisfies the condition of the proposition. If $\mathcal{E}$ is not elliptic, there exists a minimal non-elliptic subset $\mathcal{E}' \subset \mathcal{E}$. The subset $\mathcal{E}'$ is not empty since any one element subset of $\mathcal{E}$ is elliptic. Let $\mathcal{E}' = \{R_1, \dots, R_t\}$ where $1 \le t \le n$. By Lemma 1.2.10, there are positive $c_1, \dots, c_t$ such that

$$R_i \cdot (c_1 D(R_1) + \cdots + c_t D(R_t)) \ge 0$$

for any $1 \le i \le t$. The same inequality is obviously true for $t + 1 \le i \le n$. We get a contradiction with the condition of the proposition. □

Definition 2.2.5. A sequence

$$C = \{R_1, \dots, R_n\}, \quad n \ge 1,$$

of extremal rays of type (I) or (II) is called a *chain* if R_i, R_j are divisorially disjoint for $j - i > 1$, and $R_i R_{i+1}$ is a non-single arrow (thus, $R_{i+1} R_i$ is an arrow too) for any $1 \le i < n$. The extremal rays R_1, R_n are called *terminal* for the chain.

Theorem 2.2.6. *Let X be a 3-fold of $\mathcal{LT}$. Let T be a divisorially connected set of extremal rays of type* (I) *or* (II) *such that any proper subset of T is elliptic. (In particular, this is valid if either T itself is elliptic or is contained in a face of $\overline{\mathrm{NE}}(X)$ of Kodaira dimension* 3 *or T is connected parabolic or Lannér or T is an E-set such that each proper subset of T is contained in a face of $\overline{\mathrm{NE}}(X)$ of Kodaira dimension* 3.*) Assume that any two extremal rays of T are different from a pair of type $\mathfrak{B}_2$.*

Then T has one of types:

(A) *All extremal rays of T have type* (II) *and T does not have a single arrow.*

(B) *$T = \{R\} \cup C_1 \cup \cdots \cup C_k$ has only extremal rays of type* (II), *where*

$$C_1 = \{R_{11}, \dots, R_{1n_1}\},\ C_2 = \{R_{21}, \dots, R_{2n_2}\}, \dots, C_k = \{R_{k1}, \dots, R_{kn_k}\}$$

are pairwise divisorially disjoint chains, and all arrows between R and extremal rays of these chains are single arrows $R_{j1}R$, $j = 1, \dots, k$; moreover, each extremal ray of the chains $C_1, \dots, C_k$ does not belong to a pair of type $\mathfrak{B}_2$.

(C) *$T = \{R_1, R_2, R_3, \dots, R_n\}$, $n \ge 3$, where all extremal rays of T have type* (II) *and R_2R_1 is a single arrow, R_1R_3, R_3R_1 and R_2R_3, R_3R_2 are non-single arrows, $R_3, \dots, R_n$ is a chain such that $R_4, \dots, R_n$ are divisorially disjoint with extremal rays R_1 and R_2; moreover, each extremal ray $R_1, R_2, R_3, \dots, R_n$ does not belong to a pair of type $\mathfrak{B}_2$.*

(D) *$T = \{R_1, R_2, \dots, R_i, \dots, R_n\}$ is a chain such that R_1 has type* (I), *and R_i has type* (II) *for all $i > 1$; moreover, each extremal ray $R_1, R_2, \dots, R_n$ does not belong to a pair of type $\mathfrak{B}_2$.*

(E) *$T = \{Q_1, Q_2, Q_3\}$ has the type of triangle: thus, Q_1Q_2 and Q_2Q_3 are single arrows, and Q_1Q_3, Q_3Q_1 are non-single arrows; moreover, each extremal ray Q_1, Q_2, Q_3 does not belong to a pair of type $\mathfrak{B}_2$.*

(E′) $T = \{Q_1, Q_2, Q_3\}$ *has the type of special triangle: thus,* Q_1Q_2, Q_2Q_3, Q_3Q_1 *are single arrows; moreover, each extremal ray* Q_1, Q_2, Q_3 *does not belong to a pair of type* $\mathfrak{B}_2$.

Moreover, for case (B), T *is contained in a face of* $\overline{\mathrm{NE}}(X)$ *of Kodaira dimension* 3. *Thus case* (B) *is impossible when* T *is not contained in a face of* $\overline{\mathrm{NE}}(X)$ *of Kodaira dimension* 3.

Proof. If T does not have an extremal ray of type (I) and a single arrow, we get case (A).

Thus we can suppose that T has either an extremal ray of type (I) or a single arrow.

Let T_1 be the set which contains all extremal rays of type (I) and all extremal rays of single arrows of T. If $T = T_1$, then T_1 is divisorially connected, and by Theorem 2.1.7, we get one of following cases: (B) with one element chains, (D) with one extremal ray R_1, (E) or (E′).

Suppose that $T \neq T_1$. By Theorem 2.1.7, T_1 has divisorially connected components of types $\mathfrak{A}_1$, $\mathfrak{C}_n$, $\mathfrak{T}_3$ or $\mathfrak{T}_3'$. Since T is divisorially connected, for each connected component of T_1 there exists an extremal ray $Q \in T - T_1$ such that Q is joined by non-single arrows with an extremal ray of this connected component. By Theorem 2.1.8, this connected component should then have type $\mathfrak{A}_1$ or $\mathfrak{C}_n$. It follows that all connected components of T_1 have type $\mathfrak{A}_1$ or $\mathfrak{C}_n$, and T_1 has a connected component of one of these types.

We consider induction on $\sharp T$. For $\sharp T = 1, 2$ the statement is clear. Suppose that $\sharp T = n > 2$. We consider several cases.

Suppose that there exists an extremal ray $R \in T$ of type (I). Let $Q \in T$ and $Q \neq R$. We claim that then $Q + R$ is a 2-dimensional face of Kodaira dimension 3 of $\overline{\mathrm{NE}}(X)$. First, suppose that $D(R) \cap D(Q) = \emptyset$. Let H be a nef element orthogonal to Q (the set of these elements H defines a face of codimension one in $\mathrm{NEF}(X)$ by the exact sequence (1-2-1)). There exists a linear function $\alpha(H) \geq 0$ such that the map $H \to H' = H + \alpha(H)D(R)$ is linear with one-dimensional kernel, and H' is orthogonal to extremal rays Q and R. Then H' is obviously nef, since R has type (I) and $C \in R$ for any curve $C \subset D(R)$. Thus $Q + R$ is a 2-dimensional face of $\overline{\mathrm{NE}}(X)$, and, by construction, the face $(Q + R)^{\perp}$ of $\mathrm{NEF}(X)$ has codimension 2. Obviously, $(H')^3 \geq H^3 > 0$ (see the proof of Lemma 1.2.6). Thus by definition of the class $\mathcal{L}T$, this face has Kodaira dimension 3. (From now on, we will not be so formal, and just show that the corresponding face α of $\overline{\mathrm{NE}}(X)$ has numerical Kodaira dimension 3; automatically, by our construction, it will have the property $\operatorname{codim} \alpha^{\perp} = \dim \alpha$, and will have Kodaira dimension 3 by definition of the class $\mathcal{L}T$.) Now suppose that $D(Q) \cap D(R) \neq \emptyset$. Since $\sharp T = n > 2$, the set $\{Q, R\}$ is elliptic. Thus there are positive a_1, a_2 such that $Q \cdot (a_1 D(Q) + a_2 D(R)) < 0$, $R \cdot (a_1 D(Q) + a_2 D(R)) < 0$. Let H be an ample element. By Lemma 1.2.8, there are positive c_1, c_2 such that $H' = H + c_1 D(Q) + c_2 D(R)$ is orthogonal to both Q, R. Since R has type (I) and $D(Q) \cap D(R) \neq \emptyset$, we then get $\overline{\mathrm{NE}}(X, D(Q)) = Q + R$. It follows that $H' \cdot C \geq 0$ for any curve C. Thus H' is nef. Obviously, H' is orthogonal to extremal rays Q, R

only and $H' \geq H^3 > 0$. It follows that $Q + R$ is a face of dimension 2 and Kodaira dimension 3 of $\overline{\mathrm{NE}}(X)$. Thus we have proven that $Q + R$ is a 2-dimensional face of $\overline{\mathrm{NE}}(X)$ for any $Q \in T$ and $Q \neq R$.

Let us consider the contraction $f: X \to X'$ of the extremal ray R. Then the image $f(Q)$ is an extremal ray of X' for $Q \neq R$ by the above claim. Obviously, if $D(Q) \cap D(R) = \emptyset$, the extremal ray $f(Q)$ has the same type ((I) or (II)) as Q. If $D(Q) \cap D(R) \neq \emptyset$, then $f(Q)$ has type (I). By Theorem 2.1.9, the extremal ray Q with this property is unique. Since T is connected, Q exists. By the projection formula and Proposition 2.2.4, any proper subset of the set $f(T) = f(T \setminus \{R\})$ is elliptic. It follows that the set $f(T)$ has the same properties as T but has one element less. By induction, we then get that $f(T)$ has type (D). Considering preimages, one can easily see that then T has the same type (D) and has the required properties.

Now, assume that R_2R_1 is a single arrow for $R_2, R_1 \in T$. At first, suppose that there does not exist $Q \in T$ such that Q is joined by non-single arrows with the terminal R_1 of the arrow R_2R_1. Thus $R_1 \cdot D(Q) = 0$ for all $Q \in T$, $Q \neq R_1$. By Lemma 2.1.2, $\overline{\mathrm{NE}}(X, D(R_2)) = R_1 + R_2$ is a 2-dimensional face of $\overline{\mathrm{NE}}(X)$ of Kodaira dimension 3. It follows that $\overline{\mathrm{NE}}(X, D(Q)) = P + Q$, $P \subset R_1 + R_2$ for an extremal ray Q which is joined by non-single arrows with R_2. By Lemma 2.1.9, this extremal ray Q is unique if it exists. Then, as above, we can prove that R_2, Q generate a 2-dimensional face $R_2 + Q$ of Kodaira dimension 3 of $\overline{\mathrm{NE}}(X)$ for each $Q \in T$, $Q \neq R_2$ (for $Q = R_1$, this follows from Lemma 2.1.2). Let us consider the contraction $f: X \to X'$ of the extremal ray R_2. Then $f(Q)$ is a divisorial extremal ray of X' for $Q \in T$, $Q \neq R_2$. If there exists Q such that QR_2 is a non-single arrow, then $f(Q)f(R_1)$ is obviously a single arrow because Q and R_1 are divisorially disjoint by Lemma 2.1.8. If there does not exist Q such that QR_2 is a non-single arrow then there exists an extremal ray $S \in T$ such that SR_1 is a single arrow and S is divisorially disjoint with R_2. This follows from Theorems 2.1.7, 2.1.8 and the above conditions on T (otherwise, $T = \{R_2, R_1\}$). Then $f(S)f(R_1)$ is a single arrow. Thus the image $f(T)$ is a divisorially connected set of extremal rays of type (I) and (II) which contains a single arrow and as in the above case any proper subset of $f(T)$ is elliptic. By induction, $f(T)$ has type (B) and only the extremal ray $f(R_1)$ may belong to a pair of type $\mathfrak{B}_2$. One can see easily using our construction, that then T has also type (B) and only R_1 may belong to a pair of type $\mathfrak{B}_2$.

Now we consider the case when R_2R_1 is a single arrow for $R_2, R_1 \in T$ and there exists $R_3 \in T$ such that R_3R_1, R_1R_3 are non-single arrows. This is the case $\mathfrak{C}_2^{(2)}(R_3)$ of Theorem 2.1.8. By Theorem 2.1.8, then R_3R_2, R_2R_3 are non-single arrows and R_2R_1 is the only single arrow in T with the terminal R_1.

Since $\sharp T \geq 3$ and any proper subset of T is elliptic, there are $a_2 > 0, a_3 > 0$ such that $R_2 \cdot (a_2D(R_2) + a_3D(R_3)) < 0$ and $R_3 \cdot (a_2D(R_2) + a_3D(R_3)) < 0$. Let H be an ample element. By Lemma 1.2.8, there exist $b_2 > 0$, $b_3 > 0$ such that $H' = H + b_2D(R_2) + b_3D(R_3)$ is orthogonal to both R_2, R_3. By Lemma 2.1.2, $\overline{\mathrm{NE}}(X, D(R_2)) = R_1 + R_2$. It follows that $C \cdot H' \geq 0$ for any curve $C \subset D(R_2)$. We have $D(R_2) \cap D(R_3)$ is a non-empty curve. By Lemma 2.1.2, $\overline{\mathrm{NE}}(X, D(R_2)) =$

$R_1 + R_2$ is a face of $\overline{\mathrm{NE}}(X)$. It follows that $\overline{\mathrm{NE}}(X, D(R_3)) = S + R_3$ where S is a ray of $R_1 + R_2$. If follows that $C \cdot H' \geq 0$ for any curve $C \in D(R_3)$. This implies that H' is nef and only the extremal rays R_2, R_3 are orthogonal to H'. Moreover, as above, $(H')^3 \geq H^3 > 0$. Thus $R_2 + R_3$ is a 2-dimensional face of $\overline{\mathrm{NE}}(X)$ of Kodaira dimension 3.

As mentioned above, $\overline{\mathrm{NE}}(X, D(R_3)) = S + R_3$, where $S \in R_1 + R_2$. Here $S \neq R_1$, since otherwise, R_3R_1 is a single arrow. We have $D(R_3) \cap D(R_1) \neq \emptyset$. Since $R_2 + R_3$ is a face of $\overline{\mathrm{NE}}(X)$, it follows that $\overline{\mathrm{NE}}(X, D(R_1)) \subset R_1 + R_2 + R_3$. Then, as above, we can see that $R_1 + R_3$ is a 2-dimensional face of $\overline{\mathrm{NE}}(X)$ of Kodaira dimension 3. By Theorems 2.1.7 and 2.1.8, each extremal ray R_1, R_2, R_3 does not belong to a pair of type $\mathfrak{B}_2$.

If $T = \{R_1, R_2, R_3\}$, we get case (C).

Suppose that T contains an extremal ray Q different from R_1, R_2, R_3. By condition, then $\{R_1, R_2, R_3\}$ is elliptic. As we have seen, $\overline{\mathrm{NE}}(X, D(R_1))$,$\overline{\mathrm{NE}}(X, D(R_2))$, $\overline{\mathrm{NE}}(X, D(R_3))$ are contained in $R_1 + R_2 + R_3$. Then as above (using Proposition 1.2.1 and Lemmas 1.2.8 and 1.2.9), we can prove that $R_1 + R_2 + R_3$ is a 3-dimensional face of Kodaira dimension 3 of $\overline{\mathrm{NE}}(X)$.

Suppose that $Q \in T$ is different from R_1, R_2, R_3. Since $R_1 + R_2 + R_3$ is a face of $\overline{\mathrm{NE}}(X)$, by statements (1) and (2) of Lemma 2.1.9, Q is divisorially disjoint with R_1 and R_2. Then, as above, using that $\overline{\mathrm{NE}}(X, D(R_1)) \subset R_1 + R_2 + R_3$, we can prove that $R_1 + Q$ is a 2-dimensional face of Kodaira dimension 3 of $\overline{\mathrm{NE}}(X)$. For $Q = R_2, R_3$ we have proven the same above. Thus this statement is valid for any $Q \in T$ different from R_1. Let us consider the contraction $f: X \to X'$ of R_1. By the statement above, $f(Q)$ is a divisorial extremal ray for any $Q \in T$, $Q \neq R_1$. Obviously, $f(R_2)$ has type (I). Thus considering $T' = f(T)$ we get the case we had considered above, and which gives rise to type (D) of the theorem. Thus T' has type (D). One can easily see that then T has type (C) and each extremal ray of T does not belong to a pair of type $\mathfrak{B}_2$.

Let us prove the last statement. Thus we consider T which has type (B). We use induction on $\sharp T$. For $\sharp T = 2$, the statement follows from Lemma 2.1.2. If $\sharp T > 2$, let us consider the contraction $f: X \to X'$ of the extremal ray R_2 which we had used above when we considered this case. Then $T' = f(T)$ has type (B) again and has the same properties as T but $\sharp T' = n - 1$. By induction, $T' = f(T)$ is contained in a face of Kodaira dimension 3 of $\overline{\mathrm{NE}}(X')$. Then, obviously, T is also contained in a face of Kodaira dimension 3 of $\overline{\mathrm{NE}}(X)$. This finishes the proof. □

In Section 5, we will need another property of divisorial extremal rays which shows the importance of quasi-Lannér sets of divisorial extremal rays.

Definition 2.2.7. A set $S = \{Q_1, \dots, Q_k\}$ of divisorial extremal rays is called *semi-elliptic* if for any non-negative $a_1, \dots, a_k$ there exists j, $1 \leq j \leq k$, such that

$$Q_j \cdot (a_1 D(Q_1) + \cdots + a_k D(Q_k)) \leq 0.$$

In particular, by Lemma 1.2.8, an elliptic set of divisorial extremal rays is semi-elliptic.

Proposition 2.2.8. *Let $\mathcal{L} = \{R_1, \dots, R_n\}$ be a set of extremal rays of type* (I) *or* (II) *having distinct divisors $D(R_1), \dots, D(R_n)$. Assume that $\mathcal{L}$ is not semi-elliptic, i. e. there are $a_1, \dots, a_n$ such that $R_i \cdot (a_1 D(R_1) + \cdots + a_n D(R_n)) \geq 0$ for all $1 \leq i \leq n$, and one of these inequalities is strict. Assume that any proper subset of $\mathcal{L}$ is semi-elliptic (i.e. $\mathcal{L}$ is minimal not semi-elliptic). Then $\mathcal{L}$ is quasi-Lannér (see Definition* 2.2.3*). Moreover, any proper subset of $\mathcal{L}$ either is elliptic or is connected parabolic with $\sharp\mathcal{L} - 1$ elements.*

Proof. We use Proposition 2.2.4 as the equivalent definition of an elliptic set of divisorial extremal rays.

If any proper subset $\mathcal{L}' \subset \mathcal{L}$ is elliptic, then the set $\mathcal{L}$ is Lannér and is then quasi-Lannér.

Assume that $\mathcal{L}$ contains a semi-elliptic subset $\mathcal{L}'$ which is not elliptic. We should prove that then $\mathcal{L}'$ is connected parabolic and $\sharp\mathcal{L}' = \sharp\mathcal{L} - 1$. Let $\mathcal{P}$ be a minimal semi-elliptic and not elliptic subset of $\mathcal{L}'$. Then any proper subset of $\mathcal{P}$ is elliptic (use Proposition 2.2.4). It follows that $\mathcal{P}$ is connected parabolic.

Let $R \in \mathcal{L} - \mathcal{P}$. Assume that there exists an arrow from R to (an element of) $\mathcal{P}$. Let $\mathcal{P} = \{Q_1, \dots, Q_r\}$ and for some positive $c_1, \dots, c_r$ we have $\mathcal{P} \cdot (c_1 D(Q_1) + \cdots + c_r D(Q_r)) = 0$. Then for sufficiently big $\lambda > 0$ we obviously have $R \cdot (\lambda(c_1 D(Q_1) + \cdots + c_r D(Q_r)) + D(R)) > 0$ and $\mathcal{P} \cdot ((c_1 D(Q_1) + \cdots + c_r D(Q_r)) + D(R)) \geq 0$. Thus $\{R\} \cup \mathcal{P}$ is not semi-elliptic. Since $\mathcal{L}$ is minimal not semi-elliptic, we get $\mathcal{L} = \{R\} \cup \mathcal{P}$ and $\mathcal{L}' = \mathcal{P}$ is connected parabolic with $\sharp\mathcal{L} - 1$ elements.

Thus we can suppose that there is no arrow from $\mathcal{L} \setminus \mathcal{P}$ to $\mathcal{P}$. If $\mathcal{L} \setminus \mathcal{P}$ is not elliptic, arguing similarly, we can find a connected parabolic subset $\mathcal{P}' \subset \mathcal{L} \setminus \mathcal{P}$. Moreover, $\mathcal{P}, \mathcal{P}'$ are divisorially disjoint to one another and there is no arrow from $\mathcal{L} \setminus (\mathcal{P} \cup \mathcal{P}')$ to $\mathcal{P} \cup \mathcal{P}'$. Continuing these considerations, we prove that $\mathcal{L}$ is a disjoint union of connected parabolic subsets $\mathcal{P}_1, \dots, \mathcal{P}_m$ and an elliptic subset $\mathcal{E}$ such that connected parabolic subsets $\mathcal{P}_1, \dots, \mathcal{P}_m$ are divisorially disjoint to one another and there does not exist an arrow from $\mathcal{E}$ to connected parabolic subsets $\mathcal{P}_1, \dots, \mathcal{P}_m$. If follows that $\mathcal{L}$ is semi-elliptic. We get a contradiction. This finishes the proof. □

From our proof we also get

Proposition 2.2.9. *Let S be a set of extremal rays of type* (I) *or* (II) *having distinct divisors. Then S is semi-elliptic if and only if S is a disjoint union of connected parabolic subsets $\mathcal{P}_1, \dots, \mathcal{P}_m$ and an elliptic subset $\mathcal{E}$ such that parabolic subsets $\mathcal{P}_1, \dots, \mathcal{P}_m$ are divisorially disjoint to one another and there does not exist an arrow from any element of $\mathcal{E}$ to any parabolic subset $\mathcal{P}_1, \dots, \mathcal{P}_m$.*

3. The refined form of the Diagram Method

Here we want to prove a stronger form of the Diagram Method Theorem which uses the results of both Sections 1 and 2. This version will be useful below for Calabi–Yau 3-folds.

We study 3-folds X of class $\mathcal{LT}$ with conditions (a), (b) and (c) below (see Theorem 1.3.2):

(a) X has a finite polyhedral Mori cone $\overline{\mathrm{NE}}(X)$;

(b) X does not have a small extremal ray;

(c) $\overline{\mathrm{NE}}(X)$ does not have a face of numerical Kodaira dimension 1 or 2 (equivalently, there are no nef elements h with $h^3 = 0$).

From (a), (b) and (c), it follows that all extremal rays of X are divisorial of type (I) or (II).

Using Theorem 2.2.6, we define constants which are analogous to the constants $d(X)$, $C_1(X)$, $C_2(X)$ of Definition 1.3.1.

Definition 3.1. We introduce some invariants of X.

The invariants k, l, l_2 are numbers of divisorially connected components of types $\mathfrak{A}_1$, $\mathfrak{C}_k$, $k \geq 2$, $\mathfrak{C}_2$ respectively of the set of all special divisorial extremal rays on X. See Theorem 2.1.7.

The invariants

$$n(X)_D = \max_{F \in (D)} \sharp F - 1; \quad n(X)_C = \max_{F \in (C)} \sharp F - 1;$$
$$n(X)_A = \max_{F \in (A)} \sharp F - 1; \quad d(X)_A = \max_{F \in (A)} \operatorname{diam} G(F).$$

Here $F \in (A)$ (respectively $F \in (C)$ or $F \in (D)$) means that F runs through all E-sets of type (A) (respectively (C) or (D)) of Theorem 2.2.6 such that any proper subset of F is extremal of Kodaira dimension 3 and F satisfies the condition (iii) of Section 1.1 (in particular, F is Lannér). These invariants are analogous to the invariant $d(X)$ of Definition 1.3.1.

Let S be a set of divisorial extremal rays. We denote by S' the subset of S obtained by removing all extremal rays of type (I) and all extremal rays of type (II) which are ends of single arrows of X (i.e. this extremal ray belongs to a component of type $\mathfrak{C}_k$ of the set of all special extremal rays on X and is the end of a single arrow of this component). We define a symmetric distance $\rho_A(R_1, R_2)$ in S by the formula

$$\rho_A(R_1, R_2) = \begin{cases} 0, & \text{if } R_1 = R_2, \\ \rho(R_1, R_2)_{S'}, & \text{if } \{R_1, R_2\} \subset S', \\ +\infty, & \text{otherwise.} \end{cases}$$

Here $\rho(R_1, R_2)_{S'}$ denotes the ordinary distance in the graph $G(S')$ using oriented paths. The set S' does not have single arrows. Thus $\rho_A(R_1, R_2)$ is symmetric.

The invariant $C(X)_A$ is defined by the condition:

$$\sharp\big\{\{R_1, R_2\} \subset \mathcal{E} - \mathcal{E}_0 \mid 1 \le \rho_A(R_1, R_2) \le 2d(X)_A + 1\big\} \le C(X)_A \sharp(\mathcal{E} - \mathcal{E}_0)$$

for any extremal set $\mathcal{E}$ of the Kodaira dimension 3 of divisorial extremal rays on X and any its divisorially connected subset $\mathcal{E}_0 \subset \mathcal{E}$ of extremal rays of type (II).

We have the following refinement of Basic Theorem 1.3.2.

Basic Theorem 3.2. *Let X be a 3-fold of class $\mathcal{LT}$ and suppose that X satisfies conditions* (a), (b) *and* (c) *above. Then we have statements* (1), (2) *and* (3) *below:*

(1) *X does not have a pair of extremal rays of type $\mathfrak{B}_2$ and the Mori cone $\overline{\mathrm{NE}}(X)$ is simplicial. Thus by Theorem* 2.1.7, *each divisorially connected component of the set of all special divisorial extremal rays has type $\mathfrak{A}_1$, $\mathfrak{C}_n$, $\mathfrak{T}_3$ or $\mathfrak{T}'_3$.*

(2) *If the set of all special divisorial extremal rays on X has k divisorially connected components of type $\mathfrak{A}_1$ and l connected components of types $\mathfrak{C}_{n_1}, \dots, \mathfrak{C}_{n_l}$ respectively, then*

$$k + (n_1 - 1) + \cdots + (n_l - 1) \le q(X)$$

(see Definition 1.3.1 *for the invariant $q(X)$). Moreover, if there exists a connected component of type $\mathfrak{T}_3$ or $\mathfrak{T}'_3$, then every extremal ray of X belongs to this connected component and the Picard number $\rho(X) = 3$ (in particular, $k = l = 0$).*

(3) *We have the inequality:*

$$\begin{aligned}\rho(X) = \dim N_1(X) &\le kn(X)_D + l_2 \max\{n(X)_C, n(X)_A\} + 8C(X)_A + 6\\ &\le q(X) \max\{n(X)_D, n(X)_C, n(X)_A\} + 8C(X)_A + 6.\end{aligned}$$

Proof. (1) follows from the statement (1) of Basic Theorem 1.3.2 and Theorem 2.1.7.

To prove second part of (2), we use the following lemmas which are important in their own right (in fact, they are contained in Section 1) and follow from the statement (1), Lemma 1.2.8, Lemma 1.1.1 and the first statement of Theorem 1.2.13.

Lemma 3.3. *Let X be a 3-fold of class $\mathcal{LT}$ and suppose that X satisfies the conditions* (a), (b) *and* (c) *above. Let $T = \{R_1, \dots, R_n\}$ be a set of extremal rays of X. Then*

(i) *If T is extremal, there are positive $a_1, \dots, a_n$ such that $R_i \cdot (a_1 D(R_1) + \cdots + a_n D(R_n)) < 0$ for all $1 \le i \le n$, and for any non-negative $b_1, \dots, b_n$, not all* 0, *there exists i, $1 \le i \le n$, such that $R_i \cdot (b_1 D(R_1) + \cdots + b_n D(R_n)) < 0$.*

(ii) *If T is not extremal, then T contains an E-subset $\mathcal{L}$ and there are non-negative $c_1, \dots, c_n$, not all* 0, *such that $c_1 D(R_1) + \cdots + c_n D(R_n)$ is nef.*

Lemma 3.4. *Let X be a 3-fold of class $\mathcal{LT}$ and suppose that X satisfies the conditions* (a), (b) *and* (c) *above. Let $T_1 = \{R_1, \dots, R_n\}$ and $T_2 = \{Q_1, \dots, Q_m\}$ be two sets* *of extremal rays on X such that* $R_i \cdot D(Q_j) = 0$ *for all* $1 \le i \le n$ *and all* $1 \le j \le m$ *(equivalently, there does not exist an arrow from* T_1 *to* T_2*). Then either* T_1 *or* T_2 *is extremal.*

If $T_1 \cup T_2$ *is the set of all extremal rays on X and* $T_2 \ne \emptyset$*, then* T_1 *is extremal and* T_2 *is not extremal.*

Proof. We explain only the last statement. Suppose that T_2 is extremal. By the conditions and Lemma 3.3, there are positive $a_1, \dots, a_m$ such that $R \cdot (a_1 D(Q_1) + \dots + a_m D(Q_m)) \le 0$ for any extremal ray $R \in T_1 \cup T_2$. By assumption, $T_1 \cup T_2$ is the set of all extremal rays on X. It follows that, $C \cdot (a_1 D(Q_1) + \dots + a_m D(Q_m)) \le 0$ for any curve C on X. Obviously, this is absurd. Thus T_2 is not extremal. By Lemma 3.3, there are non-negative $c_1, \dots, c_m$ such that $D(Q) = c_1 D(Q_1) + \dots + c_m D(Q_m)$ is nef. By condition, $T_1 \cdot D(Q) = 0$. It follows that T_1 is extremal. □

From the last statement of Lemma 3.4, it follows

Lemma 3.5. *Let X be a 3-fold of class* $\mathcal{LT}$ *and suppose that X satisfies the conditions* (a), (b) *and* (c) *above. Then the set of all extremal rays on X is divisorially connected and is not extremal.*

Now let us prove the statement (2) of Theorem 3.2. Assume that the set of special extremal rays has a connected component P of type $\mathfrak{T}_3$ or $\mathfrak{T}_3'$. Let R be an extremal ray and $R \notin P$. By Theorem 2.1.8, then $R \cdot D(Q) = 0$ for any extremal ray $Q \in P$. We get a contradiction with Lemma 3.5. Thus any extremal ray of X belongs to P. Since P contains exactly 3 extremal rays, $\rho(X) \le 3$. By definition, P contains extremal rays Q_1, Q_2, Q_3 which define single arrows $Q_1 Q_2$ and $Q_2 Q_3$. By Lemma 2.1.2, $Q_1 + Q_2$ and $Q_2 + Q_3$ are 2-dimensional faces of $\overline{\mathrm{NE}}(X)$ which are obviously different. It follows that $\rho(X) > 2$. Thus $\rho(X) = 3$.

Let us prove the first part of (2). This is similar to the proof of the statement (2) of Basic Theorem 1.3.2. Let $P_1 = \{R_1\}, \dots, P_k = \{R_k\}$ be connected components of type $\mathfrak{A}_1$ and $S_1, \dots, S_l$ be connected components of types $\mathfrak{C}_{n_1}, \dots, \mathfrak{C}_{n_l}$ of the set of special extremal rays on X, where $S_t = \{Q_{t1}, Q_{t2}, \dots, Q_{tn_t}\}$ and $Q_{t2}Q_{t1}, \dots, Q_{tn_t}Q_{t1}$ are single arrows.

The set of all special extremal rays on X obviously satisfies the condition (i) of Lemma 3.3 and is then extremal. Let $\mathcal{E}$ be a maximal extremal set of extremal rays on X containing the set of all special extremal rays and such that each divisorially connected component of $\mathcal{E}$ contains at least one special extremal ray.

By Theorem 2.2.6, each connected component of $\mathcal{E}$ contains exactly one connected component of the set of special extremal rays. Thus $\mathcal{E}$ contains $k + l$ connected components $\mathcal{E}_1, \dots, \mathcal{E}_k, \mathcal{E}_{k+1}, \dots, \mathcal{E}_{k+l}$. Here $R_i \in \mathcal{E}_i$ and $\mathcal{E}_i$ has type (D) for $1 \le i \le k$. And $S_j \subset \mathcal{E}_{k+j}$ and $\mathcal{E}_{k+j}$ has either type (B) or (C) for $1 \le j \le l$. Renumbering, we can suppose that $\mathcal{E}_{k+j}$ has type (C) for $1 \le j \le r_1$ (in particular, $n_1 = \dots = n_{r_1} = 2$) and $\mathcal{E}_{k+j}$ has type (B) for $r_1 < j \le l$. For $r_1 < j \le l$, we additionally denote as

$\mathcal{E}_{k+j,2}, \ldots, \mathcal{E}_{k+j,n_j}$ the corresponding chains containing $Q_{j2}, \ldots, Q_{jn_j}$ respectively. These chains give divisorially connected components of $\mathcal{E}_{k+j} - \{Q_{j1}\}$.

Since $\mathcal{E}$ is extremal, by Lemma 3.3, there are positive $a(R)$ for $R \in \mathcal{E}$ such that

$$R' \cdot \sum_{R\in\mathcal{E}} a(R)D(R) < 0$$

for any $R' \in \mathcal{E}$.

Let us consider pairwise divisorially disjoint sets F_v, $v \in V$, of extremal rays which are equal to one of sets $\mathcal{E}_1, \ldots, \mathcal{E}_{k+r_1}$ or $\mathcal{E}_{k+j,u}$, $r_1 < j \leq l$, $2 \leq u \leq n_j$, and the corresponding divisors

$$D(F_v) = \sum_{R\in F_v} a(R)D(R)$$

(here $\sharp V = k + (n_1 - 1) + \cdots + (n_l - 1)$). One can easily see that $C \cdot D(F_v) \leq 0$ for any $R \in F_v$ and any curve $C \subset D(R)$.

Similarly to Lemma 1.3.4 and the proof of the statement (2) of Basic Theorem 1.3.2, we can find an extremal set $A = \{U_v \mid v \in V\}$ containing extremal rays U_v of type (II) such that we have the property: $U_v \cdot D(F_v) > 0$ but $U_v \cdot D(F_{v'}) = 0$ for $v' \neq v$. As in the the proof of the statement (2) of Theorem 1.3.2, one proves that the graph $G(A)$ is full. Thus $\sharp A = \sharp V = k + (n_1 - 1) + \cdots + (n_l - 1) \leq q(X)$. This finishes the proof of the statement (2) of Theorem 3.2.

The proof of (3) is based on the following result.

Lemma 3.6. *Let X be a 3-fold of class $\mathcal{LT}$ and suppose that X satisfies conditions* (a), (b) *and* (c) *above. Assume that $\rho(X) > 3$. Then there exists an extremal divisorially connected set $\mathcal{E}_0$ containing only extremal rays of type* (II) *and such that any E-set $\mathcal{L}$ does not contain extremal rays of type* (I) *and extremal rays of type* (II) *which are terminal vertices of single arrows on X (in particular, $\mathcal{L}$ has type* (A) *of Theorem* 2.2.6) *if $\mathcal{L}$ has at least two elements which do not belong to $\mathcal{E}_0$ and $\mathcal{L}' \cup \mathcal{E}_0$ is extremal for any proper subset $\mathcal{L}' \subset \mathcal{L}$.*

Moreover,

$$\sharp\mathcal{E}_0 \leq kn(X)_D + l_2 \max\{n(X)_C, n(X)_A\}.$$

Proof. We renumber as $R_1, \ldots, R_k$ the whole set of divisorial extremal rays of type (I) and as $R_{k+1}, \ldots, R_{k+l_2}$ the whole set of divisorial extremal rays of type (II) such that R_{k+i}, $1 \leq i \leq l_2$, belongs to a connected component of type $\mathfrak{C}_2$ of the set of all special extremal rays and R_{k+i} is the terminal vertex of the single arrow of this component.

We construct $\mathcal{E}_0$ in $k + l_2$ steps as a sequence

$$\emptyset = (\mathcal{E}_0)_0 \subset \cdots \subset (\mathcal{E}_0)_{k+l_2} = \mathcal{E}_0.$$

Here $(\mathcal{E}_0)_t$, $1 \leq t \leq k+l_2$, has the property that there does not exist an E-set $\mathcal{L}$ which contains one of extremal rays $R_1, \dots, R_t$, and $\mathcal{L}$ contains at least two elements which do not belong to $(\mathcal{E}_0)_t$, and $(\mathcal{E}_0)_t \cup \mathcal{L}'$ is extremal for any proper subset $\mathcal{L}' \subset \mathcal{L}$.

Suppose that we have constructed $(\mathcal{E}_0)_t$ with the above properties and $t < k + l_2$. Assume that there exists an E-set U which contains R_{t+1}, and U contains at least two elements which do not belong to $(\mathcal{E}_0)_t$, and $U' \cup (\mathcal{E}_0)_t$ is extremal for any proper subset $U' \subset U$. Then we set

$$(\mathcal{E}_0)_{t+1} = (\mathcal{E}_0)_t \cup (U \setminus \{R_{t+1}\}).$$

Then $(\mathcal{E}_0)_{t+1}$ is extremal by the conditions on $(\mathcal{E}_0)_t$ and U above. Suppose that there exists an E-set $\mathcal{L}$ which contains R_i, $1 \leq i \leq t+1$, and $\mathcal{L}$ contains at least two elements which do not belong to $(\mathcal{E}_0)_{t+1}$, and $\mathcal{L}' \cup (\mathcal{E}_0)_{t+1}$ is extremal for any proper subset $\mathcal{L}' \subset \mathcal{L}$. If $1 \leq i \leq t$, we have similar properties for $(\mathcal{E}_0)_t$ and get a contradiction. Thus $i = t+1$. By our conditions, there exists $Q \in \mathcal{L} \setminus \{R_{t+1}\}$ such that $Q \notin (\mathcal{E}_0)_{t+1}$ and $(\mathcal{E}_0)_{t+1} \cup (\mathcal{L} \setminus \{Q\})$ is extremal. Since $R_{t+1} \in \mathcal{L} \setminus \{Q\}$ and $U \setminus \{R_{t+1}\} \subset (\mathcal{E}_0)_{t+1}$, it follows that $U \subset (\mathcal{E}_0)_{t+1} \cup (\mathcal{L} \setminus \{Q\})$ is extremal because $(\mathcal{E}_0)_{t+1} \cup (\mathcal{L} \setminus \{Q\})$ is extremal. We get a contradiction since U is an E-set. Thus $(\mathcal{E}_0)_{t+1}$ has the required properties. Here, by construction and Theorem 2.2.6, the set U has type (D) if $1 \leq t+1 \leq k$, and the set U has type (C) or (A) if $k < t+1 \leq k+l_2$. If there does not exist the E-set U above, we just put $(\mathcal{E}_0)_{t+1} = (\mathcal{E}_0)_t$.

By our construction, $\sharp\mathcal{E}_0 \leq kn(X)_D + l_2 \max\{n(X)_C, n(X)_A\}$. By our construction, $\mathcal{E}_0$ is divisorially connected since any two different E-sets are connected by arrows (by Lemma 3.4) and extremal rays $R_1, \dots, R_{k+l_2}$ are divisorially disjoint.

Let $R_{k+l_2+1}, \dots, R_{k+l}$ be terminal vertices of arrows of all components of types $\mathfrak{C}_k$, $k \geq 3$. By our construction, any E-set $\mathcal{L}$ does not contain any of the extremal rays $R_1, \dots, R_{k+l_2}$ if $\mathcal{L}$ has at least two elements which do not belong to $\mathcal{E}_0$ and $\mathcal{L}' \cup \mathcal{E}_0$ is extremal for any proper subset $\mathcal{L}' \subset \mathcal{L}$. We claim that we even have more: the E-set $\mathcal{L}$ above does not contain any of the extremal rays $R_1, \dots, R_{k+l}$. Actually, if the $\mathcal{E}$-set $\mathcal{L}$ contains R_i, $k + l_2 < i \leq k + l$, we get a contradiction with Theorems 2.1.8 and 2.2.6: the set $\mathcal{L}$ should be of type (B), but then it is extremal and cannot be an E-set.

Since $\rho(X) > 3$, by the statement (2), any terminal vertex of a single arrow on X is one of the extremal rays $R_{t+1}, \dots, R_{t+l}$. This finishes the proof of the lemma. □

We now continue the proof of Theorem 3.2. Consider the face $\gamma \subset \mathcal{M}(X) = \mathrm{NEF}(X)/\mathbb{R}^+$ orthogonal to $\mathcal{E}_0$. Let us consider constants $C_1(X)'$ and $C_2(X)'$ which are defined by the properties:

$$\sharp\big\{(R_1, R_2) \in (\mathcal{E} - \mathcal{E}_0) \times (\mathcal{E} - \mathcal{E}_0) \bigm| 1 \leq \rho(R_1, R_2) \leq d(X)_A\big\} \leq C_1(X)'\sharp(\mathcal{E} - \mathcal{E}_0);$$

and

$$\sharp\big\{(R_1, R_2) \in (\mathcal{E} - \mathcal{E}_0) \times (\mathcal{E} - \mathcal{E}_0) \bigm| d(X)_A + 1 \leq \rho(R_1, R_2) \leq 2d(X)_A + 1\big\} \leq C_2(X)'\sharp(\mathcal{E} - \mathcal{E}_0).$$

for any extremal set $\mathcal{E}$ which contains $\mathcal{E}_0$ and distance ρ in the graph $G(\mathcal{E})$.

Directly applying Theorem 1.2 from [N8] to γ, we get the inequality

$$\dim \gamma < (16/3)C_1(X)' + 4C_2(X)' + 6, \tag{3-1}$$

which is less than we want.

To get the required estimate

$$\dim \gamma < 8C(X)_A + 6, \tag{3-2}$$

one needs to change a little the proof of Theorem 1.2 from [N8] for our case (obviously, $C(X)_A \le (C_1(X)' + C_2(X)')/2$ and (3-2) implies (3-1)).

Let $\angle$ be an oriented (plane) angle of γ. Let $\mathcal{R}(\angle)$ be the set of all extremal rays of $\overline{\mathrm{NE}}(X)$ which are orthogonal to the vertex of $\angle$. The set $\mathcal{R}(\angle)$ is the disjoint union

$$\mathcal{R}(\angle) = \mathcal{R}(\angle^{\perp}) \cup \{R_1(\angle)\} \cup \{R_2(\angle)\},$$

where $\mathcal{R}(\angle^{\perp})$ contains all rays orthogonal to the plane of the angle $\angle$, the rays $R_1(\angle)$ and $R_2(\angle)$ are orthogonal to the first and second side of the oriented angle $\angle$, respectively. Obviously, the set $\mathcal{R}(\angle)$ and the ordered pair of rays $(R_1(\angle), R_2(\angle))$ define the oriented angle $\angle$ uniquely. We define the weight $\sigma_A(\angle)$ by the following formula (for the proof of Theorem 1.2 from [N8], the definition of the weight $\sigma(\angle)$ was different):

$$\sigma_A(\angle) = \begin{cases} 1/2, & \text{if } 1 \le \rho_A(R_1(\angle), R_2(\angle)) \le 2d(X)_A + 1, \\ 0, & \text{if } 2d(X)_A + 1 < \rho_A(R_1(\angle), R_2(\angle)). \end{cases} \tag{3-3}$$

Similarly to the proof of Theorem 1.2 from [N8], one can check the conditions of Lemma 1.4 (Vinberg's Lemma) from [N8] for the polyhedron $\mathcal{M} = \gamma$ with the constants $C = C(X)_A$ and $D = 0$. The proof is even simpler because the weight $\sigma_A(\angle')$ of the angle $\angle'$ with the opposite orientation is equal to $\sigma_A(\angle)$. The inequality (3-2) follows. For the convenience of the reader, we recall Vinberg's Lemma.

Lemma 3.7. *Let $\mathcal{M}$ be a convex simple (i.e. dual to simplicial) polyhedron of dimension n. Let C and D be some numbers. Suppose that oriented angles (2-dimensional, plane) of $\mathcal{M}$ are given weights and the following conditions* (1) *and* (2) *hold:*

(1) *The sum of weights of all oriented angles at any vertex of $\mathcal{M}$ is $\le Cn + D$.*

(2) *The sum of weights of all oriented angles of any 2-dimensional face of $\mathcal{M}$ is at least $5 - k$ where k is the number of vertices of the 2-dimensional face.*

Then

$$n < 8C + 5 + \begin{cases} 1 + 8D/n & \text{if } n \text{ is even,} \\ (8C + 8D)/(n-1) & \text{if } n \text{ is odd.} \end{cases}$$

In particular, for $C \ge 0$ and $D = 0$, we have

$$n < 8C + 6.$$

Since $\dim \gamma = \dim N_1(X) - \sharp\mathcal{E}_0 - 1$ and $\sharp\mathcal{E}_0 \le kn(X)_D + l_2 \max\{n(X)_C, n(X)_A\}$, we get the estimate $\dim N_1(X) \le kn(X)_D + l_2 \max\{n(X)_C, n(X)_A\} + \dim \gamma + 1$.

By the statement (2), $k + l_2 \le q(X)$. Thus by (3-2),

$$\dim N_1(X) \le q(X) \max\{n(X)_D, n(X)_C, n(X)_A\} + 8C(X)_A + 6.$$

This finishes the proof of Theorem 3.2. □

Remark 3.8. Let us consider Fano 3-folds X with $\mathbb{Q}$-factorial terminal singularities and which satisfy the conditions (a), (b) and (c). Then the constants $q(X) \le 1$, $n(X)_D \le 1$, $n(X)_C = -1$, $n(X)_A \le 1$, $d(X)_A = 1$ and $C(X)_A = 0$ (see [N8]). Thus by Theorem 3.2, we get the main result of [N8]: $k + (n_1 - 1) + \cdots + (n_l - 1) \le 1$ and $\rho(X) \le 7$.

In Sections 4–5, we apply Basic Theorems 1.3.2 and 3.2 to Calabi–Yau 3-folds.

4. Application to Calabi–Yau 3-folds

4.1. Reminder. Here we recall general facts about Calabi–Yau 3-folds we shall use.

A projective algebraic 3-dimensional manifold X over $\mathbb{C}$ is called *Calabi–Yau* if the canonical class $K_X = 0$ and $h^1(X, \mathcal{O}_X) = h^2(X, \mathcal{O}_X) = 0$. Then $N^1(X) = H^2(X; \mathbb{R})$.

It was shown by Wilson [W1] and [W2] that the nef cone NEF(X) is locally rational polyhedral away from the cubic intersection hypersurface

$$\mathcal{W}_X = \{x \in H^2(X; \mathbb{R}) \mid x^3 = 0\}.$$

The main tool here is to use results of Kawamata [Ka1] and Shokurov [Sh] (which generalize results of Mori [Mo1]) about "polyhedrality" of the Mori cone of algebraic varieties with log terminal singularities. Thus we can speak about faces $\gamma \subset$ NEF(X) considering only faces of the locally polyhedral cone NEF(X) away from the cone $\mathcal{W}$. Then the corresponding face $\gamma' = \gamma^\perp$ of the Mori cone $\overline{\mathrm{NE}}(X)$ is finite polyhedral and $\dim \gamma' = \mathrm{codim}(\gamma')^\perp$ where $(\gamma')^\perp = \gamma$. Obviously, these faces γ' are exactly finite polyhedral faces of $\overline{\mathrm{NE}}(X)$ of numerical Kodaira dimension 3 and with the property $\mathrm{codim}(\gamma')^\perp = \dim \gamma'$. It is known (see [W1], [W2], [O]) that these faces are contractible and have Kodaira dimension 3 (one should apply theory of log terminal extremal contractions of [Ka1] and [Sh]; in fact, one need these results to prove the results of Wilson mentioned above). Thus we can speak about extremal rays of Kodaira dimension 3 (they are orthogonal to faces of NEF(X) of highest dimension) and extremal rays of types (I), (II) and (III) (small). Applying results of [Ka1] and [Sh], we get that a sequence $X \to X'$ of contractions of divisorial extremal rays gives rise to a 3-fold with $\mathbb{Q}$-factorial (canonical) singularities and X' inherits all properties of X we mentioned above. In particular, we get that X and X' belong to the class $\mathcal{LT}$.

The following fact is very important for us (see [W2] and also appendix of Shokurov): For an extremal ray R of type (II) of a Calabi–Yau 3-fold X, there exists a curve

$C \in R$ such that

$$C \cdot D(R) = -2. \tag{4-1-1}$$

This curve C is the general fiber of the ruling $f\colon D(R) \to f(D(R))$ for the contraction $f\colon X \to X'$ of the extremal ray R. But, we don't need the last fact in this paper.

Let Q be another divisorial extremal ray of X. Using a curve $C \in R$ satisfying (4-1-1), we assign to the arrow RQ the weight $C \cdot D(Q)$. Thus from now on, arrows of the graph $G(T)$ of a set T of extremal rays of type (II) have corresponding weights. We don't indicate the corresponding weight only if this weight is equal to 1.

For an extremal ray R of type (I) we choose some curve $C \in R$. This gives rise to corresponding weights of arrows too. As shown by Shokurov (see Appendix), there exists a curve $C \in R$ such that $C \cdot D(R) = -1, -2$ or -3. We shall use this result later. Now we fix the choice of C showing the weight $-k = C \cdot D(R)$ of the vertex corresponding to R. Thus from now on, arrows and black vertices of the graph $G(T)$ of any set T of extremal rays of type (I) or (II) have given weights. For a manifold X these weights are integers.

Here we want to apply the theory developed above to the study of Calabi–Yau 3-folds.

4.2. Sets of divisorial extremal rays on Calabi–Yau 3-folds. Here we describe elliptic, parabolic and Lannér (some quasi-Lannér too) sets of extremal rays of type (I) or (II) on Calabi–Yau 3-folds. By Theorem 2.2.6, they have types (A), (B), (C), (D) or (E), (E′). By the results of Section 1.2 and Theorem 2.2.6, this is enough for the description of sets of extremal rays of type (I) or (II) which are either extremal of Kodaira dimension 3 or are E-sets $\mathcal{L}$ such that each proper subset of $\mathcal{L}$ is extremal of Kodaira dimension 3 and $\mathcal{L}$ satisfies the condition (iii) of Section 1.1.

4.2.1. Sets of type (A). We have the following results:

Theorem 4.2.1.1. *Let X be a Calabi–Yau 3-fold. Suppose that $T = \{R_1, \dots, R_n\}$ is a set of extremal rays of type* (II) *on X such that all the divisors $D(R_1), \dots, D(R_n)$ are different and the graph $G(T)$ does not have a single arrow (thus, it has type* (A)*) and is connected.*

Then:

(a) *T is elliptic if and only if $G(T)$ is a Dynkin diagram of the root system type*

$$\mathbf{A}_n,\ \mathbf{B}_n,\ \mathbf{C}_n,\ \mathbf{D}_n,\ \mathbf{E}_6,\ \mathbf{E}_7,\ \mathbf{E}_8,\ \mathbf{F}_4,\ \mathbf{G}_2$$

of Table 4.1.

(b) *T is parabolic if and only if $G(T)$ is an extended Dynkin diagram of the root system type*

$$\widetilde{\mathbf{A}}_n,\ \widetilde{\mathbf{B}}_n,\ \widetilde{\mathbf{BC}}_n,\ \widetilde{\mathbf{C}}_n,\ \widetilde{\mathbf{BD}}_n,\ \widetilde{\mathbf{CD}}_n,\ \widetilde{\mathbf{D}}_n,\ \widetilde{\mathbf{E}}_6,\ \widetilde{\mathbf{E}}_7,\ \widetilde{\mathbf{E}}_8,\ \widetilde{\mathbf{BF}}_4,\ \widetilde{\mathbf{CF}}_4,\ \widetilde{\mathbf{AG}}_2,\ \widetilde{\mathbf{GA}}_2$$

of Table 4.2, where the weights of vertices show the coefficients a_i of Definition 2.2.2.

(c) *T is Lannér if and only if $G(T)$ is one of the diagrams of Table 4.3.*

Proof. Applying (4-1-1), this is a standard and well known fact. □

4.2.2. Sets of type (B).

Theorem 4.2.2.1. *Let X be a Calabi–Yau 3-fold. Suppose that $T = \{R_1, \dots, R_n\}$ is a set of extremal rays of type (II) on X such that all the divisors $D(R_1), \dots, D(R_n)$ are different and the graph $G(T)$ has type (B) of Theorem 2.2.2. Thus $T = \{R\} \cup C_1 \cup \dots \cup C_k$, $k \geq 1$, where*

$$C_1 = \{R_{11}, \dots, R_{1n_1}\},\ C_2 = \{R_{21}, \dots, R_{2n_2}\}, \dots,\ C_k = \{R_{k1}, \dots, R_{kn_k}\}$$

are pairwise divisorially disjoint chains, and all arrows between R and extremal rays of these chains are single arrows $R_{j1}R$, $j = 1, \dots, k$.

Then T cannot be parabolic or Lannér. T is elliptic if and only if each chain $C_1, \dots, C_k$ is one of the chains

$$\mathbf{A}_n,\ \mathbf{B}_n,\ \mathbf{C}_n,\ \mathbf{F}_4,\ \mathbf{G}_2$$

of Theorem 4.2.1.1, and the weights of single arrows $R_{j1}R$, $j = 1, \dots, k$, are arbitrary natural numbers.

Proof. If T is either elliptic, or parabolic or Lannér, the chains $C_1, \dots, C_k$ are elliptic because they give proper subsets of T. By Theorem 4.2.1.1, they have types $\mathbf{A}_n$, $\mathbf{B}_n$, $\mathbf{C}_n$, $\mathbf{F}_4$ or $\mathbf{G}_2$. One can easily show (see the proof of Theorem 2.2.6) that then T is elliptic for arbitrary weights of arrows $R_{j1}R$ which should be natural numbers. This finishes the proof. □

4.2.3. Sets of type (C).

Theorem 4.2.3.1. *Let X be a Calabi–Yau 3-fold. Suppose that $T = \{R_1, \dots, R_n\}$ is a set of extremal rays of type (II) on X such that all the divisors $D(R_1), \dots, D(R_n)$ are different and the graph $G(T)$ has type (C) of Theorem 2.2.6.*

Then T is elliptic, parabolic, Lannér or quasi-Lannér if and only if $G(T)$ is one of the elliptic, parabolic, Lannér or quasi-Lannér diagrams respectively of Table 4.4 below.

Proof. The corresponding calculations are very simple using (4-1-1). □

4.2.4. Sets of type (D).

Theorem 4.2.4.1. *Let X be a Calabi–Yau 3-fold. Suppose that $T = \{R_1, \ldots, R_n\}$ is a set of extremal rays of type* (I) *or* (II) *on X such that the graph $G(T)$ has type* (D) *of Theorem* 2.2.6.

Then T is elliptic, parabolic, Lannér or quasi-Lannér if and only if $G(T)$ is one of the elliptic, parabolic, Lannér or quasi-Lannér diagrams respectively of Table 4.5 *below (we recall that a black vertex corresponds to an extremal ray of type* (I)*).*

Proof. The corresponding calculations are very simple using (4-1-1). We remark that a graph $G(T)$ without a black vertex should have one of the types $\mathbf{A}_n$, $\mathbf{B}_n$, $\mathbf{C}_n$, $\mathbf{F}_4$ or $\mathbf{G}_2$ by Theorem 4.2.1.1, (a). □

4.2.5. Sets of type (E).

Theorem 4.2.5.1. *Let X be a Calabi–Yau 3-fold. Suppose that $T = \{R_1, R_2, R_3\}$ is a set of extremal rays of type* (II) *on X such that the graph $G(T)$ has type* (E) *(triangle or special triangle) of Theorem* 2.2.6.

Then T is elliptic, parabolic, Lannér or quasi-Lannér if and only if $G(T)$ is one of the elliptic, parabolic, Lannér or quasi-Lannér diagrams respectively of Table 4.6 *below.*

Proof. The corresponding calculations are very simple using (4-1-1). □

Table 4.1. Calabi–Yau elliptic diagrams without single arrows (classical Dynkin diagrams).

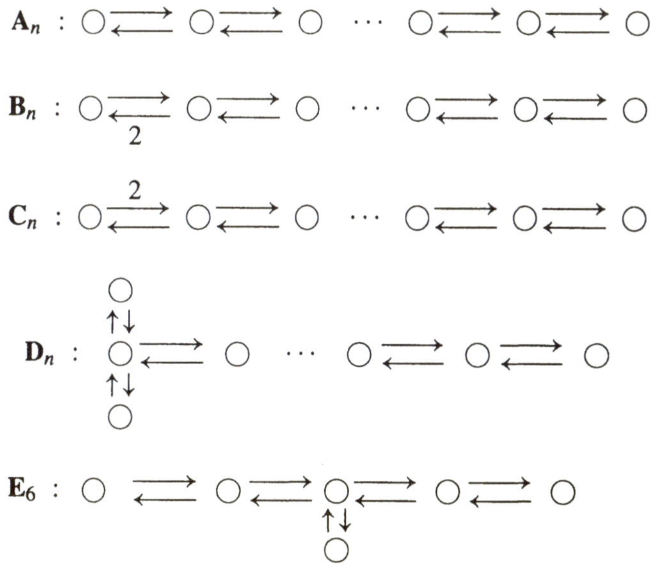

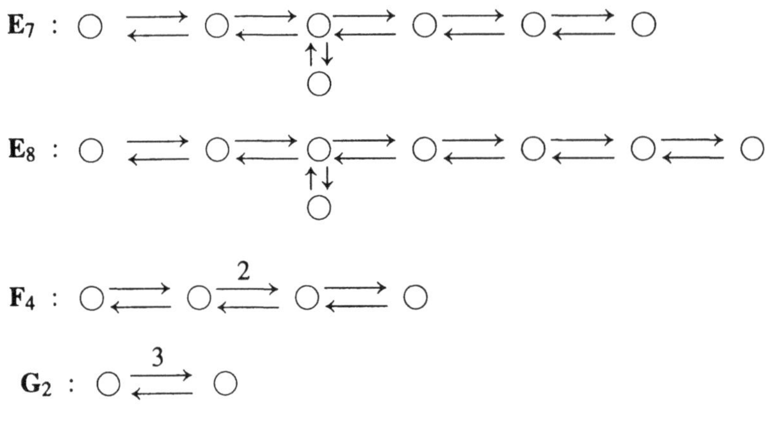

Table 4.2. Calabi–Yau parabolic diagrams without single arrows (classical extended Dynkin diagrams).

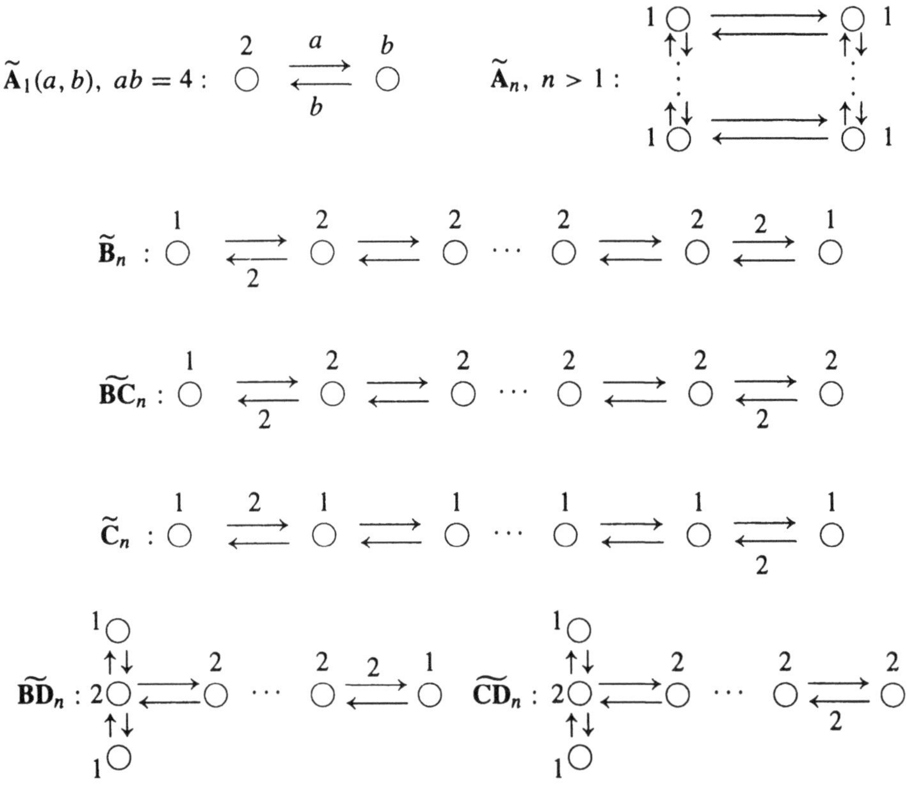

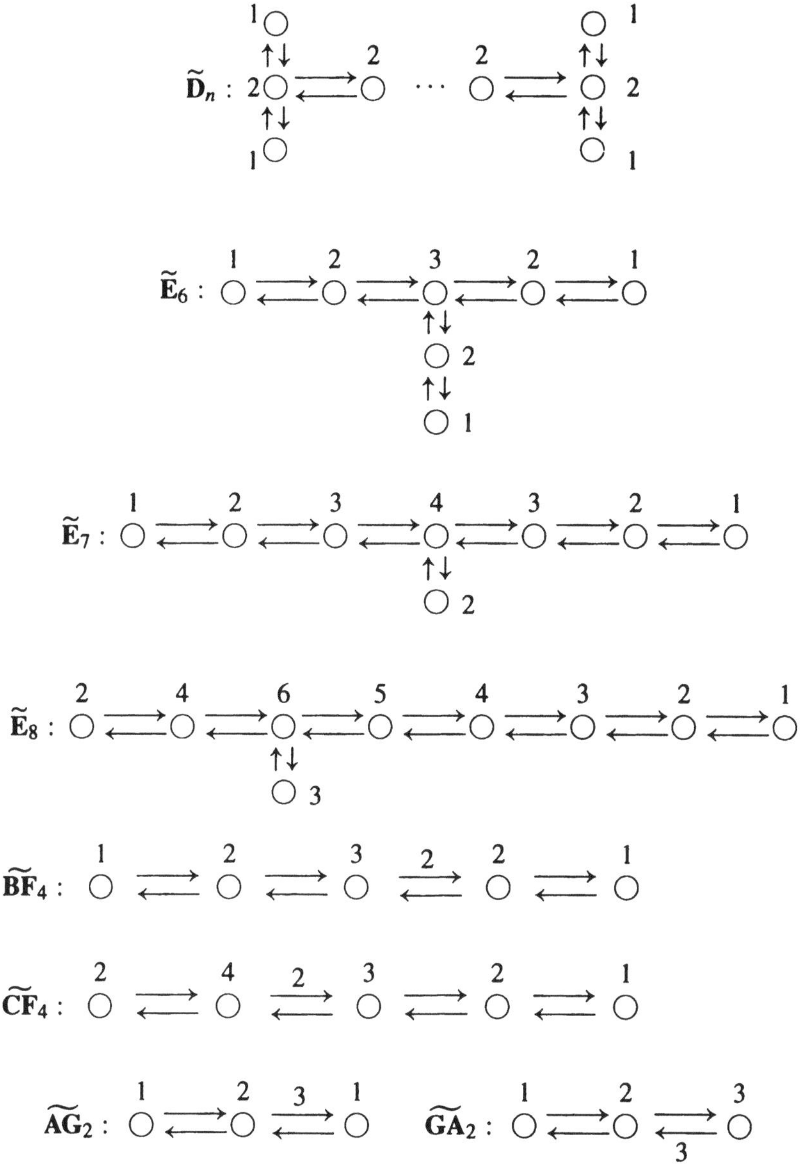

Table 4.3. Calabi–Yau Lannér diagrams without single arrows (classical Lannér diagrams).

$$\bigcirc \underset{t_{21}}{\overset{t_{12}}{\rightleftarrows}} \bigcirc \qquad \text{where } t_{12}t_{21} > 4$$

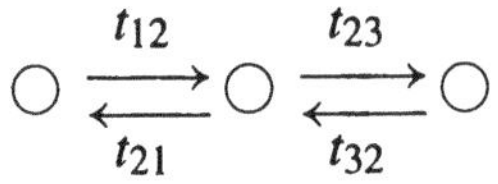

where $0 < t_{12}t_{21} < 4,\ 0 < t_{23}t_{32} < 4,\ t_{12}t_{21} + t_{23}t_{32} > 4$

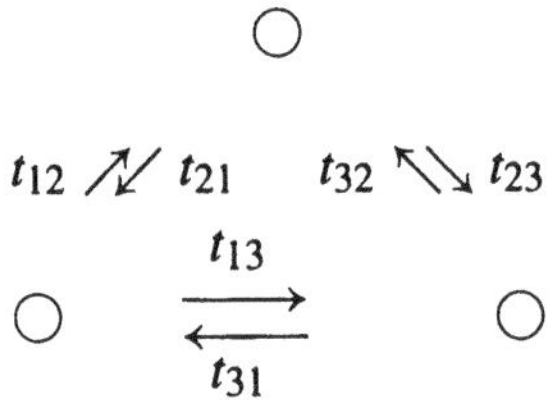

where $0 < t_{12}t_{21} < 4,\ 0 < t_{23}t_{32} < 4,$
$0 < t_{31}t_{13} < 4,\ t_{12}t_{21} + t_{23}t_{32} + t_{31}t_{13} > 3$

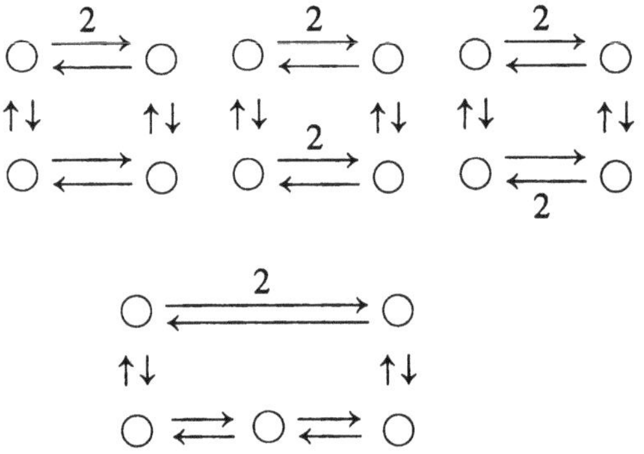

Table 4.4. Calabi–Yau diagrams of type (C).

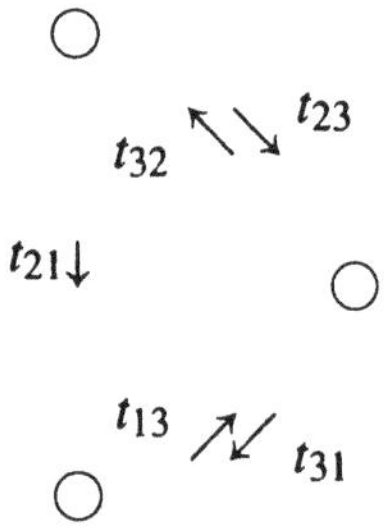

Elliptic diagram where
$t_{21}t_{13}t_{32} + 2t_{13}t_{31} + 2t_{23}t_{32} < 8$

Parabolic diagram where
$t_{21}t_{13}t_{32} + 2t_{13}t_{31} + 2t_{23}t_{32} = 8$

Quasi-Lannér diagram where
$t_{13}t_{31} \leq 4, t_{23}t_{32} \leq 4$
and $t_{21}t_{13}t_{32} + 2t_{13}t_{31} + 2t_{23}t_{32} > 8$

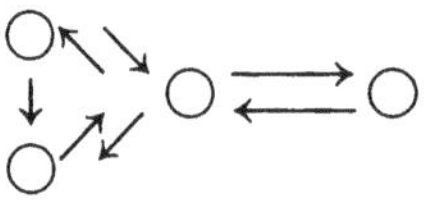
Elliptic diagram

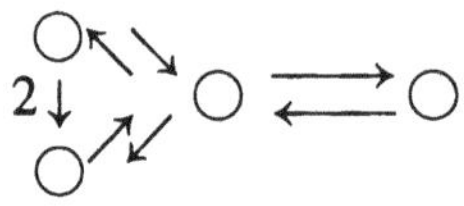

Parabolic diagram

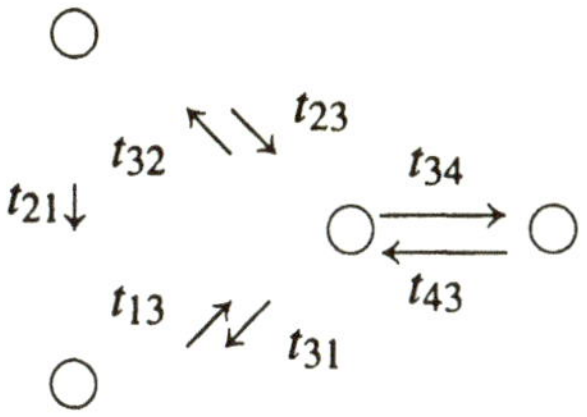

Quasi-Lannér diagram where
$t_{13}t_{31} + t_{34}t_{43} \leq 4$, $t_{23}t_{32} + t_{34}t_{43} \leq 4$,
$t_{21}t_{13}t_{32} + 2t_{13}t_{31} + 2t_{23}t_{32} \leq 8$,
$t_{21}t_{13}t_{32} + 2t_{13}t_{31} + 2t_{23}t_{32} + 2t_{34}t_{43} > 8$

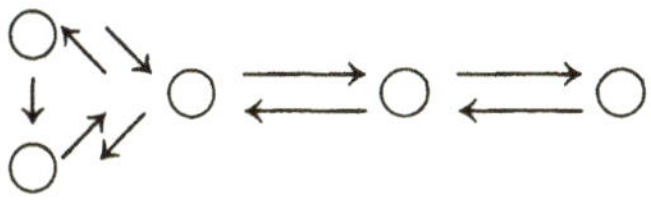

Elliptic diagram

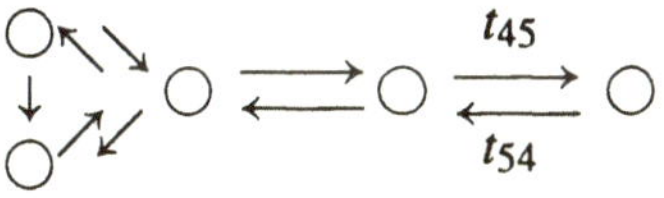

Lannér diagram
where $t_{45}t_{54} = 2$

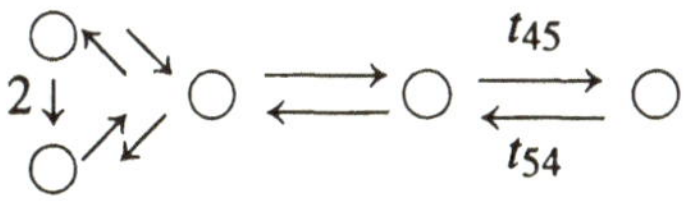

Quasi-Lannér diagram
where $1 \leq t_{45}t_{54} \leq 2$

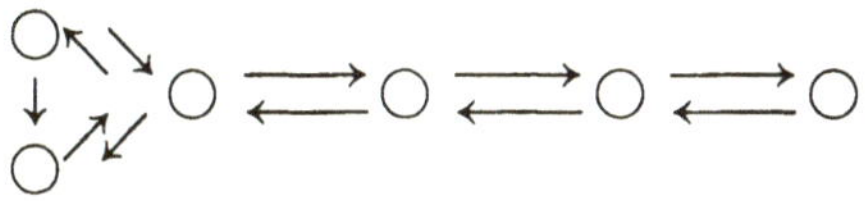

Elliptic diagram

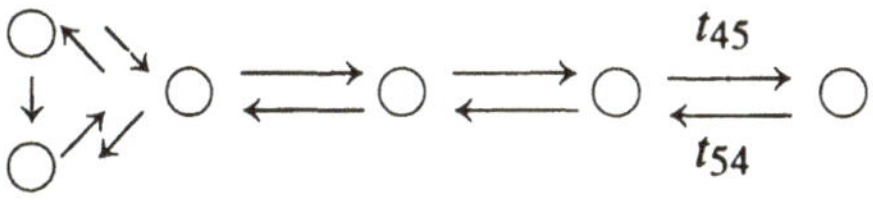

Lannér diagram where $t_{45}t_{54} = 2$

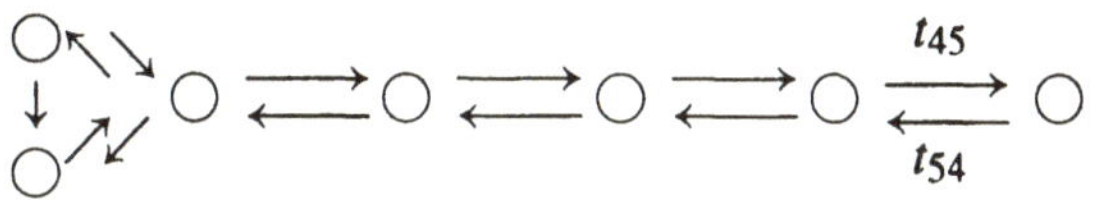

Quasi-Lannér diagram
where $t_{45}t_{54} = 2$

Table 4.5. Calabi–Yau diagrams of type (D) (n is equal to the number of white vertices).

$\mathbf{A}_n^\bullet(k; a, b)$:

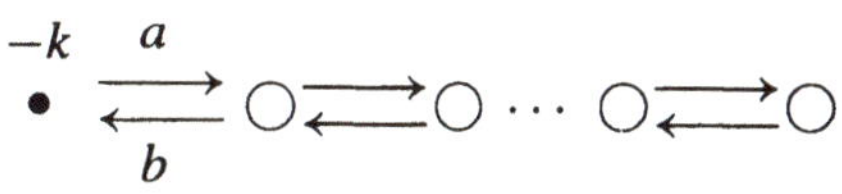

Elliptic diagram iff $ab < k(n+1)/n$. Parabolic diagram iff $ab = k(n+1)/n$. Quasi-Lannér diagram iff $kn/(n-1) \geq ab > k(n+1)/n$; in particular, $n \leq k+1$. Lannér diagram iff $kn/(n-1) > ab > k(n+1)/n$; in particular, $n \leq 2$ for $1 \leq k \leq 3$.

$\mathbf{B}_n^\bullet(k; a, b)_1$: 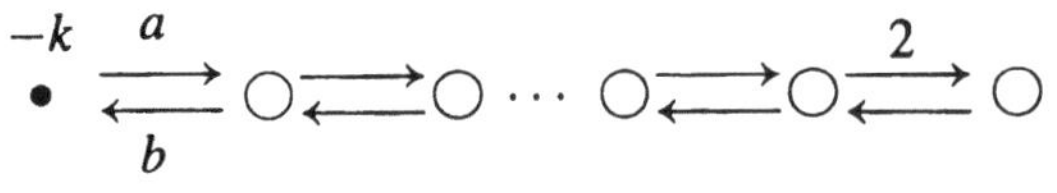

$\mathbf{C}_n^\bullet(k; a, b)_1$: 

Elliptic diagram iff $ab < k$. Parabolic diagram iff $ab = k$. Quasi-Lannér diagram iff $kn/(n-1) \geq ab > k$; in particular, $n \leq k+1$. Lannér diagram iff $kn/(n-1) > ab > k$; in particular, $n < k+1$.

$\mathbf{B}_n^\bullet(k; a, b)_2$: 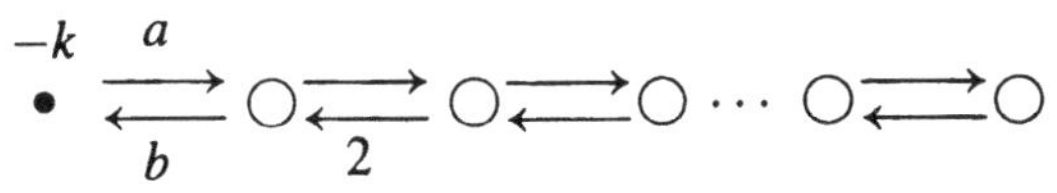

$\mathbf{C}_n^\bullet(k; a, b)_2$: 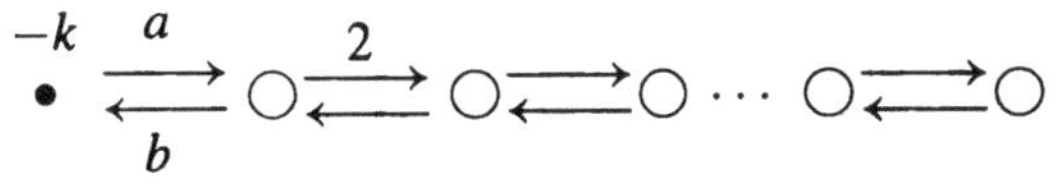

Elliptic diagram iff $ab < 2k/n$. Parabolic diagram iff $ab = 2k/n$. Quasi-Lannér diagram iff $2k/(n-1) \geq ab > 2k/n$; in particular, $n \leq 2k+1$. Lannér diagram iff $2k/(n-1) > ab > 2k/n$; in particular, $n \leq 2$ for $1 \leq k \leq 3$.

$\mathbf{F}_4^\bullet(k; a, b)_1$: 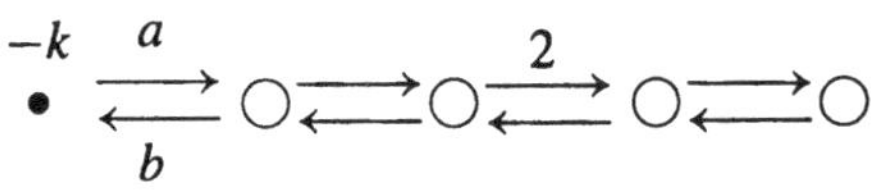

$\mathbf{F}_4^\bullet(k; a, b)_2$: 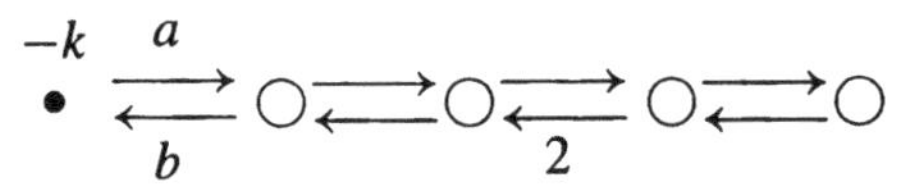

Elliptic diagram iff $ab < k/2$. Parabolic diagram iff $ab = k/2$. Quasi-Lannér diagram iff $k \geq ab > k/2$. Lannér diagram iff $k > ab > k/2$.

$\mathbf{G}_2^\bullet(k; a, b)_1$: 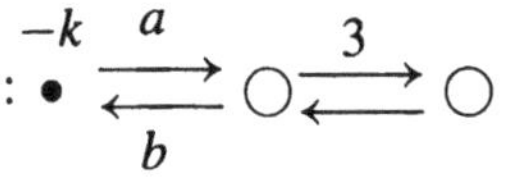 $\mathbf{G}_2^\bullet(k; a, b)_2$: 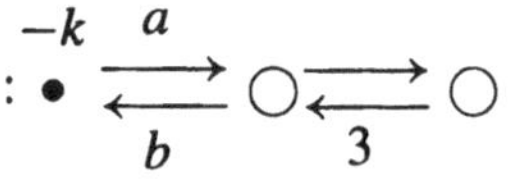

Elliptic diagram iff $ab < k/2$. Parabolic diagram iff $ab = k/2$. Quasi-Lannér diagram iff $k \geq ab > k/2$. Lannér diagram iff $k > ab > k/2$.

Table 4.6. Calabi–Yau triangle and special triangle diagrams.

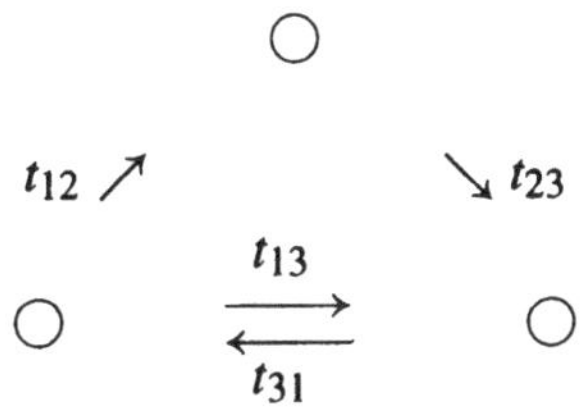

Elliptic where $t_{12}t_{23}t_{31} + 2t_{13}t_{31} < 8$

Parabolic where $t_{12}t_{23}t_{31} + 2t_{13}t_{31} = 8$

Lannér where
$t_{13}t_{31} < 4$ and $t_{12}t_{23}t_{31} + 2t_{13}t_{31} > 8$

Quasi-Lannér where
$t_{13}t_{31} \leq 4$ and $t_{12}t_{23}t_{31} + 2t_{13}t_{31} > 8$

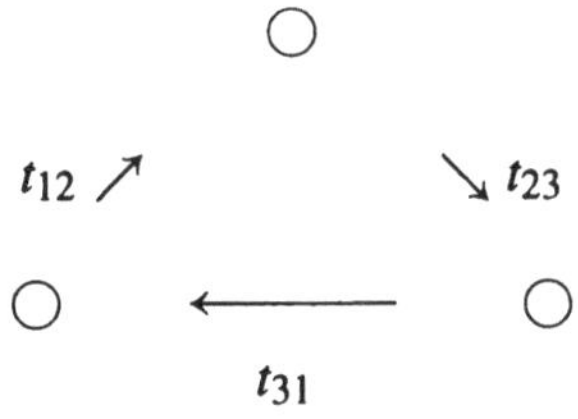

Elliptic where $t_{12}t_{23}t_{31} < 8$

Parabolic where $t_{12}t_{23}t_{31} = 8$

Lannér where $t_{12}t_{23}t_{31} > 8$

4.3. Basic results on Calabi–Yau 3-folds

To prove our main results about Calabi–Yau 3-folds, we use our results above (especially, Basic Theorems 1.3.2 and 3.2 and Theorem 2.2.6) and the result of V. V. Shokurov about the length of extremal rays of type (I) on Calabi–Yau manifolds (see Appendix of V. V. Shokurov).

Theorem 4.3.1 (joint work with V. V. Shokurov). *Let X be a Calabi–Yau 3-fold and suppose that $\rho(X) > 40$.*

Then one of two cases (i) *or* (ii) *below hold:*

(i) *There exists a small extremal ray on X.*

(ii) *There exists a nef element h such that $h^3 = 0$ (thus, the nef cone* NEF(X) *and the cubic intersection hypersurface $\mathcal{W}_X$ have a common point; here, we don't claim that h is rational!).*

Proof. Assume that X does not have a nef element h with $h^3 = 0$. Since, by Wilson [W1] and [W2], NEF(X) is locally rational polyhedral away from $\mathcal{W}_X$, using compactness arguments, we get that NEF(X) is rational finite polyhedral. It follows that $\overline{\mathrm{NE}}(X)$ is rational finite polyhedral. Thus Theorem 4.3.1 follows from the next result.

Theorem 4.3.1′. *Let X be a Calabi–Yau 3-fold. Assume that*

(a) $\overline{\mathrm{NE}}(X)$ *is finite polyhedral;*

(b) X *does not have a small extremal ray;*

(c) X *does not have a rational nef element h with $h^3 = 0$.*

Then we have the following statements (1), (2) *and* (3) *for X:*

(1) X *does not have a pair of extremal rays of type $\mathfrak{B}_2$ and $\overline{\mathrm{NE}}(X)$ is simplicial;*

(2) X *has ≤ 2 extremal rays of type* (I) *(more generally, $k+(n_1-1)+\cdots+(n_l-1) \le 2$ in the notation of Theorem* 3.1);

(3) *(joint work with V. V. Shokurov) The Picard number $\rho(X) = \dim N_1(X) \le 4k + 5l_2 + 30 \le 40$.*

Proof. From the statement (1) of Theorem 1.3.2, the statement (1) follows.

Let $\mathcal{E}$ be an extremal set of Kodaira dimension 3 of extremal rays of type (I) or (II) on X. By Proposition 1.2.1, $\mathcal{E}$ is elliptic. By Theorem 2.2.6 and calculations of Section 4.2, the graph $G(\mathcal{E})$ is one of elliptic graphs of the Tables 4.1, 4.4, 4.5, 4.6 and Theorem 4.2.2.1.

In particular, $\sharp\mathcal{E} \le 2$ if $\mathcal{E}$ contains extremal rays of type (II) and the graph $G(\mathcal{E})$ is full (i.e. any two vertices are joined by non-single arrows). Thus for the constant $q(X)$ of Basic Theorems 1.3.2 and 3.2 (see Definition 1.3.1), we have $q(X) \le 2$. By Theorems 1.3.2, (2) and 3.2, (2) we get the statement (2).

Now let us estimate $\rho(X)$. To illustrate how Basic Theorems 1.3.2, (3) and 3.2, (3) work, we first give the worse estimates for $\rho(X)$ given by these general theorems.

First, let us apply Theorem 1.3.2, (3).

Considering Tables 4.1, 4.4, 4.5 and 4.6, one can easily see that for $d \ge 2$, we have (where we take the distance in the graph $G(\mathcal{E})$):

$$\sharp\{(R_1, R_2) \in \mathcal{E} \times \mathcal{E} \mid 1 \le \rho(R_1, R_2) \le d\} \le C_1(d)\sharp\mathcal{E};$$

and

$$\sharp\{(R_1, R_2) \in \mathcal{E} \times \mathcal{E} \mid d+1 \le \rho(R_1, R_2) \le 2d+1\} \le C_2(d)\sharp\mathcal{E}.$$

where

$$C_1(d) \le 2d, \quad C_2(d) \le 2(d+1). \tag{4-3-1}$$

Let $\mathcal{L}$ be an E-set on X. Since all the extremal rays on X are of type (I) or (II) and X does not have a pair of extremal rays of type $\mathfrak{B}_2$, $\mathcal{L}$ is Lannér by Lemma 1.2.11. Using the calculations of Section 4.2, we then get that the diameter

$$\mathrm{diam}(G(\mathcal{L})) \le 4.$$

Here the maximum $\mathrm{diam}(G(\mathcal{L})) = 4$ we get for the Lannér diagram with 6 vertices of the Table 4.4 and the Lannér diagrams $\mathbf{F}_4^{\bullet}(k; a, b)_1$ and $\mathbf{F}_4^{\bullet}(k; a, b)_2$ of type (D) of the Table 4.5. Thus the constant $d(X)$ of Basic Theorem 1.3.2 is bounded by $d(X) \le 4$. Here, for Lannér diagrams of type (D) (Table 4.5), we use the result of V. V. Shokurov (see Appendix): for an extremal ray R of type (I) on X, there exists a curve $C \in R$ such that $C \cdot D(R) = -k$ where $1 \le k \le 3$. By (4-3-1), we can apply Theorem 1.3.2 with the constants $C_1(X) = 8$ and $C_2(X) = 10$. By Theorem 1.3.2, we get $\rho(X) = \dim N_1(X) \le (16/3)C_1(X)+4C_2(X)+6 = (16/3)\cdot 8+4\cdot 10+6 = 88+2/3 < 89$. Thus Basic Theorem 1.3.2 gives the estimate

$$\rho(X) \le 88.$$

Now let us apply Basic Theorem 3.2, (3). Similar considerations give constants:

$$n(X)_D \le 4, n(X)_C \le 5, n(X)_A \le 4, d(X)_A \le 2, \quad C(X)_A \le 5.$$

Thus by Basic Theorem 3.2, (3),

$$\rho(X) \le 4k + 5l_2 + 8 \cdot 5 + 6 = 4k + 5l_2 + 46,$$

and

$$\rho(X) \le 56$$

(since $k + l \le q(X) = 2$).

To get the strong estimate

$$\rho(X) \le 4k + 5l_2 + 30 \le 40,$$

we should change for this case the proof of the statement (3) of Basic Theorem 3.2. We should change the definition (3-3) of the weight $\sigma_A(\angle)$ of oriented angles of γ according to É.B. Vinberg [V1, §6, Sect. 1]. We denote this new weight by $\sigma_{AV}(\angle)$.

This is the following: We consider the extremal set $\mathcal{R}(\angle)'$ (see Definition 3.1) and connected components of the graph $G(R(\angle)')$. This graph contains non-single arrows only and its connected components have types

$$\mathbf{A}_n,\ \mathbf{B}_n,\ \mathbf{C}_n,\ \mathbf{D}_n,\ \mathbf{E}_6,\ \mathbf{E}_7,\ \mathbf{E}_8,\ \mathbf{F}_4,\ \mathbf{G}_2$$

by Theorem 3.2.1.1, (a).

The weight $\sigma_{AV}(\angle)$ is non-negative. It is not equal to zero only if both extremal rays $R_1(\angle)$, $R_2(\angle)$ belong to one connected component S of the graph $G(\mathcal{R}(\angle)')$ and we have one of five cases below:

(I) $\rho_A(R_1(\angle), R_2(\angle)) = 1$ (i.e. $R_1(\angle)$ and $R_2(\angle)$ are adjoined in $G(S)$);

(II) the connected component S contains ≤ 7 vertices;

(III) the connected component S is classical (i.e. it has type $\mathbf{A}_n$, $\mathbf{B}_n$, $\mathbf{C}_n$ or $\mathbf{D}_n$) and $R_1(\angle)$, $R_2(\angle)$ belong to the terminal interval of the order ≤ 6 of $G(S)$ (for S of type $\mathbf{D}_n$, $n \ge 5$, by definition, the terminal interval is a connected subgraph of $G(S)$ which contains a terminal vertex of S and is invariant relative to the involution automorphism of $\mathbf{D}_n$).

(IV) S has type $\mathbf{E}_8$ and the pair $R_1(\angle)$, $R_2(\angle)$ is different from pairs of vertices of $\mathbf{E}_8$ below marked by ⊗:

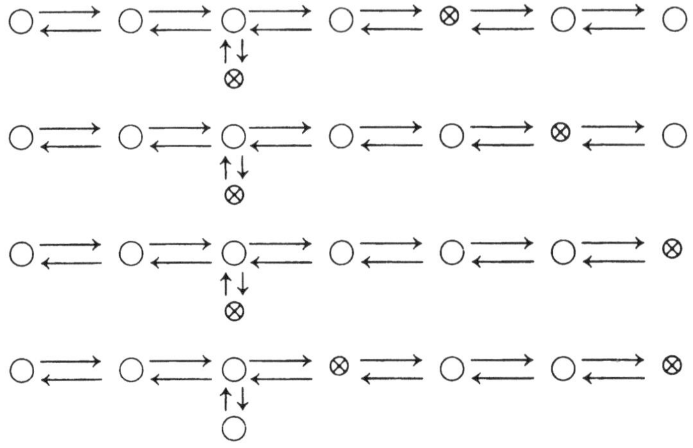

The weight $\sigma_{AV}(\angle) = 1$ if

(V) S has ≤ 4 vertices.

In all other cases (I)–(IV) above, $\sigma_{AV}(\angle) = 1/2$.

Now we should check the conditions of Vinberg's Lemma 3.7 with the constants $C = 3$ and $D = 0$ which gives the required estimate: $\dim \gamma < 30$.

The proof of the condition (1) is very similar to Vinberg [V1, §6, Sect. 2] and is not difficult.

To prove the condition (2), one should also follow to Vinberg [V1, §6, Sects. 3–12]. For our case, the elliptic diagrams are only "crystallographic", i.e. of types $\mathbf{A}_n$, $\mathbf{B}_n$, $\mathbf{C}_n$, $\mathbf{D}_n$, $\mathbf{E}_6$, $\mathbf{E}_7$, $\mathbf{E}_8$, $\mathbf{F}_4$, $\mathbf{G}_2$, and the E-sets (or Lannér diagrams) which take part in the proof have only types of Table 4.3. One should follow Vinberg step by step (we should say that his considerations are long and very delicate, and it is hard work), and check that for our "crystallographic case" he uses only two arguments which work for our situation: There do not exist a configuration of extremal rays of type (II) with one of extended Dynkin diagram of Table 4.2, because then some linear combination of divisors of these rays gives a nef element h with $h^3 = 0$. Moreover, two non-extremal sets of extremal rays of type (I) or (II) cannot be orthogonal (Lemma 3.4).

Only for the "non-crystallographic case", Vinberg uses *superhyperbolicity* arguments which probably do not work for our situation.

This finishes the proof of the theorem. □

5. $\mathbb{Q}$-factorial models of Calabi–Yau 3-folds

Applying the Diagram Method to an arbitrary Calabi–Yau 3-fold X, we have to avoid two problems: the Mori cone $\overline{\mathrm{NE}}(X)$ may not be finite polyhedral, and X may have small extremal rays. Here we want to discuss one possibility to avoid these problems and involve non-polyhedral case and small extremal rays in the game.

One can make several transformations (i) and (ii) below:

(i) Contraction of a divisorial extremal ray.

(ii) Flop in a small extremal ray.

Repeating this operations, we get some 3-fold Y. The 3-fold Y still has the most important properties: Y has $\mathbb{Q}$-factorial (canonical) singularities and belongs to the class $\mathcal{LT}$. See Kawamata [Ka2] and Shokurov [Sh2].

Definition 5.1. A 3-fold Y obtained starting from a Calabi–Yau 3-fold X and repeating transformations (i) and (ii) is called a *$\mathbb{Q}$-factorial model* of the Calabi–Yau manifold X.

Considering $\mathbb{Q}$-factorial models gives us a chance to avoid the cases when either the Mori cone is not finite polyhedral or there exists a small extremal ray. If the $\mathbb{Q}$-factorial model Y has a finite polyhedral Mori cone and does not have a small extremal ray, we can apply the Diagram Method to Y to find a rational nef element h on Y with $h^3 = 0$. By the way, using the statement (1) of Basic Theorem 1.3.2, we can prove that transformations (i) and (ii) do not make the situation worse from the point of view of the Diagram Method.

Lemma 5.2. *Let Y be a $\mathbb{Q}$-factorial model such that we have properties* (a), (b) *and* (c) *below:*

(a) $\overline{\mathrm{NE}}(Y)$ *is finite polyhedral;*

(b) *Y does not have a small extremal ray;*

(c) *Y does not have a rational nef element h with $h^3 = 0$.*

Then, starting from Y, repeating transformations (i) *and* (ii) *preserves properties* (a), (b) *and* (c).

Proof. Let us consider a contraction $f: Y \to Y'$ of a divisorial extremal ray R on Y. One easily can see that Y' has properties (a) and (c) if Y has these properties. Moreover, if Y does not have a small extremal ray, then Y' has a small extremal ray only if the divisor $D(R)$ contains another divisorial extremal ray Q. The image of Q then gives a small extremal ray on Y'. Thus R and Q define a pair of type $\mathfrak{B}_2$ on Y. This is impossible by Theorem 1.3.2, (1).

If Y does not have a small extremal ray, then Y does not have a flop, and one does not have the transformation (ii). □

We suggest the following

Conjecture 5.3. There are absolute constants $q, d, C_1, C_2, n_D, n_C, n_A, d_A, C_A$ such that for any Calabi–Yau 3-fold X, and any $\mathbb{Q}$-factorial model X of Y we have estimates $q(Y) \leq q, d(Y) \leq d, C_1(Y) \leq C_1$ and $C_2(Y) \leq C_2$ for invariants of Definition 1.3.1, and $n(Y)_D \leq n_D, n(Y)_C \leq n_C, n(Y)_A \leq n_A, d(X)_A \leq d_A$ and $C(X)_A \leq C_A$ for invariants of Definition 3.1.

If this conjecture holds, applying Theorems 1.3.2 and 3.2, we get an absolute estimate for $\rho(Y)$ if Y satisfies conditions (a), (b) and (c). Equivalently, if X has a $\mathbb{Q}$-factorial model Y with a finite polyhedral Mori cone, without small extremal rays and with big $\rho(Y)$, then Y has a rational nef element h with $h^3 = 0$. One can think that the existence of such en element h with $h^3 = 0$ for a $\mathbb{Q}$-factorial model Y of X is of similar importance as for X.

Unfortunately, now we can prove Conjecture 5.3 only for very special $\mathbb{Q}$-factorial models Y.

We can consider part of results of Section 4, as the proof of Conjecture 5.3 for non-singular models (i.e. for Calabi–Yau 3-folds).

It is possible to extend these results to a larger class of models which are still very special.

Definition 5.4. A $\mathbb{Q}$-factorial model Y of a Calabi–Yau 3-fold is called *very good* if there exists a Calabi–Yau 3-fold X and an extremal set $\mathcal{R} = \{R_1, \dots, R_k\}$ of divisorial extremal rays on X having distinct divisors $D(R_1), \dots, D(R_k)$ such that we have the following properties (i) and (ii):

(i) Y is a contraction $f: X \to Y$ of the face $\gamma = R_1 + \cdots + R_k$ of $\overline{\text{NE}}(X)$;

(ii) For any divisorial extremal ray R of Y there exists a divisorial extremal ray $\widetilde{R}$ of X such that $R = f(\widetilde{R})$. This means that for the contraction $f_R: Y \to Y'$ of R, the composition $f_R \cdot f: X \to Y'$ is the contraction of the face $R_1 + \cdots + R_k + \widetilde{R}$.

We have

Theorem 5.5. *Let Y be a very good $\mathbb{Q}$-factorial model of a Calabi–Yau 3-fold. Then Y has constants*

$$q(Y) \leq 3,\ d(Y) \leq 8,\ C_1(Y) \leq 16,\ C_2(Y) \leq 18,$$

and

$$n(Y)_D \leq 8,\ n(Y)_C \leq 9,\ n(Y)_A \leq 9,\ d(Y)_A \leq 8,\ C(Y)_A \leq 17.$$

In particular, by Theorem 1.2.3, Y has ≤ 3 extremal rays of type (I) *and*

$$\rho(Y) \leq (16/3)C_1(Y) + 4C_2(Y) + 6 < 164$$

if Y satisfies conditions (a), (b) *and* (c).

Considering the preimage of a nef element h on Y with $h^3 = 0$, we get the result.

Corollary 5.6. *Let X be a Calabi–Yau 3-fold which has a very good $\mathbb{Q}$-factorial model Y such that Y has a finite polyhedral Mori cone $\overline{\mathrm{NE}}(Y)$ and Y does not have a small extremal ray.*

Then X has a rational nef element h with $h^3 = 0$ if $\rho(Y) \geq 164$.

Sketch proof of Theorem 5.5. We use the notation of Definition 5.4. For a set Q of divisorial extremal rays on Y, we denote $f^{-1}(Q) = \{R_1, \dots, R_k\} \cup \widetilde{Q}$, where $\widetilde{Q} = \{\widetilde{R} \mid R \in Q\}$. The set $f^{-1}(Q)$ is called the preimage of Q and $\widetilde{Q}$ is called the proper preimage of Q.

The next general statements will be very useful.

Lemma 5.7. *Let $\mathcal{E}$ be an elliptic set of divisorial extremal rays on Y. Then $f^{-1}(\mathcal{E})$ is elliptic.*

Proof. Let $\mathcal{E} = \{Q_1, \dots, Q_t\}$. Since $\mathcal{E}$ is elliptic, there are positive $a_1, \dots, a_t$ such that $Q_j \cdot (a_1 D(Q_1) + \cdots + a_t D(Q_t)) < 0$ for all $1 \leq j \leq t$. By Lemma 1.2.2, the set $\{R_1, \dots, R_k\}$ is elliptic. Hence, there are positive $b_1, \dots, b_k$ such that $R_i \cdot (b_1 D(R_1) + \cdots + b_k D(R_k)) < 0$ for all $1 \leq i \leq k$. Obviously,

$$f^*(a_1 D(Q_1)+\cdots+a_t D(Q_t)) = c_1 D(R_1)+\cdots+c_k D(R_k)+a_1 D(\widetilde{Q}_1)+\cdots+a_t D(\widetilde{Q}_t)$$

with non-negative $c_1, \dots, c_k$. By the projection formula,

$$R_i \cdot (c_1 D(R_1) + \cdots + c_k D(R_k) + a_1 D(\widetilde{Q}_1) + \cdots + a_t D(\widetilde{Q}_t)) = 0$$

for all $1 \leq i \leq k$, and

$$\begin{aligned}\widetilde{Q}_j \cdot (c_1 D(R_1)+ \cdots + c_k D(R_k) + a_1 D(\widetilde{Q}_1) + \cdots + a_t D(\widetilde{Q}_t))\\ = Q_j \cdot (a_1 D(Q_1) + \cdots + a_t D(Q_t))) < 0.\end{aligned}$$

It follows that for a small $\varepsilon > 0$ the divisor

$$\begin{aligned}D(f^{-1}(Q)) = \varepsilon(b_1 D(R_1) + \cdots + b_k D(R_k))\\ + c_1 D(R_1) + \cdots + c_k D(R_k) + a_1 D(\widetilde{Q}_1) + \cdots + a_t D(\widetilde{Q}_t)\end{aligned}$$

has positive coefficients and has negative intersection with each element of $f^{-1}(Q)$. Thus $f^{-1}(Q)$ is elliptic. □

Lemma 5.8. *Let $\mathcal{L}$ be a Lannér set of divisorial extremal rays on Y. Then $f^{-1}(\mathcal{L})$ contains a quasi-Lannér subset $\mathcal{L}'$ such that $\widetilde{\mathcal{L}} \subset \mathcal{L}'$ and such that for any non-empty subset $U \subset \widetilde{\mathcal{L}}$ the set $\{R_1, \dots, R_k\} \cup (\mathcal{L}' \setminus U)$ is elliptic.*

Proof. Let $\mathcal{L} = \{Q_1, \dots, Q_t\}$. Since $\mathcal{L}$ is Lannér, there are positive $a_1, \dots, a_t$ such that $Q_j \cdot (a_1 D(Q_1) + \cdots + a_t D(Q_t)) \geq 0$ for all $1 \leq j \leq t$ and one of these inequalities is strict. Obviously, $f^*(a_1 D(Q_1) + \cdots + a_t D(Q_t)) = b_1 D(R_1) + \cdots +$

$b_kD(R_k) + a_1D(\widetilde{Q}_1) + \cdots + a_tD(\widetilde{Q}_t)$ where $b_j \geq 0$. By the projection formula,

$$R \cdot (b_1D(R_1) + \cdots + b_kD(R_k) + a_1D(\widetilde{Q}_1) + \cdots + a_tD(\widetilde{Q}_t)) \geq 0$$

for any $R \in f^{-1}(\mathcal{L})$, and one of these inequalities is strict. Thus the set $f^{-1}(\mathcal{L})$ is not semi-elliptic. By Proposition 2.2.8, $f^{-1}(\mathcal{L})$ contains a quasi-Lannér subset $\mathcal{L}'$. Since each proper subset of $\mathcal{L}$ is elliptic, by Lemma 5.7, any subset of $f^{-1}(\mathcal{L})$ is elliptic if it does not contain $\widetilde{\mathcal{L}}$. It follows that $\widetilde{\mathcal{L}} \subset \mathcal{L}'$ and $\{R_1, \ldots, R_k\} \cup (\mathcal{L}' \setminus U)$ is elliptic if U is a non-empty subset of $\widetilde{\mathcal{L}}$. □

Let us continue the proof of the theorem.

Let $\mathcal{L}$ be an E-set of divisorial extremal rays on Y such that any proper subset of $\mathcal{L}$ is extremal of Kodaira dimension 3 and $\mathcal{L}$ satisfies the condition (iii) of Section 1.1. By results of Section 1.2, the set $\mathcal{L}$ is Lannér. By Lemma 5.8, $\tilde{\mathcal{L}} \subset \mathcal{L}'$ where $\mathcal{L}'$ is a quasi-Lannér set of divisorial extremal rays on X. Considering the image of $\mathcal{L}'$ by the morphism f, one can see that

$$\operatorname{diam} \mathcal{L} \leq \operatorname{diam} \mathcal{L}'.$$

Any proper subset of $\mathcal{L}'$ is either elliptic or connected parabolic by Proposition 2.2.8. Using Theorem 2.2.6 and description in Section 4 of elliptic and connected parabolic sets of divisorial extremal rays on Calabi–Yau 3-folds, one can describe possible graphs of quasi-Lannér sets of divisorial extremal rays on X. In particular, one can see that

$$\sharp\mathcal{L}' \leq 10, \quad \operatorname{diam} \mathcal{L}' \leq 8$$

for any quasi-Lannér set $\mathcal{L}'$ of divisorial extremal rays on a Calabi–Yau 3-fold X (compare [P, Fig. 1]). It follows that $n(Y)_A \leq 9$ and

$$d(Y) \leq 8.$$

By Theorem 2.2.6, we also get $n(Y)_D \leq 8$ and $n(Y)_C \leq 9$.

Now let us consider an extremal set $\mathcal{E}$ of Kodaira dimension 3 of divisorial extremal rays on Y. The set $f^{-1}(\mathcal{E})$ is also extremal of Kodaira dimension 3 and contains $\mathcal{R} = \{R_1, \ldots, R_k\}$. For $Q_1, Q_2 \in \mathcal{E}$ we have

$$\rho(Q_1, Q_2) = \rho_{\mathcal{R}}(\widetilde{Q}_1, \widetilde{Q}_2),$$

where $\rho(Q_1, Q_2)$ is the distance in the oriented graph $G(\mathcal{E})$ and

$$\rho_{\mathcal{R}}(\widetilde{Q}_1, \widetilde{Q}_2) = \min_{\gamma} (\rho(\gamma) - v(\gamma \cap \mathcal{R})),$$

where γ is an oriented path in $G(f^{-1}(\mathcal{E}))$ joining $\widetilde{Q}_1$ and $\widetilde{Q}_2$, and $v(\gamma \cap \mathcal{R})$ is the number of vertices of γ which belong to $\mathcal{R}$.

Using the description in Section 4 of graphs of elliptic (in particular, extremal of Kodaira dimension 3) sets of divisorial extremal rays on X, one obtains that $q(Y) \leq 3$, $C_1(Y) \leq 16$, $C_2(Y) \leq 18$, $C(Y) \leq 17$. This finishes the proof. □

We hope to describe more exactly quasi-Lannér sets of divisorial extremal rays on Calabi–Yau 3-folds and give better estimates for Theorem 5.5 and Corollary 5.6 later.

We give another interesting application of Theorem 5.5.

Corollary 5.9. *Let X be a Calabi–Yau 3-fold and suppose that the Mori cone $\overline{\mathrm{NE}}(X)$ is generated by a finite set of divisorial extremal rays. Let γ be a face of the nef cone* $\mathrm{NEF}(X)$ *and $R(\gamma^{\perp})$ the set of all extremal rays orthogonal to γ. Assume that $R(\gamma^{\perp})$ does not contain an extremal ray which belongs to a pair of type $\mathfrak{B}_2$. Then γ contains a rational nef element h with $h^3 = 0$ if* $\dim \gamma > 163$.

Proof. Let $\mathcal{R} = R(\gamma^{\perp})$ and $f\colon X \to Y$ the contraction of the face $\gamma^{\perp} \subset \overline{\mathrm{NE}}(X)$ generated by $\mathcal{R}$. Since $\mathcal{R}$ does not have pairs of extremal rays of type $\mathfrak{B}_2$, the morphism f is a sequence of contractions of divisorial extremal rays which are images of extremal rays from $\mathcal{R}$. Then Y has $\mathbb{Q}$-factorial singularities (canonical) and $\gamma = f^*(\mathrm{NEF}(Y))$. Since $\overline{\mathrm{NE}}(X)$ is generated by a finite set of divisorial extremal rays and $\mathcal{R}$ does not have an extremal ray which belongs to a pair of type $\mathfrak{B}_2$, one sees that Y does not have a small extremal ray and is a very good $\mathbb{Q}$-factorial model of X with Mori cone $\overline{\mathrm{NE}}(Y)$ generated by a finite set of divisorial extremal rays. If we additionally assume that $\mathrm{NEF}(Y)$ (equivalently γ) does not have a nef element h with $h^3 = 0$, $\dim \gamma = \dim \mathrm{NEF}(Y) < 164$ by Theorem 5.5. □

Finally, we mention the following conjecture by Morrison which is connected with the condition (a) above on the Mori cone.

Conjecture 5.10 (D. Morrison, [Mor2]). For a Calabi–Yau manifold X, the nef cone $\mathrm{NEF}(X)$ is rational finite polyhedral modulo the group $\mathrm{Aut}\, X$ of biregular automorphisms of X. In particular, $\mathrm{NEF}(X)$ and the Mori cone $\overline{\mathrm{NE}}(X)$ are rational finite polyhedral if $\mathrm{Aut}\, X$ is finite.

For example, the automorphism group $\mathrm{Aut}\, X$ of a Calabi–Yau 3-fold X is finite if the cubic intersection hypersurface $\mathcal{W}_X$ is non-singular.

6. Concluding remarks

We want to give several remarks about importance of existence of rational nef elements h with $h^3 = 0$ for Calabi–Yau 3-folds.

I. I. Piatetski–Shapiro and I. R. Shafarevich [PŠ-Š] proved that a K3 surface X has a rational nef element $h \in \mathrm{Pic}\, X$ with $h^2 = 0$ if and only if the Picard lattice $\mathrm{Pic}\, X$ represents 0, i.e. there exists $0 \neq x \in \mathrm{Pic}\, X$ such that $x^2 = 0$. In particular, this is true if $\rho(X) = \dim \mathrm{Pic}\, X \geq 5$. Moreover, they proved that the linear system $|h|$ defines an elliptic fibration $|h|: X \to \mathbb{P}^1$ (i.e. the general fiber is an elliptic curve) if $h \in \operatorname{Pic} X$ is nef and $h^2 = 0$.

One can ask about similar fact for Calabi–Yau 3-folds:

Question 6.1. Does exist a rational nef element h with $h^3 = 0$ for a Calabi–Yau 3-fold X if $\rho(X) = \dim N_1(X)$ is sufficiently big?

An affirmative answer to this question would be important because of two results below.

Theorem 6.2 (P. M. H. Wilson, [W1, (3.2)′]). *Let X be a Calabi–Yau 3-fold and h a rational nef element with $h^3 = 0$. Assume that $h^2 \not\equiv 0$ and $h \cdot c_2(X) > 0$. Then for big N, the linear system $|Nh|$ defines an elliptic fibration $|Nh| : X \to S$ (i.e. S is a surface and the general fiber is an elliptic curve).*

Here the condition $h^2 \not\equiv 0$ is equivalent to saying that $\mathbb{C}h$ is not a singular point of the cubic intersection hypersurface $\mathbb{P}\mathcal{W}_X$. By Miyaoka [Mi], we always have the inequality $\mathrm{NEF}(X) \cdot c_2(X) \geq 0$. Thus for the "general case" when the cubic intersection hypersurface $\mathbb{P}\mathcal{W}_X$ is non-singular and $\mathrm{NEF}(X) \cdot c_2(X) > 0$, the existence of a rational nef element h with $h^3 = 0$ gives the existence of an elliptic fibration. Thus for this "general case", Theorems 4.3.1, 4.3.1′ and Theorem 5.5 give the existence of an elliptic fibration on Calabi–Yau manifolds under appropriate conditions. If either the cubic intersection hypersurface $\mathbb{P}\mathcal{W}_X$ is singular or one only has the non-strict condition $\mathrm{NEF}(X) \cdot c_2(X) \geq 0$, Corollary 5.9 is useful to satisfy conditions of Theorem 6.2, since Corollary 5.9 gives existence of rational nef elements h with $h^3 = 0$ in "many" faces of the nef cone $\mathrm{NEF}(X)$. We mention that it is conjectured that the condition $h \cdot c_2(X) > 0$ is not essential for the existence of the elliptic fibration $|Nh|$, see [W3]. By Oguiso [O], the linear system $|Nh|$ defines either an elliptic filtration, or a K3 or Abelian surface fibration if $|Nh|$ is not empty and N is big.

Existence of an elliptic fibration for a Calabi–Yau 3-fold (this Calabi–Yau 3-fold is called *elliptic*) is important because of the following result:

Theorem 6.3 (M. Gross, [Gro]). *There exists a finite set of families $\mathcal{X}_1 \to \mathcal{M}_1, \dots, \mathcal{X}_n \to \mathcal{M}_n$ of elliptic Calabi–Yau 3-folds with $\mathbb{Q}$-factorial terminal singularities such that each elliptic Calabi–Yau 3-fold X with $\mathbb{Q}$-factorial terminal singularities is birationally isomorphic to a fiber of one of these families. (Here all families and isomorphisms preserve elliptic fiber structure.)*

See also connected results by I. Dolgachev and M. Gross [D-Gro] and A. Grassi [Gra].

Because of these Theorems 6.2 and 6.3 and results of this paper, one can think that maybe the classification of Calabi–Yau 3-folds is simpler for high Picard number.

References

[D-Gro] I. Dolgachev and M. Gross, Elliptic threefolds I: Ogg-Shafarevich theory, J. Algebraic Geom. 3 (1994) 39–80.

[Gra] A. Grassi, Log contractions and equidimensional models of elliptic threefolds, J. Algebraic Geom. 4 (1995), 255–276.

[Gro] M. Gross, A finiteness theorem for elliptic Calabi–Yau threefolds, Duke Math. J. 74 (1994), 271–299.

[Hu] B. Hunt, A bound on the Euler number of certain Calabi–Yau threefolds, J. Reine Angew. Math. 411 (1990), 137–170.

[Ka1] Yu. Kawamata, The cone of curves of algebraic varieties, Ann. of Math. 119 (1984), 603–633.

[Ka2] —–, Crepant blowing-up of 3-dimensional canonical singularities and its application to degeneration of surfaces, Ann. of Math. 127 (1988), 93–163.

[Ka3] —–, Boundedness of Q-Fano threefolds, in: Proc. Internat. Conference on Algebra (Novosibirsk 1989), Contemp. Math. 131, Part 3 (1992), 439–445.

[Mi] Y. Miyaoka, The Chern classes and Kodaira dimension of a minimal variety, Adv. Stud. Pure Math. 10 (1987), 449–476.

[Mo1] S. Mori, Threefolds whose canonical bundles are not numerically effective, Ann. of Math. 116 (1982), 133–176.

[Mo2] —–, Flip theorem and the existence of minimal models for 3-folds, J. Amer. Math. Soc. 1 (1988), 117–253.

[MM] S. Mori and S. Mukai, On Fano 3-folds with $B_2 \geq 2$, Algebraic varieties and Analytic varieties (Tokyo, 1981), Adv. Stud. Pure Math. 1, Kinokuniya, Tokyo 1983, 101–129.

[Mor1] D. R. Morrison, The birational geometry of surfaces with rational double points, Math. Ann. 271 (1985), 415–438.

[Mor2] —–, Compactification of moduli spaces inspired by mirror symmetry, Astérisque 218 (1993), 243–271.

[N1] V. V. Nikulin, On the classification of arithmetic groups generated by reflections in Lobachevsky spaces, Izv. Akad. Nauk SSSR Ser. Mat. 45 (1) (1981), 113–142; English transl. in Math. USSR Izv. 18 (1982).

[N2] —–, Del Pezzo surfaces with log terminal singularities, Mat. Sb. 180 (2) (1989), 226–243; English transl. in Math. USSR Sbornik 66 (1990).

[N3] —–, Del Pezzo surfaces with log terminal singularities. II, Izv. Akad. Nauk SSSR Ser. Mat. 52 (1988), 1032–1050; English transl. in Math. USSR Izv. 33 (1989).

[N4] —–, Del Pezzo surfaces with log terminal singularities. III, Izv. Akad. Nauk SSSR Ser. Mat. 53 (1989), 1316–1334; English transl. in Math. USSR Izv. 35 (1990).

[N5] —–, Basis of the diagram method for generalized reflection groups in Lobachevsky spaces and algebraic surfaces with nef anticanonical class, Internat. J. Math. 7 (1996), no. 1.

[N6] —–, Algebraic 3-folds and the diagram method, Math. USSR Izv. 37 (1991), 157–189.

[N7] —–, Algebraic 3-folds and the diagram method. II, preprint Max-Planck-Institut für Mathematik Bonn, 90–104, 1990.

[N8] —–, On the Picard number of Fano 3-folds with terminal singularities, J. Math. Kyoto Univ. 34 (1994), 495–529.

[N9] —–, Discrete reflection groups in Lobachevsky spaces and algebraic surfaces, in: Proc. Internat. Congr. Math. Berkeley 1986, vol. 1, 654–669.

[N10] —–, A lecture on Kähler cone and Picard number of Calabi–Yau 3-folds, preprint MSRI 061-93 1993.

[O] K. Oguiso, On algebraic fiber space structure on a Calabi–Yau 3-fold, Internat. J. Math. 4 (1993), 439–465.

[PŠ-Š] I. I. Piateckiĭ-Šapiro and I. R. Šafarevič, A Torelli theorem for algebraic surfaces of type K3, Izv. Akad. Nauk SSSR Ser. Mat. 35 (1971), 530–572; English transl. in Math. USSR Izv. 5 (1971).

[P] M. N. Prokhorov, The absence of discrete reflection groups with non-compact fundamental polyhedron of finite volume in Lobachevsky spaces of large dimension, Izv. Akad. Nauk SSSR Ser. Mat. 50 (1986), 413–424; English transl. in Math. USSR Izv. 28 (1987).

[R1] M. Reid, Canonical 3-folds, in: Journées de geométrie algébrique d'Angers (A. Beauville, ed.), Alphen aan den Rijn, Sijhoof and Noordhoff 1980, 273–310.

[R2] —–, Minimal models of canonical 3-folds, in: Algebraic varieties and Analytic varieties, Tokyo 1981 (S.Iitaka, ed.), Adv. Stud. Pure Math. 1, Kinokuniya, Tokyo 1983, 395–418

[R3] —–, Young person's guide to canonical singularities, Proc. Sympos. Pure Math. (Summer Research Inst. Bowdoin, 1985) 46 (1987), 345–414.

[Sh] V. V. Shokurov, The nonvanishing theorem, Izv. Akad. Nauk SSSR Ser. Mat. 49 (1985), 635–651; English transl. in Math. USSR Izv. 26 (1986).

[Sh2] —–, 3-fold log flips. Izv. Russ. Akad. Nauk Ser. Mat. 56 (1992); English transl. in Russian Acad. Sci. Izv. Math. 40 (1993), 95–202.

[V1] É. B. Vinberg, The absence of crystallographic reflection groups in Lobachevsky spaces of large dimension, Trudy Moscow. Mat. Obshch. 47 (1984), 68–102; English transl. in Trans. Moscow Math. Soc. 47 (1985).

[V2] —–, Discrete reflection groups in Lobachevsky spaces, in: Proc. Internat. Congr. Math. Warszawa 1983, vol. 1, 593–601.

[W1] P. M. H. Wilson, Calabi–Yau manifolds with large Picard number, Invent. Math. 98 (1989), 139–155.

[W2] —–, The Kähler cone on Calabi–Yau threefolds, Invent. Math. 107 (1992), 561–583.

[W3] —–, The existence of elliptic fibre space structures on Calabi–Yau threefolds, Math. Ann. 300 (1994), 693–703.

Appendix by V. V. Shokurov: Anticanonical boundedness for curves

The purpose of this note is to generalize slightly the results of [K].

Theorem. *Let $f: X \to S$ be a projective morphism of normal algebraic varieties over a field of characteristic zero, and D an effective $\mathbb{R}$-divisor on X. Suppose that $K_X + D$ is $\mathbb{R}$-Cartier and divisorially log terminal near the generic points of a subvariety E, consisting of components of the* degenerate locus

$$\mathrm{Exc}(f) := \{x \in X \mid f \text{ is not finite at } x\}.$$

Then E is covered by a family (possibly disconnected) of effective 1-cycles $\{C_\lambda\}/S$ with $(-K_X - D \cdot C_\lambda) \leq 2n$ where $n = \dim E/S$ (and even $< 2n$ if $K_X + D$ is Kawamata log terminal in the generic points of E and $X \neq E$). Moreover, we could assume that the generic 1-cycles C_λ are curves, *i.e., reduced and irreducible, when $K_X + D$ is numerically definite, and the curves C_λ with $(K_X + D \cdot C_\lambda) < 0$ (resp. ≤ 0) are rational.*

The Theorem implies the following results.

Corollary 1. *If, in addition, F is an $\mathbb{R}$-Cartier divisor such that $K_X + D + F$ is also divisorially log terminal near the generic points of E, $D + F$ is effective, and $K_X + D$ is numerically semi-negative with respect to f, then E is covered by a family (possibly disconnected) of effective 1-cycles $\{C_\lambda\}/S$ (resp. with curves as generic members when $K_X + D + F$ is numerically definite) with*

$$(F \cdot C_\lambda) \geq -2n$$

(resp. $> -2n$ if $K_X + D + F$ is Kawamata log terminal in the generic points of E, and $X \neq E$).

For example, let f be an extremal divisorial contraction of a normal variety X with exceptional $\mathbb{Q}$-Cartier divisor $F = E = \mathrm{Exc}(f)$, and suppose that K_X is numerically semi-negative for f. Then we cover F by a family of rational curves $\{C_\lambda\}/S$ with

$$(F \cdot C_\lambda) \geq -2 \dim X + 2.$$

In the low dimensional case, i.e., when $\dim X \leq 3$ we have the sharper inequality $\geq -\dim X$. This is the best bound, but it is only known in that case. (See Remark 3 below.)

Proof. According to the Theorem, take a family of 1-cycles $\{C_\lambda\}/S$ with respect to the second log divisor $K_X + D + F$. Then

$$(-K_X - D \cdot C_\lambda) - (F \cdot C_\lambda) = (-K_X - D - F \cdot C_\lambda) \leq 2n$$

(resp. $< 2n$ if $K_X + D + F$ is Kawamata log terminal in the generic points of E, and $X \neq E$). But we assume that $(-K_X - D \cdot C_\lambda) \geq 0$, which gives the required inequality. □

Corollary 2 ([Ko1], cf. also [K]). *Let $f: X \to S$ be a projective morphism of normal algebraic varieties over a field of characteristic zero, and D be an effective $\mathbb{R}$-divisor on X such that $K_X + D$ is $\mathbb{R}$-Cartier. Let H be an f-ample $\mathbb{R}$-Cartier divisor and $\varepsilon > 0$. Then the number of extremal contractions* cont_R *and corresponding rays R such that $K_X + D$ is*

(*) *divisorially log terminal near a generic point of the degenerate fibers of* cont_R,

and such that $(K_X + D + \varepsilon H \cdot R) < 0$, is finite.

Thus the half-cone $(K_X + D \cdot R) < 0$ of the Kleiman–Mori cone $\overline{\mathrm{NE}}(X/S)$ is locally polyhedral when (*) holds for all extremal rays in it, including the existence of the extremal contractions cont_R.

In particular, if X is normal projective with isolated $\mathbb{Q}$-Gorenstein singularities, we have at most a finite number of extremal contractions which are negative for $K_X + \varepsilon H$ (cf. [Ko2]). Note that this does not mean that the half-cone $(K_X + D \cdot R) < 0$ is always locally polyhedral, only that, whenever this fails, there exists an extremal *non-contracted* ray R with $(K_X + D \cdot R) < 0$.

Proof. See [K]. □

Conjecture. In the Theorem we may replace the divisorially log terminal property near the generic points by the log canonical property at them, or even in them [Sh3, Corollary 1.5], and always find a family $\{C_\lambda\}/S$ of curves.

Note that, even in the definite case either we consider a covering of the generic points of E, or we admit effective 1-cycles as curves C_λ for *degenerations*. Nevertheless in this case, we abuse terminology and say a covering by curves.

We should have even more.

Theorem′. *The Conjecture implies Corollaries 1–4 with the divisorially log terminal property near the generic points replaced by the weaker assumption that $K_X + D$ is log canonical in the generic points of E, and effective 1-cycles replaced by curves.*

If $\dim X \leq 3$ the Conjecture and the improved corollaries hold.

Heuristic Arguments. Here we deduce the Conjecture and, in particular, the Theorem using the LMMP (Log Minimal Model Program). Thus this proves the Theorem and Theorem′ for $\dim X \leq 3$ [Sh3]. For all dimensions, a proof of the Theorem refines [K] and [MM], and will be given later.

First, we discuss what it means for X to be log canonical (or log terminal) in a generic point of a subvariety $E \subseteq X$. A generic point P is one of the irreducible components of E. The log divisor $K_X + D$ is log canonical (resp. log terminal, purely log terminal, or Kawamata log terminal) in P if $K_X + D$ possesses that property in a generic point of P in a naive sense. A more rigorous point of view means that for the log discrepancies a_i of the divisors E_i such that the center $c(E_i) = P$, we have the log canonical property $a_i \geq 0$ (resp. in addition, $a_i > 0$ for such exceptional divisors E_i of one log nonsingular resolution over a neighborhood of the generic point P; $a_i > 0$ for all such exceptional divisors E_i; or $a_i > 0$ for all such divisors E_i) [Sh2]. Note that purely log terminal coincides with Kawamata log terminal if P has codimension ≥ 2.

A property near means in a neighborhood.

We may assume that E and S are irreducible. For this, add also that $\dim E/S = \dim E - \dim f(E)$ if E is irreducible, and the maximum of such dimensions for irreducible components of E in general.

Now we reduce the Theorem to the case when S is a point, i.e., X is a projective variety with the constant morphism $f: X \to \mathrm{pt}$. Taking a generic hyperplane section H of S and replacing X by its inverse image $f^{-1}H$, we reduce the problem to the case when $f(E) = p$ is a point. Note that then $n = \dim E$.

If $S \neq p$ we can add a non-negative multiple of f^*H, where H is a generic hyperplane section on S through $f(E) = p$, in such a way that $K_X + D$ will be maximally log canonical in the generic point of E. This means that $K_X + D$ is log canonical in the generic point of E, and for any $\varepsilon > 0$, $K_X + D + \varepsilon f^*H$ is not so. In other words, there exists a divisor F with log discrepancy 0 and with center E.

In addition, if $S \neq p$, we may move and split D into a boundary preserving the log canonical property along E. So $K_X + D$ will be log canonical in a neighborhood of the generic point of E.

If $F = E$ is a non-exceptional divisor on X, then D has multiplicity 1 in E, and we use the Adjunction Formula and Effectiveness [Sh2, 3.1 and 3.2.2]. This reduces to the case when $X = E^\nu$ and $S = p$.

Using the LMMP, we can do the same in general. For this, we should replace X by a strict log minimal model $g: Y \to X$ of a neighborhood of E with respect to the log divisor $K_X + B$ where the boundary B has multiplicities $b_i = \min\{1, d_i\}$ (cf. [Sh2, 3.4]). According to the above perturbation of D, $D = B$ in a neighborhood of the generic point of E. Perhaps after an additional monoidal transform, we can also assume that $F = E$ is not exceptional on $X := Y$. Thereafter we replace $K_X + D$ by $f^*(K_X + D) = K_Y + D'$, i.e., D is replaced by a divisor D'. Then D' is effective by a construction of log models and by the Negativity of Birational Contractions [Sh2, 1.1]. The multiplicity of E in D equals 1 by the construction. Here we meet one annoyance: $\dim E/S$ may be higher than $n = \dim g(E)/S$ when E is exceptional for a model g. However in that case we have an additional structure, namely, a projective *Iitaka* morphism $i = g|_E: E \to g(E)$. This means that i is a projective contraction, and $K_E + D_E = (K_X + D)|E$ is numerically trivial for i. Note that

$n = \dim E/S = \dim i(E)$. The divisor E itself is a projective variety with the constant morphism $f: E \to \text{pt}$. According to the Adjunction and Projection formula, the above properties of i follow from the assumption that $K_X + D$ is maximally log canonical in E, i.e., D has multiplicity 1 in E. These imply also that a family of curves of the required type on E induces one on the subvariety $g(E)$ of the original X.

Take now $X = E$. By construction X is again a normal, projective variety with the constant morphism $f: X \to \text{pt}$. In addition, we have an Iitaka morphism $i: X \to E$ with respect to $K_X + D$. Also by the construction and the Adjunction formula, $K_X + D$ is log canonical over the generic point of E. After additional blowups, we can assume that X is $\mathbb{Q}$-factorial and $K_X + D$ is log terminal over the generic point of E. Since D has multiplicities $d_i > 1$ only for prime divisors D_i with $i(D_i) \neq E$, we can drop these components and suppose that $K_X + D$ is strictly log terminal, in particular, D is a boundary on X. However, this can spoil the Iitaka morphism i, namely, there may appear curves C in fibers of i such that $(K_X + D \cdot C) \neq 0$. Nevertheless, i will be again the Iitaka morphism over the generic point of E. This is enough, and coherent to the following arguments.

Thus we should cover X by a family of curves $\{C_\lambda\}$ with $(-K_X - D \cdot C_\lambda) \leq 2n$, where $n = \dim E$. Since $K_X + D$ is strictly log terminal and X is projective, we can apply the LMMP to X with the log divisor $K_X + D$. If $K_X + D$ is nef, then we take any family of curves $\{C_\lambda\}$ covering X. Otherwise we have an extremal contraction $g: X \to Z$. If g has fiber type, then the generic fiber F of g intersects the generic fiber of i in a finite set. Therefore $\dim F \leq n = \dim E$ and we can reduce the problem to fibers in this case. Note also that divisorial contractions and flips only decrease the intersection $(K_X + D \cdot C_\lambda)$ for the generic curve of a covering family of X. Thus from now on we can consider only birational contractions.

If g is birational, we make a divisorial contraction or a flip of X. However, the exceptional locus F of g intersects the generic fiber of i at most in a finite set. If it does not intersect, we can make such transform and preserve birationally i. Otherwise $i(F) = E$ and F is finite over the generic point of E. In this case, we use the above arguments for $E = F$ and the induction on the dimension of X, since the above arguments restrict the proof to a divisor. The corresponding Iitaka morphism is induced by i.

According to the termination of the LMMP, this completes the reduction to the case when $X = E$ and $S = p$ is a point. Thus X is normal and projective. After additional blowups, we can assume that X is also $\mathbb{Q}$-factorial. Now $n = \dim X$, and we should cover X by a family of curves $\{C_\lambda\}$ with $(-K_X - D \cdot C_\lambda) \leq 2n$. Dropping D, and after additional blowups, we can assume that $D = 0$ and X has only log terminal or even terminal singularities. In the last case we use the MMP.

If K_X is nef, then we take any family of curves $\{C_\lambda\}$ covering X. Otherwise we apply the MMP. According to the termination, after a finite number of extremal transformations we reduce to the case when X possesses a Fano fibering. Taking the generic fiber, we get the case when X is a Fano variety having only terminal singularities. Then the required covering family exists according to Miyaoka and Mori [MM].

If $K_X + D$ is Kawamata log terminal in the generic points of E and $X \neq E$, we could replace f^*H by an effective Cartier divisor which is numerically negative on E, for instance, we could take a generic anti-hyperplane section of E over a neighborhood of $p = f(E)$. Then subsequent reductions give the stronger inequality $< 2n$. □

Below in the proof of the Theorem, the result of [MM] plays the same role. The previous heuristic arguments can be replaced by the following refinement of [K, Lemma]. We lose some properties stated in the conjecture. This is the price of using homological methods (cf. [Sh3]).

Lemma. *Let $f: X \to Y$ be a projective morphism of normal algebraic varieties over a field of characteristic zero, H an $\mathbb{R}$-Cartier divisor on X, D an effective $\mathbb{R}$-divisor, E an irreducible component of the degenerate locus* Exc(f) *of f, $n =$* dim E, *and $\nu: E^\nu \to E$ the normalization. Suppose that f is finite over the generic point of Y, H is nef with respect to f, $K_X + D$ is $\mathbb{R}$-Cartier and divisorially log terminal near the generic point of E and $f(E)$ is a point. Then*

$$(H^{n-1} \cdot K_X + D \cdot E) \geq ((\nu^* H)^{n-1} \cdot K_{E^\nu})$$

(and even $>$ if H is ample and $K_X + D$ is Kawamata log terminal in the generic point of E).

Proof. For the non-strict inequality we can approximate H by $\mathbb{Q}$-Cartier ample divisors. Moreover, we prove the following multilinear inequality

$$(H_1 \dots H_{n-1} \cdot K_X + D \cdot E) \geq (\nu^* H_1 \dots \nu^* H_{n-1} \cdot K_{E^\nu})$$

where all H_i are nef (and even $>$ if all H_i are ample and $K_X + D$ is Kawamata log terminal in the generic points of E. Since any ample $\mathbb{R}$-divisor is a sum of an ample $\mathbb{R}$-divisor and an ample $\mathbb{Q}$-divisor, we may assume that all ample H_i are $\mathbb{Q}$-Cartier, or even Cartier.)

Thus as in [K], we suppose that Y is affine, H_i are very ample, and $n = 1$, i.e., we have no H_i and E is a curve.

Since $K_X + D$ is divisorially log terminal near the generic point of E, then Supp D will be log nonsingular near the generic point of E unless $K_X + D$ is purely log terminal near the generic point of E.

In the log nonsingular case, D is an $\mathbb{R}$-divisor in the generic point of E. Then adding some small effective Cartier divisor trivial near the generic point of E we preserve the statement and may assume that $D - \varepsilon F \geq 0$ where $\varepsilon > 0$ and F is an effective Cartier divisor that coincides with D near the generic point of E. This makes $K_X + D$ Kawamata log terminal near the generic point of E and, in particular, purely log terminal near the generic point of E. As a limit for $\varepsilon \to 0$ we get the required inequality.

Thus perturbing D, we may assume also that $K_X + D$ is purely and even Kawamata log terminal near the generic point of E except in the case when E is a divisor, i.e., X is a surface. The above arguments also work for the last exception.

Since we are working over a field of characteristic zero, we can replace it by $\mathbb{C}$ and replace f by an analytic contraction over a small analytic neighborhood of $p = f(E)$ in Y.

Suppose that $(K_X + D \cdot E) \leq \deg K_{E^\nu}$. Then there exists a Cartier divisor A_0 on E such that $\deg A_0 \geq (K_X + D \cdot E)$ and $H^0(E^\nu, K_{E^\nu} - \nu^* A_0) \neq 0$. As in [K], we have $H^0(E, \omega(-A_0)) \neq 0$, and we can extend A_0 to a Cartier divisor A on X. Moreover, we may assume that $A - K_X - D$ is nef $/Y$ and A is sufficiently ample$/Y$ on components of $\mathrm{Exc}(f)$ except E. "Sufficiently ample" means sufficiently ample for the vanishing assertions below to work.

According to our assumptions $K_X + D$ has nonpositive log discrepancies only for divisors F whose centers intersect Y in finite sets.

Since $A - K_X - D$ is nef, we have $R^1 f_* \mathcal{I}_X(A) = 0$ where $\mathcal{I}_X$ is an ideal sheaf of a subscheme $S \subset X$ which intersects E in a finite set. This is by the proof of [KMM, 1.2.5] and the Kawamata log terminal property of $K_X + D$ near the generic point of E. It is also important that D is effective, and X is normal. So any function regular in codimension 2 on X will be regular everywhere.

According to the construction A is sufficiently ample on S/Y, i.e., $R^i f_* \mathcal{O}_S(A) = 0$ for $i \geq 1$ by Serre vanishing. Therefore the exact sequence

$$0 \to \mathcal{I}_X(A) \to \mathcal{O}_X(A) \to \mathcal{O}_S(A) \to 0$$

implies the vanishing $R^1 f_* \mathcal{O}_X(A) = 0$.

Now base change gives $H^1(f^{-1}p, A|_{f^{-1}p}) = 0$. Again, the restriction sequence

$$0 \to \mathcal{J}(A|_{f^{-1}p}) \to \mathcal{O}_{f^{-1}p}(A|_{f^{-1}p}) \to \mathcal{O}_E(A_0) \to 0$$

and Serre vanishing imply the vanishing $H^1(E, A_0) = 0$, and we obtain the required inequality as in [K]. For this, note that the support of the ideal sheaf $\mathcal{J}$ intersects E at most in one-dimensional set E and we may choose A such that it is sufficiently ample$/Y$, i.e., $H^2(f^{-1}p, \mathcal{J}(A|_{f^{-1}p})) = 0$. □

Proof of the Theorem. If $M = -\nu^*(K_X + D)$ is not f-nef on E^ν we have a curve C/S on E^ν with $(M \cdot C) < 0$. Moreover, for the 1-cycle $N\nu(C)/S$ with $N \gg 0$,

$$(-K_X - D \cdot N\nu(C)) = (M \cdot NC) \ll 0.$$

Then it is easy to add to $N\nu(C)$ an effective family of curves$/S$ covering E and such that the 1-cycles C_λ/S we obtain satisfy $(-K_X - D \cdot C_\lambda) < 0 \leq 2n$.

Therefore, we suppose from now on that $M = -\nu^*(K_X + D)$ is f-nef on E. Then we can use the arguments of [K, Proof of the Theorem 1]. The only difference is that now M is nef. Note that we can use [MM] even for a nef divisor M, because the required inequalities are not strict. Indeed, if M is a limit of ample $\mathbb{Q}$-Cartier divisors M_n for $n \to \infty$ then the results of [MM] for M_n give in a limit the same for M. □

Corollary 3 (cf. [K, Th. 2]). *For $f: X \to S$ and D as in the Theorem, let E be a subvariety, consisting of components of*

$$\mathrm{Exc}(f) := \left\{ x \in X \,\middle|\, \begin{array}{l} \textit{an irreducible component of a fiber of } f \\ \textit{through } x \textit{ has dimension} > d = \dim X/S \end{array} \right\}.$$

Then E is covered by a family (possibly disconnected) of effective 1-cycles $\{C_\lambda\}/S$ with $(-K_X - D \cdot C_\lambda) \le 2(n-d)$ where $n = \dim E/S$ (and even $< 2(n-d)$ if $K_X + D$ is Kawamata log terminal in the generic points of E). Moreover, we could assume that the generic 1-cycles C_λ are curves when $K_X + D$ is numerically definite, and these curves C_λ with $(K_X + D \cdot C_\lambda) < 0$ (resp. ≤ 0) are rational.

Proof. Take general hyperplane sections of S (cf. the Heuristic Arguments) and then use the proof of [K, Th. 2]. □

Corollary 4. *If, in addition to Corollary 3, F is an $\mathbb{R}$-Cartier divisor such that $K_X + D + F$ is also divisorially log terminal near the generic points of E, $D + F$ is effective, and $K_X + D$ is numerically semi-negative with respect to f, then E is covered by a family (possibly disconnected) effective 1-cycles (curves when $K_X + D + F$ numerically definite) $\{C_\lambda\}/S$ with*

$$(F \cdot C_\lambda) \ge -2(n-d)$$

(resp. $> -2(n-d)$ if $K_X + D + F$ is Kawamata log terminal in the generic points of E).

Remarks. (1) Conjecturally, we may suppose that the generic members of $\{C_\lambda\}/S$ in Corollaries 3 and 4 should always be curves. This is true when $n - d \le 3$.

(2) If f is a projective morphism as in the Theorem we can introduce its D-length as the maximum of $(-K_X - D \cdot C_\lambda)$ for families $\{C_\lambda\}/S$ covering X. Of course, the length is 0 when f is finite over the generic point of S, and by the theorem the length is $\ge 2n$ if $K + D$ has mild singularities in generic points, where $n = \dim X/S$.

According to [MM], a smooth projective variety X is uniruled if and only if its length (= 0-length) is positive.

Here we could state problems for the length similar to that for Fano indexes and minimal discrepancies [Sh1].

(3) The estimate $\le 2n$ (respectively $< 2n$) is essentially derived from that of [MM] for Fano n-folds with terminal and even $\mathbb{Q}$-factorial singularities. For $n \le 3$ and in the nonsingular case, according to the classification of such Fanos, we can improve the estimate to $\le n + 1$ (cf. [M]). Since terminal singularities are nonsingular for $n \le 2$ this gives the same improvement in general for $n \le 2$.

So we have the following problems. Is the estimate $\le 2n$ sharp for $n \ge 3$? If not, try to find the right one. Why not $n + 1$?

(4) Perhaps the results proved here and the Conjecture hold in any characteristic.

Acknowledgment. Partial financial support was provided by the NSF under the grant number DMS-9200933.

References

[K] Kawamata Y., On the length of an extremal rational curve, Invent. Math. 105 (1991), 609–611.

[KMM] Kawamata Y., Matsuda K. and Matsuki, Introduction to the minimal model problem, in: Algebraic Geometry Sendai 1985, Adv. Stud. Pure Math. 10 (1987), 283–360.

[Ko1] Kollár J., The Cone theorem: Note to Kawamata's "The cone of curves of algebraic varieties", Ann. of Math. 120 (1984), 1–5.

[Ko2] Kollár J., Cone theorems and cyclic covers, in: Algebraic geometry and analytic geometry (Tokyo 1990), Springer Verlag 1991, 101–110.

[M] Mori S., Threefolds whose canonical bundles are not numerically effective, Ann. of Math. 116 (1982), 133–176.

[MM] Miyaoka Y. and Mori S., A numerical criterion of uniruledness, Ann. of Math. 124 (1986), 65–69.

[Sh1] Shokurov V. V., Problems about Fano varieties, in: Birational Geometry of Algebraic Varieties, Open Problems, The XXIII International Symposium, Division of Mathematics, The Taniguchi Foundation, August 22–August 27, 1988, Katata, 30–32.

[Sh2] Shokurov V. V., 3-fold log flips. Izv. Russ. Akad. Nauk Ser. Mat. 56 (1992); English transl. in Russian Acad. Sci. Izv. Math. 40 (1993), 95–202.

[Sh3] Shokurov V. V., 3-fold log models, Johns Hopkins Univ. preprint, (1995).

On the complete classification of Calabi–Yau threefolds of type III_0

Keiji Oguiso

Introduction

By a minimal threefold, we means a normal complex projective threefold with only $\mathbb{Q}$-factorial terminal singularities and with nef canonical divisor. A Calabi–Yau threefold means a simply connected minimal threefold X with $\mathcal{O}_X(K_X) \simeq \mathcal{O}_X$. A Calabi–Yau threefold X is called of type III_0 if there exists a nef big divisor D on X with $D.c_2(X) = 0$. The name given here fits our previous classification in [O1].

The purpose of this paper is to classify all the Calabi–Yau threefolds of type III_0 up to isomorphisms. This is an application of a remarkable structure theorem (Theorem 3) due to Shepherd-Barron and Wilson ([SW]).

In order to state our main theorem, we first construct two Calabi–Yau threefolds of type III_0. These are studied by Beauville, Roan and Yau, and the author ([B, RY, R, O2]) from a somewhat different point of view. In the sequel, we denote the primitive n-th root of unity $\exp(2\pi i/n)$ by ζ_n. By global coordinates of an abelian threefold A around $P \in A$, we mean coordinates of the holomorphic tangent space $T_{A,P}$ with respect to some fixed basis. This is nothing but global coordinates of the universal cover of A.

Example 1 (cf. [B, RY, R, O2]). Let A_3 be the product abelian threefold of three copies of the elliptic curve $E_{\zeta_3} := \mathbb{C}/(\mathbb{Z} \oplus \mathbb{Z}\zeta_3)$. Let g_3 be the automorphism of A_3 by scalar multiplication by ζ_3. Then the quotient threefold $A_3/\langle \zeta_3 \rangle$ has 27 singular points of type $\frac{1}{3}(1, 1, 1)$ and their toric resolution, which replaces each of these singular points by a single $\mathbb{P}^2$, gives a smooth Calabi–Yau threefold of type III_0. We denote this Calabi–Yau threefold by X_3. It is easy to calculate that $h^{1,1}(X_3) = 36$, $h^{2,1}(X_3) = 0$ and $\chi_{\mathrm{top}}(X_3) = 72$. In particular, X_3 is rigid.

Example 2 (cf. [RY, R]). Let $A_7 := H^0(C, \Omega^1_C)^*/H_1(C, \mathbb{Z})$ be the Jacobian variety of the Klein quartic curve $C := \{x_0x_1^3 + x_1x_2^3 + x_2x_0^3 = 0\}$ in $\mathbb{P}^2$. This is an abelian threefold. Let g_7 be the automorphism of order seven of A_7 induced by the automorphism g_C of C defined by $[x_0 : x_1 : x_2] \mapsto [x_0 : \zeta_7 x_1 : \zeta_7^3 x_2]$. Under appropriate global coordinates of A_7, the action of g_7 is represented by the diagonal matrix $\mathrm{diag}(\zeta_7, \zeta_7^2, \zeta_7^4)$. Then the quotient threefold $A_7/\langle g_7\rangle$ has the 7 singular points of type $\frac{1}{7}(1, 2, 4)$ and their toric resolution, which replaces each of these singular points by three $\mathbb{F}_2$'s crossing normally along the negative sections and fibers in a cyclic way, gives a smooth Calabi–Yau threefold of type III_0. We denote this Calabi–Yau threefold by X_7. It is easy to calculate that $h^{1,1}(X_7) = 24$, $h^{2,1}(X_7) = 0$ and $\chi_{\mathrm{top}}(X_7) = 48$. In particular, X_7 is rigid and is not isomorphic to X_3.

Our main theorem is as follows.

Main Theorem. *Conversely, any Calabi–Yau threefold of type* III_0 *is isomorphic to either* X_3 *or* X_7. *In particular there are only two different Calabi–Yau threefolds of type* III_0 *and each of them is smooth and rigid.*

We explain outline of the proof.

We call a minimal threefold Y a partial crepant resolution of a threefold Z with only canonical singularities if there is a birational morphism $f: Y \to Z$ such that $f^*(K_Z) = K_Y$. We call an automorphism, or an automorphism group, of an abelian variety Gorenstein if a non-zero global holomorphic three-form is kept invariant by its action.

The first key step is the following remarkable structure theorem for Calabi–Yau threefolds of type III_0. This is due to Shepherd-Barron and Wilson:

Theorem 3 ([SW]). *A Calabi–Yau threefold of type* III_0 *is a partial crepant resolution of the quotient* A/G *of an abelian threefold* A *by a finite Gorenstein group* G *acting freely in codimension two but with at least one fixed point. Moreover, we may assume that* G *has no non-trivial translations.*

The last statement may not be explicitely written in the literature. But this is obvious because $H := G \cap \{\text{translations}\}$ is a normal subgroup of G and $A/G = (A/H)/(G/H)$ in which A/H is also an abelian threefold. Note that the pairs $(A_3, \langle g_3\rangle)$ and $(A_7, \langle g_7\rangle)$ in Examples 1 and 2 satisfy conditions in Theorem 3 and that X_3 and X_7 is a (partial) crepant resolution of the quotients $A_3/\langle g_3\rangle$ and $A_7/\langle g_7\rangle$ respectively.

Now our strategy is more or less clear and Main Theorem is reduced to the following two lemmas:

Key Lemma. *Any pair* (A, G) *of an abelian threefold* A *and its finite automorphism group* G *satisfying condition in Theorem* 3 *is isomorphic to either* $(A_3, \langle g_3\rangle)$ *or* $(A_7, \langle g_7\rangle)$ *(under the assumption that* A/G *has a simply connected partial crepant resolution).*

Lemma 4. *Each of the quotient varieties $A_3/\langle g_3\rangle$ and $A_7/\langle g_7\rangle$ has a unique partial crepant resolution.*

It is easy to prove Lemma 4: In fact, the so called flop theorem ([Ka, Ko]) shows that any two partial crepant resolutions, say Y and Y', of Z are connected by a composite of flops over the base Z and the exceptional locus of $Y - -- \to Y'$ consists of isolated rational curves in the exceptional locus of $Y \to Z$. However, the explicit description given in Examples 1 and 2 shows that there are no such curves in the exceptional locus of each of $X_3 \to A_3/\langle g_3\rangle$ and $X_7 \to A_7/\langle g_7\rangle$.

Thus, we may prove the Key Lemma. This will be done in the proceeding sections. Here, instead of proof, we make a few remarks on the Key Lemma. At a first glance, the Key Lemma seems to be a problem on group theory. However, our proof is free from any classification results concerning groups, such as the classification of the finite groups in $SL_3(\mathbb{C})$ ([YY]) and so on. A priori, it is not at all trivial whether G can be embedded into $SL_3(\mathbb{C})$ or not, or it is even not at all trivial whether each element of G has a fixed point in A or not. This is the most crucial difference between our work and several previous works done by [RY], [R] and [YY], and also a new difficulty not appearing in previous literatures. In fact, our proof owes much more to the arithmetic of the cyclotomic integers (Lemma (1.1)) and to geometry (Lemma (3.1) and Claim (3.2)), than to group theory itself.

Acknowlegement. The author would like to express his thanks to Professor Doctor Thomas Peternell for inviting him to Universität Bayreuth in September 1994 with financial support, where he found the first idea of this paper. The author would like to express his thank to Dimitrios I. Dais for useful discussions.

§1. Abelian threefolds with a special automorphism.

Lemma (1.1). *Let (A, G) be a pair of an abelian threefold A and its finite automorphism group G satisfying the same conditions as in Theorem 3. Let 0 be an element of $A^G := \{x \in A \mid g(a) = a$ for some $g \in G - \{1\}\}$ and $g \neq 1$ be an element of the stabilizer $G_0 := \{g \in G \mid g(0) = 0\}$. Then*

(1) *the order of g, denoted by $\mathrm{ord}(g)$, is equal to either 3 or 7,*

(2) *$(A, \langle g\rangle)$ is isomorphic, in the obvious sense, to $(A_3, \langle g_3\rangle)$ if $\mathrm{ord}(g) = 3$, and*

(3) *$(A, \langle g\rangle)$ is isomorphic to $(A_7, \langle g_7\rangle)$ if $\mathrm{ord}(g) = 7$.*

Remark (1.2). The statement (1) is proved by Roan ([R]) by a geometric method. But his proof does not seem to imply the remaining two statements directly. Here we give another proof for (1), which is more algebraic and implies the remaining two statements almost automatically. The author learned this kind of method (cf. [CC]) based on cyclotomic fields and their integers from Catanese when he wrote [O2].

Proof. Put $\operatorname{ord}(g) = d$ and write A as V/Λ, where V is a three dimensional complex linear space and Λ is a discrete $\mathbb{Z}$-submodule of V of rank six. This space V is regarded as the holomorphic tangent space of A at 0 as well as the universal covering space of A.

Using appropriate global coordinates (z_1, z_2, z_3) of A, we can write the action of g by the following diagonal matrix:

$$g = \operatorname{diag}(\zeta_d^{k_1}, \zeta_d^{k_2}, \zeta_d^{k_3}),$$

where k_1, k_2, k_3 are integers satisfying

(1) $(k_1, d) = (k_2, d) = (k_3, d) = 1$ and $1 \le k_1, k_2, k_3 \le d$,

(2) $k_1 + k_2 + k_3 \equiv 0 \pmod d$,

because $\langle g \rangle$ is a Gorenstein group with only finitely many fixed points. In particular, $d \equiv 1 \pmod 2$.

By replacing g by another generator of $\langle g \rangle$ and adjusting the order of the coordinates (z_1, z_2, z_3), if necessary, we may assume without loss of generality that g has the following matrix representation:

(1.3)
$$g = \operatorname{diag}(\zeta_d, \zeta_d^{k_2}, \zeta_d^{k_3}),$$

where k_2 and k_3 are some integers such that $1 \le k_2 \le k_3 \le d$, $(k_2, d) = (k_3, d) = 1$, and $1 + k_2 + k_3 = d$.

Let $\mathbb{Z}[\langle g \rangle]$ be the group algebra defined by the cyclic group $\langle g \rangle$ over $\mathbb{Z}$. Since $g(\Lambda) = \Lambda$, the $\mathbb{Z}$-module Λ has a natural $\mathbb{Z}[\langle g \rangle]$-module structure defined through the action of g on V: $F(g)(\lambda) = \operatorname{diag}(F(\zeta_d), F(\zeta_d^{k_2}), F(\zeta_d^{k_3})) \cdot \lambda$.

Note that $\mathbb{Z}[\langle g \rangle] \simeq \mathbb{Z}[x]/(x^d - 1)$, by $g \leftrightarrow x \pmod{x^d - 1}$. Let λ be an element of either Λ or V and denote by $\Phi_d(x)$ the d-th cyclotomic polynomial. Using condition $(k_2, d) = (k_3, d) = 1$ and the explicit description above, we easily see that the following three statements are equivalent to each other:

(1) $F(g)\lambda = 0$ for some $\lambda \ne 0$,

(2) $F(g)\lambda = 0$ for all λ,

(3) $F(x) \in (\Phi_d(x))$.

This shows that $\operatorname{Ann}(\Lambda) = (\Phi_d(g))$ and Λ is a torsion free $\mathbb{Z}[\langle g \rangle]/(\Phi_d(\langle g \rangle))$-module. This also holds for V. In the sequel, we identify $\mathbb{Z}[\langle g \rangle]/(\Phi_d(\langle g \rangle)) = \mathbb{Z}[\zeta_d]$ by $g \pmod{\Phi_d(g)} \leftrightarrow \zeta_d$. Note that, under this identification, $\langle g \rangle = \langle \zeta_d \rangle$ and the action of ζ_d on Λ (as well as V) is same as that of g.

Operating $\otimes_{\mathbb{Z}}\mathbb{Q}$, to the chain of modules $\mathbb{Z} \curvearrowright \mathbb{Z}[\zeta_d] \curvearrowright \Lambda$, we get the chain of linear space structures, $\mathbb{Q} \curvearrowright \mathbb{Q}[\zeta_d] \curvearrowright \Lambda \otimes_{\mathbb{Z}} \mathbb{Q}$. Counting dimensions, we get, $\phi(d) \cdot \dim_{\mathbb{Q}[\zeta_d]}(\Lambda \otimes_{\mathbb{Z}} \mathbb{Q}) = 6$, where ϕ is the Euler function. Thus, $\phi(d)$ is either

2, 3, or 6. Combining this with $d \equiv 1 \pmod 2$, we get

(1.4) d is either 3, 7, or 9.

This shows that the possible matrix in (1.3), which is same as the action of ζ_d on V, is one of the following:

$$(1.5)\quad \begin{aligned} &(1)\quad \mathrm{diag}(\zeta_3, \zeta_3, \zeta_3) \quad \text{if } d = 3,\\ &(2)\quad \mathrm{diag}(\zeta_7, \zeta_7^2, \zeta_7^4),\ \mathrm{diag}(\zeta_7, \zeta_7^3, \zeta_7^3),\ \text{or } \mathrm{diag}(\zeta_7, \zeta_7, \zeta_7^5) \quad \text{if } d = 7,\\ &(3)\quad \mathrm{diag}(\zeta_9, \zeta_9, \zeta_9^7),\ \text{or } \mathrm{diag}(\zeta_9, \zeta_9^4, \zeta_9^4) \quad \text{if } d = 9. \end{aligned}$$

Note that the space $\Lambda \otimes_{\mathbb{Z}} \mathbb{C}$ is regarded as the complexified real tangent space of A at 0 and has the decomposition $\Lambda \otimes_{\mathbb{Z}} \mathbb{C} = V \oplus \bar{V}$, where $\bar{V}$ denotes the complex conjugate of V with respect to the natural real structure of $\Lambda \otimes_{\mathbb{Z}} \mathbb{C}$. Thus, in each case of (1.5), V is the direct sum of some eigen-subspaces corresponding to the indicated eigen-values of the action of $\zeta_d \otimes 1$ on the space $\Lambda \otimes_{\mathbb{Z}} \mathbb{C}$. For example, in the third case, V is the direct sum of some one dimensional eigen-(sub)space of the eigen-value ζ_7 and some two dimensional eigen-(sub)space of the eigen-value ζ_7^3 of the action of $\zeta_d \otimes 1$ on the space $\Lambda \otimes_{\mathbb{Z}} \mathbb{C}$.

Recall the following theorem:

Theorem (1.6) ([MH]). *The class number of the cyclotomic field $\mathbb{Q}(\zeta_m)$ (m is odd) is one if and only if m is one of the following numbers:* 3, 5, 7, 9, 11, 13, 15, 17, 19, 21, 25, 27, 33, 35, 45.

This shows that in each case of (1.4), $\mathbb{Z}[\zeta_d]$ is PID. In particular, in each case, Λ must be a free $\mathbb{Z}[\zeta_d]$-module of rank $6/\phi(d)$. This gives

$$(1.7)\quad \begin{aligned} &(1)\quad \phi_3\colon \Lambda \simeq \mathbb{Z}[\zeta_3]^{\oplus 3} \quad \text{if } d = 3,\\ &(2)\quad \phi_7\colon \Lambda \simeq \mathbb{Z}[\zeta_7] \quad \text{if } d = 7, \text{ and}\\ &(3)\quad \phi_9\colon \Lambda \simeq \mathbb{Z}[\zeta_9] \quad \text{if } d = 9. \end{aligned}$$

Here, in each case, the right hand side is the standard $\mathbb{Z}[\zeta_d]$-module and ϕ_d is an isomorphism as $\mathbb{Z}[\zeta_d]$-modules. In particular, in each case, the actions ζ_d on both sides commute with the isomorphism ϕ_d.

Observe that the eigen-values of the action $\zeta_d \otimes 1$ on the space $\mathbb{Z}[\zeta_d] \otimes_{\mathbb{Z}} \mathbb{C}$ are ζ_d^m, where m is an integer such that $0 < m < d$ and $(m, d) = 1$, and the corresponding eigen-space decomposition is as follows:

$$\mathbb{Z}[\zeta_d] \otimes_{\mathbb{Z}} \mathbb{C} = \bigoplus_{\substack{0<m<d\\ (m,d)=1}} V(m),$$

where $V(m)$ denotes the eigen-space of the eigen-value ζ_d^m. This is one dimensional.

In particular, in each case of (1.7), we get the following ϕ_d-equivariant eigen space decompositions:

(1.8)

(1) $\mathbb{Z}[\zeta_3]^{\oplus 3} \otimes_{\mathbb{Z}} \mathbb{C} = V(1)^{\oplus 3} \oplus V(2)^{\oplus 3}$ and $\Lambda \otimes_{\mathbb{Z}} \mathbb{C} = V'(1) \oplus V'(2)$ if $d = 3$,

(2) $\mathbb{Z}[\zeta_7] \otimes_{\mathbb{Z}} \mathbb{C} = \oplus_{m=1}^{6} V(m)$ and $\Lambda \otimes_{\mathbb{Z}} \mathbb{C} = \oplus_{m=1}^{6} V'(m)$ if $d = 7$, and

(3) $\mathbb{Z}[\zeta_9] \otimes_{\mathbb{Z}} \mathbb{C} = V(1) \oplus V(2) \oplus V(4) \oplus V(5) \oplus V(7) \oplus V(8)$ and $\Lambda \otimes_{\mathbb{Z}} \mathbb{C} = V'(1) \oplus V'(2) \oplus V'(4) \oplus V'(5) \oplus V'(7) \oplus V'(8)$ if $d = 9$.

Now comparing (1.5), (1.7), (1,8), we conculde that $d \neq 9$ and

(1) $\phi_3\colon (A, g) = (V'(1)/\Lambda, \zeta_3) \simeq (V(1)^{\oplus 3}/\mathbb{Z}[\zeta_3], \zeta_3)$ if $d = 3$,

(2) $\phi_7\colon (A, g) = ((V'(1) \oplus V'(2) \oplus V'(4))/\Lambda, \zeta_7) \simeq ((V(1) \oplus V(2) \oplus V(4))/\mathbb{Z}[\zeta_7], \zeta_7)$ if $d = 7$.

In particular,

(1) $(A_3, g_3) \simeq (V(1)^{\oplus 3}/\mathbb{Z}[\zeta_3], \zeta_3)$ if $d = 3$,

(2) $(A_7, g_7) \simeq ((V(1) \oplus V(2) \oplus V(4))/\mathbb{Z}[\zeta_7], \zeta_7)$ if $d = 7$.

Thus, we get the result. □

Lemma (1.9). $A_3 \not\simeq A_7$.

Proof. Assume the contrary that $A_3 \simeq A_7$. We may assume without loss of generality that this isomorphism preserves the origin, 0. Let h_3 be the automorphism of A_7 corresponding to g_3 under this isomorphism. Since g_3 is scalar multiplication by ζ_3, so is h_3. Thus, $h_3 \circ g_7 = g_7 \circ h_3$, $H := \langle h_3, g_7 \rangle = \langle h_3 \circ g_7 \rangle$ and $h_3 \circ g_7 = \operatorname{diag}(\zeta_3\zeta_7, \zeta_3\zeta_7^2, \zeta_3\zeta_7^4)$ under appropriate coordinates of A_7. This shows H is a finite Gorenstein automorphism group of A_7 with only finitely many fixed points. But this contradicts Lemma (1.1) because $h_3 \circ g_7 \in H_0$ while $\operatorname{ord}(h_3 \circ g_7) = 21$. □

Now, the Key Lemma is reduced to the following,

Key Claim.

(1) *Let G_3 be a finite Gorenstein automorphism group of A_3 containing $\langle g_3 \rangle$. Suppose that G_3 contains no non-trivial translations and that $A_3^{G_3}$ is a finite set. Then $G_3 = \langle g_3 \rangle$ provided that some partial crepant resolution of A_3/G_3 is simply connected.*

(2) *Let G_7 be a finite Gorenstein automorphism group of A_7 containing $\langle g_7 \rangle$. Suppose that G_7 contains no non-trivial translations and that $A_7^{G_7}$ is a finite set. Then $G_7 = \langle g_7 \rangle$.*

We shall prove this claim in the proceeding two sections. We conclude this section with a few remarks needed for proof of Key Claim.

The next Lemma is a simple combination of the proof (1.1) and Lemma (1.9).

Lemma (1.10).

(1) *Suppose that a pair (A_3, G_3) satisfies conditions in Key Claim* (1). *Let P be a point in $A_3^{G_3}$ and $g \neq 1$ be an element of $(G_3)_P$. Then g acts on $T_{A_3,P}$ by scalar multiplication by either ζ_3 or ζ_3^2.*

(2) *Suppose that a pair (A_7, G_7) satisfies conditions in Key Lemma* (2). *Let P be a point in $A_7^{G_7}$ and $g \neq 1$ be an element of $(G_7)_P$. Then the action of g on $T_{A_7,P}$ has three distinct eigen values. Moreover, the pair of these three distinct eigen values is $(\zeta_7^i, \zeta_7^{2i}, \zeta_7^{4i})$ for some integer i in $\{1, 2, \ldots, 6\}$.*

Lemma (1.11).

(1) *Suppose that (A_3, G_3) satisfies conditions in Key Claim* (1). *Let P be a point in $A_3^{G_3}$. Then the stabilizer $(G_3)_P$ is isomorphic to $\mathbb{Z}/3\mathbb{Z}$.*

(2) *Suppose that (A_7, G_7) satisfies conditions in Key Claim* (2). *Let P be a point in $A_7^{G_7}$. Then the stabilizer $(G_7)_P$ is isomorphic to $\mathbb{Z}/7\mathbb{Z}$.*

Proof. Let d be either 3 or 7. Assume the contrary that $(G_d)_P$ is not isomorphic to $\mathbb{Z}/d\mathbb{Z}$. By (1.10), each element of $(G_d)_P - \{1\}$ is of order d. Thus, $(G_d)_P$ contains a subgroup $H_d = \langle a, b\rangle$ isomorphic to $(\mathbb{Z}/d\mathbb{Z})^{\oplus 2}$. Since $ab = ba$, a and b are represented by diagonal matrices, say A and B, under some common appropriate global coordinates around P. Assume $d = 3$. Then, by (1.10)(1), after replacing a and b by other generators of $\langle a\rangle$ and $\langle b\rangle$ respectively if necessary, we get $A = B = \mathrm{diag}(\zeta_3, \zeta_3, \zeta_3)$. But this is absurd. Assume $d = 7$. Then, by (1.10)(2), after replacing a and b by other generators of $\langle a\rangle$ and $\langle b\rangle$ respectively if necessary, we get $A = \mathrm{diag}(\zeta_7, \zeta_7^2, \zeta_7^4)$ and $B = \mathrm{diag}(\zeta_7, \zeta_7^4, \zeta_7^2)$. Then ab is represented by $AB = \mathrm{diag}(\zeta_7^2, \zeta_7^6, \zeta_7^6)$. But this contradicts (1.10)(2). This completes the proof. □

§2. Proof of Key Claim (1)

In this section, we prove Key Claim (1).

Lemma (2.1). *Suppose that $h \neq 1 \in G_3$ has a fixed point. Then $h \in \langle g_3\rangle$.*

Proof. Let P be a fixed point of h. Then (1.10)(1) shows that $h = t_a \circ g_3^i$, where t_a is a translation by some $a \in A_3$. Thus, $h \circ g_3^{-i} = t_a$. Since G contains no non-trivial translations, $a = 0$. This implies $h = g_3^i$. □

Lemma (2.2). *$\langle g_3\rangle$ is a normal subgroup of G_3. Moreover, the quotient group $G/\langle g_3\rangle$ acts freely on $A_3/\langle g_3\rangle$.*

Proof. Let $h \in G_3$. Take a point $Q \in (A_3)^{\langle g_3\rangle}$ and put $P := h^{-1}(Q)$. Then $h^{-1} \circ g_3 \circ h(P) = h^{-1} \circ g_3(Q) = h^{-1}(Q) = P$. Thus, $h^{-1} \circ g_3 \circ h \in \langle g_3\rangle$ by (2.1). This gives the first half part of (2.2). Assume the contrary that $G/\langle g_3\rangle$ has a fixed point on $A_3/\langle g_3\rangle$. Then there exist a point $P \in A_3$ and an element $h \in G_3 - \langle g_3\rangle$ such that $h(P) \in \{P, g_3(P), g_3^2(P)\}$. Thus, there is an integer i such that $g_3^{-i} \circ h(P) = P$. Thus, by (2.1), $g_3^{-i} \circ h \in \langle g_3\rangle$ and then $h \in \langle g_3\rangle$. This contradicts our choice of h. □

Proof of Key Claim (1). Let X be a partial crepant resolution of A_3/G_3. Since X_3 is the unique partial crepant resolution of $A_3/\langle g_3\rangle$ as was observed in Lemma 4, the quotient group $G_3/\langle g_3\rangle$ acts on X_3. This action is free by (2.2). Thus, one of the partial crepant resolutions of A_3/G_3 is given by $X_3/(G_3/\langle g_3\rangle)$. Thus, by the flop theorem, we have $\pi_1(X) \simeq \pi_1(X_3/(G_3/\langle g_3\rangle)) = G_3/\langle g_3\rangle$. Thus, $G_3 = \langle g_3\rangle$ provided that A_3/G_3 has a simply connected partial crepant resolution. Now we are done. □

§3. Proof of Key Claim (2)

In this section, we prove Key Claim (2).

Lemma (3.1). *Take global coordinates (z_1, z_2, z_3) of A_7 so that $g_7 = \mathrm{diag}(\zeta_7, \zeta_7^2, \zeta_7^4)$. Let h be an element of G_7. Then the matrix representation of h with respect to this global coordinates is given by*

$$h\begin{pmatrix} z_1 \\ z_2 \\ z_3 \end{pmatrix} = \begin{pmatrix} a_{11} & 0 & 0 \\ 0 & a_{22} & 0 \\ 0 & 0 & a_{33} \end{pmatrix}\begin{pmatrix} z_1 \\ z_2 \\ z_3 \end{pmatrix} + \begin{pmatrix} a \\ b \\ c \end{pmatrix}.$$

Proof. We need a geometric argument. First, we prove,

Claim (3.2). $H^2(A_7, \mathbb{C})^{G_7} (:= \{w \in H^2(A_7, \mathbb{C}) \mid g^*(w) = w$ *for every* $g \in G_7\}) = \oplus_{i=1}^3 \mathbb{C} dz_i \wedge d\bar{z}_i$.

Proof. Write $|G_7| = 7k$. Then k is an integer. Consider the quotient map $\pi\colon A_7 \to A_7/G_7$ and put $\mathrm{Sing}(A_7/G_7) = \{Q_i \mid i = 1, \ldots, n\}$. By Lemma (1.11)(2), $\pi^{-1}(Q_i)$ consists of k points for each i. We denote $\pi^{-1}(Q_i) = \{P_{ij} \mid j = 1, \ldots, k\}$. Then $A_7^{G_7} = \{P_{ij} \mid i = 1, \ldots, n, j = 1, \ldots, k\}$. Since the map $\pi\colon A_7 - A_7^{G_7} \to A_7/G_7 - \mathrm{Sing}(A_7/G_7)$ is unramified and of degree $7k$, we get $\chi_{\mathrm{top}}(A_7/G_7 - \mathrm{Sing}(A_7/G_7)) = (0 - nk)/7k = -n/7$. Therefore we have $n = 7m$ for some positive integer m and $\chi_{\mathrm{top}}(A_7/G_7 - \mathrm{Sing}(A_7/G_7)) = -m$. On the other hand, since by (1.11)(2),

$G_{P_{ij}} = \langle \mathrm{diag}(\zeta_7, \zeta_7^2, \zeta_7^4)\rangle$ under appropriate coordinates around P_{ij}, each singular point Q_i is of type $\frac{1}{7}(1,2,4)$. Thus, their toric resolution, which replaces each Q_i by three $\mathbb{F}_2$'s crossing normally along the negative sections and fibers in a cyclic way, gives a crepant resolution X of A_7/G_7. Let E_i be the exceptional divisor over Q_i. Then $\chi_{\mathrm{top}}(E_i) = 7$ and we get,

(3.3) $$\chi_{\mathrm{top}}(X) = -m + 7m \times 7 = 48m.$$

Note that any quotient singularities are rational singularities. Hence we have $h^i(\mathcal{O}_X) = h^i(\mathcal{O}_{A_7/G_7}) \le \dim(H^i(A_7,\mathbb{C})^{G_7}) \le \dim(H^i(A_7,\mathbb{C})^{\langle g_7\rangle})$.

Since $H^1(A_7,\mathbb{C}) = \mathbb{C}\langle dz_i, d\bar{z}_i\rangle_{i=1,2,3}$, we get $H^1(A_7,\mathbb{C})^{\langle g_7\rangle} = 0$ by an explicit calculation. Thus, $h^1(\mathcal{O}_X) = \{0\}$. Since X is a smooth projective threefold with $h^1(\mathcal{O}_X) = 0$ and with $\mathcal{O}_X(K_X) \simeq \mathcal{O}_X$, we have $h^2(\mathcal{O}_X) = 0$ and then,

(3.4) $$\chi_{\mathrm{top}}(X) = 2(h^{1,1}(X) - h^{2,1}(X)).$$

Since $h^2(\mathcal{O}_X) = 0$ and since A_7/G_7 is $\mathbb{Q}$-factorial, we calculate,

(3.5) $$h^{1,1}(X) = \rho(X) = 21m + \rho(A_7/G_7),$$

where $21m$ is nothing but the number of the exceptional divisors of $X \to A_7/G_7$ and ρ denotes Picard number.

An explicit calculation shows $H^2(A_7,\mathbb{C})^{\langle g_7\rangle} = \oplus_{i=1}^3 \mathbb{C}dz_i \wedge d\bar{z}_i$. Thus,

(3.6) $$\rho(A_7/G_7) \le \dim H^2(A_7,\mathbb{C})^{G_7} \le \dim H^2(A_7,\mathbb{C})^{\langle g_7\rangle} = 3.$$

Combining (3.3),(3.4),(3.5) and (3.6), we get

(3.7) $$48m \le 2(21m + 3 - h^{2,1}(X)) \le 42m + 6.$$

Thus, $m = 1$ and all the inequalities in (3.7) and (3.6) must be equalities. This implies that $|\mathrm{Sing}(A_7/G_7)| = 7$, $h^{1,1}(X) = 24$, $h^{2,1}(X) = 0$ (cf. Example 2) and $H^2(A_7,\mathbb{C})^{G_7} = H^2(A_7,\mathbb{C})^{\langle g_7\rangle} = \oplus_{i=1}^3 \mathbb{C}dz_i \wedge d\bar{z}_i$. This completes the proof of claim (3.2). □

We return to the proof of (3.1). Take $h \in G_7$ and write

$$h\begin{pmatrix} z_1 \\ z_2 \\ z_3 \end{pmatrix} = \begin{pmatrix} a_{11} & a_{12} & a_{13} \\ a_{21} & a_{22} & a_{23} \\ a_{31} & a_{32} & a_{33} \end{pmatrix}\begin{pmatrix} z_1 \\ z_2 \\ z_3 \end{pmatrix} + \begin{pmatrix} a \\ b \\ c \end{pmatrix}.$$

By (3.2), $dz_i \wedge d\bar{z}_i = (h^{-1})^*(dz_i \wedge d\bar{z}_i) = (a_{i1}dz_1 + a_{i2}dz_2 + a_{i3}dz_3) \wedge (\bar{a}_{i1}d\bar{z}_1 + \bar{a}_{i2}d\bar{z}_2 + \bar{a}_{i3}d\bar{z}_3)$ for $i = 1, 2, 3$.

Comparing both sides, we get $a_{ii}\bar{a}_{ii} = 1$ and $a_{ij}\bar{a}_{ii} = 0$ for $j \ne i$. Thus, $a_{ij} = 0$ provided that $i \ne j$. This proves (3.1). □

Lemma (3.8). *Under the same notation as in* (3.1), $a_{ii} \ne 1$ *for each* $i = 1, 2, 3$ *provided that* $h \ne 1$. *In particular, each element of* G_7 *has a fixed point.*

Proof. Assume the contrary that there is an element $h \in G - \{1\}$ such that $a_{ii} = 1$ for some i. For simplicity of notation, we assume $i = 1$. Then

$$h\begin{pmatrix} z_1 \\ z_2 \\ z_3 \end{pmatrix} = \begin{pmatrix} 1 & 0 & 0 \\ 0 & a_{22} & 0 \\ 0 & 0 & a_{33} \end{pmatrix}\begin{pmatrix} z_1 \\ z_2 \\ z_3 \end{pmatrix} + \begin{pmatrix} a \\ b \\ c \end{pmatrix}.$$

Note that $a_{22}a_{33} = 1$ but $a_{22} \neq 1$ (and hence $a_{33} \neq 1$) because G_7 is Gorenstein and has no non-trivial translations.

Calculate

$$h \circ g_7^i \begin{pmatrix} z_1 \\ z_2 \\ z_3 \end{pmatrix} = \begin{pmatrix} \zeta_7^i & 0 & 0 \\ 0 & \zeta_7^{2i}a_{22} & 0 \\ 0 & 0 & \zeta_7^{4i}a_{33} \end{pmatrix}\begin{pmatrix} z_1 \\ z_2 \\ z_3 \end{pmatrix} + \begin{pmatrix} a \\ b \\ c \end{pmatrix}.$$

Since $\zeta_7^{2i} \neq \zeta_7^{2j}$ and $\zeta_7^{4i} \neq \zeta_7^{4j}$ if $1 \leq i \neq j \leq 6$, there is an integer k $(1 \leq k \leq 6)$ such that $\zeta_7^{2k}a_{22} \neq 1$ and $\zeta_7^{4k}a_{33} \neq 1$. Since $h \circ g_7^k$ has a fixed point for this k, $a_{22} = \zeta_7^p$ and hence $a_{33} = \zeta_7^{7-p}$ for some $p \in \{1, 2, \ldots, 6\}$ by (1.10)(2). But, then,

$$h \circ g_7^p \begin{pmatrix} z_1 \\ z_2 \\ z_3 \end{pmatrix} = \begin{pmatrix} \zeta_7^p & 0 & 0 \\ 0 & \zeta_7^{3p} & 0 \\ 0 & 0 & \zeta_7^{3p} \end{pmatrix}\begin{pmatrix} z_1 \\ z_2 \\ z_3 \end{pmatrix} + \begin{pmatrix} a \\ b \\ c \end{pmatrix}.$$

Thus, $h \circ g_7^p$ has a fixed point, say Q, while the action $h \circ g_7^p$ on $T_{A_7,Q}$ has only two distinct eigen-values. This contradicts (1.10)(2). This finishes the proof. □

Proof of Key Claim (2). Let h be an element of $G_7 - \{1\}$. Then by (3.8) and (1,11)(2), h is of order 7. Thus, by (3.1) and (3.8), we have,

$$h\begin{pmatrix} z_1 \\ z_2 \\ z_3 \end{pmatrix} = \begin{pmatrix} \zeta_7^i & 0 & 0 \\ 0 & \zeta_7^j & 0 \\ 0 & 0 & \zeta_7^k \end{pmatrix}\begin{pmatrix} z_1 \\ z_2 \\ z_3 \end{pmatrix} + \begin{pmatrix} a \\ b \\ c \end{pmatrix},$$

where i, j, k are elements in $\{1, 2, \ldots, 6\}$. Thus,

$$h \circ g_7^{-i} \begin{pmatrix} z_1 \\ z_2 \\ z_3 \end{pmatrix} = \begin{pmatrix} 1 & 0 & 0 \\ 0 & \zeta_7^{j-2i} & 0 \\ 0 & 0 & \zeta_7^{k-4i} \end{pmatrix}\begin{pmatrix} z_1 \\ z_2 \\ z_3 \end{pmatrix} + \begin{pmatrix} a \\ b \\ c \end{pmatrix}.$$

By (3.8) again, we get $h \circ g_7^{-i} = 1$. This implies, $G_7 = \langle g_7 \rangle$. Now we are done. □

References

[B] A. Beauville, Some remarks on Kähler manifolds with $c_1 = 0$, in: Classification of algebraic and analytic manifolds (K. Ueno ed.), Progr. Math. 39 (1983), 1–26.

[CC] F. Catanese and C. Ciliberto, On the irregularity of cyclic coverings of algebraic surfaces, in: Geometry of complex projective varieties, Cetrano (Italy) (A. Lanteri, M. Palleschi, D. C.Struppa eds.), Mediterranean Press, 89–115, 1993.

[Ka] Y. Kawamata, The crepant blowing-ups of 3-dimensional canonical singularities and its application to degeneration of surfaces, Ann. of Math. 127 (1988), 93–163.

[KMM] Y. Kawamata, K. Matsuda, and K. Matsuki, Introduction to the minimal model problem, Adv. Stud. Pure Math. 10 (1987), 283–360.

[Ko] J. Kollár, Flops, Nagoya Math. J. 113 (1989), 15–36.

[MM] J. M. Masley and H. L. Montgomery, Cyclotomic fields with unique factorization, J. Angew. Reine Math. 286 (1976), 248–256.

[O1] K. Oguiso, On algebraic fiber space structures on a Calabi–Yau 3-fold, Internat. J. Math. 4 (1993), 439–465.

[O2] K. Oguiso, On certain rigid fibered Calabi–Yau threefolds, Math. Z., to appear.

[R] M. Roan, On the generalization of Kummer surfaces, J. Diff. Geom. 30 (1989), 523–537.

[RY] M. Roan and S. Yau, On Ricci flat 3-fold, Acta Math. Sinica 3 (1987), 256–288.

[SW] N. I. Shepherd-Barron and P. M. H. Wilson, Singular threefolds with numerically trivial first and second Chern class, J. Alg. Geom. 3 (1994), 265–281.

[W] P. M. H. Wilson, The role of c_2 in Calabi–Yau classification — a preliminary survey, preprint, 1994.

[YY] S. T.Yau and Yung Yu, Gorenstein quotient singularities in dimension three, Mem. Amer. Math. Soc. 505, 1993.

On Halphen's speciality theorem

Roberto Paoletti

1. Introduction

The effort to understand how the postulation of algebraic varieties reflects itself in their geometry is one of the most classical veins of algebraic geometry. For space curves, some fundamental work in this sense goes back to the nineteenth century and is due to Halphen. In his *Mémoire sur la classification des courbes gauches algébriques* ([Hal 82]), published in 1882, Halphen exposed various results of this flavour, among which his celebrated speciality theorem (see Theorem 1.2 below). Through the work of Gruson and Peskine this has then been generalized to a statement holding for codimension two subvarieties of any $\mathbb{P}^n$ ([GP 77], [GP 77-1]; see Theorem 1.3). For $n = 2$, in particular, an especially sharp result exists, which is due to Ellia and Peskine ([EP 88]; cfr Theorem 1.1 below). In the present paper we revisit these questions by means of vector bundle techniques, and show that they can be addressed in a very simple and geometric way by means of Bogomolov's instability theorem on a surface. In a few words, the basic idea is the following: first, we use techniques from [La 86], [Ty 87] and [GH 78] together with Bogomolov's instability theorem to extend Ellia and Peskine's result about points in the plane to any surface with $\mathrm{Pic}(S) \simeq \mathbb{Z}$; then we reduce the general case of $X^r \subset \mathbb{P}^{r+2}$ to this one by cutting X with general complete intersection surfaces of very large degree.

It is certainly well-known to the experts that in the case of space curves the speciality theorem can be approached to a large extent using reflexive sheaves, following Hartshorne's correspondence ([Har 80]); for example, in [GP 83] various classical statement about the postulation of space curves are obtained along these lines. It should be possible, similarly, to extend this correspondence to subvarieties of codimension two of any $\mathbb{P}^n$, $n \geq 3$, and to prove the speciality theorem along these lines for $n \geq 3$. The present approach is arguably more direct and elementary, in that it uses the notion of stability for a finite scheme to establish the theorem for finite schemes on a surface, and then reduces the general case to a consequence of the statement holding for surfaces. An advantage of this line argument is that the speciality theorem generalizes to a geometric statement holding in any variety X such that $\mathrm{Pic}(X) \cong \mathbb{Z}$.

We work throughout on an algebraically closed field of characteristic zero.

To start with, we consider the following statements:

Theorem 1.1. (a) *Let $\xi \subset \mathbb{P}^2$ be an arbitrary finite subscheme of degree d, and denote by τ the largest k for which $h^1(\mathbb{P}^2, \mathcal{I}_\xi(kH)) \neq 0$. Suppose that there exists a real number s with $0 < s \leq \sqrt{d}$ such that*

$$\tau + 3 \geq s + \frac{d}{s}. \tag{1.1}$$

Then there exists a curve D of degree $y \leq s$ containing a subscheme ξ_s of ξ of degree d_s with

$$y\left[\tau + \frac{1}{2}(5 - y)\right] \geq d_s \geq y(\tau + 3 - y). \tag{1.2}$$

If inequality is strict in (1.1), *then the curve D is unique. If $y = s$, furthermore, ξ_s is a complete intersection of type $(s, \frac{d}{s})$.*

(b) *More generally, let S be a smooth complete surface with* $\mathrm{Pic}(S) \simeq \mathbb{Z}$ *and let H be the ample generator of* $\mathrm{Pic}(S)$. *Define the integer a by $-\omega_S \equiv aH$. Let $\xi \subset S$ be a reduced finite subscheme of degree d and define τ as in* (a). *Suppose that $\tau + a > s + \frac{d}{sH^2}$ for some s and that $s^2 \leq d$. Then there exists an integer $y \leq s$ and a curve $D \in |yH|$ containing a subscheme ξ_s of ξ of degree d_s satisfying $d_s \geq y(\tau + a - y)H^2$. Suppose furthermore that $H^1(S, \mathcal{O}_S(x)) = 0$ for $x \geq 0$; if $y = s$, then ξ_s is a complete intersection of type $(s, \frac{d}{sH^2})$. If furthermore $\pi_1(S) = 0$ then the same conclusions hold if $\tau + a \geq s + \frac{d}{sH^2}$.*

Part (a) is a statement of Ellia and Peskine ([EP 88]), who derived it from a more general result concerning the geometry of a finite scheme in the plane having a numerical character with certain properties. This same result has also been obtained by Mark Green, who has reinterpreted numerical characters in the setting of the theory of initial ideals. Both arguments are of a deeply algebraic nature, and it is rather natural to ask whether there is a concrete geometric interpretation of this kind of results.

One hint in this direction is given by the following observation. As shown in [LP 93], Ellia and Peskine's statement implies that any free pencil of degree $< \frac{1}{4}d^2$ on an integral plane curve C of degree d is the moving part of the pencil cut out on C by a pencil of plane curves of a suitable degree $r < \frac{d}{2}$ (various results of this kind had already been obtained by Coppens: see [Cop 92]). On the other hand, in [Pa 93] the same result about linear series on plane curves is obtained by applying Bogomolov's instability theorem to the rank two bundle on $\mathbb{P}^2$ that one associates to the free pencil on C by the standard elementary transformation, as follows. Let V be a pencil of sections globally generating the line bundle A on C; then, following Lazarsfeld ([La 86]), we can define a rank two bundle $\mathcal{E}$ on $\mathbb{P}^2$ by the exactness of the sequence $0 \to \mathcal{E} \to V \otimes \mathcal{O}_{\mathbb{P}^2} \to A \to 0$. Note that $c_1(\mathcal{E}) = -d$, $c_2(\mathcal{E}) = \deg(A)$, and if $k \leq \frac{d^2}{4}$ then $\mathcal{E}$ is not stable. This elementary transformation, in turn, generalizes to a procedure developped by Griffiths and Harris ([GH], [GH 78]) and by Tyurin ([Ty 87]) that on a smooth surface associates a vector bundle to a finite scheme which is suitably "special". The basic fact in this sense is that if $\xi \subset S$ is a reduced finite scheme and L is a line bundle on S, then ξ is the zero scheme of a section of a rank two bundle $\mathcal{E}$ on S with $\wedge^2\mathcal{E} \simeq L$ if and only if ξ satisfies the Cayley–Bacharach property with respect to the adjoint line bundle $L \otimes \omega_S$. It is therefore tempting to guess that it should be possible to reprove Ellia and Peskine's statement using vector bundle methods along these lines, applying Bogomolov's theorem to this construction. This is precisely the content of Section 2.

On the other hand, it is not obvious that there should be any algebraic argument for a proof of part (b) of Theorem 1.1. Our geometric argument, however, works for any surface with $\mathrm{Pic}(S) \simeq \mathbb{Z}$. Furthermore, (b) is very useful in the proof of Halphen's theorem that we give in Section 3, and that we shall discuss shortly; in fact, in the way to the speciality theorem we only need to worry about reduced finite subschemes of a surface.

As a consequence of the proof of Theorem 1.1, furthermore, we shall also be able to identify ξ_s as the "τH-stable part" of ξ in the sense of [Ty 87]. We are going to explain exactly what this means in the next section, but let us note that in the case of a reduced finite subscheme ξ of the plane this has a very pictorial interpretation, as follows. There exists a smooth curve C of degree $\tau + 3$ through ξ, and the τ-stable part ξ_s is just $\xi - B$, where B is the base locus of the linear series $|\xi|$ on C.

Briefly, the argument is as follows. To the subscheme $\xi_s \subset \xi$ one associates, according to the above observations, a rank two vector bundle $\mathcal{E}$ on S, which sits in a short exact sequence

$$0 \to \mathcal{O}_S \to \mathcal{E} \to \mathcal{I}_{\xi_s}(\tau + a) \to 0.$$

Then the hypothesis of Theorem 1.1 imply that $\mathcal{E}$ is Bogomolov-unstable, so that there is a maximal destabilizing line bundle $\mathcal{O}_S(x) \subset \mathcal{E}$. This yields a curve of degree $y = \tau + a - x$ through ξ_s, and then one shows directly that the unstability conditions force y to satisfy the required inequalities.

Part (b) of Theorem 1.1 is useful in proving the speciality theorem almost inductively, in a sense that we are now going to make more precise. The precise statement is part (a) of the following:

Theorem 1.2. (a) (Halphen) *Let $C \subset \mathbb{P}^3$ be an integral curve of degree d and define*

$$e = \max\{k \mid h^1(C, \mathcal{O}_C(k)) \neq 0\}. \tag{1.3}$$

Let $f(x) = x^2 - (e+4)x + d$, and let s be the smallest real root of $f(x)$. Then there exists a hypersurface $V \subset \mathbb{P}^3$ of degree $y \leq s$ containing C. If $y = s$, then C is a complete intersection of type $(s, \frac{d}{s})$. More generally, let C be a finite union of reduced irreducible curves $C_1, \ldots, C_n$, where the C_i's are all nondegenerate. For each i, let $e_i = e(C_i)$ be defined as above, and let $e = \min\{e_i\}$. Then the same conclusions hold.

(b) *More generally, let P be a smooth projective threefold with* $\mathrm{Pic}(P) \cong \mathbb{Z}H$, *where H is the ample generator, and let c be the integer such that $\omega_P = -cH$. Let $C \subset P$ be an integral curve, and set $d = C \cdot H$, $e = \max\{k \mid h^2(P, \mathcal{I}_C(k)) \neq 0\}$ and $f(x) = x^2 - (e+c)x + \frac{d}{H^3}$. Then:* i) *if $f(X)$ has two distinct real roots $s_1 < s_2$, then there exists an integer $y > 0$ such that $y \leq s_1$ and $d \geq y(e + c - y)H^3$, and* *a surface* $D \in |yH|$ *such that* $D \supset C$; ii) *if* $H^1(P, \mathcal{O}_P(x)) = 0$ *for* $x \geq 0$, *and if* $y = s$, *then C is a complete intersection in P, of type* $(s, \frac{d}{sH^3})$; iii) *if* $\pi_1(P) = 0$, *the same holds if* $s_1 = s_2$.

The way this is reduced to Theorem 1.1 is very simple: for $S \subset \mathbb{P}^3$ a general surface of very large degree m with $\mathrm{Pic}(S) \simeq \mathbb{Z}$, let $C_S = C \cap S$ be the corresponding section of C; then by uniform position C_S is a stable finite subscheme of S. Define $\tau = \tau_{C_S}$ as before. Then $\tau \geq e + m$, and we are in the situation of Theorem 1.1. Hence we obtain a curve $D \in |\mathcal{O}_S(yH)|$ that contains C_S, where $y \leq s$. However, for $m \gg 0$ we also have an isomorphism $H^0(\mathbb{P}^3, \mathcal{I}_C(y)) \simeq H^0(S, \mathcal{J}_{C_S}(y))$, where $\mathcal{J}_{C_S} \subset \mathcal{O}_S$ is the ideal sheaf of $C_S \subset S$. This gives us the desired hypersurface. Note that, in addition, this also proves the inequality $d \geq y(\tau + 3 - y)$.

Finally, in the last section, we show that essentially the same argument also proves the generalized speciality theorem of Gruson and Peskine ([GP 77], [GP 77-1]):

Theorem 1.3. *Let* $X^r \subset \mathbb{P}^{r+2}$ *be an integral codimension two subvariety, and define*

$$\tau_X =: \max\{k \mid H^{r+1}(\mathbb{P}^{r+2}, \mathcal{I}_X(k)) \neq 0\}.$$

Consider the polynomial $f(x) = x^2 - (\tau_X + r + 3)x + d$. *If s is the smallest real root of* $f(x)$, *then there exists a hypersurface* $V \subset \mathbb{P}^{r+2}$ *through X having degree* $y \leq s$. *If* $y = s$, *X is a complete intersection of type* $(s, \frac{d}{s})$. *Furthermore,* $d \geq y(\tau_X + r + 3 - y)$.

Obviously, this statement also generalizes to arbitrary projective manifolds such that $\mathrm{Pic}(X) \cong \mathbb{Z}$, exactly as in part (b) of Theorem 1.2. We leave it to the interested reader to make this explicit.

Acknowledgments. I am very grateful to Angelo Lopez and Pietro Pirola for some fruitful discussions that have stimulated this paper, and to Christian Peskine for some interesting remarks. I am also endebted to the Department of Mathematics of the University of Pavia, and especially to Maurizio Cornalba, for its hospitality and to the Istituto Nazionale di Alta Matematica "F. Severi" for its support during Spring and Summer 1993.

2. Preliminaries

We collect here some facts about vector bundles and zero-cycles that will be handy in the sequel.

Definition 2.1. Let S be any smooth, regular projective surface, L a line bundle on S and consider a finite subscheme $\xi \subset S$. The *index of specialty of* ξ *w.r.t.* L is defined to be

$$\delta(\xi, L) =: h^1(S, \mathcal{J}_\xi(L)). \tag{2.4}$$

$\delta(\xi, L)$ measures the failure of ξ to impose independent conditions on L.

Definition 2.2. ξ is said to be *L-stable* if for every proper subscheme $\xi' \subset \xi$ we have $\delta(\xi', L) < \delta(\xi, L)$. ([Ty 87])

The following fact is proved in [Ty 87]:

Lemma 2.1. *There is a unique L-stable subscheme $\xi_s \subset \xi$ such that that $\delta(\xi_s, L) = \delta(\xi, L)$ and which is maximal with respect with this property.*

Definition 2.3. Let $\xi_s \subset \xi$ be the subscheme described in Lemma 2.1. We shall refer to ξ_s as the *maximal L-stable subscheme of ξ*.

Remark 2.1. To fix ideas, suppose that ξ is reduced; then, we uniquely determine a decomposition $\xi = \xi_s + \xi_r$, where $\delta(\xi_s, L) = \delta(\xi, L)$ and the residual subscheme ξ_r imposes independent conditions on the linear system $|\mathcal{J}_{\xi_s}(L)|$. In general, to an L-stable 0-cycle ξ there is associated a vector bundle $\mathcal{E}$ on the surface S, sitting in a short exact sequence

$$0 \to H^1(S, \mathcal{J}_\xi(L)) \otimes \mathcal{O}_S \to \mathcal{E} \to \mathcal{J}_\xi(L \otimes \omega_S^{-1}) \to 0. \tag{2.5}$$

This is constructed as follows. Let $V = H^1(S, \mathcal{J}_\xi(L))$. Then by Serre duality we have

$$\mathrm{Ext}^1(\mathcal{J}_\xi(L \otimes \omega_S^{-1}), V \otimes \mathcal{O}_S) \simeq \mathrm{Hom}(V, V),$$

and the identity map id_V gives an intrinsically defined extension. As shown in [Ty 87], this extension is locally free if and only if ξ is L-stable.

Definition 2.4. Let S be a smooth complete surface, $\xi \subset S$ a finite subscheme of degree d and L a line bundle on S. ξ is said to satisfy the Cayley–Bacharach property with respect L if every global section of L vanishing along a subscheme $\xi' \subset \xi$ of degree $d-1$ also vanishes along ξ. In other words, if $\xi' \subset \xi$ and $\deg(\xi') = d-1$ then $H^0(S, \mathcal{J}_{\xi'}(L)) = H^0(S, \mathcal{J}_\xi(L))$ ([GH 78]).

For a proof of the following theorem see [GH] or [GH 78]:

Theorem 2.1. *Let S be a smooth complete surface and let $\xi = P_1 + \cdots + P_d$ be a reduced finite subscheme of S. Let L be a line bundle on S. Then there exists a rank two vector budle $\mathcal{E}$ on S such that $\wedge^2\mathcal{E} = L$, and a section $s \in H^0(S, \mathcal{E})$ whose zero-scheme is ξ if and only if ξ satisfies the Cayley–Bacharach property with respect to the adjoint line bundle $L + K_S$.*

The following fact is easy but useful:

Lemma 2.2. *Let $\xi \subset S$ be L-stable. Then ξ satisfies the Cayley–Bacharach condition with respect to L.*

Proof. Suppose, to the contrary, that there exists a subscheme $\xi' \subset \xi$ such that $[\xi] = [\xi'] + [P]$ (where $[\xi]$ denotes the associated 0-cycle), and not all global sections of L vanishing along ξ' vanish along ξ. This implies that $\delta(\xi', L) = \delta(\xi, L)$, which contradicts the assumption that ξ is L-stable. □

Remark 2.2. Let then ξ be L-stable. Following the construction outlined in Remark 2.1, we associate to ξ a vector bundle on S of rank $h^1(S, \mathcal{J}_\xi(L)) + 1$; on the other hand, by Lemma 2.2, we can also associate to ξ a rank two vector bundle on S, living in an exact sequence whose extension class is some $e \in V^*$. It is easy to convince oneself that all the rank two bundles that one obtains in this way through Cayley–Bacharach are just quotients of the higher rank bundle arising from Tyurin's argument.

Definition 2.5. In the sequel, we shall be interested in surfaces such that $\mathrm{Pic}(S) \simeq \mathbb{Z}$, and we shall denote by H the ample generator of the Picard group. Given a finite subscheme $\xi \subset S$, we shall define an integer τ_ξ by

$$\tau_\xi = \max\{k \mid h^1(S, \mathcal{J}_\xi(kH)) \neq 0\}. \tag{2.6}$$

In other terms, τ_ξ is the largest integer k for which ξ fails to impose independent conditions on $\mathcal{O}_S(kH)$. Furthermore, we shall say that ξ is *stable* if it is $\tau_\xi H$-stable, and in general we shall denote by ξ_s the maximal stable subscheme of ξ with respect to $\tau_\xi H$.

Lemma 2.3. *Let $\xi \subset S$ be a reduced finite subscheme in uniform position. Then ξ is stable.*

Proof. If ξ is not stable, we can decopose it as $\xi = \xi_s + \xi_r$, where ξ_s is the maximal stable part of ξ and ξ_r imposes independent conditions on the linear series $|\mathcal{I}_{\xi_s}(\tau)|$. Pick $P \in \xi_r$, and consider $\xi' = \xi_s + P$. For $Q \in \xi_s$ let $\xi_s' = \xi_s - Q$. Then ξ_s and $\xi_1 = \xi_s' + P$ have the same degree. However, because P imposes independent conditions on $|\mathcal{I}_{\xi_s}(\tau)|$ it also imposes independent conditions on $|\mathcal{I}_{\xi_s'}(\tau)|$, and therefore $\delta(\xi_1, \mathcal{O}(\tau)) = \delta(\xi_s', \mathcal{O}(\tau)) < \delta(\xi_s, \mathcal{O}(\tau))$, where the last inequality holds by stability. But this contradicts the assumption of uniform position. □

Remark 2.3. Putting together Lemmas 2.2 and 2.3, we clearly have that if ξ is L-stable, then it satisfies the Cayley–Bacharach propery w.r.t. L; on the other hand, it is clear that if $\xi = \sum_i \xi_i$, where the ξ_is are disjoint and satisfy Caley–Bacharach w.r.t. L, then so does ξ. A trivial but handy consequence is then that if the ξ_is are in uniform position and if for each i we have $h^1(S, \mathcal{J}_\xi(L)) \neq 0$, then ξ satisfies Cayley–Bacharach with respect to L.

Let us consider the decomposition $\xi = \xi_s + \xi_r$, where ξ_s is the maximal stable subscheme of ξ. Since ξ imposes independent conditions on lH for $l > \tau$, so does ξ_s; hence $\tau_{\xi_s} = \tau_\xi$. Suppose that ξ_s is reduced, or at least a local complete intersection.

Then by Lemma 2.2 we obtain a short exact sequence

$$0 \to \mathcal{O}_S \to \mathcal{E} \to \mathcal{J}_{\xi,s}(L \otimes \omega_S^{-1}) \to 0, \tag{2.7}$$

where $\mathcal{E}$ is a rank two vector bundle with $c_1(\mathcal{E}) = L - K_S$ and $c_2(\mathcal{E}) = d$.

The following statement is a consequence of Theorem 3.1 in [DG 88]; we include an alternative proof for the convenience of the reader.

Lemma 2.4. *Let $\xi = P_1 + \cdots + P_d \subset \mathbb{P}^2$ be a reduced subscheme. Denote by P_ξ the blow-up of $\mathbb{P}^2$ along ξ, and by $E = f^{-1}(\xi)$ the exceptional divisor. Then the divisor $(\tau + 3)H - E$ is globally generated.*

Proof. From the exact sequence $0 \to \mathcal{I}_P((\tau+3)H - E) \to \mathcal{O}_{P_\xi}((\tau+3)H - E)) \to \mathbb{C}_P \to 0$, we only need to check $H^1(P_\xi, \mathcal{I}_P((\tau+3)H - E)) = 0$ for every $P \in P_\xi$. Now every $P \in P_\xi$ can be written as a complete intersection: if $P \in P_\xi \setminus E \simeq \mathbb{P}^2 \setminus \xi$, then we can write $P = L_1 \cap L_2$, where L_1 and $L_2 \subset \mathbb{P}^2$ are lines disjoint from ξ; if $P \in E_1 = f^{-1}P_1$, P represents a tangent direction at P_1 and if L is a line through P_1 with direction P then $P = E_1 \cap \tilde{L}$, where $\tilde{L}$ is the proper transform of L. In the first case, we have a Koszul resolution

$$0 \to \mathcal{O}_{P_\xi}(-2) \to \mathcal{O}_{P_\xi}(-1)^2 \to \mathcal{I}_P \to 0.$$

Hence we obtain an exact sequence

$$\begin{gathered} 0 \to \mathcal{O}_{P_\xi}((\tau+1)H - E) \\ \to \mathcal{O}_{P_\xi}((\tau+2)H - E)^2 \to \mathcal{I}_P((\tau+3)H - E) \to 0. \end{gathered}$$

From this we get the exact sequence

$$\begin{gathered} H^1(P_\xi, \mathcal{O}_\xi((\tau+2)H - E))^2 \to H^1(P_\xi, \mathcal{J}_P((\tau+3)H - E)) \\ \to H^2(P_\xi, \mathcal{O}_{P_\xi}((\tau+1) - E)); \end{gathered}$$

from the definition of τ we see that the left cohomology group vanishes, while by Serre duality $H^2((\tau+1)H - E) = 0 \simeq H^0(P_\xi, \mathcal{O}_{P_\xi}(2E - (\tau+4))^* \simeq H^0(\mathbb{P}^2, \mathcal{O}_{\mathbb{P}^2}(-\tau - 4))^* = 0$.

In the second case, suppose to keep notation simple that the line L through P_1 with direction P does not meet $\xi \setminus P_1$. Then we have a Koszul resolution

$$\begin{gathered} 0 \to \mathcal{O}_{P_\xi}((\tau+2)H - E) \\ \to \mathcal{O}_{P_\xi}((\tau+2)H - \sum_{i \geq 2} E_i) \oplus \mathcal{O}_{P_\xi}((\tau+3)H - E - E_1) \\ \to \mathcal{I}_{P_1}((\tau+3)H - E) \to 0. \end{gathered}$$

So we need to establish the following vanishings:

$$H^1(P_\xi, \mathcal{O}((\tau+2)H - \sum_{i \geq 2} E_i)) = 0 \tag{2.8}$$

$$H^1(P_\xi, \mathcal{O}((\tau+3)H - E - E_1)) = 0 \tag{2.9}$$

and

$$H^2(P_\xi, \mathcal{O}((\tau+2)H - E)) = 0. \tag{2.10}$$

As to (2.10), it easily follows from Serre duality. (2.8) follows by taking the long exact sequence of the short exact sequence

$$0 \to \mathcal{O}((\tau+2)H - E) \to \mathcal{O}((\tau+2)H - \sum_{i\geq 2} E_i) \to \mathcal{O}_{E_1} \to 0.$$

For (2.9), we consider a smooth conic $Q \subset \mathbb{P}^2$ passing through P_1 but disjoint from the rest of ξ, and let $\tilde{Q}$ be its proper transform. Then we have an exact sequence

$$0 \to \mathcal{O}((\tau+1)H - E) \to \mathcal{O}((\tau+3)H - E - E_1)$$
$$\to \mathcal{O}_Q((\tau+4)H) \to 0,$$

and (2.9) follows from the long exact cohomology sequence. The case where L meets ξ in more than a point is dealt with similarly. □

Corollary 2.1. *Let $\xi \subset \mathbb{P}^2$ be a reduced finite subscheme and let $\tau = \tau_\xi$ be as in Definition 2.5. Then there exists a smooth curve C of degree $\tau+3$ containing ξ.*

Proof. The general element of the linear system $|(\tau+3)H - E|$ is smooth and, as one easily checks, irreducible and therefore it maps down in $\mathbb{P}^2$ to an integral curve which is non-singular away from ξ. Furthermore, since $[(\tau+3)H - E]\cdot E_i = 1$ for every i, we see that such a curve has to be non-singular also along ξ. □

Remark 2.4. In the hypothesis of Theorem 1.1, $[(\tau+3)H - E]^2 \geq 3d$, so that $(\tau+3)H - E$ is a big and nef divisor on P_ξ.

Lemma 2.5. *Let $\xi \subset \mathbb{P}^2$ be a finite reduced subscheme. If C is a smooth curve of degree $\tau+3$ through ξ, then ξ is stable (i.e., τ-stable: see Definition 2.5) if and only if the linear series $|\xi|$ on C is base point free.*

The proof is straightforward and can be found in [Ty 87].

Let $\xi \subset \mathbb{P}^2$ be any stable finite subscheme, and let $\mathcal{E}$ be the vector bundle associated to ξ sitting in the exact sequence (2.5). Set $V =: H^1(\mathbb{P}^2, \mathcal{J}_\xi(\tau))$; (2.5) can be rewritten as:

$$0 \to V \otimes \mathcal{O}_{\mathbb{P}^2} \to \mathcal{E} \to \mathcal{I}_\xi(\tau+3) \to 0. \tag{2.11}$$

Lemma 2.6. *We have $H^i(\mathbb{P}^2, \mathcal{E}(-2)) = H^i(\mathbb{P}^2, \mathcal{E}(-3)) = 0$, for $i = 1, 2$. In particular, $\mathcal{E}$ is globally generated.*

Proof. Observe that $H^2(\mathbb{P}^2, \mathcal{J}_\xi(x)) = 0$, for $x > -3$. Twisting (2.11) by $\mathcal{O}_{\mathbb{P}^2}(-2)$, and using $H^1(\mathbb{P}^2, \mathcal{J}_\xi(\tau+1)) = 0$, we obtain the desired vanishing for $\mathcal{E}(-2)$. Next twist (2.11) by $\mathcal{O}_{\mathbb{P}^2}(-3)$, and observe that the coboundary map $H^1(\mathbb{P}^2, \mathcal{J}_\xi(\tau)) \to V \otimes H^2(\mathbb{P}^2, \mathcal{O}_{\mathbb{P}^2}(-3)) \simeq V$ is the identity, by the construction of $\mathcal{E}$ (Remark 2.1). As to global generation, for $P \in \mathbb{P}^2$ look at the exact sequence $0 \to \mathcal{I}_P(\mathcal{E}) \to \mathcal{E} \to \mathbb{C}^2_P \to 0$. We want to show that $H^1(\mathbb{P}^2, \mathcal{I}_P(\mathcal{E})) = 0$. Representing P as a complete intersection of two lines we have an exact sequence

$$0 \to \mathcal{E}(-2) \to \mathcal{E}(-1)^2 \to \mathcal{I}_P(\mathcal{E}) \to 0.$$

Hence we have an exact sequence

$$H^1(\mathcal{E}(-1))^2 \to H^1(\mathcal{I}_P(\mathcal{E})) \to H^2(\mathcal{E}(-2)). \tag{2.12}$$

□

Corollary 2.2. *Let $\xi \subset \mathbb{P}^2$ be an arbitrary finite stable subscheme. Then there exists a flat family ξ_t, $t \in \Delta$ of subschemes of $\mathbb{P}^2$, such that ξ_t is stable and reduced for all $t \neq 0$, $\xi_0 = \xi$ and $\tau_{\xi_t} = \tau_\xi$ for all t. Furthermore, $H^1(\mathbb{P}^2, \mathcal{J}_{\xi_t}(\tau)) \simeq H^1(\mathbb{P}^2, \mathcal{J}_\xi(\tau))$, $\forall t \in \Delta$.*

Proof. Since $\mathcal{E}$ is globally generated, for a general linear subspace W of dimension $h^1(\mathbb{P}^2, \mathcal{I}_\xi(\tau))$ of $H^0(\mathbb{P}^2, \mathcal{E})$ the morphism $W \otimes \mathcal{O}_{\mathbb{P}^2} \to \mathcal{E}$ drops rank along a reduced finite subscheme ξ_W of $\mathbb{P}^2$. We then just need to take V_t to be a family of such subspaces of $H^0(\mathbb{P}^2, \mathcal{E})$ with $\xi_t =: \xi_{V_t}$ reduced for $t \neq 0$, $V_0 = H^1(\mathbb{P}^2, \mathcal{I}_\xi(\tau))$. For each t, we then have a short exact sequence

$$0 \to V_t \otimes \mathcal{O}^2_{\mathbb{P}} \to \mathcal{E} \to \mathcal{J}_{\xi_t}(\tau + 3) \to 0. \tag{2.13}$$

To complete the proof we only need to show that $h^1(\mathbb{P}^2, \mathcal{J}_{\xi_t}(\tau + 1)) = 0$, $h^1(\mathbb{P}^2, \mathcal{J}_{\xi_t}(\tau)) \neq 0$. To check the first condition, look at the exact sequence $0 \to V_t \otimes \mathcal{O}_{\mathbb{P}^2}(-2) \to \mathcal{E}(-2) \to \mathcal{I}_{\xi_t}(\tau + 1) \to 0$, that one obtains by twisting (2.13) by $\mathcal{O}_{\mathbb{P}^2}(-2)$. Using the vanishings in Lemma 2.6, we have the desired vanishing. Similarly, twisting (2.13) by $\mathcal{O}_{\mathbb{P}}(-3)$ and again invoking Lemma 2.6 we get the second condition. □

Remark 2.5. Observe that from the construction of Corollary 2.2 it follows that for every t there is an exact sequence $0 \to \mathcal{O}_{\mathbb{P}^2} \to \mathcal{E} \to \mathcal{J}_{\xi_t}(\tau + 3) \to 0$, where $\mathcal{E}$ *is the same vector bundle for all* t's.

Finally, we recall the content of Bogomolov's instability theorem ([Bo 78]). Recall first of all that for a rank two vector bundle $\mathcal{E}$ on a smooth surface S the *discriminant* of $\mathcal{E}$ is $\Delta(\mathcal{E}) = c_1(\mathcal{E})^2 - 4c_2(\mathcal{E})$.

Theorem 2.2. *Let S be a smooth projective surface, $\mathcal{E}$ a rank two vector bundle. If $\Delta(\mathcal{E}) > 0$, then there is a saturated line bundle $\mathcal{L} \subset \mathcal{E}$ such that $2c_1(\mathcal{L}) - c_1(\mathcal{E}) \in N^+(S)$, and in fact*

$$[2c_1(\mathcal{L}) - c_1(\mathcal{E})]^2 \geq \Delta(\mathcal{E}). \tag{2.14}$$

Remark 2.6. The inequality (2.14) is elementary; since we shall need to tell when it is an equality, let us derive it explicitly. Since $\mathcal{L} \subset \mathcal{E}$ is saturated, it sits in an exact sequence $0 \to \mathcal{L} \to \mathcal{E} \to \mathcal{Q} \otimes \mathcal{J}_Z \to 0$, where $Z \subset S$ is some finite local complete intersection. Then $c_2(\mathcal{E}) = c_1(\mathcal{L}) \cdot c_1(\mathcal{Q}) + \deg(Z)$, and an easy computation then shows that $[2c_1(\mathcal{L}) - c_1(\mathcal{E})]^2 = \Delta(\mathcal{E}) + 4\deg(Z)$. Hence, equality holds in (2.14) if and only if $Z = \emptyset$, i.e. if and only if $\mathcal{L} \to \mathcal{E}$ never drops rank.

Furthermore, in the case $S = \mathbb{P}^2$ if $\Delta(\mathcal{E}) \geq 0$ then $\mathcal{E}$ is not stable.

3. Proof of Theorem 1.1

The reader interested only in the shortest way to the speciality theorem may as well follow the argument only in the case of a reduced stable finite scheme.

Proof of Theorem 1.1. Observe that (a) is clearly a special case of (b), except that some more specific conclusions are drawn. We shall only give the detailed argument for (a), since the proof of (b) is essentially identical, up to notation. Let then $\xi \subset \mathbb{P}^2$ be a finite subscheme and let us suppose to start with that ξ is stable and reduced. By Theorem 2.1 and Lemma 2.2, we have an exact sequence

$$0 \to \mathcal{O} \to \mathcal{E} \to \mathcal{I}_\xi(\tau + a) \to 0, \tag{3.15}$$

where $\mathcal{E}$ is a rank two vector bundle.

Lemma 3.1. *If $\tau + 3 \geq s + \frac{d}{s}$, then there exists a curve $D \subset \mathbb{P}^2$ through ξ and of degree y with $y \leq d^2$ and $2y \leq \tau + 3$ (strict inequalities hold if $\tau + 3 > s + \frac{d}{s}$).*

Proof. We start with the following claim:

Claim 3.1. *$\mathcal{E}$ is not stable.*

Proof. The Chern classes of $\mathcal{E}$ are then $c_1(\mathcal{E}) = \tau + 3$ and $c_2(\mathcal{E}) = d$; its discriminant is therefore

$$\Delta(\mathcal{E}) = (\tau + 3)^2 - 4d. \tag{3.16}$$

By condition (1.1), we have $(\tau + 3)^2 \geq s^2 + (\frac{d}{s})^2 + 2d$ and this together with (3.16) yields

$$\Delta(\mathcal{E}) \geq \left(\frac{d}{s} - s\right)^2 \geq 0. \tag{3.17}$$

Therefore $\mathcal{E}$ is not stable. □

We have therefore a maximal destabilizing line bundle $\mathcal{O}(x) \subset \mathcal{E}$, and the destabilizing condition says that

$$x \geq \frac{1}{2}(\tau + 3). \tag{3.18}$$

In particular, the induced composition $\mathcal{O}(x) \to \mathcal{E} \to \mathcal{I}_\xi(\tau+3)$ is non-zero, and therefore we have a non-zero morphism

$$\mathcal{O} \to \mathcal{I}_\xi(\tau+3-x), \tag{3.19}$$

i.e. we have produced a curve of degree $y = \tau + 3 - x$ containing the scheme ξ. This curve is in fact uniquely detemined by the above construction: it is easy to check that $H^0(\mathbb{P}^2, \mathcal{I}_\xi(y)) \simeq H^0(\mathcal{E}(-x)) \simeq \mathbb{C}$. Furthermore, the condition (2.14) gives $x^2 - x(\tau+3) + d \geq 0$. This can be rewritten $d \geq x[\tau+3-x]$, and therefore we have

$$d \geq y(\tau+3-y). \tag{3.20}$$

Now $x > \frac{1}{2}(\tau+3)$ can be rewritten as $\tau+3 > 2y$, while (3.20) can be rewritten $y + \frac{d}{y} \geq \tau+3$. Hence we have $y + \frac{d}{y} \geq 2y$, and so

$$d \geq y^2. \tag{3.21}$$

If $\tau + 3 > s + \frac{d}{s}$, then the inequalities (3.18) and in (3.21) are strict.

Lemma 3.2. *Suppose that $s^2 \leq d$. Then $y \leq s$. If equality holds, then ξ is a complete intersection of type $(s, \frac{d}{s})$.*

Proof. For $x^2 \leq d$ the function $x + \frac{d}{x}$ is decreasing in x; therefore since $y + \frac{d}{y} \geq \tau + 3 \geq s + \frac{d}{s}$ we must have $y \leq s$.

If $y = s$, then we have $\tau + 3 \geq s + \frac{d}{s}$ by assumption, and on the other hand $\tau + 3 \leq s + \frac{d}{s}$ by (3.20); therefore $\tau + 3 = s + \frac{d}{s}$. Hence, we are in the case were equality holds in (2.14); it follows that $\mathcal{O}_{\mathbb{P}^2}(x) \to \mathcal{E}$ never drops rank (Remark 2.6). Hence we have an exact sequence $0 \to \mathcal{O}(x) \to \mathcal{E} \to \mathcal{O}(\tau+3-x) \to 0$, and from this we obtain a splitting $\mathcal{E} \simeq \mathcal{O}(s) \oplus \mathcal{O}(\frac{d}{s})$. From the resulting exact sequence $0 \to \mathcal{O} \to \mathcal{O}(s) \oplus \mathcal{O}(\frac{d}{s}) \to \mathcal{I}_\xi(s + \frac{d}{s}) \to 0$ it is then easy to conclude that ξ is a complete intersection of a curve of degree s and a curve of degree $\frac{d}{s}$. □

Lemma 3.3. *Let ξ' and $\tilde{\xi}$ be finite schemes in $\mathbb{P}^2$, such that $\tilde{\xi} \supset \xi'$ and* $\deg(\tilde{\xi}) = \deg(\xi') + 1$. *Then for every integer k we have*

$$h^1(\mathcal{I}_{\tilde{\xi}}(k)) \geq h^1(\mathcal{I}_{\xi'}(k)) \geq h^1(\mathcal{I}_{\tilde{\xi}}(k)) - 1.$$

Proof. We have an exact sequence $0 \to \mathcal{I}_{\tilde{\xi}}(k) \to \mathcal{I}_{\xi'}(k) \to \mathcal{I}^{\tilde{\xi}}_{\xi'}(k) \to 0$. Now the sheaf $\mathcal{I}^{\tilde{\xi}}_{\xi'}$, the ideal sheaf of ξ' in $\tilde{\xi}$ has finite support, and therefore its higher cohomology vanishes. On the other hand, $H^0(\mathcal{I}^{\tilde{\xi}}_{\xi'}(k))$ is the space of functions on $\tilde{\xi}$ that vanish along ξ', and is therefore isomorphic to $\mathbb{C}$. Hence we have an exact sequence $\mathbb{C} \to H^1(\mathcal{I}_{\tilde{\xi}}(k)) \to H^1(\mathcal{I}_{\xi'}(k)) \to 0$, and therefore

$$h^1(\mathcal{I}_{\tilde{\xi}}(k)) \geq h^1(\mathcal{I}_{\xi'}(k)) \geq h^1(\mathcal{I}_{\tilde{\xi}}(k)) - 1. \tag{3.22}$$

□

By Corollary 2.1, there exists a smooth curve $C \subset \mathbb{P}^2$ of degree $\tau + 3$ through ξ, and by Lemma 2.5 and the stability assumption the linear series $|\xi|$ on C is base point free. Let $\tilde{\xi}$ denote the complete intersection of the curve D of degree y produced above and C, so that in particular $\xi \subset \tilde{\xi}$.

On the other hand, by the very definition of τ we have $h^1(\mathbb{P}^2, \mathcal{I}_\xi(\tau + 1)) = 0$. By Lemma 3.3, we then have

$$\deg(\xi) \leq \deg(\tilde{\xi}) - h^1(\mathbb{P}^2, \mathcal{I}_{\tilde{\xi}}(\tau + 1)), \tag{3.23}$$

where $\deg(\tilde{\xi}) = y(\tau + 3)$. Since $\mathcal{I}_C \simeq \mathcal{O}_{\mathbb{P}^2}(-\tau - 3)$, we have an exact sequence

$$0 \to \mathcal{O}_{\mathbb{P}^2}(-2) \to \mathcal{I}_{\tilde{\xi}}(\tau + 1) \to \mathcal{O}_C(\tau + 1 - y) \to 0,$$

from which we get an isomorphism $H^1(\mathbb{P}^2, \mathcal{I}_\xi(\tau + 1)) \simeq H^1(C, \mathcal{O}_C(\tau + 1 - y))$. Since $\omega_C \simeq \mathcal{O}_C(\tau)$, by applying Serre duality on C we then have $h^1(\mathbb{P}^2, \mathcal{J}_\xi(\tau+1)) = h^0(C, \mathcal{O}_C(y - 1))$ and it is easy to see that this is $\geq h^0(\mathbb{P}^2, \mathcal{O}_{\mathbb{P}^2}(y - 1)) = \frac{1}{2}(y + 1)y$. Together with (3.23) this finally implies

$$d \leq y(\tau + 3) - \frac{1}{2}(y + 1)y = y\Big[\tau + \frac{1}{2}(5 - y)\Big]. \tag{3.24}$$

This completes the proof of the case where ξ is reduced and stable. If ξ is stable but not reduced, let us apply Corollary 2.2 to find a family ξ_t of reduced stable schemes degenerating to ξ and with the same value of τ for all $t \in T$. Furthermore, the above construction also provides (perhaps after restricting T) a flat family of curves D_t of degree y containing ξ_t. By taking the limit for $t \to 0$ we obtain a limit curve of degree y containing ξ ([Fu 84], chapter 11).

Finally, let us consider the case where ξ is not stable. Let $\xi_s \subset \xi$ be the stable part of ξ, d_s its degree. We have $\tau + 3 \geq s + \frac{d}{s} > s + \frac{d_s}{s}$, and the argument goes as before to give a curve of degree y containing ξ_s. We have that $y^2 < d_s < d$; since $y + \frac{d}{y} > y + \frac{d_s}{y} \geq \tau + 3 \geq s + \frac{d}{s}$ we conclude that $y < s$; the rest of the argument goes over as before.

The argument for part (b) is essentially the same, and therefore we shall only give a brief description. Suppose first that $\tau + a > s + \frac{d}{sH^2}$. As before, we have a short exact sequence (3.15), and under the hypothesis of the theorem $\mathcal{E}$ is not stable. Therefore, we obtain a maximal destabilizing line bundle $\mathcal{O}(x) \subset \mathcal{E}$. Since in our situation $\tau + a > 0$, we have a non-zero section $\mathcal{O}_S \to \mathcal{I}_\xi(y)$, where $y = \tau + a - x$. The argument then proceed exactly as before. If $\tau + a = s + \frac{d}{sH^2}$, and if $\Delta(\mathcal{E}) = 0$, then from the equality $c_1(\mathcal{E})^2 = 4c_2(\mathcal{E})$ and the hypothesis that $c_2(\mathcal{E})$ divides H^2

we have in particular that $c_1(\mathcal{E}) = 2kH$, for some k. Twisting by $\mathcal{O}_S(-k)$, we get $c_1(\mathcal{E}(-k)) = 0$, $c_2(\mathcal{E}(-k)) = 0$; by a theorem of Mehta and Ramanathan ([MR 84]), $\mathcal{E}(-k)$ comes from a representation of the fundamental group, and therefore it is trivial if $\pi_1(S) = 0$. □

4. Proof of Halphen's speciality theorem

We now show how the above arguments can be used to give a simple proof of Halphen's theorem, by essentially deriving it from Theorem 1.1.

Proof of Theorem 1.2. Let $S \subset \mathbb{P}^3$ be a general hypersurface of degree $m \gg s$; in particular we may assume that S is smooth and that $\mathrm{Pic}(S) \simeq \mathbb{Z}$. Denote by $C_S = C \cap S$ the corresponding section. Suppose to fix ideas that C is irreducible. Then C_S is a reduced finite subscheme of S of degree dm, and by uniform position it is stable (Lemma 2.3). Let $\tau = \tau_{C_S}$ be as in Theorem 1.1, and let $a = 4 - m$. Then there exists a rank two vector bundle $\mathcal{E}$ on S with $\wedge^2(\mathcal{E}) \simeq \mathcal{O}_S(\tau + a)$ and a section $s \in H^0(S, \mathcal{E})$ vanishing exactly along C_S. In particular, $c_2(\mathcal{E}) = d \cdot m$ is a multiple of m. Furthermore, by the Lefshetz hyperplane theorem S is simply connected. Hence we are in the hypothesis of Theorem 1.1, part (b).

Claim 4.1. *If $m \gg 0$, then $\tau = e + m$.*

Proof. Let $\mathcal{I}_C$ and $\mathcal{J}_{C_S}$ denote, respectively, the ideal sheaves of C in $\mathbb{P}^3$ and of C_S in S. Then we have an exact sequence $0 \to \mathcal{I}_C(-m) \to \mathcal{I}_C \to \mathcal{J}_{C_S} \to 0$. For $m \gg 0$, we may assume that $H^i(\mathbb{P}^3, \mathcal{I}_C(e+m)) = 0$, for $i > 0$. Taking cohomology, we find the exact sequence

$$\begin{aligned} 0 = H^1(\mathbb{P}^3, \mathcal{I}_C(e+m)) &\to H^1(S, \mathcal{J}_{C_S}(e+m)) \\ \to H^2(\mathbb{P}^3, \mathcal{I}_C(e)) &\to H^2(\mathbb{P}^3, \mathcal{I}_C(e+m)) = 0. \end{aligned} \tag{4.25}$$

Hence, we get the isomorphism $H^1(S, \mathcal{J}_{C_S}(e+m)) \cong H^2(\mathbb{P}^3, \mathcal{I}_C(e)) \neq 0$, which shows that $\tau \geq e + m$. The same sequence with e replaced by $e + 1$ shows that $H^1(S, \mathcal{J}_{C_S}(e+1+m)) = 0$, and this completes the proof. □

Let now $s \in \mathbb{R}$ be the smallest root of $f(x) = x^2 - x(e+4) + d$. Clearly, $s > 0$. We have $\tau + a = e + 4 = s + \frac{d}{s}$. Furthermore, for $0 < x < s$ we have $f(x) > 0$, i.e. $x + \frac{d}{x} > e + 4$. Hence, $x + \frac{d}{x}$ is decreasing in x, for $x \leq s$, and so $s^2 \leq d$. □

On S we have $H^2 = m$ and so $\frac{1}{H^2}\deg(C_S) = \frac{1}{m}dm = d$. By the second part of Theorem 1.1, it follows that there exist $y \leq s$ and $D \in |\mathcal{O}_S(y)|$ containing C_S. The proof is then reduced to the following

Claim 4.2. *For $m \gg 0$, we have $H^0(\mathbb{P}^3, \mathcal{I}_C(y)) \simeq H^0(S, \mathcal{J}_{C_S}(y))$.*

Proof. We have an exact sequence

$$0 \to H^0(\mathbb{P}^3, \mathcal{I}_C(y-m)) \to H^0(\mathbb{P}^3, \mathcal{I}_C(y)) \\ \to H^0(S, \mathcal{J}_{C_S}(y)) \to H^1(\mathbb{P}^3, \mathcal{I}_C(y-m)). \tag{4.26}$$

Hence we only need to show that the first and last cohomology groups vanish when $m \gg s$. The first vanishing is obvious. As to the second, look at the exact sequence $0 = H^0(C, \mathcal{O}_C(y-m)) \to H^1(\mathbb{P}^3, \mathcal{I}_C(y-m)) \to H^1(\mathbb{P}^3, \mathcal{O}_{\mathbb{P}^3}(y-m)) = 0$. □

If C is not irreducible, then — for general S — C_S will be a finite disjoint union of finite schemes ξ_s, each in uniform position; invoking Remark 2.3, we can then argue in the same way as before.

This completes the proof of Halphen's theorem.

As to part (b), the argument is the same since, in the notation of Theorem 1.1, we have $f(x) = x^2 - (e+c)x + \frac{d}{H^3} = x^2 - (\tau + a)x + \frac{\deg(C \cdot S)}{H \cdot sH}$; just observe that if $H^1(P, \mathcal{O}_P(x)) = 0$ for $x \geq 0$, then $H^1(S, \mathcal{O}_S(x)) = 0$ for $x \geq 0$, when $S \in |mH|$ and $m \gg 0$; furthermore, if P is simply connected then so is S, by the Lefschetz hyperplane theorem. □

5. The generalized speciality theorem

The argument in the proof of Theorem 1.2 does not use in an essential way the fact that we are dealing with a curve in $\mathbb{P}^3$; what is really relevant is rather that we have a codimension two subvariety of projective space, and that the right information about the postulation of C puts us in the hypothesis of Theorem 1.1 when we look at the hypersurface sections of C. It is therefore clear that the same construction should also give a proof of the generalized speciality theorem of Gruson and Peskine ([GP 77], [GP 77-1]).

Proof of Theorem 1.3. We first generalize Claim 4.1:

Lemma 5.1. *Let $X^r \subset \mathbb{P}^{r+2}$ be as in the statement of Theorem* 1.3. *Let $V \subset \mathbb{P}^{r+2}$ be a general hypersurface of degree $m \gg 0$, and let $X_V = X \cap V$ the corresponding hypersurface section of X, so that $X_V \subset V$ is an integral codimension two subvariety. Let $\tau_{X_V} = \max\{k \mid H^r(V, \mathcal{J}_{X_V}(k)) \neq 0\}$. Then $\tau_{X_V} = \tau_X + m$.*

Proof. If $\mathcal{J}_{X_V}$ is the ideal sheaf of X_V in V, then we have an exact sequence $0 \to \mathcal{I}_X(-m) \to \mathcal{I}_X \to \mathcal{J}_{X_V} \to 0$. Proceeding as in the proof of Claim 4.1 we have the statement. □

Corollary 5.1. *In the hypothesis of Lemma* 5.1, *let* $S \subset \mathbb{P}^{r+2}$ *be the complete intersection of r general hypersurfaces of degree m, and let* $\xi = X \cap S$. *Then* ξ *is a stable* 0*-cycle in S, and* $\tau_\xi = \tau_X + rm$.

Intersect X with a general complete intersection surface of type $(m, \dots, m)$, $m \gg 0$. Using Lemma 5.1, argue as in the proof of Halphen's theorem. □

References

[Bo 78] F. Bogomolov, Unstable Vector Bundles and Curves on Surfaces, Proc. ICM Helsinki 1978.

[Cop 92] M. Coppens, Free Linear Systems On Integral Gorenstein Curves, J. Algebra 145 (1992), 209–218.

[DG 88] E. D. Davis, A. V. Geramita, Birational Morphisms to $\mathbb{P}^2$; an ideal theoretic perspective, Math. Ann. 279 (1988), 435–448.

[EP 88] Ph. Ellia and Ch. Peskine, Groupes de Points de P^2: caractère et position uniforme, in: Algebraic Geometry, L'Aquila 1988 (A. Biancofiore, E. L. Livorni and A. J. Sommese, eds.), Lecture Notes in Math. 1417, Springer-Verlag 1990.

[Fu 84] W. Fulton, Intersection Theory, Springer-Verlag 1984.

[GH] Ph. Griffiths and J. Harris, Principles of Algebraic Geometry, Wyley Publ. Co..

[GH 78] P. Griffiths and J. Harris, Residues and Zero-cycles on an Algebraic Surface, Ann. of Math. 108 (1978), 461–505.

[GP 77] L. Gruson and C. Peskine, Théorème de Spécialité, Sém. E.N.S. (1977–1978), Astérisque 71–72.

[GP 77-1] L. Gruson and C. Peskine, Genre des Courbes dans l'Espace Projectif, in: Lecture Notes in Math. 687, Springer-Verlag 1977.

[GP 83] L. Gruson and C. Peskine, Postulation des courbes gauches, in: Lecture Notes in Math. 997, 218–227, Springer-Verlag 1983.

[Hal 82] G. Halphen, Mémoire sur la classification des courbes gauches algébriques, J. Ecole Polyt. 52 (1882), 1–200, .

[Har 80] R. Hartshorne, Stable Reflexive Sheaves, Math. Ann. 254 (1980), 121–176.

[La 86] R. Lazarsfeld, Brill–Noether–Petri without degenerations, J. Differential. Geom. 23 (1986), 299–307.

[LP 93] A. Lopez and G. P. Pirola, On the Curves through a general point of a smooth surface in $\mathbb{P}^3$, preprint.

[MR 84] V. B. Mehta and A. Ramanathan, Restriction of Stable Sheaves and Representations of the Fundamental Group, Invent. Math. 77 (1984), 163–172.

[Pa 93] R. Paoletti, Free Pencils on Divisors, Math. Ann. 303 (1995), 109–123.

[Ty 87] A. N. Tyurin, Cycles, Curves and Vector Bundles on an Algebraic Surface, Duke Math. J. 54 (1987) 1–26.

Threefolds with extremal Chern classes

Th. Peternell and P. M. H. Wilson

Introduction

The famous Miyaoka inequality in the theory of algebraic threefolds states that the Chern classes of a minimal threefold X satisfy $c_1^2(X) \leq 3c_2(X)$, meaning that the 1-cycle $3c_2 - c_1^2$ is in the closure of the cone of effective curves. A minimal threefold is a normal projective $\mathbb{Q}$-factorial threefold with at most terminal singularities such that K_X is nef. This nefness condition is equivalent to the non-uniruledness of X. A natural question is therefore to determine the structure of those threefolds for which equality occurs. In this paper we shall investigate this problem for Gorenstein minimal threefolds. A Gorenstein model will be a normal Gorenstein $\mathbb{Q}$-factorial projective variety with at most terminal singularities. Our first result states that every 3-dimensional Gorenstein model X with $\kappa(X) \geq 0$ and $c_1^2 = 3c_2$ is necessarily minimal, i.e. K_X is nef. An essential point is that in passing to a minimal model we will not leave the category of Gorenstein models. The structure of Gorenstein minimal threefolds with $c_1 = 3c_2$ and $\kappa(X) = 0$ has already been determined in [SBW94]: X is the smooth quotient of an abelian threefold by a finite group. The case $\kappa(X) = 3$ does not occur, at least if X is supposed to be smooth from the beginning. The reason is the sharper inequality $c_1^3 \geq \frac{8}{3}c_1 \cdot c_2$ which holds because K_X is big and nef. In the singular case this inequality is still unknown but expected to be true. Concerning the intermediate cases $\kappa = 1, 2$ we prove in the Sections 2 and 3:

Theorem. *Let X be a Gorenstein minimal threefold with $\kappa = 1$ or 2. Then X is automatically smooth, the Iitaka fibration is almost smooth in the sense that the only singular fibers are multiples of abelian surfaces, resp. elliptic curves, and finally X does not contain any rational curves.*

For uniruled threefolds our problem is not of the same interest since Miyaoka's inequality does not hold here; and it is apparently not possible to get reasonable results. However Miyaoka's inequality implies that $c_2(X)$ is always in the closure of the cone of effective curves for a minimal threefold X, and it is interesting to ask for the position of c_2 in the Mori cone in the non-minimal and in the uniruled case, too. This is the subject of Section 4. If $-c_2(X)$ is in the Mori cone, we shall write $c_2(X) \leq 0$. We shall prove that the condition $c_2(X) \leq 0$ but $c_2(X) \neq 0$ implies $\kappa(X) = -\infty$. The case $c_2(X) = 0$ is of course possible if $\kappa(X) = 0$ or 1, but then X arises from a smooth minimal model (those are classified before) by blowing up points. If $\kappa(X) = -\infty$ and if $c_2(X) \leq 0$, then X is birational via a finite sequence of Mori contractions to a $\mathbb{P}_1$-bundle or a conic bundle over a smooth surface, and a lot about these fibrations can be said (see 4.6).

Acknowledgements. We would like to thank the Deutsche Forschungsgemeinschaft for financial subsistance of our work and M. Andreatta and G. Bolondi for organising a very stimulating and enjoyable meeting in Trento.

0. Preliminaries

0.1. Notations

(a) A minimal variety is throughout this paper a complex projective variety X which is $\mathbb{Q}$-factorial and has at worst terminal singularities, and such that K_X is nef; see for instance [KMM87] for details. A Gorenstein model is a complex projective variety which is Gorenstein, $\mathbb{Q}$-factorial and with at worst terminal singularities. By (5.1) of [Ka88], any 3-dimensional Gorenstein model has only factorial cDV singularities.

(b) $\overline{NE}(X)$ denotes the closed cone of (classes) of effective 1-cycles on X.

(c) Suppose X is a projective variety with canonical singularities. Let $\pi : \hat{X} \to X$ be a desingularisation which is crepant in codimension 2 and let $c_i(\hat{X}) = c_i(\bigwedge^i T_{\hat{X}})$. Then we define the i-th (numerical) Chern class ($i = 1, 2$) of X by $c_i(X) = \pi_*(c_i(\hat{X})) \in N^i(X)$, the group of i-cycles modulo numerical equivalence. Throughout this paper, the equality $c_1^2 = 3c_2$ will always be interpreted as equality of the numerical equivalence classes.

0.2. Proposition. *Let X be a projective 3-fold with at most terminal singularities. Then the following Riemann–Roch formula holds:*

$$\chi(X, \mathcal{O}_X) = \frac{1}{24}c_1(X) \cdot c_2(X) + \frac{1}{24}\sum\left(r(x) - \frac{1}{r(x)}\right),$$

the sum being taken over all virtual terminal quotient singularities x of X with index $r(x) > 1$.

For a proof see [Fl87, Re87].

1. Minimality of threefolds with $c_1^2 = 3c_2$ and $\kappa \geq 0$

In his paper [Mi87] Miyaoka proved that on a minimal projective variety the 1-cycle $3c_2 - c_1^2$ is always semi-positive in a suitable sense (often we write c_i instead of $c_i(X)$). For threefolds this can be stated in the following way.

1.1. Theorem (Miyaoka). *Let X be a minimal projective 3-fold. Then* $3c_2 - c_1^2 \in \overline{NE}(X)$.

The aim of these notes is to investigate the "extremal" case $3c_2 - c_1^2 = 0$ for Gorenstein 3-folds X (with at most terminal singularities). Because of (1.1) this is of most interest if $\kappa(X) \geq 0$. The case $\kappa(X) = -\infty$ will be discussed later (Sect. 4).

1.2. Theorem. *Let X be a 3-dimensional Gorenstein model with $\kappa(X) \geq 0$. If $c_1^2(X) = 3c_2(X)$, then X is minimal, i.e. K_X is nef.*

The strategy of the proof of (1.2) is, assuming K_X not nef, to pass to a minimal model by Mori contractions and then to apply (1.1) in order to derive a contradiction. Fortunately we will see that we do not leave the category of Gorenstein models and therefore we need not take flips into account; this considerably simplifies matters. The following lemma is essential.

1.3. Lemma. *Let X be a 3-dimensional Gorenstein model and $\varphi : X \to Y$ a birational Mori contraction. Then*

(a) *φ is divisorial, i.e. there is a unique prime divisor $E \subset X$ such that* $\dim \varphi(E) \leq 1$ *and $\varphi|X \setminus E$ is biholomorphic;*

(b) *if* $\dim \varphi(E) = 1$, *then Y is Gorenstein;*

(c) *if* $\dim \varphi(E) = 0$ *and if $c_1^2(X) = 3c_2(X) + A$ with $A \in \overline{NE}(X)$, then Y is Gorenstein unless $A \cdot E = 10$.*

Proof. (a) follows from [Be85], since X is Gorenstein.

(b) E is Cartier by [Ka88, 5.1], so K_Y is Cartier, i.e. Y is Gorenstein by virtue of the formula $K_X = \varphi^*(K_Y) + E$ (which follows from the fact that generically φ is a blow-up).

(c) Assume that Y is not Gorenstein. Then the classification of divisorial Mori contractions in dimension 3 ([Mo82] in the smooth case and [Cu88] in general) gives $E \simeq \mathbb{P}_2$ with normal bundle $N_{E|X} = \mathcal{O}(-2)$. So $p = \varphi(E)$ is an index 2 point on Y. Moreover, locally around p (in the analytic sense) Y is given by the quotient of $\mathbb{C}^3$ by $(z_1, z_2, z_3) \to (-z_1, -z_2, -z_3)$. From $c_1^2 = 3c_2 + A$ we obtain

$$K_X^2 \cdot E = 3E \cdot c_2(X) + A \cdot E.$$

Since $K_X^2 \cdot E = 1$, we must show $E \cdot c_2(X) = -3$. In fact, using $K_X = \varphi^*(K_Y) + \frac{1}{2}E$, we compute

$$c_1(Y) \cdot c_2(Y) = \varphi^*(c_1(Y)) \cdot c_2(X) = c_1(X) \cdot c_2(X) + \frac{1}{2}E \cdot c_2(X).$$

Since $\chi(X, \mathcal{O}_X) = \chi(Y, \mathcal{O}_Y)$, the local term R corresponding to the singularity p in the Riemann–Roch formula (0.2) for $24\chi(Y, \mathcal{O}_Y)$ is $-\frac{1}{2}E \cdot c_2(X)$, and $R = \frac{3}{2}$ since the index $r(p) = 2$. Alternatively, one may reduce to the case X smooth and compute in this case $E \cdot c_2(X) = -3$ by virtue of the exact sequence

$$0 \to T_E \to T_X|E \to N_E \to 0.$$

This finishes the proof of (1.3). □

Proof of 1.2. Assume that K_X is not nef and let $\varphi : X \to Y$ be a Mori contraction. By (1.3) Y is again Gorenstein. Since $\kappa(X) \neq -\infty$, φ is birational and in fact a divisorial contraction by 1.3(a) with exceptional prime divisor E.

(1) Assume $\dim \varphi(E) = 1$ and let $C = \varphi(E)$. We compute easily

$$c_1^2(Y) = 3c_2(Y) + 4C.$$

In fact, φ is generically a blow-up (outside the singularities of C and of Y), so it is sufficient to prove the formula for blow-ups in manifolds. But this is easy by first desingularising Y and C, then using the formulas in the smooth case (see e.g. [Fu84,15.4])

$$c_1(X) = \varphi^*(c_1(Y)) - E$$

$$c_2(X) = \varphi^* c_2(Y) - E^2 + (\deg K_C)l,$$

$$E^2 = -\varphi^*(C) + c_1(N_C)l,$$

where l is a curve contracted by φ, then computing from these formulas

$$\varphi^*(c_1^2(Y) - 3c_2(Y)) = 4\varphi^*(C) - 4c_1(N_C)l - 2(K_Y \cdot C)l$$

and finally going back to the original X resp. Y.

(2) If $\dim \varphi(E) = 0$, then first note $c_1^2(Y) = 3c_2(Y)$. From 1.3(c) we see that Y is Gorenstein.

In summary, Y is always Gorenstein and $c_1^2(Y) = 3c_2(Y) + 4C$, with C possibly empty.

(3) We now consider Y and assume that C is non-empty. By (1.1) K_Y is not nef, so let $\psi : Y \to Y'$ be a Mori contraction. Again ψ is of divisorial type; let E' be the exceptional divisor. If $\dim \psi(E') = 1$, we see as in (1) that Y' is Gorenstein with

$$c_1^2(Y') = 3c_2(Y') + 4\tilde{C} + 4C',$$

where $\tilde{C} = \psi_*(C)$ and $C' = \psi(E')$. Now if dim $\psi(E') = 0$, then first Y' is Gorenstein: if not, apply 1.3(c) with $A = 4C$ and deduce $A.E = 10$, a contradiction as E is Cartier. Moreover $c_1^2(Y') = 3c_2(Y') + 4\tilde{C}$.

In summary, Y' is also Gorenstein with $c_1^2(Y') = 3c_2(Y') + 4A$ with A an effective cycle. Proceeding in this way, we obtain a sequence of divisorial contractions

$$Y \to Y' \to \cdots \to Y^k = Z,$$

where K_Z is nef, since $\kappa(X) \geq 0$. But then $c_1^2(Z) = 3c_2(Z) + 4B$ with an effective cycle $B \neq 0$ which contradicts (1.1). The assertion $B \neq 0$ is seen as follows. Assuming $B = 0$ and letting $W = Y^{k-1}$ for short, we have $c_1^2(Z) = 3c_2(Z)$ and will prove that $c_1^2(W) - 3c_2(W)$ is not in $\overline{NE}(W)$. In fact, the map $\psi : W \to Z$ contracts the exceptional divisor, say D, to a point (otherwise we would have $B \neq 0$.) From the equation

$$c_1(W) = \psi^*(c_1(Z)) - \lambda D, \quad (\lambda = 1, 2)$$

and the fact that Z is Gorenstein, we deduce that $c_2(W) \cdot D = 0$, and hence that $c_2(W) = \psi^*(c_2(Z))$. Therefore $c_1^2(W) - 3c_2(W) = \lambda^2 D^2$, which is not in the cone of effective curves, contradicting our previous conclusions. □

1.4. Remark. Theorem 1.2 suggests that $c_1^2 - 3c_2$ should never be in $\overline{NE}(X) \setminus \{0\}$ if $\kappa(X) \geq 0$, even if X is not Gorenstein. The proof of (1.2) does not work however in this case except if $c_1^2 = 3c_2 + 4A$, where $A = \sum n_i C_i$ with positive integers n_i. To get the general statement it would be sufficient that the condition $c_1^2 - 3c_2 \in \overline{NE}(X)$ is invariant under flips.

Another possible generalisation would be the following: if $\kappa(X) \geq 0$ and if $3c_2(X) - c_1^2(X) \in \partial\overline{NE}(X)$, then X is minimal.

Next we want to obtain more specific information about 3-folds X with $c_1^2 = 3c_2$ for each given value of $\kappa(X)$. The case $\kappa = 0$ has been settled by Shepherd-Barron and Wilson [SBW94]:

1.5. Theorem. *Let X be a 3-dimensional Gorenstein model with $\kappa(X) = 0$ and $c_1^2 = 3c_2$. Then X is a smooth quotient of an abelian 3-fold by a finite group.*

For the proof just note that $K_X \equiv 0$, hence $c_2(X) = 0$, hence by [SBW94] X is a quotient of an abelian 3-fold by a finite group. Since the local fundamental group of a potential singularity of X vanishes, X being Gorenstein with at most terminal singularities, X must be smooth.

The other extreme case is $\kappa = 3$:

1.6. Theorem. *There is no minimal smooth 3-fold X of general type with $c_1^2 = 3c_2$.*

We shall use the Miyaoka–Yau inequality

$$c_1^3(X) \geq \frac{8}{3} c_1(X) \cdot c_2(X).$$

In the case K_X is ample this is proved by Yau [Yau77], in the case K_X is only big and nef by Tsuji [Ts88]. From $c_1^2 = 3c_2$ we conclude $c_1(X) \cdot c_2(X) = c_1^3(X) = 0$, which is impossible.

Of course one would like to have Theorem 1.6 in case X a Gorenstein model, but the corresponding Miyaoka–Yau inequality seems not yet to be available. Note that for $0 \leq \kappa \leq 2$, we show that a Gorenstein model with $c_1^2 = 3c_2$ is in fact smooth, and so we would expect (1.6) to remain true for Gorenstein models.

2. Threefolds with $c_1^2 = 3c_2$: Case $\kappa = 1$

In this section we investigate 3-dimensional Gorenstein models X of Kodaira dimension 1 with $c_1^2 = 3c_2$. Since $c_1^2 = 0$, we have by (1.2) (equivalently) $c_2(X) = 0$ and X is minimal. The structure of those threefolds is described by

2.1. Theorem. *Let X be a 3-dimensional Gorenstein model with $\kappa(X) = 1$ and $c_1^2 = 3c_2$. Then the following assertions hold.*

(1) *X is smooth (and minimal).*

(2) *The Iitaka fibration $f : X \to C$ is almost smooth, i.e. the reduced fibres are all smooth. The only singular fibers are multiples of abelian or hyperelliptic surfaces.*

(3) $c_3(X) = \chi_{\text{top}}(X) = 0$.

(4) *X does not contain rational curves.*

Proof. By (1.2) K_X is nef. Note that by [Ka88,5.1] X is factorial, so every Weil divisor on X is Cartier. We define $\tilde{\Omega}_X^1 = (\Omega_X^1)^{**}$. Then there is a natural injection $f^*(\Omega_C^1) \to \tilde{\Omega}_X^1$; we define $\mathcal{O}_X(D)$ to be the saturation of $f^*(\Omega_C^1)$ in $\tilde{\Omega}_X^1$. We distinguish the cases whether C is rational or not.

(a) First we assume C to be irrational. Then D is effective and of course $\dim f(D) = 0$. We have an exact sequence

$$0 \to \mathcal{O}_X(D) \to \tilde{\Omega}_X^1 \to V \to 0 \qquad (S)$$

with V a torsion free sheaf of rank 2. Let $W = V^{**}$. Then W is reflexive, hence locally free outside a finite set. Let $Q = W/V$.

We claim that $W_H = W|H$ is Bogomolov unstable for a general hyperplane section $H \subset X$ (such that $H \cap \text{Sing}(W) = \emptyset$) except if $D^2 = 0$ and $Q_H = 0$.

In fact, let $V_H = V|H$. Then V_H is torsion free. From the restricted sequence (S_H) we obtain (let $D_H = D|H$ and observe $c_2(X) = 0$) that

$$c_1(V_H) = K_X|H - D_H$$

$$c_2(V_H) = -K_X \cdot H \cdot D_H + D_H^2 = D_H^2.$$

Thus

$$c_1(W_H) = K_X \cdot H - D_H$$

$$c_2(W_H) = c_2(V_H) + c_2(Q_H) = D_H^2 + c_2(Q_H).$$

It follows that $c_1^2(W_H) - 4c_2(W_H) = -3D_H^2 - 4c_2(Q_H)$, and since $D_H^2 \leq 0, c_2(Q_H) \leq 0$, we deduce that W_H is Bogomolov unstable as claimed, unless $D^2 = 0$ and $Q_H = 0$. The inequalities $D_H^2 \leq 0$ and $c_2(Q_H) \leq 0$ are seen as follows. The first one just follows from the fact that $f|H$ contracts $D \cap H$, for the second first notice that $\mathcal{O}_H(D)$ is saturated in Ω_H^1, H being general. Hence the cokernel of the inclusion map is of the form $I_W \otimes \mathcal{O}_H(K_H + D)$, with W a subscheme supported in codimension 2 (i.e. on a finite set of points). This cokernel is on the other hand nothing else than the cokernel of the morphism $\mathcal{O}_H(H) \to V_H$. Consequently

$$\text{Coker}(\mathcal{O}_H(H) \to W_H) = I_{W'}\mathcal{O}_H(K_H + D)),$$

with $W' \subset W$ a subscheme. We conclude that $c_2(Q_H) = \deg W' - \deg W \leq 0$, proving also the second inequality (with strict inequality unless $Q_H = 0$).

We suppose now that not both $D^2 = 0$ and $Q_H = 0$. Then in particular we conclude that W_H is unstable in Mumford's sense with respect to any polarisation. By Bogomolov instability there exists (see [Re77]) a line bundle $L \subset W_H$ such that

$$(2L + K_X|H - D_H) \cdot A > 0 \tag{$*$}$$

for every ample divisor A on H; moreover

$$c_1(2L + K_X|H - D_H)^2 > 0. \tag{$*'$}$$

Let $L' \subset W_H$ be a maximal H_H destabilising subsheaf. We claim $L' - L$ is effective: consider the sequence

$$0 \to L' \to W_H \to I_Z \otimes W_H \otimes L'^{-1} \to 0$$

with Z a finite subscheme to conclude that in case of non-effectivity we would have an injection

$$L \subset I_Z \otimes \det W_H \otimes L'^{-1}.$$

This contradicts $L' \cdot H^2 \geq L \cdot H^2 > \frac{1}{2} W_H \cdot H^2$ from $(*)$, proving the claim.

We are aiming for

$$K_X \cdot H \cdot L' > 0. \tag{$**$}$$

In fact, from $(*)$ and $(*')$ we deduce $(2L + K_X|H - D_H) \cdot F > 0$ for a fiber F of f. Thus

$$0 < (2L + K_X|H - D_H) \cdot K_X|H = 2L \cdot K_X \cdot H.$$

$L' - L$ being effective, we obtain $L' \cdot K_X \cdot H > 0$, i.e. $(**)$.

If $H \in |2^m B|$, with B ample and m sufficiently large, then by Mehta–Ramanathan [MR82] L' extends to a line bundle on X, denoted by the same symbol, which is the maximal destabilising subsheaf of W. Now consider a general smooth fiber F of f. Then in a neighbourhood of F, we can view L' as a subsheaf of the relative cotangent sheaf $\Omega^1_{X|C}$. Thus L'_F is a subsheaf of Ω^1_F. From $(**)$ we have $L' \cdot H \cdot F > 0$, so Ω^1_F is unstable with respect to H_F, a contradiction.

So we have proved that $D^2 = 0$ and $Q_H = 0$, i.e. $V = W$ outside a finite set. We want to show that these two conditions imply claim (2). For a smooth hyperplane section H, we have $D_H^2 = 0$. Since $\operatorname{supp} D_H$ is contained in the fibers of $H \to C$, it follows from Zariski's lemma that $D_H = \sum \lambda_i \tilde{F}_i$, where the $\tilde{F}_i$ are scheme-theoretic fibers of $H \to C$. Consequently

$$D = \sum \lambda_i F_i$$

with the F_i being scheme-theoretic fibers of f. Since Q is supported at a finite set, the map $\mathcal{O}_X(D) \to \tilde{\Omega}^1_X$ has zeros only at finitely many points. Now the singular set $\operatorname{sing}(V)$ contains

$$\{x \in X \setminus \operatorname{sing}(X) \mid \operatorname{red}(f^{-1}f(x)) \text{ is singular at } x\},$$

this statement following immediately from the expression $D = \sum \lambda_i F_i$ by a local base change around multiple fibers ($t \mapsto t^{\lambda_i}$). Hence red F is normal for every fiber F of f. In order to see that red F is actually smooth, first assume that F is reduced and singular. Then F is a normal (connected) Gorenstein surface whose canonical bundle is given by $K_F = K_X|F$, the normal bundle N_F being trivial. So K_F is torsion. In [Um81] it is proved that in the case $K_F \simeq \mathcal{O}_F$ that we have the cohomology vanishing

$$H^1(\mathcal{O}_F) = 0.$$

This remains true also in the torsion case: either adapt the proof or pass to a cover of F. The normal bundle N_F being trivial, we deduce

$$H^1(\mathcal{O}_{\hat{F}}) = 0$$

for the formal completion of F. But this contradicts the fact that $R^1 f_*(\mathcal{O}_X)$ has rank 1 or 2. In case F is non-reduced either argue with λF or make a local base change $t \to t^\lambda$ and then apply the arguments given above.

So we have seen that the only singular fibers of f are multiples of smooth surfaces. Performing a base change on the base, we obtain a covering $\tilde{X} \to X$ such that the induced map $\tilde{f} : \tilde{X} \to \tilde{C}$ has smooth fibers. Hence $\tilde{X}$ is smooth, and thus X itself has only quotient singularities. On the other hand, X is Gorenstein with only terminal singularities, hence the local fundamental groups of the singularities are zero. So X

must be smooth. Since $c_2(X) = 0$, we have $c_2(\text{red } F) = 0$, so red F is an abelian surface or hyperelliptic and (4) follows. Finally, (3) is an immediate consequence of (2), and so the proof is complete in case $g(C) \geq 1$.

(b) Now let $C = \mathbb{P}_1$. Choose a ramified $2:1$ covering $g : \tilde{C} \to C$, branched over at least 4 points, all the branch points contained in the set over which f is smooth. We obtain a diagram

$$\begin{array}{ccc} \tilde{X} & \xrightarrow{\tilde{g}} & X \\ \downarrow & & \downarrow \\ \tilde{C} & \xrightarrow{g} & C \end{array}$$

with a minimal projective Gorenstein variety $\tilde{X}$ and an induced morphism $\tilde{f} : \tilde{X} \to \tilde{C}$ from the original morphism $f : X \to C$. Since $\tilde{\Omega}^1_{\tilde{X}|X}$ is a line bundle living on the ramification locus of $\tilde{g}$, its c_2 vanishes, so we can compute

$$c_2(\tilde{X}) = c_2(X) + \tilde{g}^* c_1(\Omega^1_X) \cdot c_1(\Omega^1_{\tilde{X}|X}) = 0.$$

Since $g(\tilde{C}) \geq 1$, we conclude by part (a) that $\tilde{X}$ is smooth and that $\tilde{f} : \tilde{X} \to \tilde{C}$ is almost smooth in the sense of the theorem. As before we see that X is smooth and obviously f is almost smooth. Now (3) is clear. Finally for (4), if X contains a rational curve, say B, then B has to be a multi-section of f. After a base change, B becomes a section. Passing perhaps to a covering of X such that all fibers of f are abelian, we deduce that there is a holomorphic map $B \to \mathcal{A}_2$, the moduli space of polarised abelian surfaces. This map has to be constant, so f is a fiber bundle. Thus X is a product $\mathbb{P}_1 \times A$ with A an abelian surface, so we conclude $\kappa(X) = -\infty$, a contradiction. □

2.2. Corollary. *Let X be as in* (2.1). *Then X admits a finite cover $\tilde{X}$ which is a smooth abelian fibration over a curve.*

We now prove a converse to Theorem 2.1.

2.3. Proposition. *Let X be a smooth projective 3-fold, $f : X \to C$ a surjective map to a compact Riemann surface. Assume that the general fiber is abelian or hyperelliptic and that the only singular fibers are multiples of smooth surfaces. Then $c_2(X) = 0$.*

Proof. The exact sequence

$$0 \to f^*(\Omega^1_C) \to \Omega^1_X \to \Omega^1_{X|C} \to 0$$

yields

$$c_2(\Omega^1_X) = c_2(\Omega^1_{X|C})$$

by virtue of

$$c_1(\Omega^1_{X|C}) \cdot c_1(f^*(\Omega^1_C)) = 0.$$

So it suffices to show

$$c := c_2(\Omega^1_{X|C}) = 0.$$

Since $c \cdot F = 0$ for all fibers F of f, the 1-cycle c is contained in finitely many fibers $F_1, \ldots, F_r$. Let $x_i = f(F_i)$. Choose a small ball B around x_i and let $U = f^{-1}(B)$. Performing a base change $B \to B, t \to t^{m_i}$ we obtain an etale cover $U' \to U$ which is smooth over B. Thus $c_2(U') = 0$, hence $c|U = 0$. This proves $c = 0$. □

3. Threefolds with $c_1^2 = 3c_2$: Case $\kappa = 2$

We now turn to 3-dimensional Gorenstein models X with $\kappa(X) = 2$ and $c_1^2(X) = 3c_2(X)$. Multiplying by c_1 we obtain $0 = c_1^3 = 3c_1c_2 = \chi(\mathcal{O}_X)$. For some multiple $m \in \mathbb{N}$, the line bundle $\mathcal{O}_X(mK_X)$ is generated by global sections [Ka85] and gives a morphism $f : X \to S$ to a normal projective surface S which is the Iitaka fibration of X.

We will prove that X is again smooth (3.5), that f has at most multiple elliptic curves as singular fibers (3.5), that S is covered by the ball (with ramification) (3.6), and finally that X does not admit any rational curve (3.7).

We start with the following elementary lemma.

3.1. Lemma. *The set $N := \{y \in S \mid X_y$ is singular but not a multiple of an elliptic curve$\}$ is finite.*

Proof. Either apply [Gr94] or argue as follows. Take a general hyperplane section $C \subset S$ and let $Y = f^{-1}(C)$. Then we have

$$0 = c_1(X)^2 \cdot Y = 3c_2(X) \cdot Y = 3c_2(Y) - 3K_Y \cdot Y|Y = 3c_2(Y),$$

hence the elliptic fibration has only multiple fibers as singular fibers (see e.g. [BPV84]). This proves our claim. □

Our aim is to prove that actually $N = \emptyset$. We let $A \subset S$ denote the closure of

$$\{s \in S \mid X_s \text{ is multiple}\}.$$

If A_i are the irreducible components of A and if m_i is the multiplicity of a general fiber X_s for $s \in A_i$, then we set

$$\Delta = \sum_i (1 - \frac{1}{m_i}) A_i.$$

3.2. Lemma. *$K_X \equiv f^*(K_S + \Delta)$ and (S, Δ) has only log-terminal singularities; in particular S has only quotient singularities.*

Proof. [Na87]. □

3.3. Proposition. *With the assumptions as stated above, we have:*

(a) *f has no 2-dimensional fibers.*

(b) *Let F be a singular fiber. Then there is no component F_1 such that $\Omega^1_X|F_1 \to \Omega^1_{F_1}$ is generically an isomorphism; hence generically* rk $\Omega^1_{F_1} \le 2$.

Proof. (1) Assume that dim $f^{-1}(s_0) = 2$ for some $s_0 \in S$ and let F be the scheme-theoretic fiber over s_0. We are going to examine the structure of the two dimensional components of the fiber. Since S has only Gorenstein singularities and, since we shall argue only locally around s_0, we perform a base change $\tilde{S} \to S$ such that the resulting fibration $\tilde{f} : \tilde{X} \to \tilde{S}$ has smooth base space $\tilde{S}$. Note that $\tilde{X}$ will again be a Gorenstein model and that $c_2(\tilde{X}) = 0$. Indeed for this last fact observe that $c_2(\tilde{X}) = c_2(\Omega_{\tilde{X}|X})$, hence $c_2(\tilde{X})$ is represented by a cycle in $\tilde{f}^{-1}(0)$. Since $\tilde{f}^{-1}(0) = f^{-1}(0)$ set-theoretically, it follows that $c_2(X)$ is represented up to a multiple by the same cycle. Therefore our claim follows and we conclude that we may assume S to be smooth from the beginning.

We assume S to be a ball say and take a general 1-form without zeros $\sigma \in H^0(\Omega^1_S)$. Let $s = f^*(\sigma) \in H^0(\Omega^1_X)$. Then the zero set $\{s = 0\}$ is contained in the singularity set Sing(f) of f. Let D the 2-dimensional part of $\{s = 0\}$. Observe that D consists exactly of those 2-dimensional components of F which have multiplicity > 1 in F, i.e. which are non-reduced. Now consider s as a section of $\Omega^1_X(-D)$ and let Z be the scheme defined by its vanishing. Then dim $Z \le 1$. We let Z_1 be the 1-dimensional part of Z, considered as a 1-cycle.

Then we obtain, using the assumption $c_1^2 = 3c_2$ which yields $c_2(X) = 0$ in our local setting:

$$c_2(\Omega^1_X(-D)) = c_2(\Omega^1_X) + D^2,$$

hence $[Z] = D^2$. Next we note that supp $Z_1 \subset F$. This is seen as follows. Let Z' be a 1-dimensional component of Z not lying in F. Note that it is sufficient to prove that Z' must be contained in a multiple fiber of f (then shrink S). If not then take any multiple fiber l meeting Z'. Since s has a zero at one point of l it follows that s vanishes at all of l. This is clear by restricting s to the elliptic curve l.

Letting H be a general hyperplane section of X, we obtain

$$Z_1 \cdot H = D^2 \cdot H.$$

Now let $F_1, \ldots, F_q$ be the irreducible components of F of dimension 2. Then we can write

$$D = \sum_i \lambda_i F_i,$$

with $\lambda_i \geq 0$, but not all 0. Since $D \cdot H = \sum_i \lambda_i F_i$ is an exceptional cycle in H, we must have

$$(\sum_i \lambda_i F_i)^2 \cdot H < 0,$$

which contradicts $Z_1.H \geq 0$. This is a contradiction: D must be 0.

But then we have $Z_1 \cdot H = 0$, thus $Z_1 = \emptyset$. In particular we note that there is no curve $C \subset F$ such that generically rk $\Omega^1_F|C = 3$. This implies that

(a) every F_i is normal and that

(b) $F_i \cap F_j = \emptyset$ if $i \neq j$ (recall that X is factorial).

Moreover we know that F is generically reduced along F_i. We fix i and shall write $G = F_i$ for short.

Note that the above arguments also work if dim $f^{-1}(s_0) = 1$ (case $D = 0$) and therefore prove statement (b).

(2) We claim that there is a map $\varphi : X \to X'$ blowing down all components F_i to points and which factors over $f : X \to S$ via a map $f' : X' \to S$. Having proved the existence of φ, we conclude that f' is equidimensional and that X' has only isolated singularities but some canonical non-terminal ones. Now Proposition 3.4 (including the remark following (3.4)) gives the contradiction.

In order to prove the existence of φ, it is sufficient to show the contractability for just one component $G = F_i$ by virtue of (b). Let $H \subset X$ be a hyperplane section and consider the generically finite map $\alpha : H \to S$. Since the curve C cut out on H by G is contracted by α, it follows $C^2 < 0$, i.e. $G \cdot C < 0$. Now X having only canonical singularities, the pair $(X, \epsilon G)$ has only log terminal singularities for sufficiently small positive rational numbers ϵ. But $K_X + \epsilon G$ is not nef since $K_X + \epsilon G \cdot C = \epsilon G \cdot C < 0$, so the log contraction theorem gives a contraction $\psi : X \to W$ factoring over f which either contracts G to a point or to a curve or which contracts only a finite numbers of curves to a certain number of terminal singularities (if X is terminal). In the first case we are done, so it remains to exclude the other two cases, which is done separately in (3) and (4).

(3) We now exclude the case that ψ contracts G to a curve B. Since G is normal by (a), the general fiber l of $p : G \to B$ is a smooth rational curve with normal bundle $N_{l|X} = \mathcal{O} \oplus \mathcal{O}(-2)$. The map f factors over $f' : W \to S$. Now consider the exact sequence

$$0 \to f'^*(\Omega^1_S) \to \Omega^1_{X'} \to \Omega^1_{X'|S} \to 0$$

and restrict this sequence to the curve $B = \psi(G)$. Now we obtain a contradiction by computing ranks (note that X' is singular along B, that B is not contained in a 2-dimensional component of $G' = f'^{-1}(s_0)$ and that G' is generically reduced along B, so that rk $\Omega^1_{X'|S}|B_0 = \text{rk}\, \Omega^1_{B_0} = 1$ on some open subset B_0 of B).

(4) Now assume that ψ is a small contraction. Then W has only terminal singularities and is also Gorenstein, so it has only cDV singularities. Let Z denote the

exceptional locus of ψ. We perform a flop $\psi' : X' \to W$ with exceptional locus Z'. The resulting divisor G' on X' fulfils $G' \cdot Z' > 0$. The two inequalities $G \cdot Z < 0$ and $G' \cdot Z' > 0$ can be translated as follows. If $h : \tilde{X} \to X, h' : \tilde{X} \to X'$ is a resolution of the birational map $X \to X'$, then $h^*(G) = h'^*(G') + E$ with E an effective divisor supported on the exceptional set of $\psi \circ h = \psi' \circ h'$ and $-E$ relatively ample with respect to both h and h'. Hence we obtain

$$G^3 = h^*(G)^2 \cdot h'^*(G') = h'^*(G')^3 + h'^*(G') \cdot E^2 < G'^3. \qquad (*)$$

Next observe that $3c_2(X') - c_1^2(X')$ is an effective 1-cycle supported on Z', consequently we have

$$c_2(X') \cdot G' \geq 0. \qquad (**)$$

From $R^i h'_*(\mathcal{O}_{\tilde{X}}(E)) = 0$ for $i \geq 1$, we obtain

$$\chi(\mathcal{O}_X(G)) = \chi(h^*(\mathcal{O}_X(G))) = \chi(h'^*(\mathcal{O}_{X'}(G')) \otimes \mathcal{O}_{\tilde{X}}(E)) = \chi(\mathcal{O}_{X'}(G')).$$

Hence Riemann–Roch gives (remembering that X, X' are factorial)

$$\frac{G^3}{6} + \frac{c_2(X) \cdot G}{12} = \frac{G'^3}{6} + \frac{c_2(X') \cdot G'}{12},$$

contradicting $(*)$ and $(**)$.

The proof of (3.3) is now complete. □

3.4. Proposition. *Assume now that X has isolated canonical rather than terminal singularities but that the Iitaka fibration $f : X \to S$ is equidimensional, with S having only quotient singularities and the other assumptions unchanged. Then f has only multiple fibers as singular fibers.*

Proof. With the same argument as in the very first sentences of the proof of (3.3) we may assume that S is smooth. Let $f^{-1}(s_0)$ be a singular fiber, by assumption purely 1-dimensional. We assume that this fiber is not multiple. We choose a general smooth curve C through s_0 and let $X_C = f^{-1}(C)$. Then X_C is an irreducible reduced surface.

(1) We claim that X_C has only isolated singularities. This is clear over $C \setminus A$ since X_C is smooth outside the singularity set of X and Sing(f). Over $A \cap C$ it just follows from Bertini by (3.3.b). In the next step we will see that X_C has only rational singularities, hence is Cohen-Macaulay so that in total X_C is normal.

(2) Let $\sigma : \hat{X}_C \to X_C$ be a minimal desingularisation. Then the composite map $\hat{X}_C \to C$ must be the Iitaka fibration of $\hat{X}_C$ and in particular X_C can have only rational singularities. We conclude, using (1), that X_C is Gorenstein, so has only rational double points. Consequently $K^2_{\hat{X}_C} = K^2_{X_C} = 0$, in particular $\hat{X}_C$ is minimal.

(3) In the next step we compute $c_2(\hat{X}_C)$. To this extent we embed the desingularisation $\sigma : \hat{X}_C \to X_C$ into a sequence of blow-ups of simple points $\rho : \hat{X} \to X$, i.e.

$\hat{X}_C \subset \hat{X}$ and $\rho|\hat{X}_C = \sigma$. Denoting the exceptional components of ρ by E_i, we have

$$K_{\hat{X}} = \rho^*(K_X) + \sum \lambda_i E_i$$

with semi-positive coefficients λ_i (X has only canonical singularities). Then we compute

$$c_2(\hat{X}) \cdot \hat{X}_C = c_2(\hat{X}_C) + c_1(\Omega^1_{\hat{X}_C}) \cdot c_1(N_{\hat{X}_C|\hat{X}})^* = c_2(\hat{X}_C) - K^2_{\hat{X}_C} + K_{\hat{X}_C} \cdot (K_{\hat{X}}|\hat{X}_C).$$

Now $K_{\hat{X}}|\hat{X}_C = \sum \lambda_i E_i|X_C$, and hence, using the fact that the singularities of X_C are only rational double points, it follows that

$$K_{\hat{X}_C} \cdot K_{\hat{X}}|\hat{X}_C = 0,$$

hence finally

$$c_2(\hat{X}) \cdot \hat{X}_C = c_2(\hat{X}_C). \tag{1}$$

By virtue of

$$c_2(\hat{X}) = \rho^*(c_2(X))$$

we obtain from $c_2(X) = 0$ and from (1) that $c_2(\hat{X}_C) = 0$. The conclusion is that $\hat{f}|\hat{X}_C$ has at most multiple singular fibers. In particular $\hat{X}_C = X_C$ and f has only multiple singular fibers. This ends the proof of (3.4). □

Notice that in case S is smooth we have proved that X_C is smooth, so that X itself must be smooth.

Coming back again to our original situation where X is a Gorenstein model we pass again to a local smooth cover $\tilde{S} \to S$. Then $\tilde{X} \to \tilde{S}$ has only multiple fibers and $\tilde{X}$ is smooth. Hence X has only quotient singularities, so X is smooth as explained in (2.1). So we may state

3.5. Theorem. *Let X be a 3-dimensional Gorenstein model with $c_1^2(X) = 3c_2(X)$. Assume that $\kappa(X) = 2$. Then X is smooth and minimal, and the Iitaka fibration has at most multiple elliptic curves as singular fibers.*

Next we want to investigate the structure of (S, Δ). For this purpose the notion of log Chern classes $\bar{c}_i(S, \Delta)$ is important. For $i = 1$ we just have $\bar{c}_1(S, \Delta) = K_S + \Delta$, while the definition of $\bar{c}_2$ is more complicated and will be explained in the proof of (3.6); we refer to [Ko90] and [Me93].

3.6. Theorem. *We have $\bar{c}_1^2(S, \Delta) = \bar{3}c_2(S, \Delta)$, in particular there is a branched covering $B \to S$, with $B \subset \mathbb{C}^2$ the unit disc, which is branched exactly along Δ.*

Proof. Once the first statement is proved, the second is just [Ko90, Theorem 2]. For the proof of the first statement we have to go into Megyesi's thesis [Me93], chap. 3. We shall use the language of Q-sheaves and Q-varieties. We take an atlas of charts as described in [Me93, p. 44], i.e. we have Zariski open sets $U_\alpha \subset S$, quasi-projective smooth surfaces S_α, finite groups G_α acting on S_α faithfully with finite étale projections $p'_\alpha : X_\alpha/G_\alpha \to U_\alpha$. Moreover the projection map $X_\alpha \to X_\alpha/G_\alpha$ is a Galois covering which is ramified over the preimage of Δ. In the setting of Q-sheaves [Me93, p. 45], the Q-sheaf $\hat{\Omega}^1_S(\log B)$ on S corresponds to the rank 2 vector bundles $\Omega^1_{S_\alpha}$ on S_α. Moreover, if X_α is defined by the pull-back diagram

$$\begin{array}{ccc} X_\alpha & \overset{h_\alpha}{\to} & X|U_\alpha \\ \downarrow & & \downarrow \\ S_\alpha & \to & U_\alpha \end{array}$$

with projection maps $f_\alpha : X_\alpha \to S_\alpha$, then we have the exact sequence of G_α-sheaves

$$0 \to f^*_\alpha(\Omega^1_{X_\alpha}) \to \Omega^1_{X_\alpha} \to \mathcal{O}_{X_\alpha} \to 0$$

since $\Omega^1_{X_\alpha|S_\alpha}$ is trivial. Note that $\Omega^1_{X_\alpha} = h^*_\alpha(\Omega^1_X|U_\alpha)$ since h_α is unramified. Now let $\tilde{S}$ be the normal surface obtained by patching together the local charts S_α as described in [Me93]; algebraically $\tilde{S}$ is the normalisation of S in a Galois extension of $\mathbb{C}(S)$ which contains all the function fields $\mathbb{C}(S_\alpha)$. Then $\hat{\Omega}^1_S(\log B)$ corresponds to a rank 2 vector bundle $\tilde{V}$ on $\tilde{S}$. The construction of $\tilde{S}$ is made because now S appears as a global quotient of $\tilde{S}$ via a map g. Letting $\tilde{X} = X \times_S \tilde{S}$, we have projection maps $h : \tilde{X} \to X$ and $q : \tilde{X} \to \tilde{S}$. From the previous exact sequence we therefore deduce the exact sequence of vector bundles on $\tilde{X}$

$$0 \to q^*(\tilde{V}) \to h^*(\Omega^1_X) \to \mathcal{O}_{\tilde{X}} \to 0.$$

We shall denote the i-th Chern class of a Q-sheaf F by $\hat{c}_i(F)$. By definition

$$\hat{c}_2(\hat{\Omega}^1_S(\log B)) = \frac{1}{|G|} c_2(\tilde{V}) \in A^2(\tilde{S}) \otimes \mathbb{Q}.$$

But

$$\hat{c}_2(\Omega^1_X) = \frac{1}{|G|} c_2(h^*\Omega^1_X)$$

as an element in $A^2(\tilde{S}) \otimes \mathbb{Q}$, and so, considered as forms on $\mathrm{Div}(X)$, we have $c_2(X) \equiv f^*\hat{c}_2(\hat{\Omega}^1_S(\log B))$. If F denotes the numerical class of a fiber f, we can therefore write $c_2(X) \equiv \hat{c}_2(\hat{\Omega}^1_S(\log B))F$. As computed on p. 50 in [Me93], we have

$$\hat{c}_1(\hat{\Omega}^1_S(\log B)) = K_S + B,$$

and so $K_X = f^*\hat{c}_1(\Omega^1_S(\log B))$. Putting things together, $K_X^2 = 3c_2(X)$ implies that $(K_S + B)^2 = 3\hat{c}_2(\hat{\Omega}^1_S(\log B))$, i.e.

$$\bar{c}_1(S, B)^2 = 3\bar{c}_2(S, B).$$

This ends the proof of 3.6. □

3.7. Corollary. *X does not contain rational curves.*

Proof. Assume that $l \subset X$ is a rational curve and let $C = f(l)$. First note that $C \cap \Delta \neq \emptyset$. In fact, if $C \cap \Delta = \emptyset$, then by (3.6) we would obtain a non-constant map $\mathbb{P}_1 \to B$ to the unit disc B in $\mathbb{C}^2$, which is absurd. Now consider the surface X_C with the multisection l. Via the base change $l \to C$ and after possibly normalising l, we obtain a smooth surface $\tilde{f} : \tilde{X}_C \to \mathbb{P}_1$ which has a section. Hence $\tilde{f}$ is a smooth fibration, hence an elliptic fiber bundle and therefore $\tilde{X}_C = \mathbb{P}_1 \times E$ with an elliptic curve E. Let $\hat{X}_C$ be the normalisation of X_C, then $\hat{X}_C$ is smooth (because $\tilde{X}_C$ is smooth) and we have a map $\tilde{X}_C \to \hat{X}_C$. Therefore we obtain a 1-dimensional family $\hat{C}_t$ of rational curves in $\hat{X}_C$. These have $\hat{C}_t^2 = 0$ because otherwise $\hat{X}_C$ would be rational, which is not the case since by [BPV84] we see that $\chi(\mathcal{O}_{\hat{X}_C}) = 0$. We conclude that $\hat{X}_C$ is a product and hence X_C is a product, too. But this implies that C cannot meet Δ, a contradiction. □

4. The position of c_2 in the Mori cone

Since Miyaoka's inequality $c_1^2 \leq 3c_2$ only holds in general for minimal threefolds X, which in particular have $\kappa(X) \neq -\infty$, it does not make much sense to ask for threefolds with $\kappa = -\infty$ satisfying $c_1 = 3c_2$. However Miyaoka's inequality says in particular that $c_2(X) \in \overline{NE}(X)$, and the previous results when $\kappa(X) \leq 1$ can also be viewed as results on threefolds with $c_2(X) = 0$. It seems interesting to investigate in general the position of $c_2(X)$ relative to the Mori cone $\overline{NE}(X)$ regardless whether X is minimal or not or the value $\kappa(X)$. In particular we ask for the structure of threefolds with $c_2(X) \in -\overline{NE}(X)$. We shall abbreviate this by

$$c_2(X) \leq 0.$$

Our first result says that this class behaves well with respect to birational Mori contractions.

4.1. Proposition. *Let X be a 3-dimensional Gorenstein model such that $c_2(X) \leq 0$. Let $\varphi : X \to Y$ be a birational Mori contraction. Then Y is again a Gorenstein model and $c_2(Y) \leq 0$.*

Proof. Let E be the prime divisor contracted by φ.

(a) We assume here that $\dim \varphi(E) = 0$. Let $\{p\} = \varphi(E)$. Since $D \cdot c_2(Y) = \varphi^*(D) \cdot c_2(X)$ it is obvious $c_2(Y) \cdot M \leq 0$ for M ample on Y. Thus we need only to show that p is a Gorenstein point. Assume p not Gorenstein, then by [Cu88], $E \simeq \mathbb{P}_2$ and p is an index 2 quotient singularity. By Riemann–Roch (0.2) we get

$$\chi(\mathcal{O}_Y) = \frac{c_1(Y) \cdot c_2(Y)}{24} + \frac{1}{16}$$

and

$$\chi(\mathcal{O}_X) = \frac{c_1(X) \cdot c_2(X)}{24}$$

(cf. proof of (1.3)). As in (1.3), we deduce that $E \cdot c_2(X) = -3$.

Write

$$c_2(X) = \varphi^*(c_2(Y)) + Z$$

with a 1-cycle Z living on E. Then it follows $Z \cdot E < 0$, so Z is numerically an effective 1-cycle on $E \simeq \mathbb{P}_2$. This clearly contradicts $c_2(X) \leq 0$ (it would be sufficient to state that $-Z$ cannot be an effective 1-cycle on E.)

(b) Now we treat the case that $\varphi(E)$ is a curve C. Let $\nu : \tilde{C} \to C$ be the normalisation. By [Cu88] C is locally a complete intersection. Hence $E \to C$ is a $\mathbb{P}_1$-bundle and the normalisation $\mu : \tilde{E} \to E$ is just obtained by pulling back the $\mathbb{P}_1$-bundle structure. Let $\tilde{C}_0 \subset \tilde{E}$ be a section with minimal self-intersection and put $C_0 = \mu(\tilde{C}_0)$. Let F be a general fiber of $E \to C$. Then there is a formula

$$0 = c_2(X) = \varphi^*(c_2(Y)) + \alpha C_0 + \beta F$$

and intersecting with E we see that $\alpha = 1$. So

$$0 = \varphi_*\varphi^*(c_2(Y)) + \varphi_*(C_0) = c_2(Y) + C.$$

This gives the first claim. Since φ is the blow-up of Y along C, it follows also that Y is Gorenstein. □

By (4.1) it is sufficient to study threefolds with $c_2(X) \leq 0$ which are either minimal or admit a non-birational Mori contraction. Before we start this we deduce from (4.1) the following general structural statement.

4.2. Proposition. *Let X be a 3-dimensional Gorenstein model with $c_2(X) \leq 0$ but $c_2(X) \neq 0$. Then $\kappa(X) = -\infty$.*

Proof. In view of (1.1) and (4.1) it is sufficient to prove the following. Let $\varphi : X \to Y$ be a divisorial Mori contraction. Then $c_2(Y) \neq 0$. In order to prove this non-vanishing, assume $c_2(Y) = 0$. Then Riemann–Roch gives $\chi(\mathcal{O}_Y) = \chi(\mathcal{O}_X) = 0$. Write $c_2(X) = -A$ with $A \in \overline{NE}(X) \setminus \{0\}$. Then we conclude $K_X \cdot A = 0$. On the other hand $\varphi_*(A) = 0$, so $K_X \cdot A < 0$, since $-K_X$ is φ-ample, contradiction. □

4.3. Corollary. *Let X be a 3-dimensional Gorenstein model with $c_2(X) \leq 0$. Then X is birationally equivalent — via a finite sequence of Mori contractions — either to a minimal Gorenstein model X' with $c_2(X') = 0$, in which case we also have $c_2(X) = 0$ or to a $\mathbb{P}_1$-bundle or a conic bundle over a smooth surface Y.*

Proof. By (4.1) we find a Gorenstein model X' by performing a finite sequence of divisorial Mori contractions on X which is either minimal or admits a Mori contraction

$\psi : X' \to Y$ with dim $Y \leq 2$. Since $c_2(X') \leq 0$, we conclude by (4.2) that $c_2(X') = 0$ in case X' is minimal, hence $c_2(X) = 0$ again by (4.2). So we now assume the existence of a non-birational Mori contraction ψ. We exclude the cases dim $Y = 0$ and dim $Y = 1$. In fact, if dim $Y = 0$, then X' is Fano, in particular $\chi(\mathcal{O}_{X'}) = 1$. On the other hand $c_2(X') \leq 0$ implies $\chi(\mathcal{O}_{X'}) \leq 0$ by Riemann–Roch. If dim $Y = 1$, then we consider a general smooth fiber F of ψ. We obtain

$$c_2(F) = c_2(X') \cdot F \leq 0.$$

However F is a del Pezzo surface, so $c_2(F) > 0$. Now that we know dim $Y = 2$, it is proved in [Cu88] that Y is smooth and that ψ is a conic bundle (in case X' is smooth this is already in [Mo82]). □

Next we have a closer look at the case $c_2(X) = 0$ and $\kappa(X) \geq 0$.

4.4. Corollary. *Let X be a 3-dimensional Gorenstein model with $c_2(X) = 0$ and $\kappa(X) \geq 0$. Then X is obtained from a smooth minimal model X' with $c_2(X') = 0$ by blowing up points.*

Proof. We know already that we can find a finite sequence of divisorial Mori contraction, say $\varphi : X \to X'$ such that X' is a minimal Gorenstein model with $c_2(X') = 0$. In order to see the smoothness of X', first observe that $\kappa(X')$ can only be 0 or 1, by (1.1) and $c_2(X') = 0$. In case $\kappa(X') = 1$, we refer for the smoothness to (2.1), in case $\kappa(X') = 0$, see the argument following (1.5). Now suppose that φ is composed say of the Mori contractions $\varphi_i : X_i \to X_{i+1}$, $1 \leq i \leq r-1$. Then we shall prove that all X_i are smooth and φ_i is the blow up of a point. Notice first that $c_2(X_i) = 0$ for all i by (4.2). If φ_i would contract a divisor to a curve, then we know that $c_2(X_{i+1}) \neq 0$, a contradiction. So every φ_i contracts its exceptional divisor to a point. But then it follows from descending induction on i that every X_i is smooth and φ_i is the blow-up of a point. □

Parts of the proof do not use the assumption on the Kodaira dimension, so we may state

4.5. Corollary. *Let X be a 3-dimensional Gorenstein model with $\kappa(X) = -\infty$ and $c_2(X) = 0$. Then X arises from a 3-dimensional Gorenstein model X' with $c_2(X') = 0$ admitting a Mori contraction to a surface by a finite sequence of divisorial Mori contractions which contract the exceptional divisor to a point.*

What remains to be done is to investigate Gorenstein models X with $c_2(X) \leq 0$ admitting a Mori contraction $\varphi : X \to Y$ onto a smooth surface Y. We know that φ is a conic bundle and let Δ denote its discriminant locus (possibly empty).

4.6. Lemma. *Let X be a 3-dimensional Gorenstein model and $\varphi : X \to Y$ a Mori contraction to a surface Y.*

(1) *If $c_2(X) \le 0$ then $2K_Y - \Delta \in \overline{NE}(Y)$. In particular $\kappa(Y) \ge 0$.*

(2) *If moreover $\chi(\mathcal{O}_X) = 0$ then $0 \le \kappa(Y) \le 1$. If $\kappa(Y) = 0$, then X is a conic bundle over Y with vanishing discriminant locus. The same holds for $\kappa(Y) = 1$ in the case Y is minimal. In particular X is smooth in these cases.*

(3) *$c_2(X) = 0$ if and only if X is a conic bundle over a torus or a hyperelliptic surface with empty discriminant locus. Moreover these conditions are equivalent to $\chi(\mathcal{O}_X) = 0$ and $\Delta \equiv 2K_Y$.*

Proof. Let $H \subset Y$ be a general hyperplane section (not passing through the singular points of Δ), meeting Δ in $(\Delta \cdot H)$ points transversally. Let $H' = \varphi^{-1}(H)$. From the exact sequence

$$0 \to N^*_{H'|X} \to \Omega^1_X|H' \to \Omega^1_{H'} \to 0$$

we deduce

$$c_2(X) \cdot H' = c_2(H') - K_{H'} \cdot N_{H'} = 4(1-g) + \Delta \cdot H + 2H^2,$$

where g is the genus of the curve H. Hence

$$\Delta \cdot H = 2K_Y \cdot H + c_2(X) \cdot H'. \qquad (*)$$

(1) is now an immediate consequence of $(*)$.

(2) Since $\chi(\mathcal{O}_X) = 0$ we have $\chi(\mathcal{O}_Y) = 0$, hence $\kappa(Y) = 0$ or 1. Assuming $\kappa(Y) = 0$ we let $g : Y \to Y'$ be the minimal model of Y. It follows from (1) that Δ is contained in the fibers of g. Now Δ has the following obvious property. Every smooth rational component of Δ meets the union of the other components in at least two points. The reason is that every irreducible component Δ_i of Δ gives rise to a two-sheeted covering of Δ_i. Now it is an easy exercise that this property implies $\Delta = 0$. If $\kappa(Y) = 1$, then Δ must be contained in the fibers of the Iitaka fibration, i.e. the composition of the map $g : Y \to Y'$ to the minimal model followed by the usual Iitaka fibration $h : Y' \to B$. Notice that h has as singular fibers only multiples of elliptic curves. As above we therefore see that Δ does not contain any rational curve. In the case Y is minimal, we now apply (4.7) below to get the claim.

(3) One direction being obvious, we shall assume $c_2(X) = 0$. Then $(*)$ gives the minimality of Y and $\Delta \equiv 2K_Y$, hence we obtain $K_Y \equiv 0$ from (2). Since $\chi(\mathcal{O}_Y) = 0$, Y is a torus or hyperelliptic. For the last statement assume that $\chi(\mathcal{O}_X) = 0$ and $\Delta \equiv 2K_Y$. Then $(*)$ yields $c_2(X) \cdot H' = 0$ and Riemann–Roch gives $K_X \cdot c_2(X) = 0$. Since K_X and $\varphi^*(\mathrm{Pic}(Y))$ generate $N^1(X)$, it follows that $c_2(X) = 0$. □

4.7. Lemma. *Let X be a 3-dimensional Gorenstein model and $\varphi : X \to Y$ a Mori contraction to a smooth minimal surface Y with $\kappa(Y) = 1$. Assume that $\chi(\mathcal{O}_Y) = 0$. Then the discriminant locus Δ of φ is not contained in the fibers of the Iitaka fibration of Y (if $\Delta \neq 0$.)*

Proof. Assume the contrary. Let $g : Y \to B$ be the Iitaka fibration of Y and denote by $h : X \to B$ the composite map. By our assumptions, g has at most multiple elliptic curves as singular fibers. Then the general fiber X_s, $s \in B$ is a ruled surface over an elliptic curve. Pick a component $C \subset \Delta$ (which is an elliptic curve) and put $X_0 = f^{-1}(g(C))$. Let $\mu : \tilde{C} \to C$ be the etale two-sheeted covering associated with the conic bundle structure and denote by $\tilde{X}_0 \to \tilde{C}$ the induced surface which is a $\mathbb{P}_1$-bundle over $\tilde{C}$. Then the resulting map $\nu : \tilde{X}_0 \to X_0$ is nothing than the normalisation of X_0. For simplicity of notations we put $S = X_0$ and $\tilde{S} = \tilde{X}_0$. Notice the formula

$$K_{\tilde{S}} = \nu^*(K_S) - C'$$

with some section C' of $\tilde{S}$. Let e be the invariant of $\tilde{S}$ in the sense of [Ha77, chap. 5.2]. Take a section C_0 of $\tilde{S}$ with $C_0^2 = -e$ (which is then by definition of e a section with minimal self-intersection) and let F denote a ruling fiber. Then

$$K_{\tilde{S}} \equiv -2C_0 - eF.$$

Furthermore we can write

$$C' \equiv C_0 + \beta F \tag{1}$$

with some non-negative integer β. Thus it follows

$$\nu^*(K_S) \equiv -C_0 + (\beta - e)F.$$

We have $K_{X_s}^2 = 0$, so $\nu^*(K_S)^2 = 0$, which yields

$$\beta = \frac{e}{2}. \tag{2}$$

From (1) we have on the other hand

$$\beta \geq e \text{ or } \beta = 0.$$

So we conclude $\beta = e = 0$ since $\beta \geq 0$.

Let $e(s) = e(X_s)$ for the general smooth X_s; of course then $e(s)$ is independent of s. We claim that $e(s) = 0$ or -1. To prove this let us exclude the inequality $e(s) > 0$. In fact, in this case we have a family (B_s) of sections $B_s \subset X_s$ with $B_s^2 = -e(s) < 0$. The (B_s) converge to an effective 1-cycle $B_0 \subset S = X_0$. Now we conclude that $B_0^2 < 0$ and, letting $\tilde{B}_0 = \nu^{-1}(B_0)$, we even have $\tilde{B}_0^2 < 0$, contradicting $e(\tilde{S}) = 0$. Thus $e(s) \in \{0, -1\}$. If $e(s) = 0$, then K_{X_s} is numerically divisible by 2, so does K_S and hence also $\nu^*(K_S)$, which is not the case. So we end up with $e(s) = -1$ and thus $-K_{X_s} \equiv 2B_s + F_s$. Going to the limit, it follows $\nu^*(-K_{\tilde{S}}) \equiv 2\tilde{B}_0 + 2F$, a contradiction. This completes the proof of (4.7). □

4.8. Example. It is easy to construct examples of projective bundles X over surfaces with $c_2(X) \leq 0$, but $c_2(X) \neq 0$. Necessarily these surfaces have to be non-minimal. Take for instance an abelian surface A and let $\hat{A} \to A$ be the blow-up of the point $p \in A$. Let $E \subset \hat{A}$ be the exceptional curve and let $V = \mathcal{O}_{\hat{A}} \oplus \mathcal{O}_{\hat{A}}(kE)$, for some $k \geq 1$. Let $X = \mathbb{P}(V)$. Then it is easy to see that $c_2(X) \leq 0$.

References

[Be85] Benveniste, X., Sur le cone des 1-cycles effectives de certaines variétés de dimension 3, Math. Ann. 272 (1985), 257–265.

[BPV84] Barth, W., Peters,C., Van de Ven, A., Compact complex surfaces, Ergeb. Math. Grenzgeb. (3) 4, Springer-Verlag 1984.

[Cu88] Cutkosky, S., Elementary contractions of Gorenstein threefolds, Math. Ann. 280 (1988), 521–525.

[Fl87] Fletcher, A. R., Contributions to Riemann–Roch on projective 3-folds with only canonical singularities and applications, Proc. Symp. Pure Math. 46 (1987), 221–231.

[Fu84] Fulton, W., Intersection theory, Ergeb. Math. Grenzgeb. (3) 2, Springer-Verlag 1984.

[Gr94] Grassi, A., On a question of J. Kollár. Cont. Math. 162, 209-214 (1994)

[Ha77] Hartshorne, R., Algebraic Geometry, Grad. Texts in Math. 52, Springer-Verlag 1977.

[Ka85] Kawamata, Y., Pluricanonical systems on minimal algebraic varieties, Invent. Math. 79 (1985), 567–588.

[Ka88] Kawamata, Y., The crepant blowing-up of 3-dimensional canonical singularities and its application to degenerations of surfaces, Ann. Math. 127 (1988), 93–163.

[KMM87] Kawamata, Y., Matsuda, K., Matsuki, K., Introduction to the minimal model program, Adv. Stud. Pure Math. 10 (1987), 283–360.

[Ko90] Kobayashi, R., Uniformisation of complex surfaces, Adv. Stud. Pure Math. 18, Part 2, 313–394 (1990).

[Me93] Megyesi, G., Inequalities between Chern numbers, Ph.D. Thesis, Cambridge 1993.

[Mi87] Miyaoka, Y., The Chern classes and Kodaira dimension of a minimal variety, Adv. Stud. Pure Math. 10, 449–476 (1987)

[Mo82] Mori, S., Threefolds whose canonical bundles are not numerically effective, Ann. Math. 116 (1982), 133–176.

[MR82] Mehta, V. B., Ramanathan, A., Semistable sheaves on projective varieties and their restriction to curves, Math. Ann. 258 (1982), 213–224.

[Na87] Nakayama, N., On Weierstrass models, in: Algebraic geometry and commutative algebra, Vol. in honour of M. Nagata, 405–431, Kinokuniya 1987.

[Re77] Reid, M., Bogomolov's theorem $c_1^2 \leq 4c_2$, Proc. Internat. Symp. Alg. Geom. Kyoto 1977, 633–642.

[Re87] Reid, M., Young person's guide to canonical singularities, Proc. Symp. Pure Math. 46 (1987), 345–414.

[SBW94] Shepherd-Barron, N. I., Wilson, P. M. H., Singular threefolds with numerically trivial first and second Chern classes, J. Alg. Geom. 3 (1994), 265–282.

[Ts88] Tsuji, H., Stability of tangent bundles of minimal algebraic varieties, Topology 27 (1988), 429–442.

[Um81] Umezu, Y., On normal projective surfaces with trivial dualizing sheaf, Tokyo J. Math. 4 (1981), 343–354.

[Yau77] Yau, S.-T., Calabi's conjecture and some new results in algebraic geometry, Proc. Nat. Acad. Sci. USA (1977), 1798–1799.

List of Contributors

Valery Alexeev
Department of Mathematics, University of Georgia, Athens, GA 30602-7403, USA
valery@math.uga.edu

E. Ballico
Department of Mathematics, University of Trento, 38050 Povo (TN), Italy
ballico@science.unitn.it

Victor V. Batyrev
Universität-Gesamthochschule-Essen, Fachbereich Mathematik, Universitätsstraße 2, 45141 Essen, Germany
victor.batyrev@aixrs1.hrz.uni-essen.de

Lev A. Borisov
Department of Mathematics, University of Michigan, Ann Arbor, MI 48109-1003, USA
lborisov@math.lsa.umich.edu

Thierry Bouche
Université de Grenoble, Laboratoire associé au CNRS (URA 188), Institut Fourier, B.P. 74, 38402 Saint-Martin d'Hères Cedex, France
bouche@fourier.grenet.fr

Robert Braun
Mathematisches Institut, Universiät Bayreuth, 95445 Bayreuth, Germany
braun@btm8x1.mat.uni-bayreuth.de

Frédéric Campana
Départment de Mathématiques, Université Nancy 1, BP 239, 54506 Vandœuvre-les-Nancy Cedex, France
Frederic.Campana@iecn.u-nancy.fr

Fabrizio Catanese
Dipartimento di Matematica, Via Buonarroti 2, 56127 Pisa, Italy
catanese@dm.unipi.it and catanese@vm.cnuce.cnr.it

Ciro Ciliberto
Dipartimento di Matematica, Università di Roma II, Tor Vergata, Viale della Ricerca Scientifica, 00133 Roma, Italy
ciliberto@mat.utovrm.it

Lawrence Ein
Department of Mathematics, University of Illinois at Chicago, Chicago, IL 60680, USA
U22425@uic.edu

Hubert Flenner
Institut für Mathematik, Ruhr-Universität Bochum, Universitätsstraße 150, Gebäude NA, 44801 Bochum, Germany
flenn@erubc.rz.ruhr-uni-bochum.de

Yukari Ito
Department of Mathematical Sciences, University of Tokyo, Tokyo 113, Japan
yukari-i@tansei.cc.u-tokyo.ac.jp

Yujiro Kawamata
Department of Mathematical Sciences, University of Tokyo, Komaba, Meguro, Tokyo 153, Japan
kawamata@ms.u-tokyo.ac.jp

Antonio Lanteri
Dipartimento di Matematica "F. Enriques", Università, Via C. Saldini 50, 20133 Milano, Italy
lanteri@vmimat.mat.unimi.it

Robert Lazarsfeld
Department of Mathematics, University of California Los Angeles, Los Angeles, CA 90024, USA
rkl@math.ucla.edu

Angelo Felice Lopez
Dipartimento di Matematica, Terza Università di Roma, Via Corrado Segre 2, 00146 Roma, Italy
lopez@matrm3.mat.uniroma3.it

Hidetoshi Maeda
Graduate School of Mathematics, Kyushu University, 4-2-1 Ropponmatsu, Chuo-ku, Fukuoka 810, Japan
hmaeda@math.kyushu-u.ac.jp

Rick Miranda
Department of Mathematics, Colorado State University, Ft. Collins, CO 80523, USA
miranda@riemann.math.colostate.edu

Stefan Müller-Stach
Universität-Gesamthochschule-Essen, Fachbereich Mathematik, Universitätsstraße 2, 45141 Essen, Germany

Michael Nakamaye
Department of Mathematics, Harvard University, Cambridge, MA 02138, USA
nakamaye@math.harvard.edu

Viacheslav V. Nikulin
Steklov Mathematical Institute, Moscow, Russia
slava@nikulin.mian.su

Keiji Oguiso
Faculty of Science, Ochanomizu University, Otsuka, Bunkyo, Tokyo, Japan
oguiso@math.ocha.ac.jp

Roberto Paoletti
Department of Pure Mathematics, University of Cambridge, 16 Mill Lane, Cambridge CB2 1SB, England
rp@pmms.cam.ac.uk

Th. Peternell
Mathematisches Institut, 95447 Bayreuth, Universitätsstraße 30, Germany,
peternel@btm8x1.mat.uni-bayreuth.de

Miles Reid
Mathematics Institute, University of Warwick, Coventry CV4 7AL, England
miles@maths.warwick.ac.uk

B. Russo
Department of Mathematics, University of Trento, 38050 Povo (TN), Italy
russo@science.unitn.it

Wolfgang Vogel
Mathematics Department, Massey University, Palmerston North, New Zealand
W.Vogel@massey.ac.nz

P. M. H. Wilson,
Department of Pure Mathematics, 16 Mill Lane, Cambridge CB2 1SB, England
pmhw@pmms.cam.ac.uk